Methods in Enzymology

Volume 272
CYTOCHROME P450
Part B

METHODS IN ENZYMOLOGY

EDITORS-IN-CHIEF

John N. Abelson   Melvin I. Simon

DIVISION OF BIOLOGY
CALIFORNIA INSTITUTE OF TECHNOLOGY
PASADENA, CALIFORNIA

FOUNDING EDITORS

Sidney P. Colowick and Nathan O. Kaplan

Methods in Enzymology

Volume 272

Cytochrome P450

Part B

EDITED BY

Eric F. Johnson

THE SCRIPPS RESEARCH INSTITUTE
LA JOLLA, CALIFORNIA

Michael R. Waterman

VANDERBILT UNIVERSITY SCHOOL OF MEDICINE
NASHVILLE, TENNESSEE

ACADEMIC PRESS

San Diego   London   Boston   New York   Sydney   Tokyo   Toronto

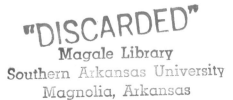

Academic Press, Inc.
525 B Street, Suite 1900, San Diego, California 92101-4495, USA
http://www.apnet.com

Academic Press Limited
24-28 Oval Road, London NW1 7DX, UK
http://www.hbuk.co.uk/ap/

International Standard Serial Number: 0076-6879

International Standard Book Number: 0-12-182173-0

PRINTED IN THE UNITED STATES OF AMERICA
96 97 98 99 00 01 EB 9 8 7 6 5 4 3 2 1

Table of Contents

CONTRIBUTORS TO VOLUME 272 ix

PREFACE xv

VOLUMES IN SERIES xvii

Section I. Heterologous Expression of P450s

1. Maximizing Expression of Eukaryotic Cytochrome   HENRY J. BARNES   3
   P450s in *Escherichia coli*

2. Construction of Plasmids and Expression in *Esche-*   CHARLES W. FISHER,
   richia coli of Enzymatically Active Fusion Pro-   MANJUNATH S. SHET, AND
   teins Containing the Heme-Domain of a P450   RONALD W. ESTABROOK   15
   Linked to NADPH-P450 Reductase

3. N-Terminal Modifications That Alter P450 Mem-   STEVEN J. PERNECKY AND
   brane Targeting and Function   MINOR J. COON   25

4. Purification of Functional Recombinant P450s   F. PETER GUENGERICH,
   from Bacteria   MARTHA V. MARTIN,
     ZUYU GUO, AND
     YOUNG-JIM CHUN   35

5. Application of Electrochemistry for P450-Cata-   RONALD W. ESTABROOK,
   lyzed Reactions   KEVIN M. FAULKNER,
     MANJUNATH S. SHET, AND
     CHARLES W. FISHER   44

6. Yeast Expression of Animal and Plant P450s in Op-   DENIS POMPON,
   timized Redox Environments   BENEDICTE LOUERAT,
     ALEXIS BRONINE, AND
     PHILIPPE URBAN   51

7. Generation of the Cytosolic Domain of Microsomal   ULRICH SCHELLER,
   P450 52A3 after High-Level Expression in *Sac-*   THOMAS JURETZEK, AND
   charomyces cerevisiae   WOLF-HAGEN SCHUNCK   65

8. Use of Heterologous Expression Systems to Study   SYLVAINE LECOEUR,
   Autoimmune Drug-Induced Hepatitis   JEAN-CHARLES GAUTIER,
     CLAIRE BELLOC,
     ALINE GAUFFRE, AND
     PHILLIPE H. BEAUNE   76

9. Recombinant Baculovirus Strategy for Coexpres- CAROLINE A. LEE,
   sion of Functional Human Cytochrome P450 and   THOMAS A. KOST, AND
   P450 Reductase   COSETTE SERABJIT-SINGH   86

Section II. Enzyme Assays

10. *In Vitro* Assessment of Various Cytochromes P450   HEYO K. KROEMER,
    and Glucuronosyltransferases Using the Antiar-   SIGRID BOTSCH,
    rhythmic Propafenone as a Probe Drug   GEORG HEINKELE, AND
                                                       MONIKA SCHICK   99

11. *In Vivo* and *in Vitro* Measurement of CYP2C19 Ac- PETER J. WEDLUND AND
    tivity   GRANT R. WILKINSON   105

12. Chlorzoxazone: An *in Vitro* and *in Vivo* Substrate DANIELE LUCAS,
    Probe for Liver CYP2E1   JEAN-FRANÇOIS MENEZ, AND
                                                         FRANCOIS BERTHOU   115

13. Assays for CYP1A2 by Testing *in Vivo* Metabolism   BING-KOU TANG AND
    of Caffeine in Humans   WERNER KALOW   124

14. Assays of Omeprazole Metabolism as a Substrate   DONALD J. BIRKETT,
    Probe for Human CYP Isoforms   TOMMY ANDERSSON, AND
                                                     JOHN O. MINERS   132

15. Use of Tolbutamide as a Substrate Probe for Human JOHN O. MINERS AND
    Hepatic Cytochrome P450 2C9   DONALD J. BIRKETT   139

16. Assays of CYP2C8- and CYP3A4-Mediated Metab-   THOMAS WALLE   145
    olism of Taxol *in Vivo* and *in Vitro*

17. Tamoxifen Metabolism by Microsomal Cytochrome   DAVID KUPFER AND
    P450 and Flavin-Containing Monooxygenase   SHANGARA S. DEHAL   152

18. Trimethadione: Metabolism and Assessment of He- EINOSUKE TANAKA AND
    patic Drug-Oxidizing Capacity   YOSHIHIKO FUNAE   163

19. Antipyrine, Theophylline, and Hexobarbital as *in* KEES GROEN AND
    Vivo P450 Probe Drugs   DOUWE D. BREIMER   169

20. Imipramine: A Model Drug for P450 Research   KIM BRØSEN,
                                                 ERIK SKJELBO, AND
                                                 KARIN KRAMER NIELSEN   177

21. Measurement of Human Liver Microsomal Cyto-   A. DAVID RODRIGUES   186
    chrome P450 2D6 Activity Using [*O-Methyl-*
    ^{14}C]Dextromethorphan as Substrate

Section III. Determination of Genotype

22. CYP2D6 Multiallelism   ANN K. DALY,
                                                  VIDAR M. STEEN,
                                                  KAREN S. FAIRBROTHER, AND
                                                  JEFFREY R. IDLE   199

23. Genetic Tests Which Identify the Principal Defects   JOYCE A. GOLDSTEIN AND
    in *CYP2C19* Responsible for the Polymorphism   JOYCE BLAISDELL   210
    in Mephenytoin Metabolism

24. Genetic Polymorphism of Human Cytochrome   VESSELA NEDELCHEVA,
    P450 2E1   IRENE PERSSON, AND
    MAGNUS INGELMAN-
    SUNDBERG   218

25. Identification of Allelic Variants of the Human   KANAME KAWAJIRI,
    CYP1A1 Gene   JUNKO WATANABE, AND
    SHIN-ICHI HAYASHI   226

Section IV. Plant P450s

26. Analysis of Herbicide Metabolism by Monocot Mi-   SZE-MEI CINDY LAU AND
    crosomal Cytochrome P450   DANIEL P. O'KEEFE   235

27. Microsome Preparation from Woody Plant Tissues   JERRY HEFNER AND
    RODNEY CROTEAU   243

28. Detection, Assay, and Isolation of Allene Oxide   ALAN R. BRASH AND
    Synthase   WENCHAO SONG   250

29. Cinnamic Acid Hydroxylase Activity in Plant Mi-   FRANCIS DURST,
    crosomes   IRÈNE BENVENISTE,
    MICHEL SCHALK, AND
    DANIÈLE WERCK-REICHHART   259

30. Isolation of Plant and Recombinant CYP79   BARBARA ANN HALKIER,
    OLE SIBBESEN, AND
    BIRGER LINDBERG MØLLER   268

31. Cloning of Novel Cytochrome P450 Gene Se-   TIMOTHY A. HOLTON AND
    quences via Polymerase Chain Reaction Ampli-   DIANE R. LESTER   275
    fication

Section V. Insect P450s

32. Preparation of Microsomes from Insects and Purifi-   JEFFREY G. SCOTT   287
    cation of CYP6D1 from House Flies

33. Quantification of Ecdysteroid Biosynthesis during   TIMOTHY J. SLITER,
    Short-Term Organ Culture   KOMSUN SUDHIVORASETH, AND
    JOHN L. MCCARTHY   292

34. Sampling P450 Diversity by Cloning Polymerase   MARK J. SNYDER,
    Chain Reaction Products Obtained with Degen-   JULIE A. SCOTT,
    erate Primers   JOHN F. ANDERSEN, AND
    RENE FEYEREISEN   304

Section VI. Analysis of P450 Structure

35. Structural Alignments of P450s and Extrapolations   SANDRA E. GRAHAM-
    to the Unknown   LORENCE AND
                                                         JULIAN A. PETERSON   315

36. Predicting the Rates and Regioselectivity of Reac-   JEFFREY P. JONES AND
    tions Mediated by the P450 Superfamily   KENNETH R. KORZEKWA   326

37. Substrate Docking Algorithms and the Prediction   JAMES J. DE VOSS AND
    of Substrate Specificity   PAUL R. ORTIZ DE
                                                         MONTELLANO   336

38. Using Molecular Modeling and Molecular Dynam-   MARK D. PAULSEN,
    ics Simulation to Predict P450 Oxidation Products   JOHN I. MANCHESTER, AND
                                                         RICK L. ORNSTEIN   347

39. Approaches to Crystallizing P450s   THOMAS L. POULOS   358

40. Crystallization Studies of NADPH–Cytochrome   JUNG-JA PARK KIM,
    P450 Reductase   DAVID L. ROBERTS,
                                                         SNEZANA DJORDJEVIC,
                                                         MIUG WANG,
                                                         THOMAS M. SHEA, AND
                                                         BETTIE SUE SILER MASTERS   368

Section VII. Regulation

41. Sphingolipid-Dependent Signaling in Regulation of   EDWARD T. MORGAN,
    Cytochrome P450 Expression   MARIANA NIKOLOVA-
                                                         KARAKASHIAN,
                                                         JIN-QIANG CHEN, AND
                                                         ALFRED H. MERRILL, JR.   381

42. Use of Human Hepatocytes to Study P450 Gene In-   STEPHEN C. STROM,
    duction   LIUBOMIR A. PISAROV,
                                                         KENNETH DORKO,
                                                         MELISSA T. THOMPSON,
                                                         JOHN D. SCHUETZ, AND
                                                         ERIN G. SCHUETZ   388

43. Cytochrome P450 mRNA Induction: Quantitation   MICHAEL J. FASCO,
    by RNA–Polymerase Chain Reaction Using Cap-   CHRISTOPHER TREANOR, AND
    illary Electrophoresis   LAURENCE KAMINSKY   401

44. Targeted Disruption of Specific Cytochromes P450   PEDRO M. FERNANDEZ-
    and Xenobiotic Receptor Genes   SALGUERO AND
                                                         FRANK J. GONZALEZ   412

AUTHOR INDEX   431

SUBJECT INDEX   457

Contributors to Volume 272

Article numbers are in parentheses following the names of contributors.
Affiliations listed are current.

JOHN F. ANDERSEN (34), *Department of Entomology, University of Arizona, Tucson, Arizona 85721*

TOMMY ANDERSSON (14), *Clinical Pharmacology, Astra Hässle AB, S-43183 Möndal, Sweden*

HENRY J. BARNES (1), *Immune Complex Corporation, La Jolla, California 92038*

PHILLIPE BEAUNE (8), *CHU NECKER-IN-SERM U 75, Biochimie Pharmacologique et Métabolique, Université René Descartes, F-75730 Paris Cedex 15, France*

CLAIRE BELLOC (8), *CHU NECKER-IN-SERM U 75, Biochimie Pharmacologique et Métabolique, Université René Descartes, F-75730 Paris Cedex 15, France*

IRÉNE BENVENISTE (29), *Department of Cellular and Molecular Enzymology, Plant Molecular Biology Institute, CNRS, Strasbourg, France*

FRANCOIS BERTHOU (12), *Labs Biochimie-Nutrition, Faculté de Médecine, F-29285 Brest, France*

DONALD J. BIRKETT (14, 15), *Department of Clinical Pharmacology, Flinders University of South Australia, Flinders Medical Center, Bedford Park, South Australia 5042, Australia*

JOYCE BLAISDELL (23), *Laboratory of Biochemical Risk Analysis, National Institute of Environmental Health Sciences, Research Triangle Park, North Carolina 27709*

SIGRID BOTSCH (10), *Dr. Margarete Fischer-Bosch-Institute für Klinische Pharmakologie, D-70341 Stuttgart, Germany*

ALAN R. BRASH (28), *Department of Pharmacology, Vanderbilt University Medical Center, Nashville, Tennessee 37232*

DOUWE D. BREIMER (19), *Division of Pharmacology, Leiden/Amsterdam Center for Drug Research, University of Leiden, 2300 RA Leiden, The Netherlands*

ALEXIS BRONINE (6), *Centre de Génétique Moléculaire, Centre National de la Recherche Scientifique, 91198 Gif-sur-yvette Cedex, France*

KIM BRØSEN (20), *Department of Clinical Pharmacology, Institute of Medical Biology, Odense University, DK-5000 Odense, Denmark*

JIN-QIANG CHEN (41), *Department of Environmental Health Sciences, Johns Hopkins University School of Public Health, Baltimore, Maryland 21218*

YOUNG-JIN CHUN (4), *Toxicology Research Center, Korean Research Institute of Chemical Technology, Taejon, South Korea*

MINOR J. COON (3), *Department of Biological Chemistry, University of Michigan Medical School, Ann Arbor, Michigan 48109*

RODNEY CROTEAU (27), *Department of Genetics and Cell Biology, Institute of Biological Chemistry, Washington State University, Pullman, Washington 99164*

ANN K. DALY (22), *Department of Pharmacological Sciences, University of Newcastle upon Tyne Medical School, Framlington Place, Newcastle upon Tyne NE2 4HH, United Kingdom*

JAMES J. DE VOSS (37), *Department of Pharmaceutical Chemistry, School of Pharmacy, University of California, San Francisco, California 94143*

SHANGARA S. DEHAL (17), *Worcester Foundation for Biomedical Research, Shrewsbury, Massachusetts 01545*

SNEZANA DJORDJEVIC (40), *Department of Biochemistry, Center for Advanced Biotechnology and Medicine, Piscataway, New Jersey 08854*

KENNETH DORKO (42), *Department of Pathology, University of Pittsburgh, Pittsburgh, Pennsylvania 15261*

FRANCIS DURST (29), *Department of Cellular and Molecular Enzymology, Plant Molecular Biology Institute, CNRS, Strasbourg, France*

RONALD W. ESTABROOK (2, 5), *Department of Biochemistry, The University of Texas Southwestern Medical Center, Dallas, Texas 75235*

KAREN S. FAIRBROTHER (22), *Department of Pharmacological Sciences, University of Newcastle upon Tyne Medical School, Framlington Place, Newcastle upon Tyne NE2 4HH, United Kingdom*

MICHAEL J. FASCO (43), *New York State Department of Health, Wadsworth Center, Albany, New York 12201*

KEVIN M. FAULKNER (5), *Department of Biochemistry, The University of Texas Southwestern Medical Center, Dallas, Texas 75235*

PEDRO M. FERNANDEZ-SALGUERO (44), *Division of Basic Science, National Cancer Institute, National Institutes of Health, Bethesda, Maryland 20892*

RENÉ FEYEREISEN (34), *Department of Entomology, University of Arizona, Tucson, Arizona 85721*

CHARLES W. FISHER (2, 5), *Department of Biochemistry, The University of Texas Southwestern Medical Center, Dallas, Texas 75235*

YOSHIHIKO FUNAE (18), *Laboratory of Chemistry, Osaka City University Medical School, Osaka 545, Japan*

ALINE GAUFFRE (8), *CHU NECKER-IN-SERM U 75, Biochimie Pharmacologique et Métabolique, Université René Descartes, F-75730 Paris Cedex 15, France*

JEAN-CHARLES GAUTIER (8), *CHU NECKER-INSERM U 75, Biochimie Pharmacologique et Métabolique, Université René Descartes, F-75730 Paris Cedex 15, France*

JOYCE A. GOLDSTEIN (23), *Laboratory of Biochemical Risk Analysis, National Institute of Environmental Health Sciences, Research Triangle Park, North Carolina 27709*

FRANK J. GONZALEZ (44), *Division of Basic Science, National Cancer Institute, National Institutes of Health, Bethesda, Maryland 20892*

SANDRA E. GRAHAM-LORENCE (35), *Department of Biochemistry, The University of Texas Southwestern Medical Center, Dallas, Texas 75235*

KEES GROEN (19), *Department of Clinical Pharmacokinetics, Janssen Research Foundation, B-2340 Beerse, Belgium*

F. PETER GUENGERICH (4), *Department of Biochemistry and Center in Molecular Toxicology, Vanderbilt University School of Medicine, Nashville, Tennessee 37232*

ZUYU GUO (4), *Rhone-Poulenc-Rorer, Department of Drug Metabolism, Collegeville, Pennsylvania 19426*

BARBARA ANN HALKIER (30), *Department of Plant Biology, Plant Biochemistry Laboratory, Royal Veterinary and Agricultural University, DK-1871 Frederiksberg C, Denmark*

SHIN-ICHI HAYASHI (25), *Department of Biochemistry, Saitama Cancer Center Research Institute, Saitama 362, Japan*

JERRY HEFNER (27), *Department of Genetics and Cell Biology, Institute of Biological Chemistry, Washington State University, Pullman, Washington 99164*

GEORG HEINKELE (10), *Dr. Margarete Fischer-Bosch-Institute für Klinische Pharmakologie, D-70341 Stuttgart, Germany*

TIMOTHY A. HOLTON (31), *Florigene Pty. Ltd., Collingwood, Victoria 3066, Australia*

JEFFREY R. IDLE (22), *Faculty of Medicine, Institute of Cancer Research and Molecular Biology, The University of Trondheim, N-7005 Trondheim, Norway*

MAGNUS INGELMAN-SUNDBERG (24), *Institute of Environmental Medicine, Karolinska Institute, S-171 77 Stockholm, Sweden*

JEFFREY P. JONES (36), *Department of Pharmacology and Physiology, University of Rochester, Rochester, New York 14642*

THOMAS JURETZEK (7), *Department of Cell Biology, Max Delbrück Center for Molecular Medicine, D-13125 Berlin, Germany*

WERNER KALOW (13), *Department of Pharmacology, University of Toronto, Toronto, Ontario M5S 1A8, Canada*

LAURENCE KAMINSKY (43), *New York State Department of Health, Wadsworth Center, Albany, New York 12201*

KANAME KAWAJIRI (25), *Department of Biochemistry, Saitama Cancer Center Research Institute, Saitama 362, Japan*

JUNG-JA PARK KIM (40), *Department of Biochemistry, Medical College of Wisconsin, Milwaukee, Wisconsin 53226*

KENNETH R. KORZEKWA (36), *Center for Clinical Pharmacology, University of Pittsburgh, Pittsburgh, Pennsylvania 15213*

THOMAS A. KOST (9), *Division of Biochemistry, Glaxo-Wellcome, Inc., Research Triangle Park, North Carolina 27709*

HEYO K. KROEMER (10), *Dr. Margarete Fischer-Bosch-Institute für Klinische Pharmakologie, D-70341 Stuttgart, Germany*

DAVID KUPFER (17), *Worcester Foundation for Biomedical Research, Shrewsbury, Massachusetts 01545*

SZE-MEI CINDY LAU (26), *Central Research and Development, The Dupont Company Experimental Station, Wilmington, Delaware 19880*

SYLVAINE LECOEUR (8), *CHU NECKER-IN-SERM U 75, Biochimie Pharmacologique et Métabolique, Université René Descartes, F-75730 Paris Cedex 15, France*

CAROLINE A. LEE (9), *Agouron Pharmaceuticals Inc., San Diego, California 92121*

DIANE R. LESTER (31), *Florigene Pty. Ltd., Collingwood, Victoria 3066, Australia*

BENEDICTE LOUERAT (6), *Centre de Génétique Moléculaire, Centre National de la Recherche Scientifique, 91198 Gif-sur-yvette Cedex, France*

DANIELE LUCAS (12), *Labs Biochimie-Nutrition, Faculté de Médecine, F-29285 Brest, France*

JOHN I. MANCHESTER (38), *Environmental Molecular Sciences Laboratory, Pacific Northwest National Laboratories, Richland, Washington 99352*

MARTHA V. MARTIN (4), *Department of Biochemistry and Center in Molecular Toxicology, Vanderbilt University School of Medicine, Nashville, Tennessee 37232*

BETTIE SUE SILER MASTERS (40), *Department of Biochemistry, The University of Texas Health Science Center, San Antonio, Texas 78284*

JOHN L. McCARTHY (33), *Department of Biological Sciences, Southern Methodist University, Dallas, Texas 75275*

JEAN-FRANÇOIS MENEZ† (12), *Labs Biochimie-Nutrition, Faculté de Médecine, F-29285 Brest, France*

ALFRED H. MERRILL, JR. (41), *Department of Biochemistry, Emory University School of Medicine, Atlanta, Georgia 30322*

JOHN O. MINERS (14, 15), *Department of Clinical Pharmacology, Flinders Medical Centre, Bedford Park, South Australia 5042, Australia*

BIRGER LINDBERG MØLLER (30), *Department of Plant Biology, Plant Biochemistry Laboratory, Royal Veterinary and Agricultural University, DK-1871 Frederiksberg C, Denmark*

EDWARD T. MORGAN (41), *Department of Pharmacology, Emory University School of Medicine, Atlanta, Georgia 30322*

VESSELA NEDELCHEVA (24), *Center of Industrial Hygiene and Occupational Diseases, National Institute of Public Health, Praha, Czech Republic*

KARIN KRAMER NIELSEN (20), *Department of Clinical Pharmacology, Institute of Medical Biology, Odense University, DK-5000 Odense, Denmark*

MARIANA NIKOLOVA-KARAKASHIAN (41), *Department of Biochemistry, Emory University School of Medicine, Atlanta, Georgia 30322*

DANIEL P. O'KEEFE (26), *Central Research and Development, The Dupont Company Experimental Station, Wilmington, Delaware 19880*

† Deceased.

RICK L. ORNSTEIN (38), *Environmental Molecular Sciences Laboratory, Pacific Northwest National Laboratory, Richland, Washington 99352*

PAUL R. ORTIZ DE MONTELLANO (37), *Department of Pharmaceutical Chemistry, School of Pharmacy, University of California, San Francisco, California 94143*

MARK D. PAULSEN (38), *Environmental Molecular Sciences Laboratory, Pacific Northwest National Laboratory, Richland, Washington 99352*

STEVEN J. PERNECKY (3), *Department of Chemistry, Eastern Michigan University, Ypsilante, Michigan 48197*

IRENE PERSSON (24), *Institute of Environmental Medicine, Karolinska Institute, S-171 77 Stockholm, Sweden*

JULIAN A. PETERSON (35), *Department of Biochemistry, The University of Texas Southwestern Medical Center, Dallas, Texas 75235*

LIUBOMIR A. PISAROV (42), *Department of Pathology, University of Pittsburgh, Pittsburgh, Pennsylvania 15261*

DENIS POMPON (6), *Centre de Génétique Moléculaire, Centre National de la Recherche Scientifique, 91198 Gif-sur-yvette Cedex, France*

THOMAS L. POULOS (39), *Departments of Biochemistry and Molecular Biology and Physiology Biophysics, University of California, Irvine, California 92717*

DAVID L. ROBERTS (40), *Department of Biochemistry, Medical College of Wisconsin, Milwaukee, Wisconsin 53226*

A. DAVID RODRIGUES (21), *Drug Metabolism Department, Abbott Laboratories, Abbott Park, Illinois 60064*

MICHEL SCHALK (29), *Department of Cellular and Molecular Enzymology, Plant Molecular Biology Institute, CNRS, Strasbourg, France*

ULRICH SCHELLER (7), *Department of Cell Biology, Max Delbrück Center for Molecular Medicine, D-13125 Berlin, Germany*

MONIKA SCHICK (10), *Dr. Margarete Fischer-Bosch-Institute für Klinische Pharmakologie, D-70341 Stuttgart, Germany*

ERIN D. SCHUETZ (42), *Department of Pharmaceutical Sciences, St. Jude Children's Research Hospital, Memphis, Tennessee 38105*

JOHN D. SCHUETZ (42), *Department of Pharmaceutical Sciences, St. Jude Children's Research Hospital, Memphis, Tennessee 38105*

WOLF-HAGEN SCHUNCK (7), *Department of Cell Biology, Max Delbrück Center for Molecular Medicine, D-13125 Berlin, Germany*

JEFFREY G. SCOTT (32), *Department of Entomology, Cornell University, Ithaca, New York 14853*

JULIE A. SCOTT (34), *Department of Entomology, University of Arizona, Tucson, Arizona 85721*

COSETTE J. SERABJIT-SINGH (9), *Division of BioAnalysis and Drug Metabolism, Glaxo-Wellcome, Inc., Research Triangle Park, North Carolina 27709*

THOMAS M. SHEA (40), *Department of Biochemistry, The University of Texas Health Science Center, San Antonio, Texas 78284*

MANJUNATH S. SHET (2, 5), *Department of Biochemistry, The University of Texas Southwestern Medical Center, Dallas, Texas 75235*

OLE SIBBESEN (30), *Department of Plant Biology, Plant Biochemistry Laboratory, Royal Veterinary and Agricultural University, DK-1871 Frederiksberg C, Denmark*

ERIK SKJELBO (20), *Department of Clinical Pharmacology, Institute of Medical Biology, Odense University, DK-5000 Odense, Denmark*

TIMOTHY J. SLITER (33), *Department of Biological Sciences, Southern Methodist University, Dallas, Texas 75275*

MARK J. SNYDER (34), *Department of Entomology, University of Arizona, Tucson, Arizona 85721*

WENCHAO SONG (28), *Department of Pharmacology and Center for Experimental Therapeutics, Stellar-Chance Laboratories, University of Pennsylvania, Philadelphia, Pennsylvania 19104*

VIDAR M. STEEN (22), *Dr. Einar Martens Research Group for Biological Psychiatry, Center for Molecular Medicine, Haukeland Hospital, N-5021 Bergen, Norway*

STEPHEN C. STROM (42), *Department of Pathology, University of Pittsburgh, Pittsburgh, Pennsylvania 15261*

KOMSUN SUDHIVORASETH (33), *Department of Biological Sciences, Southern Methodist University, Dallas, Texas 75275*

EINOSUKE TANAKA (18), *Institute of Community Medicine, University of Tsukuba, Tsukuba 305, Japan*

BING-KOU TANG (13), *Department of Pharmacology, University of Toronto, Toronto, Ontario M5S 1A8, Canada*

MELISSA T. THOMPSON (42), *Department of Pathology, University of Pittsburgh, Pittsburgh, Pennsylvania 15261*

CHRISTOPHER TREANOR (43), *New York State Department of Health, Wadsworth Center, Albany, New York 12201*

PHILIPPE URBAN (6), *Centre de Génétique Moléculaire, Centre National de la Recherche Scientifique, 91198 Gif-sur-yvette Cedex, France*

THOMAS WALLE (16), *Department of Cell and Molecular Pharmacology and Experimental Therapeutics, Medical University of South Carolina, Charleston, South Carolina 29425*

MIUG WANG (40), *Department of Biochemistry, Medical College of Wisconsin, Milwaukee, Wisconsin 53226*

JUNKO WATANABE (25), *Department of Biochemistry, Saitama Cancer Center Research Institute, Saitama 362, Japan*

PETER J. WEDLUND (11), *College of Pharmacy, University of Kentucky, Lexington, Kentucky 40536*

DANIÉLE WERCK-REICHHART (29), *Department of Cellular and Molecular Enzymology, Plant Molecular Biology Institute, CNRS, Strasbourg, France*

GRANT R. WILKINSON (11), *Department of Pharmacology, Vanderbilt University, Nashville, Tennessee 37232*

Preface

The study of cytochrome P450s continues to grow at a rapid pace. The Preface to Volume 206 of *Methods in Enzymology* on Cytochrome P450, published in 1991, noted 160 distinct P450s largely of mammalian origin. A quick check at this time indicates a count of 477 P450 sequences distributed among 127 subfamilies comprising 70 families of which only 26 and 14, respectively, are mammalian. This reflects a rapid increase in the identification and characterization of plant, insect, and microbial P450s made possible by the techniques of molecular biology.

This volume includes several chapters that focus on methods for the investigation of P450s in these areas as well as on advances in general techniques for the heterologous expression of P450s from cloned sequences. These approaches have also expanded our knowledge of the human P450 enzymes, and a number of chapters that address the enzymatic characterization of variations among individuals in the expression of the human enzymes, as well as procedures for characterizing associated genotypes, are included. Moreover, three-dimensional structures are now available for three additional P450s (compared to the single structure available in 1991), providing an understanding of conserved features of P450s and significant refinements in the prediction of P450 structures and substrate specificities.

Once again we have asked contributors to provide their detailed experimental procedures, and they did so with enthusiasm and promptness. We hope readers will find this compilation of protocols useful in their work. We thank the authors for their contributions and the staff of Academic Press for their help in the preparation of this volume.

ERIC F. JOHNSON
MICHAEL R. WATERMAN

METHODS IN ENZYMOLOGY

VOLUME I. Preparation and Assay of Enzymes
Edited by SIDNEY P. COLOWICK AND NATHAN O. KAPLAN

VOLUME II. Preparation and Assay of Enzymes
Edited by SIDNEY P. COLOWICK AND NATHAN O. KAPLAN

VOLUME III. Preparation and Assay of Substrates
Edited by SIDNEY P. COLOWICK AND NATHAN O. KAPLAN

VOLUME IV. Special Techniques for the Enzymologist
Edited by SIDNEY P. COLOWICK AND NATHAN O. KAPLAN

VOLUME V. Preparation and Assay of Enzymes
Edited by SIDNEY P. COLOWICK AND NATHAN O. KAPLAN

VOLUME VI. Preparation and Assay of Enzymes (*Continued*)
Preparation and Assay of Substrates
Special Techniques
Edited by SIDNEY P. COLOWICK AND NATHAN O. KAPLAN

VOLUME VII. Cumulative Subject Index
Edited by SIDNEY P. COLOWICK AND NATHAN O. KAPLAN

VOLUME VIII. Complex Carbohydrates
Edited by ELIZABETH F. NEUFELD AND VICTOR GINSBURG

VOLUME IX. Carbohydrate Metabolism
Edited by WILLIS A. WOOD

VOLUME X. Oxidation and Phosphorylation
Edited by RONALD W. ESTABROOK AND MAYNARD E. PULLMAN

VOLUME XI. Enzyme Structure
Edited by C. H. W. HIRS

VOLUME XII. Nucleic Acids (Parts A and B)
Edited by LAWRENCE GROSSMAN AND KIVIE MOLDAVE

VOLUME XIII. Citric Acid Cycle
Edited by J. M. LOWENSTEIN

VOLUME XIV. Lipids
Edited by J. M. LOWENSTEIN

VOLUME XV. Steroids and Terpenoids
Edited by RAYMOND B. CLAYTON

VOLUME XVI. Fast Reactions
Edited by KENNETH KUSTIN

VOLUME XVII. Metabolism of Amino Acids and Amines (Parts A and B)
Edited by HERBERT TABOR AND CELIA WHITE TABOR

VOLUME XVIII. Vitamins and Coenzymes (Parts A, B, and C)
Edited by DONALD B. MCCORMICK AND LEMUEL D. WRIGHT

VOLUME XIX. Proteolytic Enzymes
Edited by GERTRUDE E. PERLMANN AND LASZLO LORAND

VOLUME XX. Nucleic Acids and Protein Synthesis (Part C)
Edited by KIVIE MOLDAVE AND LAWRENCE GROSSMAN

VOLUME XXI. Nucleic Acids (Part D)
Edited by LAWRENCE GROSSMAN AND KIVIE MOLDAVE

VOLUME XXII. Enzyme Purification and Related Techniques
Edited by WILLIAM B. JAKOBY

VOLUME XXIII. Photosynthesis (Part A)
Edited by ANTHONY SAN PIETRO

VOLUME XXIV. Photosynthesis and Nitrogen Fixation (Part B)
Edited by ANTHONY SAN PIETRO

VOLUME XXV. Enzyme Structure (Part B)
Edited by C. H. W. HIRS AND SERGE N. TIMASHEFF

VOLUME XXVI. Enzyme Structure (Part C)
Edited by C. H. W. HIRS AND SERGE N. TIMASHEFF

VOLUME XXVII. Enzyme Structure (Part D)
Edited by C. H. W. HIRS AND SERGE N. TIMASHEFF

VOLUME XXVIII. Complex Carbohydrates (Part B)
Edited by VICTOR GINSBURG

VOLUME XXIX. Nucleic Acids and Protein Synthesis (Part E)
Edited by LAWRENCE GROSSMAN AND KIVIE MOLDAVE

VOLUME XXX. Nucleic Acids and Protein Synthesis (Part F)
Edited by KIVIE MOLDAVE AND LAWRENCE GROSSMAN

VOLUME XXXI. Biomembranes (Part A)
Edited by SIDNEY FLEISCHER AND LESTER PACKER

VOLUME XXXII. Biomembranes (Part B)
Edited by SIDNEY FLEISCHER AND LESTER PACKER

VOLUME XXXIII. Cumulative Subject Index Volumes I–XXX
Edited by MARTHA G. DENNIS AND EDWARD A. DENNIS

VOLUME XXXIV. Affinity Techniques (Enzyme Purification: Part B)
Edited by WILLIAM B. JAKOBY AND MEIR WILCHEK

VOLUME XXXV. Lipids (Part B)
Edited by JOHN M. LOWENSTEIN

VOLUME XXXVI. Hormone Action (Part A: Steroid Hormones)
Edited by BERT W. O'MALLEY AND JOEL G. HARDMAN

VOLUME XXXVII. Hormone Action (Part B: Peptide Hormones)
Edited by BERT W. O'MALLEY AND JOEL G. HARDMAN

VOLUME XXXVIII. Hormone Action (Part C: Cyclic Nucleotides)
Edited by JOEL G. HARDMAN AND BERT W. O'MALLEY

VOLUME XXXIX. Hormone Action (Part D: Isolated Cells, Tissues, and Organ Systems)
Edited by JOEL G. HARDMAN AND BERT W. O'MALLEY

VOLUME XL. Hormone Action (Part E: Nuclear Structure and Function)
Edited by BERT W. O'MALLEY AND JOEL G. HARDMAN

VOLUME XLI. Carbohydrate Metabolism (Part B)
Edited by W. A. WOOD

VOLUME XLII. Carbohydrate Metabolism (Part C)
Edited by W. A. WOOD

VOLUME XLIII. Antibiotics
Edited by JOHN H. HASH

VOLUME XLIV. Immobilized Enzymes
Edited by KLAUS MOSBACH

VOLUME XLV. Proteolytic Enzymes (Part B)
Edited by LASZLO LORAND

VOLUME XLVI. Affinity Labeling
Edited by WILLIAM B. JAKOBY AND MEIR WILCHEK

VOLUME XLVII. Enzyme Structure (Part E)
Edited by C. H. W. HIRS AND SERGE N. TIMASHEFF

VOLUME XLVIII. Enzyme Structure (Part F)
Edited by C. H. W. HIRS AND SERGE N. TIMASHEFF

VOLUME XLIX. Enzyme Structure (Part G)
Edited by C. H. W. HIRS AND SERGE N. TIMASHEFF

VOLUME L. Complex Carbohydrates (Part C)
Edited by VICTOR GINSBURG

VOLUME LI. Purine and Pyrimidine Nucleotide Metabolism
Edited by PATRICIA A. HOFFEE AND MARY ELLEN JONES

VOLUME LII. Biomembranes (Part C: Biological Oxidations)
Edited by SIDNEY FLEISCHER AND LESTER PACKER

VOLUME LIII. Biomembranes (Part D: Biological Oxidations)
Edited by SIDNEY FLEISCHER AND LESTER PACKER

VOLUME LIV. Biomembranes (Part E: Biological Oxidations)
Edited by SIDNEY FLEISCHER AND LESTER PACKER

VOLUME LV. Biomembranes (Part F: Bioenergetics)
Edited by SIDNEY FLEISCHER AND LESTER PACKER

VOLUME LVI. Biomembranes (Part G: Bioenergetics)
Edited by SIDNEY FLEISCHER AND LESTER PACKER

VOLUME LVII. Bioluminescence and Chemiluminescence
Edited by MARLENE A. DELUCA

VOLUME LVIII. Cell Culture
Edited by WILLIAM B. JAKOBY AND IRA PASTAN

VOLUME LIX. Nucleic Acids and Protein Synthesis (Part G)
Edited by KIVIE MOLDAVE AND LAWRENCE GROSSMAN

VOLUME LX. Nucleic Acids and Protein Synthesis (Part H)
Edited by KIVIE MOLDAVE AND LAWRENCE GROSSMAN

VOLUME 61. Enzyme Structure (Part H)
Edited by C. H. W. HIRS AND SERGE N. TIMASHEFF

VOLUME 62. Vitamins and Coenzymes (Part D)
Edited by DONALD B. MCCORMICK AND LEMUEL D. WRIGHT

VOLUME 63. Enzyme Kinetics and Mechanism (Part A: Initial Rate and Inhibitor Methods)
Edited by DANIEL L. PURICH

VOLUME 64. Enzyme Kinetics and Mechanism (Part B: Isotopic Probes and Complex Enzyme Systems)
Edited by DANIEL L. PURICH

VOLUME 65. Nucleic Acids (Part I)
Edited by LAWRENCE GROSSMAN AND KIVIE MOLDAVE

VOLUME 66. Vitamins and Coenzymes (Part E)
Edited by DONALD B. MCCORMICK AND LEMUEL D. WRIGHT

VOLUME 67. Vitamins and Coenzymes (Part F)
Edited by DONALD B. MCCORMICK AND LEMUEL D. WRIGHT

VOLUME 68. Recombinant DNA
Edited by RAY WU

VOLUME 69. Photosynthesis and Nitrogen Fixation (Part C)
Edited by ANTHONY SAN PIETRO

VOLUME 70. Immunochemical Techniques (Part A)
Edited by HELEN VAN VUNAKIS AND JOHN J. LANGONE

VOLUME 71. Lipids (Part C)
Edited by JOHN M. LOWENSTEIN

VOLUME 72. Lipids (Part D)
Edited by JOHN M. LOWENSTEIN

VOLUME 73. Immunochemical Techniques (Part B)
Edited by JOHN J. LANGONE AND HELEN VAN VUNAKIS

VOLUME 74. Immunochemical Techniques (Part C)
Edited by JOHN J. LANGONE AND HELEN VAN VUNAKIS

VOLUME 75. Cumulative Subject Index Volumes XXXI, XXXII, XXXIV–LX
Edited by EDWARD A. DENNIS AND MARTHA G. DENNIS

VOLUME 76. Hemoglobins
Edited by ERALDO ANTONINI, LUIGI ROSSI-BERNARDI, AND EMILIA CHIANCONE

VOLUME 77. Detoxication and Drug Metabolism
Edited by WILLIAM B. JAKOBY

VOLUME 78. Interferons (Part A)
Edited by SIDNEY PESTKA

VOLUME 79. Interferons (Part B)
Edited by SIDNEY PESTKA

VOLUME 80. Proteolytic Enzymes (Part C)
Edited by LASZLO LORAND

VOLUME 81. Biomembranes (Part H: Visual Pigments and Purple Membranes, I)
Edited by LESTER PACKER

VOLUME 82. Structural and Contractile Proteins (Part A: Extracellular Matrix)
Edited by LEON W. CUNNINGHAM AND DIXIE W. FREDERIKSEN

VOLUME 83. Complex Carbohydrates (Part D)
Edited by VICTOR GINSBURG

VOLUME 84. Immunochemical Techniques (Part D: Selected Immunoassays)
Edited by JOHN J. LANGONE AND HELEN VAN VUNAKIS

VOLUME 85. Structural and Contractile Proteins (Part B: The Contractile Apparatus and the Cytoskeleton)
Edited by DIXIE W. FREDERIKSEN AND LEON W. CUNNINGHAM

VOLUME 86. Prostaglandins and Arachidonate Metabolites
Edited by WILLIAM E. M. LANDS AND WILLIAM L. SMITH

VOLUME 87. Enzyme Kinetics and Mechanism (Part C: Intermediates, Stereochemistry, and Rate Studies)
Edited by DANIEL L. PURICH

VOLUME 88. Biomembranes (Part I: Visual Pigments and Purple Membranes, II)
Edited by LESTER PACKER

VOLUME 89. Carbohydrate Metabolism (Part D)
Edited by WILLIS A. WOOD

VOLUME 90. Carbohydrate Metabolism (Part E)
Edited by WILLIS A. WOOD

VOLUME 91. Enzyme Structure (Part I)
Edited by C. H. W. HIRS AND SERGE N. TIMASHEFF

VOLUME 92. Immunochemical Techniques (Part E: Monoclonal Antibodies and General Immunoassay Methods)
Edited by JOHN J. LANGONE AND HELEN VAN VUNAKIS

VOLUME 93. Immunochemical Techniques (Part F: Conventional Antibodies, Fc Receptors, and Cytotoxicity)
Edited by JOHN J. LANGONE AND HELEN VAN VUNAKIS

VOLUME 94. Polyamines
Edited by HERBERT TABOR AND CELIA WHITE TABOR

VOLUME 95. Cumulative Subject Index Volumes 61–74, 76–80
Edited by EDWARD A. DENNIS AND MARTHA G. DENNIS

VOLUME 96. Biomembranes [Part J: Membrane Biogenesis: Assembly and Targeting (General Methods; Eukaryotes)]
Edited by SIDNEY FLEISCHER AND BECCA FLEISCHER

VOLUME 97. Biomembranes [Part K: Membrane Biogenesis: Assembly and Targeting (Prokaryotes, Mitochondria, and Chloroplasts)]
Edited by SIDNEY FLEISCHER AND BECCA FLEISCHER

VOLUME 98. Biomembranes (Part L: Membrane Biogenesis: Processing and Recycling)
Edited by SIDNEY FLEISCHER AND BECCA FLEISCHER

VOLUME 99. Hormone Action (Part F: Protein Kinases)
Edited by JACKIE D. CORBIN AND JOEL G. HARDMAN

VOLUME 100. Recombinant DNA (Part B)
Edited by RAY WU, LAWRENCE GROSSMAN, AND KIVIE MOLDAVE

VOLUME 101. Recombinant DNA (Part C)
Edited by RAY WU, LAWRENCE GROSSMAN, AND KIVIE MOLDAVE

VOLUME 102. Hormone Action (Part G: Calmodulin and Calcium-Binding Proteins)
Edited by ANTHONY R. MEANS AND BERT W. O'MALLEY

VOLUME 103. Hormone Action (Part H: Neuroendocrine Peptides)
Edited by P. MICHAEL CONN

VOLUME 104. Enzyme Purification and Related Techniques (Part C)
Edited by WILLIAM B. JAKOBY

VOLUME 105. Oxygen Radicals in Biological Systems
Edited by LESTER PACKER

VOLUME 106. Posttranslational Modifications (Part A)
Edited by FINN WOLD AND KIVIE MOLDAVE

VOLUME 107. Posttranslational Modifications (Part B)
Edited by FINN WOLD AND KIVIE MOLDAVE

VOLUME 108. Immunochemical Techniques (Part G: Separation and Characterization of Lymphoid Cells)
Edited by GIOVANNI DI SABATO, JOHN J. LANGONE, AND HELEN VAN VUNAKIS

VOLUME 109. Hormone Action (Part I: Peptide Hormones)
Edited by LUTZ BIRNBAUMER AND BERT W. O'MALLEY

VOLUME 110. Steroids and Isoprenoids (Part A)
Edited by JOHN H. LAW AND HANS C. RILLING

VOLUME 111. Steroids and Isoprenoids (Part B)
Edited by JOHN H. LAW AND HANS C. RILLING

VOLUME 112. Drug and Enzyme Targeting (Part A)
Edited by KENNETH J. WIDDER AND RALPH GREEN

VOLUME 113. Glutamate, Glutamine, Glutathione, and Related Compounds
Edited by ALTON MEISTER

VOLUME 114. Diffraction Methods for Biological Macromolecules (Part A)
Edited by HAROLD W. WYCKOFF, C. H. W. HIRS, AND SERGE N. TIMASHEFF

VOLUME 115. Diffraction Methods for Biological Macromolecules (Part B)
Edited by HAROLD W. WYCKOFF, C. H. W. HIRS, AND SERGE N. TIMASHEFF

VOLUME 116. Immunochemical Techniques (Part H: Effectors and Mediators of Lymphoid Cell Functions)
Edited by GIOVANNI DI SABATO, JOHN J. LANGONE, AND HELEN VAN VUNAKIS

VOLUME 117. Enzyme Structure (Part J)
Edited by C. H. W. HIRS AND SERGE N. TIMASHEFF

VOLUME 118. Plant Molecular Biology
Edited by ARTHUR WEISSBACH AND HERBERT WEISSBACH

VOLUME 119. Interferons (Part C)
Edited by SIDNEY PESTKA

VOLUME 120. Cumulative Subject Index Volumes 81–94, 96–101

VOLUME 121. Immunochemical Techniques (Part I: Hybridoma Technology and Monoclonal Antibodies)
Edited by JOHN J. LANGONE AND HELEN VAN VUNAKIS

VOLUME 122. Vitamins and Coenzymes (Part G)
Edited by FRANK CHYTIL AND DONALD B. MCCORMICK

VOLUME 123. Vitamins and Coenzymes (Part H)
Edited by FRANK CHYTIL AND DONALD B. MCCORMICK

VOLUME 124. Hormone Action (Part J: Neuroendocrine Peptides)
Edited by P. MICHAEL CONN

VOLUME 125. Biomembranes (Part M: Transport in Bacteria, Mitochondria, and Chloroplasts: General Approaches and Transport Systems)
Edited by SIDNEY FLEISCHER AND BECCA FLEISCHER

VOLUME 126. Biomembranes (Part N: Transport in Bacteria, Mitochondria, and Chloroplasts: Protonmotive Force)
Edited by SIDNEY FLEISCHER AND BECCA FLEISCHER

VOLUME 127. Biomembranes (Part O: Protons and Water: Structure and Translocation)
Edited by LESTER PACKER

VOLUME 128. Plasma Lipoproteins (Part A: Preparation, Structure, and Molecular Biology)
Edited by JERE P. SEGREST AND JOHN J. ALBERS

VOLUME 129. Plasma Lipoproteins (Part B: Characterization, Cell Biology, and Metabolism)
Edited by JOHN J. ALBERS AND JERE P. SEGREST

VOLUME 130. Enzyme Structure (Part K)
Edited by C. H. W. HIRS AND SERGE N. TIMASHEFF

VOLUME 131. Enzyme Structure (Part L)
Edited by C. H. W. HIRS AND SERGE N. TIMASHEFF

VOLUME 132. Immunochemical Techniques (Part J: Phagocytosis and Cell-Mediated Cytotoxicity)
Edited by GIOVANNI DI SABATO AND JOHANNES EVERSE

VOLUME 133. Bioluminescence and Chemiluminescence (Part B)
Edited by MARLENE DELUCA AND WILLIAM D. MCELROY

VOLUME 134. Structural and Contractile Proteins (Part C: The Contractile Apparatus and the Cytoskeleton)
Edited by RICHARD B. VALLEE

VOLUME 135. Immobilized Enzymes and Cells (Part B)
Edited by KLAUS MOSBACH

VOLUME 136. Immobilized Enzymes and Cells (Part C)
Edited by KLAUS MOSBACH

VOLUME 137. Immobilized Enzymes and Cells (Part D)
Edited by KLAUS MOSBACH

VOLUME 138. Complex Carbohydrates (Part E)
Edited by VICTOR GINSBURG

VOLUME 139. Cellular Regulators (Part A: Calcium- and Calmodulin-Binding Proteins)
Edited by ANTHONY R. MEANS AND P. MICHAEL CONN

VOLUME 140. Cumulative Subject Index Volumes 102–119, 121–134

VOLUME 141. Cellular Regulators (Part B: Calcium and Lipids)
Edited by P. MICHAEL CONN AND ANTHONY R. MEANS

VOLUME 142. Metabolism of Aromatic Amino Acids and Amines
Edited by SEYMOUR KAUFMAN

VOLUME 143. Sulfur and Sulfur Amino Acids
Edited by WILLIAM B. JAKOBY AND OWEN GRIFFITH

VOLUME 144. Structural and Contractile Proteins (Part D: Extracellular Matrix)
Edited by LEON W. CUNNINGHAM

VOLUME 145. Structural and Contractile Proteins (Part E: Extracellular Matrix)
Edited by LEON W. CUNNINGHAM

VOLUME 146. Peptide Growth Factors (Part A)
Edited by DAVID BARNES AND DAVID A. SIRBASKU

VOLUME 147. Peptide Growth Factors (Part B)
Edited by DAVID BARNES AND DAVID A. SIRBASKU

VOLUME 148. Plant Cell Membranes
Edited by LESTER PACKER AND ROLAND DOUCE

VOLUME 149. Drug and Enzyme Targeting (Part B)
Edited by RALPH GREEN AND KENNETH J. WIDDER

VOLUME 150. Immunochemical Techniques (Part K: *In Vitro* Models of B and T Cell Functions and Lymphoid Cell Receptors)
Edited by GIOVANNI DI SABATO

VOLUME 151. Molecular Genetics of Mammalian Cells
Edited by MICHAEL M. GOTTESMAN

VOLUME 152. Guide to Molecular Cloning Techniques
Edited by SHELBY L. BERGER AND ALAN R. KIMMEL

VOLUME 153. Recombinant DNA (Part D)
Edited by RAY WU AND LAWRENCE GROSSMAN

VOLUME 154. Recombinant DNA (Part E)
Edited by RAY WU AND LAWRENCE GROSSMAN

VOLUME 155. Recombinant DNA (Part F)
Edited by RAY WU

VOLUME 156. Biomembranes (Part P: ATP-Driven Pumps and Related Transport: The Na,K-Pump)
Edited by SIDNEY FLEISCHER AND BECCA FLEISCHER

VOLUME 157. Biomembranes (Part Q: ATP-Driven Pumps and Related Transport: Calcium, Proton, and Potassium Pumps)
Edited by SIDNEY FLEISCHER AND BECCA FLEISCHER

VOLUME 158. Metalloproteins (Part A)
Edited by JAMES F. RIORDAN AND BERT L. VALLEE

VOLUME 159. Initiation and Termination of Cyclic Nucleotide Action
Edited by JACKIE D. CORBIN AND ROGER A. JOHNSON

VOLUME 160. Biomass (Part A: Cellulose and Hemicellulose)
Edited by WILLIS A. WOOD AND SCOTT T. KELLOGG

VOLUME 161. Biomass (Part B: Lignin, Pectin, and Chitin)
Edited by WILLIS A. WOOD AND SCOTT T. KELLOGG

VOLUME 162. Immunochemical Techniques (Part L: Chemotaxis and Inflammation)
Edited by GIOVANNI DI SABATO

VOLUME 163. Immunochemical Techniques (Part M: Chemotaxis and Inflammation)
Edited by GIOVANNI DI SABATO

VOLUME 164. Ribosomes
Edited by HARRY F. NOLLER, JR., AND KIVIE MOLDAVE

VOLUME 165. Microbial Toxins: Tools for Enzymology
Edited by SIDNEY HARSHMAN

VOLUME 166. Branched-Chain Amino Acids
Edited by ROBERT HARRIS AND JOHN R. SOKATCH

VOLUME 167. Cyanobacteria
Edited by LESTER PACKER AND ALEXANDER N. GLAZER

VOLUME 168. Hormone Action (Part K: Neuroendocrine Peptides)
Edited by P. MICHAEL CONN

VOLUME 169. Platelets: Receptors, Adhesion, Secretion (Part A)
Edited by JACEK HAWIGER

VOLUME 170. Nucleosomes
Edited by PAUL M. WASSARMAN AND ROGER D. KORNBERG

VOLUME 171. Biomembranes (Part R: Transport Theory: Cells and Model Membranes)
Edited by SIDNEY FLEISCHER AND BECCA FLEISCHER

VOLUME 172. Biomembranes (Part S: Transport: Membrane Isolation and Characterization)
Edited by SIDNEY FLEISCHER AND BECCA FLEISCHER

VOLUME 173. Biomembranes [Part T: Cellular and Subcellular Transport: Eukaryotic (Nonepithelial) Cells]
Edited by SIDNEY FLEISCHER AND BECCA FLEISCHER

VOLUME 174. Biomembranes [Part U: Cellular and Subcellular Transport: Eukaryotic (Nonepithelial) Cells]
Edited by SIDNEY FLEISCHER AND BECCA FLEISCHER

VOLUME 175. Cumulative Subject Index Volumes 135–139, 141–167

VOLUME 176. Nuclear Magnetic Resonance (Part A: Spectral Techniques and Dynamics)
Edited by NORMAN J. OPPENHEIMER AND THOMAS L. JAMES

VOLUME 177. Nuclear Magnetic Resonance (Part B: Structure and Mechanism)
Edited by NORMAN J. OPPENHEIMER AND THOMAS L. JAMES

VOLUME 178. Antibodies, Antigens, and Molecular Mimicry
Edited by JOHN J. LANGONE

VOLUME 179. Complex Carbohydrates (Part F)
Edited by VICTOR GINSBURG

VOLUME 180. RNA Processing (Part A: General Methods)
Edited by JAMES E. DAHLBERG AND JOHN N. ABELSON

VOLUME 181. RNA Processing (Part B: Specific Methods)
Edited by JAMES E. DAHLBERG AND JOHN N. ABELSON

VOLUME 182. Guide to Protein Purification
Edited by MURRAY P. DEUTSCHER

VOLUME 183. Molecular Evolution: Computer Analysis of Protein and Nucleic Acid Sequences
Edited by RUSSELL F. DOOLITTLE

VOLUME 184. Avidin–Biotin Technology
Edited by MEIR WILCHEK AND EDWARD A. BAYER

VOLUME 185. Gene Expression Technology
Edited by DAVID V. GOEDDEL

VOLUME 186. Oxygen Radicals in Biological Systems (Part B: Oxygen Radicals and Antioxidants)
Edited by LESTER PACKER AND ALEXANDER N. GLAZER

VOLUME 187. Arachidonate Related Lipid Mediators
Edited by ROBERT C. MURPHY AND FRANK A. FITZPATRICK

VOLUME 188. Hydrocarbons and Methylotrophy
Edited by MARY E. LIDSTROM

VOLUME 189. Retinoids (Part A: Molecular and Metabolic Aspects)
Edited by LESTER PACKER

VOLUME 190. Retinoids (Part B: Cell Differentiation and Clinical Applications)
Edited by LESTER PACKER

VOLUME 191. Biomembranes (Part V: Cellular and Subcellular Transport: Epithelial Cells)
Edited by SIDNEY FLEISCHER AND BECCA FLEISCHER

VOLUME 192. Biomembranes (Part W: Cellular and Subcellular Transport: Epithelial Cells)
Edited by SIDNEY FLEISCHER AND BECCA FLEISCHER

VOLUME 193. Mass Spectrometry
Edited by JAMES A. MCCLOSKEY

VOLUME 194. Guide to Yeast Genetics and Molecular Biology
Edited by CHRISTINE GUTHRIE AND GERALD R. FINK

VOLUME 195. Adenylyl Cyclase, G Proteins, and Guanylyl Cyclase
Edited by ROGER A. JOHNSON AND JACKIE D. CORBIN

VOLUME 196. Molecular Motors and the Cytoskeleton
Edited by RICHARD B. VALLEE

VOLUME 197. Phospholipases
Edited by EDWARD A. DENNIS

VOLUME 198. Peptide Growth Factors (Part C)
Edited by DAVID BARNES, J. P. MATHER, AND GORDON H. SATO

VOLUME 199. Cumulative Subject Index Volumes 168–174, 176–194

VOLUME 200. Protein Phosphorylation (Part A: Protein Kinases: Assays, Purification, Antibodies, Functional Analysis, Cloning, and Expression)
Edited by TONY HUNTER AND BARTHOLOMEW M. SEFTON

VOLUME 201. Protein Phosphorylation (Part B: Analysis of Protein Phosphorylation, Protein Kinase Inhibitors, and Protein Phosphatases)
Edited by TONY HUNTER AND BARTHOLOMEW M. SEFTON

VOLUME 202. Molecular Design and Modeling: Concepts and Applications (Part
A: Proteins, Peptides, and Enzymes)
Edited by JOHN J. LANGONE

VOLUME 203. Molecular Design and Modeling: Concepts and Applications (Part
B: Antibodies and Antigens, Nucleic Acids, Polysaccharides, and Drugs)
Edited by JOHN J. LANGONE

VOLUME 204. Bacterial Genetic Systems
Edited by JEFFREY H. MILLER

VOLUME 205. Metallobiochemistry (Part B: Metallothionein and Related Molecules)
Edited by JAMES F. RIORDAN AND BERT L. VALLEE

VOLUME 206. Cytochrome P450
Edited by MICHAEL R. WATERMAN AND ERIC F. JOHNSON

VOLUME 207. Ion Channels
Edited by BERNARDO RUDY AND LINDA E. IVERSON

VOLUME 208. Protein–DNA Interactions
Edited by ROBERT T. SAUER

VOLUME 209. Phospholipid Biosynthesis
Edited by EDWARD A. DENNIS AND DENNIS E. VANCE

VOLUME 210. Numerical Computer Methods
Edited by LUDWIG BRAND AND MICHAEL L. JOHNSON

VOLUME 211. DNA Structures (Part A: Synthesis and Physical Analysis of
DNA)
Edited by DAVID M. J. LILLEY AND JAMES E. DAHLBERG

VOLUME 212. DNA Structures (Part B: Chemical and Electrophoretic Analysis
of DNA)
Edited by DAVID M. J. LILLEY AND JAMES E. DAHLBERG

VOLUME 213. Carotenoids (Part A: Chemistry, Separation, Quantitation, and Antioxidation)
Edited by LESTER PACKER

VOLUME 214. Carotenoids (Part B: Metabolism, Genetics, and Biosynthesis)
Edited by LESTER PACKER

VOLUME 215. Platelets: Receptors, Adhesion, Secretion (Part B)
Edited by JACEK J. HAWIGER

VOLUME 216. Recombinant DNA (Part G)
Edited by RAY WU

VOLUME 217. Recombinant DNA (Part H)
Edited by RAY WU

VOLUME 218. Recombinant DNA (Part I)
Edited by RAY WU

VOLUME 219. Reconstitution of Intracellular Transport
Edited by JAMES E. ROTHMAN

VOLUME 220. Membrane Fusion Techniques (Part A)
Edited by NEJAT DÜZGÜNEŞ

VOLUME 221. Membrane Fusion Techniques (Part B)
Edited by NEJAT DÜZGÜNEŞ

VOLUME 222. Proteolytic Enzymes in Coagulation, Fibrinolysis, and Complement Activation (Part A: Mammalian Blood Coagulation Factors and Inhibitors)
Edited by LASZLO LORAND AND KENNETH G. MANN

VOLUME 223. Proteolytic Enzymes in Coagulation, Fibrinolysis, and Complement Activation (Part B: Complement Activation, Fibrinolysis, and Nonmammalian Blood Coagulation Factors)
Edited by LASZLO LORAND AND KENNETH G. MANN

VOLUME 224. Molecular Evolution: Producing the Biochemical Data
Edited by ELIZABETH ANNE ZIMMER, THOMAS J. WHITE, REBECCA L. CANN, AND ALLAN C. WILSON

VOLUME 225. Guide to Techniques in Mouse Development
Edited by PAUL M. WASSARMAN AND MELVIN L. DEPAMPHILIS

VOLUME 226. Metallobiochemistry (Part C: Spectroscopic and Physical Methods for Probing Metal Ion Environments in Metalloenzymes and Metalloproteins)
Edited by JAMES F. RIORDAN AND BERT L. VALLEE

VOLUME 227. Metallobiochemistry (Part D: Physical and Spectroscopic Methods for Probing Metal Ion Environments in Metalloproteins)
Edited by JAMES F. RIORDAN AND BERT L. VALLEE

VOLUME 228. Aqueous Two-Phase Systems
Edited by HARRY WALTER AND GÖTE JOHANSSON

VOLUME 229. Cumulative Subject Index Volumes 195–198, 200–227

VOLUME 230. Guide to Techniques in Glycobiology
Edited by WILLIAM J. LENNARZ AND GERALD W. HART

VOLUME 231. Hemoglobins (Part B: Biochemical and Analytical Methods)
Edited by JOHANNES EVERSE, KIM D. VANDEGRIFF, AND ROBERT M. WINSLOW

VOLUME 232. Hemoglobins (Part C: Biophysical Methods)
Edited by JOHANNES EVERSE, KIM D. VANDEGRIFF, AND ROBERT M. WINSLOW

VOLUME 233. Oxygen Radicals in Biological Systems (Part C)
Edited by LESTER PACKER

VOLUME 234. Oxygen Radicals in Biological Systems (Part D)
Edited by LESTER PACKER

VOLUME 235. Bacterial Pathogenesis (Part A: Identification and Regulation of Virulence Factors)
Edited by VIRGINIA L. CLARK AND PATRIK M. BAVOIL

VOLUME 236. Bacterial Pathogenesis (Part B: Integration of Pathogenic Bacteria with Host Cells)
Edited by VIRGINIA L. CLARK AND PATRIK M. BAVOIL

VOLUME 237. Heterotrimeric G Proteins
Edited by RAVI IYENGAR

VOLUME 238. Heterotrimeric G-Protein Effectors
Edited by RAVI IYENGAR

VOLUME 239. Nuclear Magnetic Resonance (Part C)
Edited by THOMAS L. JAMES AND NORMAN J. OPPENHEIMER

VOLUME 240. Numerical Computer Methods (Part B)
Edited by MICHAEL L. JOHNSON AND LUDWIG BRAND

VOLUME 241. Retroviral Proteases
Edited by LAWRENCE C. KUO AND JULES A. SHAFER

VOLUME 242. Neoglycoconjugates (Part A)
Edited by Y. C. LEE AND REIKO T. LEE

VOLUME 243. Inorganic Microbial Sulfur Metabolism
Edited by HARRY D. PECK, JR., AND JEAN LeGALL

VOLUME 244. Proteolytic Enzymes: Serine and Cysteine Peptidases
Edited by ALAN J. BARRETT

VOLUME 245. Extracellular Matrix Components
Edited by E. RUOSLAHTI AND E. ENGVALL

VOLUME 246. Biochemical Spectroscopy
Edited by KENNETH SAUER

VOLUME 247. Neoglycoconjugates (Part B: Biomedical Applications)
Edited by Y. C. LEE AND REIKO T. LEE

VOLUME 248. Proteolytic Enzymes: Aspartic and Metallo Peptidases
Edited by ALAN J. BARRETT

VOLUME 249. Enzyme Kinetics and Mechanism (Part D: Developments in Enzyme Dynamics)
Edited by DANIEL L. PURICH

VOLUME 250. Lipid Modifications of Proteins
Edited by PATRICK J. CASEY AND JANICE E. BUSS

VOLUME 251. Biothiols (Part A: Monothiols and Dithiols, Protein Thiols, and Thiyl Radicals)
Edited by LESTER PACKER

VOLUME 252. Biothiols (Part B: Glutathione and Thioredoxin; Thiols in Signal Transduction and Gene Regulation)
Edited by LESTER PACKER

VOLUME 253. Adhesion of Microbial Pathogens
Edited by RON J. DOYLE AND ITZHAK OFEK

VOLUME 254. Oncogene Techniques
Edited by PETER K. VOGT AND INDER M. VERMA

VOLUME 255. Small GTPases and Their Regulators (Part A: Ras Family)
Edited by W. E. BALCH, CHANNING J. DER, AND ALAN HALL

VOLUME 256. Small GTPases and Their Regulators (Part B: Rho Family)
Edited by W. E. BALCH, CHANNING J. DER, AND ALAN HALL

VOLUME 257. Small GTPases and Their Regulators (Part C: Proteins Involved in Transport)
Edited by W. E. BALCH, CHANNING J. DER, AND ALAN HALL

VOLUME 258. Redox-Active Amino Acids in Biology
Edited by JUDITH P. KLINMAN

VOLUME 259. Energetics of Biological Macromolecules
Edited by MICHAEL L. JOHNSON AND GARY K. ACKERS

VOLUME 260. Mitochondrial Biogenesis and Genetics (Part A)
Edited by GIUSEPPE M. ATTARDI AND ANNE CHOMYN

VOLUME 261. Nuclear Magnetic Resonance and Nucleic Acids
Edited by THOMAS L. JAMES

VOLUME 262. DNA Replication
Edited by JUDITH L. CAMPBELL

VOLUME 263. Plasma Lipoproteins (Part C: Quantitation)
Edited by WILLIAM A. BRADLEY, SANDRA H. GIANTURCO, AND JERE P. SEGREST

VOLUME 264. Mitochondrial Biogenesis and Genetics (Part B)
Edited by GIUSEPPE M. ATTARDI AND ANNE CHOMYN

VOLUME 265. Cumulative Subject Index Volumes 228, 230–262

VOLUME 266. Computer Methods for Macromolecular Sequence Analysis
Edited by RUSSELL F. DOOLITTLE

VOLUME 267. Combinatorial Chemistry
Edited by JOHN N. ABELSON

VOLUME 268. Nitric Oxide (Part A: Sources and Detection of NO; NO Synthase)
Edited by LESTER PACKER

VOLUME 269. Nitric Oxide (Part B: Physiological and Pathological Processes)
Edited by LESTER PACKER

VOLUME 270. High Resolution Separation and Analysis of Biological Macromolecules (Part A: Fundamentals)
Edited by BARRY L. KARGER AND WILLIAM S. HANCOCK

VOLUME 271. High Resolution Separation and Analysis of Biological Macromolecules (Part B: Applications)
Edited by BARRY L. KARGER AND WILLIAM S. HANCOCK

VOLUME 272. Cytochrome P450 (Part B)
Edited by ERIC F. JOHNSON AND MICHAEL R. WATERMAN

VOLUME 273. RNA Polymerase and Associated Factors (Part A) (in preparation)
Edited by SANKAR ADHYA

VOLUME 274. RNA Polymerase and Associated Factors (Part B) (in preparation)
Edited by SANKAR ADHYA

VOLUME 275. Viral Polymerases and Related Proteins (in preparation)
Edited by LAWRENCE C. KUO, DAVID B. OLSEN, AND STEVEN S. CARROLL

VOLUME 276. Macromolecular Crystallography (Part A) (in preparation)
Edited by CHARLES W. CARTER, JR. AND ROBERT M. SWEET

Section I

Heterologous Expression of P450s

[1] Maximizing Expression of Eukaryotic Cytochrome P450s in *Escherichia coli*

By Henry J. Barnes

The high level, heterologous expression of eukaryotic cytochrome P450 enzymes presents several problems to the host cell. These proteins are tightly associated with cellular membranes which places an increased burden on this subcellular compartment. Furthermore, active enzyme requires the proper coordination of noncovalently bound heme which must be produced and incorporated into the nascent polypeptide. Several heterologous expression systems have been used successfully to express cytochrome P450s.[1,2] The bacterium *Escherichia coli* has been harnessed for generating large quantities of enzymatically active protein.[3-9] Numerous cytochrome P450s have been produced in this manner and expression levels have been impressive, exceeding 500 nmol/liter in several instances.[3,9] This chapter concentrates on various aspects that have been found to be important for achieving high expression levels of these enzymes in *E. coli*.

Background

Successful synthesis of cytochrome P450s in *E. coli* requires the cDNA encoding a particular P450 to be cloned downstream of an efficient promoter and ribosome-binding site situated on a multicopy number plasmid. The promoter directs the synthesis of hybrid mRNA molecules that contain signals immediately upstream of the initiation codon that directs *E. coli* ribosomes to initiate translation of the inserted cDNA sequence. Once

---

[1] M. R. Waterman and E. F. Johnson, *Methods Enzymol.* **206**, Section II, 85–145 (1991).

[2] F. J. Gonzalez and K. R. Korzekwa, *Annu. Rev. Pharmacol. Toxicol.* **35**, 369 (1995).

[3] T. H. Richardson, F. Jung, K. J. Griffin, M. Wester, J. L. Raucy, B. Kemper, L. M. Bornheim, C. Hassett, C. J. Omiecinski, and E. F. Johnson, *Arch. Biochem. Biophys.* **323**, 87 (1995).

[4] B. A. Halkier, H. C. Nielsen, B. Koch, and B. L. Møller, *Arch. Biochem. Biophys.* **322**, 369 (1995).

[5] E. M. J. Gilliam, Z. Guo, M. V. Martin, C. M. Jenkins, and F. P. Guengerich, *Arch. Biochem. Biophys.* **319**, 540 (1995).

[6] G. H. John, J. A. Hasler, Y. He, and J. R. Halpert, *Arch. Biochem. Biophys.* **314**, 367 (1994).

[7] J. F. Anderson, J. G. Utermohlen, and R. Feyereisen, *Biochemistry* **33**, 2171 (1994).

[8] T. Imai, H. Globerman, J. M. Gertner, N. Kagawa, and M. R. Waterman, *J. Biol. Chem.* **268**, 19681 (1993).

[9] C. W. Fisher, D. L. Caudle, C. Martin-Wixtrom, L. C. Quattrochi, R. H. Tukey, M. R. Waterman, and R. W. Estabrook, *FASEB J.* **6**, 759 (1992).

translated, the nascent polypeptides associate with *E. coli* membranes, bind a molecule of heme, and fold into a final tertiary structure. The events that culminate in proper protein folding and membrane insertion are complex and poorly understood. Clearly, the elevated synthesis of a membrane-bound cytochrome P450 in *E. coli* would elicit drastic changes in cellular metabolism related to heme and lipid biosynthesis. In addition, any proteins required for the folding (Gro ES or Gro EL) or membrane targeting (e.g., products of the *E. coli* Sec genes) may need to be present in greater amounts than found under normal physiological conditions. If the expression rate of recombinant cytochrome P450 exceeds the capability of the cell to synthesize these ancillary elements, the recombinant P450 would fail to fold correctly and irreversibly accumulate in inclusion bodies.[10] Such aggregated protein does not exhibit a characteristic cytochrome P450 spectrum and is not enzymatically active. Many of the specific requirements for the expression of active membrane-bound cytochrome P450 described in this chapter are probably conditions that allow the bacterial cell to make the appropriate metabolic adjustments for proper folding and membrane integration.

Minimum Requirements for Expression

The minimum requirements for cytochrome P450 expression in *E. coli* were delineated during the development of a heterologous expression system for bovine cytochrome P450 17α-hydroxylase (CYP17).[11,12] Comparative examination of several expression constructs revealed a number of factors that had a fundamental impact on the intracellular distribution and solubility of the recombinant P450. These factors involve specific elements contained in the expression plasmid (the promoter and *lac* repressor gene), the structure or sequence of the mRNA surrounding the initiation codon of the cytochrome P450, and idiosyncrasies of the *E. coli* strain and culture conditions. Failure to adhere to these basic requirements results in either the expression of exceedingly low amounts of protein or the production of significant amounts of recombinant protein that accumulates in inclusion bodies.[12]

Transcription: Promoter Characteristics

Plasmid-based *E. coli* expression systems involve the placement of a cDNA at an appropriate distance downstream of a bacterial promoter and

[10] D. C. Williams, R. M. Van Frank, W. L. Munt, and J. P. Burnett, *Science* **215**, 687 (1982).
[11] H. J. Barnes, M. P. Arlotto, and M. R. Waterman, *Proc. Natl. Acad. Sci. U.S.A.* **88**, 5597 (1991).
[12] H. J. Barnes, Ph.D. Thesis, The University of Texas Southwestern Medical Center at Dallas (1992).

Shine–Dalgarno (S-D) element.[13,14] One of the most popular and successful series of expression vectors utilizes the powerful T7 phage promoter.[15] T7 phage promoter-based vectors have been shown to be capable of synthesizing large amounts of recombinant bovine CYP17 protein. However, subcellular fractionation and immunoblot analysis demonstrated that the vast majority of the recombinant protein produced by T7 promoter expression plasmids was found in inclusion bodies.[12] This suggests that exceptionally strong promoters, such as the T7 phage promoter (which is approximately 15 times more powerful than the *lac* promoter[16]), may not be compatible with the production of enzymatically active, membrane-associated cytochrome P450 in *E. coli*.

Currently, the *lac* promoter, or derivatives thereof, is preferred for cytochrome P450 expression. The *in vivo* basal promoter activity can be effectively controlled by sufficient quantities of the *lac* repressor protein, most conveniently provided by a plasmid-encoded *lac* I gene. Transcription can be initiated with the addition of isopropyl β-D-thiogalactoside (IPTG).

Two *E. coli* expression vectors that have been successfully used to produce high levels of enzymatically active membrane-bound cytochrome P450 are shown in Fig. 1. Both contain *lac* or a *lac*-derived promoter(s), a unique 5′ *Nde*I restriction site, and several other unique 3′ restriction sites that facilitate the directional cloning of DNA sequences.

The vector backbone of pCW Ori+ (a derivative of plasmid pHSe5[17,18]) contains a pBR322 origin of plasmid replication, the *lac* I^q gene, the β-lactamase gene for conferring ampicillin resistance, and a bacteriophage origin of replication. The transcription/translation region contains the following: a *lacUV5* promoter and two copies of a *tac* promoter cassette followed by a translation initiation region derived from the phage T4 lysozyme gene. This DNA sequence contains a 7-bp S-D element[19] located three nucleotides 5′ of the initiation condon contained within an *Nde*I restriction site. A *trp* A transcription termination cassette is located downstream of the unique 3′ restriction sites (Fig. 1A).

[13] P. Balbas and F. Bolivar, *Methods Enzymol.* **185,** 14 (1990).
[14] L. Gold and G. D. Stormo, *Methods Enzymol.* **185,** 89 (1990).
[15] F. W. Studier, A. Rosenberg, J. J. Dunn, and J. W. Dubendorff, *Methods Enzymol.* **185,** 60 (1990).
[16] P. J. Lopez, I. Iost, and M. Dreyfus, *Nucleic Acids Res.* **22,** 1186 (1994).
[17] J. A. Gegner and F. W. Dahlquist, *Proc. Natl. Acad. Sci. U.S.A.* **88,** 750 (1991).
[18] D. C. Muchmore, I. P. McIntosh, C. B. Russell, D. E. Anderson, and F. W. Dahlquist, *Methods Enzymol.* **177,** 44 (1989).
[19] L. Gold and G. Stormo, in "*Escherichia coli* and *Salmonella typhimurium*: Cellular and Molecular Biology" (F. C. Neidhardt and J. L. Ingraham, eds.), p. 1302. American Society for Microbiology, Washington, DC, 1987.

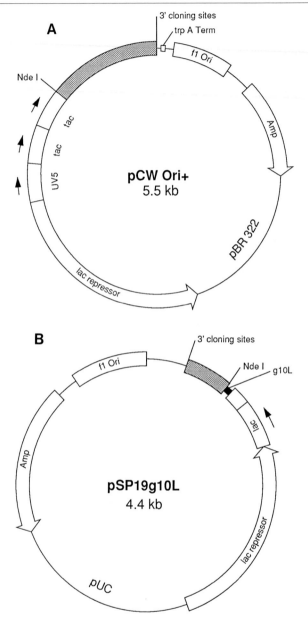

FIG. 1. Plasmid maps of pCW Ori+ and pSP19g10L. Plasmid elements are not drawn to scale. See text for details.

pSP19g10L is a derivative of pSPORT-1 (GIBCO-BRL) in which the sequence between the *Xba*I and *Hin*dIII restriction sites is replaced by 304 nucleotides consisting of a 40-bp *Xba*I–*Nde*I fragment upstream of the *Nde*I cloning site that contains the S-D element and the leader sequence from T7 phage gene 10 (22924 to 22963 of the T7 genome). This is followed by a 264-bp (183 to 447 of pUC19) *Nde*I–*Hin*dIII fragment from plasmid pUC19 that contains the polylinker of that plasmid. The vector contains a pUC origin of replication, a plasmid-encoded *lac* I gene, the β-lactamase gene, and a phage origin of replication. Transcription initiates from the *lac* promoter contained within the 5' end of a *lac* Z gene fragment (Fig. 1B).

Translation: Inhibitory Secondary Structures

Nucleotide alterations in the vicinity of the initiation codon have been shown to facilitate translation of a particular cDNA sequence. This is consistent with the idea that a primary inhibitory effect on translation initiation results from the formation of mRNA secondary structures within this region.[20,21] Joining the promoter/RBS sequence of an *E. coli* expression vector to a cDNA sequence represents the fusion of nucleic acid sequences from divergent sources (i.e., prokaryotic and eukaryotic) and unrelated genes (the *lac* operon and a cytochrome P450). Incompatibility between these sequences is a major cause for the failure of many cDNAs to be expressed in *E. coli*.[22] In many instances, modest alterations within the amino-terminal codons lead to drastic increases in the level of protein expression.[23–26]

Although the pCW plasmid contains all of the elements required for recombinant protein expression, certain nucleotide changes in the amino-terminal codons of the *CYP17* cDNA were shown to promote high level expression (~400 nmol/liter) as constructs containing the unmodified cDNA failed to produce any immuno detectable CYP17 protein.[11] The modifications of the amino-terminal codons that proved helpful for CYP17 expression were based on the information reviewed by Gold and Stormo:[19] namely, an S-D sequence of four to five nucleotides, a spacing of seven to

[20] M. H. de Smit, and J. van Duin, *Proc. Natl. Acad. Sci. U.S.A.* **87,** 7668 (1990).
[21] A. C. Looman, J. Bodlaender, M. de Gruyter, A. Vogelaar, and P. H. van Knippenberg, *Nucleic Acids Res.* **14,** 5481 (1986).
[22] L. Gold, *Methods Enzymol.* **185,** 11 (1990).
[23] N. Watson and E. R. Olson, *Gene* **86,** 137 (1990).
[24] U. S. Bucheler, D. Werner, and R. H. Schirmer, *Gene* **96,** 271 (1990).
[25] G. Morelle, R. Frank, and A. Meyerhans, *Biochim. Biophys. Acta* **1089,** 320 (1991).
[26] S. H. Rangwala, R. F. Finn, C. E. Smith, S. A. Berberich, W. J. Salsgiver, W. C. Stallings, G. I. Glover, and P. O. Olins, *Gene* **122,** 263 (1992).

nine nucleotides between the S-D sequence and the initiation codon, GCU as the second codon, the sequence UUAA embedded within the fourth and fifth codons, and avoidance of secondary structures around the ribosome-binding site. These nucleotide alterations and the corresponding amino acid changes are indicated in bold.

Native	ATG	TGG	CTG	CTC	CTG	GCT	GTC	TTT
	Met	Trp	Leu	Leu	Leu	Ala	Val	Phe
Modified	ATG	**GCT**	CTG	**TTA**	**TTA**	GCA	GTT	TTT
	Met	**Ala**	Leu	Leu	Leu	Ala	Val	Phe

A variety of eukaryotic cytochrome P450s have been successfully expressed using the pCW vector and the first eight codons of the modified bovine *CYP17* amino terminus. These methods take advantage of the fact that the leader sequence (i.e., the sequence between the promoter and the initiation codon) of the pCW vector in combination with the modified amino-terminal sequence of the bovine *CYP17* cDNA apparently creates an mRNA that is efficiently translated by *E. coli* ribosomes. This successful combination has been achieved by either of two different methods. The first consists of the direct replacement of the first eight codons of the cytochrome P450 with the amino-terminal codons of the modified bovine *CYP17* cDNA.[27] Alternatively, protein sequence alignments of several microsomal cytochrome P450 proteins have been used to first identify an N-terminal region of high similarity between bovine CYP17 and the cytochrome P450 to be expressed.[9] Even though these procedures entail a drastic alteration of the N terminus, these techniques have generated constructs that produce large quantities of recombinant cytochrome P450 whose enzymatic activity is identical or very similar to that of the native enzyme.[5,6,9,27–29] The retention of catalytic characteristics in these enzymes is consistent with the belief that the amino terminus of these proteins functions as a membrane-anchoring domain, and that this eukaryotic domain is recognized as such in *E. coli.*[30]

When these molecules are expressed in *E. coli,* the nucleotide sequence of this membrane-anchoring domain also forms part of the prokaryotic translation initiation sequence.[19] Thus, at the protein level this sequence must function as a signal sequence/membrane anchor while at the nucleo-

[27] T. H. Richardson, M.-H. Hsu, T. Kronbach, H. J. Barnes, G. Chan, M. R. Waterman, B. Kemper, and E. F. Johnson, *Arch. Biochem. Biophys.* **300,** 510 (1993).

[28] Z. Guo, E. M. J. Gilliam, S. Ohmori, R. H. Tukey, and F. P. Guengerich, *Arch. Biochem. Biophys.* **312,** 436 (1994).

[29] E. M. J. Gilliam, T. Baba, B.-R. Kim, S. Ohmori, and F. P. Guengerich, *Arch. Biochem. Biophys.* **305,** 123 (1993).

[30] R. J. Edwards, B. P. Murray, A. M. Singleton, and A. R. Boobis, *Biochemistry* **30,** 71 (1991).

tide level it must function as an efficient translation initiation sequence. Although there is subtle species specific variation in signal peptide design,[31] prokaryotic and eukaryotic secretory proteins are readily translocated across heterologous membranes.[32,33] The requirement that this sequence fulfill both of these functions may explain the results of several investigators that have found either poor expression levels or high levels of apoprotein when this sequence is altered.[4,34,35]

In contrast to results obtained with pCW, the use of the vector pSP19g10L allows the high level expression ($\sim$500 nmol/liter) of the bovine *CYP17* cDNA regardless of whether the native or modified sequence is used.[12] The ability of this vector to efficiently translate both of these cDNA sequences is in all likelihood due to the gene 10 leader sequence. This sequence has been demonstrated to direct the high level translation of a wide variety of foreign genes in *E. coli*.[36] It has a 6-bp S-D element situated an optimal distance of eight nucleotides[37] from the ATG initiation codon. Also, the A/T content of this sequence is 80% and thus would *not* be expected to form highly stable mRNA secondary structures that severely inhibit translation initiation or elongation.[38,39] Although this vector may not effectively translate all nucleotide sequences, it is notably more tolerant of a wider variety of amino-terminal sequences than is pCW.[12,40]

Method

Figure 2 is an outline of a cloning strategy for obtaining a cytochrome P450 *E. coli* expression plasmid that reduces the likelihood of sequence artifacts being introduced during the cloning. The starting material is, ideally, a cloned full-length cDNA of determined sequence and whose protein product has demonstrated enzymatic activity. The polymerase chain reaction (PCR)-amplified coding sequence is double digested and cloned into the corresponding sites of the expression vector. The amount of DNA to

[31] G. von Heijne and L. Abrahmsen, *FEBS Lett.* **244,** 439 (1989).
[32] P. D. Garcia, J. Ghrayeb, M. Inouye, and P. Walter, *J. Biol. Chem.* **262,** 9463 (1987).
[33] I. T. Fecycz and G. Blobel, *Proc. Natl. Acad. Sci. U.S.A.* **84,** 3723 (1987).
[34] E. M. J. Gilliam, Z. Guo, and F. P. Guengerich, *Arch. Biochem. Biophys.* **312,** 59 (1994).
[35] P. Sandhu, Z. Guo, T. Baba, M. Martin, R. H. Tukey, and F. P. Guengerich, *Arch. Biochem. Biophys.* **309,** 168 (1994).
[36] P. O. Olins and S. H. Rangwala, *Methods Enzymol.* **185,** 115 (1990).
[37] S. Ringquist, S. Shinedling, D. Barrick, L. Green, J. Binkley, G. D. Stormo, and L. Gold, *Mol. Microbiol.* **6,** 1219 (1992).
[38] B. Schauder and J. E. G. McCarthy, *Gene* **78,** 59 (1989).
[39] G. Buell and N. Panayotatos, *in* "Maximizing Gene Expression" (W. Reznikoff and L. Gold, eds.), p. 345. Butterworth, Boston, 1986.
[40] H. J. Barnes, unpublished data (1995).

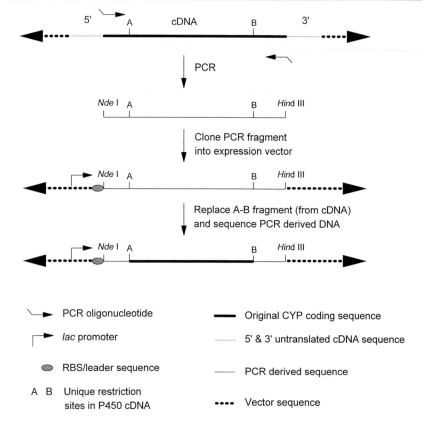

FIG. 2. Strategy for the construction of cytochrome P450 expression plasmids. See text for details.

be sequenced is minimized by replacement of most of the coding sequence with a restriction fragment from the original cDNA sequence. Thus, only the PCR-derived segments and cloning junctions of the final expression plasmid need be sequenced.

Primer Design

The guidelines for basic PCR primer design have been described,[41,42] as have the PCR conditions for plasmid DNA templates.[43] Ideally, the 3′

[41] R. K. Saiki, in "PCR Technology: Principals and Applications for DNA Application" (H. A. Erlich, ed.), p. 7. Stockton Press, New York, 1989.
[42] M. A. Innis and D. Gelfand, in "PCR Protocols: Guide to Methods and Applications" (M. A. Innis, D. H. Gelfand, J. J. Sninsky, and T. J. White, eds), p. 3. Academic Press, San Diego, CA, 1990.
[43] R. Higuchi, B. Krummel, and R. K. Saiki, Nucleic Acids Res. 16, 7351 (1988).

ends of the primers should have a perfect 15- to 25-bp homology to the cDNA template in order to assure efficient priming. Two or three G or C residues are added to the 5' end of the primers to allow efficient restriction enzyme cleavage of the adjacent sites engineered into the PCR product. High fidelity reaction conditions[44,45] should be employed when using *Taq* DNA polymerase to minimize the occurrence of spurious mutations. For example, the denaturation temperature can usually be lowered to below 90° after the first several PCR cycles and only 15–20 cycles are adequate to generate sufficient amounts of PCR product. Alternatively, thermostable DNA polymerases (e.g., Vent and *Pfu* DNA polymerases) which exhibit lower error rates than *Taq* polymerase are commercially available.

5'-Sense Primer

The 5'-sense primer introduces an *Nde*I restriction site (CA'TATG) that contains the initiation codon as well as any desired nucleotide modifications to the amino-terminal codons. The nucleotide sequence of the first seven or eight codons may be modified so as to conform to either the nucleotide preferences observed in *E. coli* genes[46,47] or to random sequences that have been shown to function as efficient translation initiation sequences when cloned into a *lac* Z reporter construct.[48] These studies identified a distinct nucleotide preference for this region (A > T > G > C) and, if desired, the degeneracy of the genetic code can be exploited to incorporate these preferences into the amino-terminal codons of the cytochrome P450 to be expressed. Also, mRNA secondary structure prediction programs[49] can be used to examine this region so as to incorporate nucleotide modifications that disrupt exceedingly stable structures, especially if these involve the S-D element or initiation codon.[37,39]

3'-Antisense Primer

A 3'-antisense oligonucleotide is used to introduce a preferably unique cloning site that is coincident with, or very near to, the stop codon of the target cytochrome P450 cDNA. The inclusion of extraneous 3'-untranlated sequence can have a detrimental effect on the level of cytochrome P450 expression in *E. coli*.[3] If possible, the preferred stop sequence (TAAT),

[44] K. Eckert and T. A. Kunkel, *Nucleic Acids Res.* **18,** 3739 (1990).
[45] L. Ling, P. Keohavong, C. Dias, and W. G. Thilly, *PCR Methods Appl.* **1,** 63 (1991).
[46] G. D. Stormo, T. D. Schneider, and L. Gold, *Nucleic Acids Res.* **10,** 2971 (1982).
[47] T. D. Schneider, G. D. Stromo, L. Gold, and A. Ehrenfeucht, *J. Mol. Biol.* **188,** 415 (1986).
[48] M. Dreyfus, *J. Mol. Biol.* **204,** 79 (1988).
[49] J. A. Jaeger, P. H. Turner, and M. Zuker, *Proc. Natl. Acad. Sci. U.S.A.* **86,** 7706 (1989).

found in highly expressed *E. coli* genes, should be incorporated into the 3′ primer.[50]

Culture Conditions

Perhaps surprisingly, the cell strain, culture media, temperature, and other factors have a dramatic effect on the expression level and/or subcellular distribution of the recombinant bovine CYP17 protein.[12] Optimization of these parameters enabled the identification of specific conditions that promoted the expression of other cytochrome P450s in *E. coli*.

E. coli Strain

Certain *E. coli* strains were shown to be very efficient hosts for the expression of the recombinant bovine CYP17 enzyme.[12] No distinguishing genetic markers were found in these stains that correlated with the ability to produce high levels of the recombinant enzyme. Cell strain specific variability in the expression level has been noted for other proteins[39,51] and a highly idiosyncratic expression has been observed with homologous proteins that are 96 and 91% identical at the amino acid and nucleotide level.[52] The commonly available *E. coli* strains JM109, DH5α, and XL-1 blue have been used to express high levels of several cytochrome P450s. As a starting point, all three of these strains should be evaluated for their ability to express a particular cytochrome P450. Protease-deficient strains may be useful for the production of proteolytically sensitive cytochrome P450s.[53]

Media and Temperature

Cultures using standard LB media at 37°, the optimal growth temperature for *E. coli*, resulted in a modest level of CYP17 expression with the majority of the recombinant protein accumulating in inclusion bodies.[12] However, use of a rich buffered media combined with a growth/induction temperature of 28–30° increased CYP17 expression to levels easily detectable in difference spectra of whole cells. Induction of cells above ~32° resulted in an accumulation of recombinant protein in inclusion bodies, while induction at temperatures below 25° resulted in a precipitous drop

---

[50] W. P. Tate and C. M. Brown, *Biochemistry* **31,** 2443 (1992).
[51] M. Nieboer, J. Kingma, and B. Witholt, *Mol. Microbiol.* **8,** 1039 (1993).
[52] T. Date, K. Tanihara, S. Yamamoto, N. Nomura, and A. Matsukage, *Nucleic Acids Res.* **20,** 4859 (1992).
[53] S. Gottesman, *Methods Enzymol.* **185,** 119 (1990).

in the expression level.[12] The media formulation that has been found to be most useful is Terrific broth (TB).[54] This phosphate-buffered media maintains the culture pH near neutrality, even at high cell densities, and contains a readily utilizable carbon source, glycerol. Detailed experimental protocols for the growth and induction of *E. coli* cells containing cytochrome P450 expression plasmids in shake flasks have been described.[9,27] Excessive autoclaving of the TB media should be avoided and the 10× phosphate salts must be autoclaved separately. As indicated earlier, the optimal temperature range during induction seems to be rather narrow and thus sensitive to drastic fluctuations in the incubator temperature. Cultures should be cooled to the induction temperature prior to the addition of IPTG (0.25–1 mM). The incubator shake speed during induction, combined with the flask geometry, also affects optimal expression levels. Erlenmeyer flasks approximately half filled and shaken at 100–200 rpm routinely produce optimal yields. Although not an absolute requirement for high level expression of all cytochrome P450s, the addition of δ-aminolevulinic acid (ALA), the first committed intermediate in heme biosynthesis, to a final concentration of 0.5 mM 30–60 min before induction can dramatically improve expression. Richardson *et al.*[3] reported a 2- to 10-fold enhancement in the expression of seven 2C enzymes, while three other 2C enzymes that expressed at high levels remained unaffected by the addition of ALA.[3] Similarly, Gilliam *et al.*[5] found a strong increase in the expression of human 1A1, 2C10, and 2E1; a modest improvement in the level of 3A4 expression; and no effect on 1A2.

Conclusions

During the past several years, the essential parameters for the production of eukaryotic P450 enzymes in *E. coli* have been established. Recombinant proteins generated in this way have yielded valuable information regarding structure/function relationships.[55–60] Issues related to purification and reconstitution of catalytic activity have also been addressed. Amino-

[54] K. D. Tartof and C. A. Hobbs, *Focus (Bethesda, Md.)* **9**, 12 (1987).

[55] J. R. Larson, M. J. Coon, and T. D. Porter, *J. Biol. Chem.* **266**, 7321 (1991).

[56] Y. Sagara, H. J. Barnes, and M. R. Waterman, *Arch. Biochem. Biophys.* **304**, 272 (1993).

[57] S. J. Pernecky, N. M. Olken, L. L. Bestervelt, and M. J. Coon, *Arch. Biochem. Biophys.* **318**, 446 (1995).

[58] A. C. Kempf, U. M. Zanger, and U. A. Meyer, *Arch. Biochem. Biophys.* **321**, 277 (1995).

[59] K. G. Ravichandran, S. S. Boddupalli, C. A. Hasemann, J. A. Peterson, and J. Deisenhofer, *Science* **261**, 731 (1993).

[60] C. A. Hasemann, R. G. Kurumbail, S. S. Boddupalli, J. A. Peterson, and J. Deisenhofer, *Structure* **2**, 41 (1995).

and carboxy-terminal polyhistidine tags that do not interfere with enzyme activity have been used effectively for affinity purification.[8,58] Also, phase separation protocols have been applied successfully to simplify purification schemes.[4,5,12] Recombinant CYP17 was used to identify a soluble flavodoxin/NADPH-flavodoxin reductase system in *E. coli* that can support catalytic activity,[61] and this protein–protein association has been exploited using Sepharose-bound flavodoxin as an affinity purification matrix.[5] The requirement for a reductase has inspired various efforts to simplify the generation of active enzyme. Catalytically competent protein fusions between various cytochrome P450s and NADPH-P450 reductase have been expressed in *E. coli*,[62–64] and inorganic compounds have been used to provide reducing equivalents for these fusion proteins.[65] These innovations are likely to expand the utility of *E. coli*-expressed P450 enzymes and enable the eventual application of recombinant P450s to such diverse tasks as biosensors, bioremediation, and industrial-scale biocatalysis. The ability to express and purify large quantities of essentially native enzyme also offers more immediate practical opportunities in the fields of pharmacology and toxicology. Purified enzyme enables detailed kinetic characterization of human P450s and provides useful predictive tools for assessing drug–drug interactions. These enzymes can also be used for the generation of sufficient quantities of specific metabolites to allow detailed study of procarcinogen activation or drug bioactivation. Clearly, the continued application of this heterologous expression system will make significant contributions to our understanding of these important enzymes.

Acknowledgment

The author is indebted to Dr. Toby Richardson for insightful comments and critical reading of this manuscript.

[61] C. M. Jenkins and M. R. Waterman, *J. Biol. Chem.* **269**, 27401 (1994).

[62] C. W. Fisher, M. S. Shet, D. L. Caudle, C. Martin-Wixtrom, and R. W. Estabrook, *Proc. Natl. Acad. Sci. U.S.A.* **89**, 10817 (1992).

[63] M. S. Shet, C. W. Fisher, P. L. Holmans, and R. W. Estabrook, *Proc. Natl. Acad. Sci. U.S.A.* **90**, 11748 (1993).

[64] M. S. Shet, C. W. Fisher, M. P. Arlotto, C. H. L. Shackleton, P. L. Holmans, C. Martin-Wixtrom, Y. Saeki, and R. W. Estabrook, *Arch. Biochem. Biophys.* **311**, 402 (1994).

[65] K. M. Faulkner, M. S. Shet, C. W. Fisher, and R. W. Estabrook, *Proc. Natl. Acad. Sci. U.S.A.* **92**, 7705 (1995).

[2] Construction of Plasmids and Expression in *Escherichia coli* of Enzymatically Active Fusion Proteins Containing the Heme-Domain of a P450 Linked to NADPH-P450 Reductase

By CHARLES W. FISHER, MANJUNATH S. SHET, and RONALD W. ESTABROOK

The demonstration[1,2] that mammalian P450s can be expressed in *Escherichia coli* for the synthesis of large amounts of enzymatically active recombinant proteins has made the extension of this methodology possible[3-5] for the expression of recombinant fusion proteins containing the heme-domain of a P450 linked to the flavin-domains of the flavoprotein, NADPH-P450 reductase. This construct serves as a cassette for expressing other P450-domains as well as other flavoprotein-domains. The availability of large amounts of these recombinant fusion proteins, which could serve as self-sufficient enzymatic units for the study of P450-catalyzed reactions, has many advantages. First, the presence of the flavin-domains of a NADPH-P450 reductase greatly facilitates the isolation and purification of these fusion proteins since they bind to an ADP-Sepharose affinity matrix[6] and serve as a convenient step for the purification of the enzyme. Second, studies for the measurement of P450-catalyzed activities using these fusion proteins avoid the need for a purified flavoprotein for reconstitution studies. Third, these P450-containing fusion proteins can function in intact bacteria,[7] permitting a simple, convenient method for assaying activities without the need for purification. This allows the screening of mutants of a fusion protein by directly measuring catalytic activities by using intact cells. Fourth, these constructs can help answer interesting questions about the mecha-

[1] H. J. Barnes, M. P. Arlotto, and M. R. Waterman, *Proc. Natl. Acad. Sci. U.S.A.* **88,** 5597 (1991).
[2] H. J. Barnes, *Methods Enzymol.* **272,** Chap. 1, 1996 (this volume).
[3] C. W. Fisher, M. S. Shet, D. L. Caudle, C. A. Martin-Wixtröm, and R. W. Estabrook, *Proc. Natl. Acad. Sci. U.S.A.* **89,** 10817 (1992).
[4] M. S. Shet, C. W. Fisher, P. L. Holmans, and R. W. Estabrook, *Proc. Natl. Acad. Sci. U.S.A.* **90,** 11748 (1993).
[5] M. S. Shet, K. M. Faulkner, P. L. Holmans, C. W. Fisher, and R. W. Estabrook, *Arch. Biochem. Biophys.* **318,** 314 (1995).
[6] Y. Yasukochi and B. S. S. Masters, *J. Biol. Chem.* **251,** 5337 (1976).
[7] M. S. Shet, C. W. Fisher, M. P. Arlotto, C. H. L. Shackleton, P. L. Holmans, C. A. Martin-Wixtröm, Y. Saeki, and R. W. Estabrook, *Arch. Biochem. Biophys.* **311,** 402 (1994).

nism(s) of electron transport from the flavin-domains of the flavoprotein to the heme-domain of a P450. Fifth, the plasmid(s) used for expressing these fusion proteins can be further modified by including additional domains of interest, e.g., a polyhistidine-domain or a biotin-binding protein domain for the immobilization of functional enzyme units in a bioreactor.

Heme proteins containing both a flavoprotein reductase-domain as well as a heme-domain are now well known in biology.[8–10] Currently there is great interest in understanding the enzymology and biological function of nitric oxide synthase,[9] a multidomain protein related to the P450s with at least two redox centers. The P450 of highest known enzymatic activity is the natural fusion protein $P450_{BM3}^{10}$ of *Bacillus megaterium*. This fusion protein has hydroxylation activities exceeding 3000 nmol of fatty acid metabolized/min/nmol of P450.

Expression of recombinant fusion proteins, containing the heme-domain of rat P450 1A1 joined to the flavin-domain of rat NADPH-P450 reductase, was first reported by Murakami et al.[11] in 1987. In this seminal study they showed the requirements for engineering the cDNAs of the two proteins (P450 1A1 and NADPH-P450 reductase) and they evaluated the limitations imposed by different constructs on the ability to express an enzymatically active fusion protein in yeast. This group has subsequently shown the usefulness of their method in the construction and expression in yeast of other fusion proteins containing the heme-domains of P450 17A[12] and P450 21[13,14] linked to yeast NADPH-P450 reductase. Other investigators have applied the protocol developed by this group for the construction of a variety of similar fusion proteins that can be expressed in yeast[15] or COS1[16] cells.

Our laboratory has focused on the expression of similar fusion proteins in *E. coli*.[3–5,7] We have introduced a nomenclature for these fusion proteins as illustrated by the following example: rF450[mHum1A2/mRatOR]L1 represents the recombinant (r) fusion protein (F) of P450 (450) composed of

[8] F. Lederer, *in* "Chemistry and Biochemistry of Flavoenzymes" (F. Muller, ed.), Vol. 2, p. 153. CRC Press, Boca Raton, FL, 1991.

[9] L. J. Roman, E. A. Sheta, P. Martasek, S. S. Gross, Q. Liu, and B. S. S. Masters, *Proc. Natl. Acad. Sci. U.S.A.* **92,** 8428 (1995).

[10] Y. Miura and A. J. Fulco, *Biochim. Biophys. Acta* **388,** 305 (1975).

[11] H. Murakami, Y. Yabusaki, T. Sakaki, M. Shibata, and H. Ohkawa, *DNA* **6,** 189 (1987).

[12] M. Shibata, T. Sakaki, Y. Yabusaki, H. Murakami, and H. Ohkawa, *DNA Cell Biol.* **9,** 27 (1990).

[13] T. Sakaki, M. Shibata, Y. Yabusaki, H. Murakami, and H. Ohkawa, *DNA Cell Biol.* **9,** 603 (1990).

[14] Y. Yabusaki, H. Murakami, T. Sakaki, M. Shibata, and H. Ohkawa, *DNA* **7,** 701 (1988).

[15] N. E. Wittekindt, F. E. Wurgler, and C. Sengstag, *DNA Cell Biol.* **14,** 273 (1995).

[16] J. A. Harikrishna, S. M. Black, G. D. Szklarz, and W. L. Miller, *DNA Cell Biol.* **12,** 371 (1993).

a modified (m) human (Hum) P450 1A2 (1A2) connected to a modified (m) rat (Rat) NADPH-P450 reductase (OR) by a designated peptide linker (L1). A listing of those artificial fusion proteins that we have successfully expressed in *E. coli*, with an indication of the level of spectrophotometrically detectable P450 observed, is summarized in Table I.

Design of Constructs

We used the plasmid pCWmod17,[2] which contains the cDNA for bovine P450 17α-hydroxylase (CYP17A), as the starting material for construction of a plasmid, pCWFBov17OR, containing the cDNA of a fusion protein composed of the heme-domain of bovine P450 17A and the flavoprotein-domain of rat NADPH-P450 reductase. As illustrated in Fig. 1, pCWmod17 was transformed into *dcm-E. coli* strain GM48 to permit the isolation of unmethylated plasmid DNA for digestion with *Stu*I (A of Fig. 1). Plasmid DNA was prepared, and a 40-bp fragment was removed by digesting the plasmid with *Stu*I and *Hin*dIII (B). The cDNA for the coding region of bovine P450 17A was polymerase chain reaction (PCR) amplified (C) with the 5′ primer described by Barnes[2] and a 3′ primer (GGGGGCGGTTGTTTGGATA<u>GTCGAC</u>GGGGTGCTACCCTCAGC) deleting the TGA stop codon and incorporating a *Sal*I site (under-

TABLE I

EXPRESSION OF FUSION PROTEINS CONTAINING A P450-DOMAIN AND THE FLAVIN-DOMAIN
OF NADPH-P450 REDUCTASE

Construct	Activity measured	Expression level (nmol/liter)
rF450[mHum1A2/mRatOR]L1	Estradiol 2-hydroxylase	150–250
	Ethoxyresorufin deethylase	
	Acetanilide hydroxylase	
	Caffiene demethylase	
	Arachidonic acid hydroxylase	
	Imipramine *N*-demethylase	
rF450[mHum3A4/mRatOR]L1	Steroid 6β-hydroxylase	200–300
	Erythromycin *N*-demethylase	
	Nifedipine hydroxylase	
	Benzphetamine *N*-demethylase	
	Imipramine *N*-demethylase	
rF450[mRat4A1/mRatOR]L1	Fatty acid omega hydroxylase	400–500
rF450[mBov17A/mRatOR]L1	Steroid 17α-hydroxylase	600–700
	Steroid 17,20-lyase	

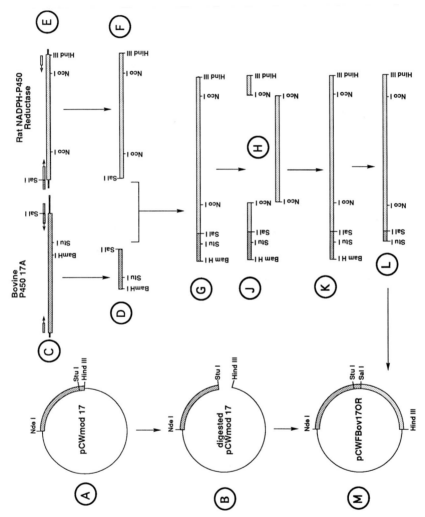

FIG. 1. The construction of the plasmid for expression of rF450[mBov17A/mRatOR]L1 from pCWmod17. Details are given in the text.

lined) which partially encodes the codons for the Ser-Thr linker segment. This PCR product was digested with *Bam*HI and *Sal*I to give the fragment (D) shown in Fig. 1. In this way a segment of the 3' portion of the cDNA for P450 17A was prepared for ligation with the modified 5' portion of the rat NADPH-P450 reductase cDNA. The plasmid for rat liver NADPH-P450 reductase described by Shen *et al.*[17](pOR263) was PCR amplified (E) using primers *deleting* the codons for the amino-terminal membrane-anchoring region of the cDNA for the reductase and incorporating a *Sal*I site encoding Ser-Thr as a linker (GTTGAGGGTAGCACCCCGTCGACTATCCAAACAACCGCCCCC) as described by Sakaki *et al.*[13] The 3' reductase primer incorporated a *Hin*dIII site after the TAG stop codon. The PCR fragment was made blunt ended using the Klenow fragment of DNA polymerase I to fill in incompletely synthesized strands and was subcloned into the *Sma*I site of pTZ19R, and the 5' and 3' ends of the amplified sequence were sequenced to the two internal *Nco*I sites. The *Sal*I–*Hin*dIII insert was removed by digestion and isolated (F). The *Sal*I–*Hin*dIII reductase domain (F) was then ligated to the *Bam*H I–*Sal*I fragment (D) containing the P450 domain and to the plasmid pTZ19R. This was digested with *Bam*HI and *Hin*dIII (G). A positive clone was sequenced from the *Stu*I site to the *Sal*I site in the P450 domain region. This positive clone (G) was then digested with *Nco*I to remove the 1.4-kb internal *Nco*I fragment (H) from the PCR-generated reductase domain. The 1.4-kb *Nco*I fragment was isolated from a plasmid preparation of pOR263, i.e., the original rat liver NADPH-P450 reductase cDNA, and the *Nco*I fragment was ligated into the *Nco*I-deleted vector construct (J) to give (K). The construct with the replaced *Nco*I fragment in the correct orientation was identified and transformed into *E. coli* GM48 and digested with *Stu*I and *Hin*dIII to give (L). The 1.9-kb *Stu*I–*Hin*dIII fragment was ligated with the digested pCWmod17 (B), and positive clones were identified by restriction digestion. We have named this plasmid pCWFBov17OR (M).

The preparation of a second fusion protein, rF450[mRat4A1/mRatOR]L1 containing rat P450 4A1 linked to rat NADPH-P450 reductase, was accomplished by modifying the plasmid pCWFBov17OR as illustrated in Fig. 2. First, the cDNA for rat P450 4A1 (A of Fig. 2) was PCR amplified using the 5' primer (TACATATGGCTCTGTTATTAGCAGTTTTTCT| GGTTCTGCTGCTGGTC). This construct (B of Fig. 2) incorporates a 27-bp fragment from the modified bovine P450 17A cDNA (5' to |), includes the *Nde*I site, and deletes a portion of the coding sequence of the amino terminus of rat P450 4A1. Alignment of the amino acid sequences for both proteins permitted the selection of the site of the junction. The carboxy

---

[17] A. L. Shen, T. D. Porter, T. E. Wilson, and C. B. Kasper, *J. Biol. Chem.* **264,** 7584 (1989).

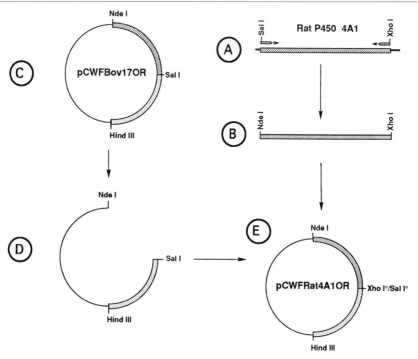

FIG. 2. The construction of plasmids containing the cDNAs for the heme-domain of other P450s and the flavoprotein-domain from human, yeast, or plant NADPH-P450 reductases. Details are given in the text.

terminus of the cDNA for the P450 4A1 was modified using the 3'-primer (AGTCCGGCTCGAGTGGAGCTTCTTGAGATA) for PCR amplification. This sequence incorporates an XhoI site (underlined) encoding Ser-Thr as a linker, replacing the TAA stop codon, to allow fusion with the modified DNA sequence of NADPH-P450 reductase, as described earlier. This PCR product was digested with NdeI and XhoI to give B of Fig. 2. The plasmid encoding rF450[mBov17A/mRatOR]L1 described earlier (C of Fig. 2) was digested with NdeI and SalI to remove the domain of P450 17A. The digested plasmid containing only the reductase domain (D of Fig. 2) was ligated with the NdeI–XhoI PCR-amplified rat P450 4A1 domain (B of Fig. 2) to give (E). The construct with the replaced NdeI–XhoI fragment in the correct orientation was identified by restriction digestion and transformed into E. coli. Positive colonies were selected as described earlier.

     Fusion proteins for other P450s, including the human P450s 1A2 and 3A4, have been constructed using a strategy similar to that described for the preparation of the fusion protein containing rat P450 4A1. Selection of the site of junction of the 5' sequence of a P450 to the nine amino acid sequence from the modified bovine P450 17A cDNA is subjective and depends on a visual estimate of the best alignment of amino acids. In addition, the design of the primer used for the 3' junction of the cDNA of the P450 which links to the 3' terminus of the cDNA for the truncated reductase may require introduction of a different restriction site other than *Sal*I since this restriction site may be present in the nucleotide sequence of the P450.

We have also introduced the cDNAs for other flavoproteins, such as human NADPH-P450 reductase as well as yeast NADPH-P450 reductase. The design of suitable primers for PCR synthesis of cDNAs to replace the nucleotide sequence of rat NADPH-P450 reductase follows essentially the same principles described earlier for replacement of the nucleotide sequence of P450 17A with cDNAs for different P450s.

Other more current protocols of mutagenesis may prove more efficient for the creation of modified sequences. The primers described earlier used in the original construction were intended for use in an overlapping PCR protocol and are longer than necessary for the described method of mutagenesis and ligation. While such overlapping primers are unnecessary since they are trimmed by restriction enzyme digestion, they have utility in identifying the correctly constructed fusion sequences as well as for subtractive screening to detect insertions in the linker region.

An important consideration in the creation of any construct by mutagenesis is to minimize the region of DNA that is subjected to mutagenesis and to confirm by sequencing that only the desired mutations have been introduced into this mutagenized region. The use of (low error rate) thermostable DNA polymerases with proofreading activities has been heralded for use in mutagenesis. The use of these enzymes has led investigators to assume that only the desired mutations have been introduced. Such assumptions may have negative consequences for expression or the validity of subsequent enzymatic studies.

Cell Growth and Harvesting

Growth conditions for the expression of P450 fusion proteins have been described in detail.[7] Briefly, *E. coli*, DH5α, are transformed with the plasmid containing the engineered cDNAs for a P450 linked to a selected NADPH-P450 reductase, and ampicillin-resistant colonies are streaked on a Luria-Bertani (LB) agar/ampicillin plate (10 g of tryptone, 5 g of yeast extract, 10 g of NaCl, 100 mg of ampicillin, and 12 g of agar per liter). For the initial screening of high producing cells, in our laboratory a number of clones are selected and inoculated into 10 ml of LB for growth overnight at 37° with shaking. A 1-ml aliquot of each is used as an inoculum for 1 liter of Terrific broth (TB) containing thiamine (1 mM final concentration), potassium phosphate buffer, pH 7.5 (100 mM), rare salt solution,[18] and ampicillin (100 μg/ml) in a 2.8-liter Fernbach flask. Cells are grown at 37° in a rotary shaker for about 4 to 5 hr (until an optical density of about 0.4 at 600 nm is attained) when isopropyl-β-D-thiogalactopyranoside (1 mM final concentration) is added using sterile conditions and the incubation temperature is shifted to about 27° for 48–72 hr. It should be noted that it is necessary to try a variety of conditions to optimize the level of expression of the recombinant protein. Each laboratory has its own "magic formula" for best yield. In this laboratory we follow the principles of (a) growth at low temperatures (24–30°); (b) use of slow rates of agitation of the growth media (125 rpm, avoiding the formation of foam on the surface of the growth medium); and (c) sampling at 24 and 48 hr to determine by spectrophotometric measurements the level of expression of P450 in each flask and selecting those flasks with the highest yield.

After 72 hr of growth the cells are harvested by centrifugation at 500 g for 10 min. The pelleted cells are washed by suspension with a Dounce homogenizer in 10 mM potassium phosphate buffer, pH 7.5, containing 0.15 M NaCl. The resuspended cells are pelleted as described earlier and the pellets are drained and weighed. The washed cells are suspended with a twofold (v/w) of TSE buffer (75 mM Tris–HCl, pH 7.5, 250 mM sucrose, and 0.25 mM EDTA) and divided into 100-ml aliquots (33 g wet weight of cells). A sample of the washed cells is assayed spectrophotometrically for P450 content, and the suspended cells are frozen at −80° where they can be stored for at least 1 year with no loss of P450 content.

The disruption of cells, isolation of membranes, solubilization, and purification of the fusion proteins have been described elsewhere.[7]

Additional Comments

Of interest is the influence of different linker groups on the enzymatic activity of a fusion protein. A series of experiments have been carried out where the serine-threonine linker (L1), first described by Sakaki,[13] was replaced by a number of different amino acid sequences. This was accomplished by using two complementary synthetic oligonucleotides which were phosphorylated and ligated into the SalI restriction site of the construct. In some instances, sequence analysis revealed that multiple copies were

[18] S. Bauer and J. Shiloach, *J. Biotechol. Bioeng.* **16,** 933 (1974).

TABLE II

EFFECT OF MODIFYING THE LINKER REGION OF rF450[mBov17A/mRatOR]Lx
ON THE RATE OF 17α-HYDROXYLATION OF PROGESTERONE[a]

Sequence of linker	Rate of progesterone metabolism (nmol 17αOH-P4/min/ nmol P450)
ST	1.0
SRGGGGGST	1.7
SKKKKKST	2.0
SKKKKKSKKKKKST	1.9
SKKKKKSKKKKKSKKKKKSTSSSSST	1.2
SKKKKKSTSSSSSTSSSSSTSSSSSTSSSSSTSSSSST	1.8
SKKKKKSTSSSSSKKKKKSKKKKKSKKKKKSKKKKKSKKKKKST	1.9

[a] Membranes from transformed *E. coli* were suspended in a buffer mixture containing 50 mM Tris–Cl (pH 7.5), 10 mM MgCl$_2$, 0.8 mM sodium isocitrate, 0.1 units isocitrate dehydrogenase/ml, and 10 μM [³H]progesterone (200,000 cpm/ml). The final concentration of P450, rF450[mBov17A/mRatOR]Lx, was 1 nmol/ml. The reaction was started by the addition of NADPH (1 mM final concentration). Aliquots were removed at 5, 10, and 15 min for analysis by HPLC of product formation.

inserted, resulting in linkers of different amino acid length. As shown in Table II, increasing the length of the linker by adding clusters of glycine molecules, lysine molecules, or serine molecules has only a modest effect on the rate of 17α-hydroxylation of progesterone, as catalyzed by the membrane-bound form of rF450[mBov17A/mRatOR]Lx. Of interest is the observation that extending the size of the linker region to include as many as 30 lysine molecules does not significantly influence the 17α-hydroxylase activity of the fusion protein. From these preliminary experiments one concludes that the number of amino acids included or the presence of charged amino acids in the linker region of the fusion protein has only a small (twofold) effect on the rate of catalysis.

One of the goals of this laboratory has been to develop P450 catalysts that might be of commerical value for chemical synthetic processes. Therefore it was of interest to engineer and express fusion proteins that might be immobilized to a matrix permitting the regenerative flow through of reactant chemicals. In addition, it was of interest to include additional domains in the fusion protein that might influence a P450 reaction, e.g., the domain for cytochrome b_5 or the presence of a second flavoprotein-domain. A number of combinations have been successfully constructed and tested and are schematically illustrated in Fig. 3. The variety of constructs that can be designed and engineered is large, indicating many opportunities

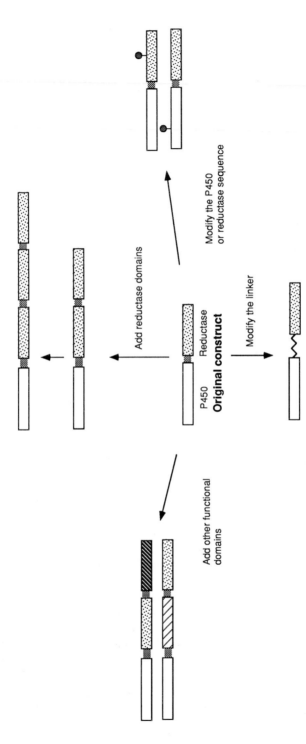

FIG. 3. Modifications of recombinant fusion proteins containing the heme-domain of a P450 and the reductase-domain of a flavoprotein, including other redox domains, sites for immobilization, mutant forms, and linkers such as cytochrome b_5, biotin-binding protein, and a polyhistidine-domain.

to apply this type of fusion protein to problems related to understanding the function of P450s.

Acknowledgment

This work was supported in part by grants from NIGMS (GM16488-25) and NIEHS (ESO7628-01).

[3] N-Terminal Modifications That Alter P450 Membrane Targeting and Function

By STEVEN J. PERNECKY and MINOR J. COON

Introduction

Several P450 cytochromes lacking the usual N-terminal region have been studied in different laboratories.[1–9] In most instances the modified enzymes were examined after heterologous expression and found to have retained some or all of the original catalytic activity and membrane-binding capability. The highly hydrophobic N-terminal region that is normally present in mammalian P450 cytochromes[10] serves as the signal peptide for cotranslational insertion of the protein into the endoplasmic reticulum,[11] and the following hydrophobic proline-rich segment promotes stabilization of the enzyme.[12] To examine in detail the role of these regions in membrane targeting and function, we have removed the signal peptide (designated S1

[1] T. Sakaki, K. Oeda, M. Miyoshi, and H. Ohkawa, *J. Biochem.* (*Tokyo*) **98**, 167 (1985).

[2] J. R. Larson, M. J. Coon, and T. D. Porter, *J. Biol. Chem.* **266**, 7321 (1991).

[3] J. R. Larson, M. J. Coon, and T. D. Porter, *Proc. Natl. Acad. Sci. U.S.A.* **88**, 9141 (1991).

[4] B. J. Clark and M. J. Waterman, *J. Biol. Chem.* **266**, 5898 (1991).

[5] Y. C. Li and Y. L. Chiang, *J. Biol. Chem.* **266**, 19186 (1991).

[6] C. Cullin, *Biochem. Biophys. Res. Commun.* **184**, 1490 (1992).

[7] L.-C. Hsu, M.-C. Hu, H.-C. Cheng, J.-C. Lu, and B. Chung, *J. Biol. Chem.* **268**, 14682 (1993).

[8] S. J. Pernecky, J. R. Larson, R. M. Philpot, and M. J. Coon, *Proc. Natl. Acad. Sci. U.S.A.* **90**, 2651 (1993).

[9] S. J. Pernecky, N. M. Olken, L. L. Bestervelt, and M. J. Coon, *Arch. Biochem. Biophys.* **318**, 446 (1995).

[10] D. A. Haugen, L. G. Armes, K. T. Yasunobu, and M. J. Coon, *Biochem. Biophys. Res. Commun.* **77**, 967 (1977).

[11] S. Monier, P. Van Luc, G. Kreibich, D. D. Sabatini, and M. Adesnik, *J. Cell Biol.* **107**, 457 (1988).

[12] S. Yamazaki, K. Sato, K. Suhara, M. Sakaguchi, K. Mihara, and T. Omura, *J. Biochem.* (*Tokyo*) **114**, 652 (1993).

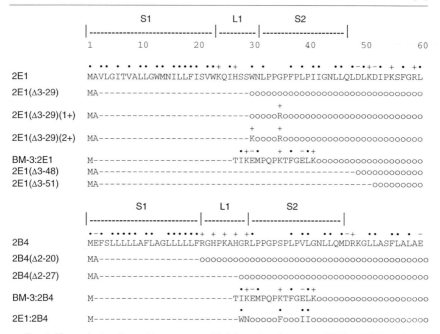

FIG. 1. N-terminal amino acid sequences of full-length and truncated 2E1 and 2B4 variants. The N-terminal sequences of 2B4 and 2E1 are shown with symbols above to indicate hydrophobic (●), positively charged (+), and negatively charged residues (−). The N-terminal sequences of the truncated variants are shown beneath that of the full-length clone; deletions in sequence are indicated with a dash, amino acid replacements are shown with the appropriate symbol above the residue as for the full-length sequence, and the retained residues are indicated by open circles. The first two hydrophobic segments (S1 and S2) and the first intervening loop region (L1) are shown above the two full-length sequences.

by Nelson and Strobel[13]) from 2B4 and 2E1 as shown in Fig. 1. The N-terminal region of 2E1 (Δ3–29) was modified by the insertion of positive charges, the N-terminal region of 2E1 (Δ3–29) and 2B4 (Δ2–27) was replaced with the N-terminal region of P450 BM-3 (CYP102), and the N-terminal region of 2B4 (Δ2–27) was replaced with that from 2E1 (Δ3–29) as shown in Fig. 1. A more extensive deletion in 2E1 included removal of the proline-rich region, S2, as well as S1.

Truncation and modification to decrease the hydrophobicity of the N-terminal region of P450 2E1 and 2B4 generally cause the cytochromes to change from a membranous to a cytosolic location when expressed in bacteria.[8,9] Related studies with purified 2E1 (Δ3–29) revealed no alteration in catalytic properties,[3] but removal of the N-terminal region from 2B4

[13] D. R. Nelson and H. W. Strobel, *J. Biol. Chem.* **263,** 6038 (1988).

lowers the catalytic activity with all substrates tested.[9] The role of the N-terminal region in promoting extensive aggregation of P450 enzymes is demonstrated by disaggregation of shortened P450s to monomers by the detergent cholate at concentrations that do not alter the multimeric aggregation state of the full-length enzymes. The retention of catalytic activity and the monomerization of P450 cytochromes at low detergent concentrations, all achieved through N-terminal modification, are being used to improve the suitability of these enzymes as candidates for crystallization attempts. The methods described here for the heterologous expression, determination of subcellular localization, and purification of shortened P450 cytochromes, as well as examination of their catalytic properties, have been described previously.[2,3,8,9]

Procedures

Bacterial Expression Vectors

Two bacterial expression vectors driven by an isopropyl β-D-thiogalactoside (IPTG)-inducible *trc* promoter are used in our laboratory to examine the role of the N-terminal region in subcellular localization and function. The Pharmacia vector, pKK223-2, which was originally modified to express full-length and shortened 2E1 variants,[2] is not sufficient to express the full-length and shortened 2B4 variants. However, when a double-stranded oligonucleotide sequence containing the translation enhancer sequence from a bacteriophage gene was cloned between the IPTG-inducible promoter and the *Nco*I site to give pJL, full-length 2B4 was expressed in pJL at sufficient levels that a CO-reduced difference spectrum could be detected with bacterial membranes. Full-length and shortened 2E1 and 2B4 cDNAs are inserted into the *Nco*I and *Hin*dIII sites of pJL. Two approaches are used to construct forms of P450 2E1 and 2B4 (Fig. 1) with modified N-terminal amino acid sequences. For construction of P450 2B4 (Δ2–27), the full-length 2B4 in pJL is digested with *Ecl*XI, blunt ends are created with the Klenow fragment of DNA polymerase, and a 10-mer *Nco*I linker (5'-AGCCATGGCT; Pharmacia) is ligated to the 2B4 sequence, thus placing the start codon Met and Ala in front of Gly-28. The 2E1 construct without codons for amino acids 3–29 is similarly obtained by inserting an *Nco*I site at a *Pvu*II site via a synthetic linker present at position 84 of the cDNA. A double-stranded synthetic linker containing *Nco*I and *Hae*II sites is ligated to 2E1 cDNA (previously digested with *Nco*I and *Hae*II) to obtain 2E1 (Δ3–51). In another approach, fragments generated by the polymerase chain reaction (PCR) and digested with *Nco*I and *Bam*HI are cloned into similarly digested pJL2E1 to yield 2E1 (Δ3–29)(1+), 2E1 (Δ3–29)(2+), and 2E1 (Δ3–48). A similar method involving a 435-bp PCR fragment (*Nco*I–*Bsm*I) is used to create the 5′-coding region of 2B4 (Δ2–20), which places a Met and an Ala codon before Arg-21. The BM–3:2E1 chimera is generated by a three-way ligation of (a) a double-stranded *Nco*I to *Pst*I fragment containing the 5′-coding region of BM-3; (b) a *Pst*I to *Bam*HI fragment encoding nucleotides 166 through 486 in 2E1 cDNA (in which PCR was previously used to introduce a *Pst*I site in the 5′ end); and (c) *Nco*I/*Bam*HI-digested 2E1 cDNA in pJL2.[9] The BM3:2B4 chimera is similarly constructed from the BM-3 linker, a PCR-generated fragment (*Pst*I–*Bsm*I) extending from nucleotides 127 to 480, and *Nco*I/*Bsm*I-digested pJL2B4. The 2E1:2B4 chimera is obtained by cloning an 81-bp PCR fragment of the 5′-coding region of 2E1 (Δ3–29), previously digested with *Nco*I and *Pst*I, into similarly digested 2B4 in pJL.

The 2B4 (Δ2–27) is also expressed fused to glutathione *S*-transferase (GST) and is constructed from pGEX-KN, a Pharmacia vector modified by Hakes and Dixon[14] to include a polyglycine kinker region encoding a sequence that improves the efficiency of thrombin cleavage immediately preceding the Met of the shortened P450. To obtain this clone, the 2B4 clone in pBluescript[15] is digested with *Hind*III, made blunt-ended with the Klenow fragment of T4 DNA polymerase, and digested at the *Not*I site of the vector; the fragment is ligated to pGEX-KN (previously digested with *Not*I and *Sma*I). Finally, the dephosphorylated double-stranded linker shown below is cloned into *Not*I and *Xma*III sites of this vector to generate pGEX-KT-2B4 (Δ2–27).

```
GGCCGCCTGGTTCCTCGTATGGCT
CGGACCAAGGAGCATACCGACCGG
```

The DNA sequence of all PCR-generated fragments is determined to assess the fidelity of *Taq* DNA polymerase (GIBCO-BRL). All plasmids are transfected into MV1304 cells, a *rec*A⁻ derivative of JM105.[9]

Expression Conditions and Preparation of Subcellular Fractions

For investigation of subcellular localization, overnight cultures of each clone in *Escherichia coli* are diluted 100-fold in LB broth containing 50 μg/ml ampicillin and grown to OD$_{600\ nm}$ 0.5–1.0 at 37° in an orbital shaker, and P450 expression is induced with 1 mM IPTG. Cells are harvested after a 4-hr induction in centrifuge bottles, resuspended in 20 mM potassium phosphate buffer, pH 7.4, containing 5 mM EDTA and 50 mM KCl, and

[14] D. J. Hakes and J. E. Dixon, *Anal. Biochem.* **202,** 293 (1992).
[15] R. Gasser, M. Negishi, and R. M. Philpot, *Mol. Pharmacol.* **32,** 22 (1988).

lysed by two passes through a French pressure cell at 18,000 p.s.i. Cell debris is removed by centrifugation at 12,000 g for 10 min, and the lysate is subjected to ultracentrifugation at 142,000 g for 1 hr at 4°; "membrane" and "cytosolic" fractions are arbitrarily defined on the basis of whether the material is sedimented or retained in the supernatant fraction under these conditions. The membrane fraction is washed once in 50 mM Tris acetate buffer, pH 7.4, containing 1 mM EDTA and 0.8 M KCl, repelleted, and then suspended in 100 mM potassium phosphate buffer, pH 7.4, containing 20% glycerol and 1 mM EDTA. Experiments with full-length and shortened P450 2E1 and 2B4 enzymes indicated that only about 1% of the expressed P450 is localized in the outer membrane and that none is present in the periplasm[2]; hence, only cytosolic and membrane fractions are prepared for determination of subcellular localization. The protein content of each fraction is determined using bicincochinic acid. Under these growth conditions, levels of expression are routinely low, i.e., less than 50 nmol of P450 per liter.

For high level expression of P450 in E. coli, TB media with 100 μg/ml ampicillin is inoculated for overnight growth at room temperature, induction of P450 expression is carried out the following morning with cells in plateau phase by the addition of IPTG to 1 mM concentration in the culture, and the cells are shaken slowly for 24 to 48 hr with periodic additions of ampicillin to promote plasmid retention. The addition of δ-aminolevulinic acid (1 mM final concentration) to the culture at the time of induction gives 300 to 400 nmol of CO-reactive P450 per liter of cells with 2B4 (Δ2–27) fused to GST. The 2E1 (Δ3–29) is expressed at a level of about 500 nmol/ liter under these conditions without a requirement for the heme precursor. Unfortunately, these expression conditions do not allow for the production of sufficient quantities of 2E1 (Δ3–29) for purification when the 5′-coding sequence is modified or deleted. To determine the P450 content in intact cells, 1 ml of cultured cells is centrifuged, and the cells are resuspended in 2 ml of 100 mM potassium phosphate buffer, pH 7.4, containing 20% glycerol and 1.0% cholate, an ample amount of sodium dithionite is added, and the sample is divided into two cuvettes. Carbon monoxide is vigorously bubbled into one cuvette, and the CO-reduced difference spectrum is determined with a double-beam recording spectrophotometer.

Determination of Subcellular Localization

Quantitative immunoblot analysis of membranes and cytosol is performed with polyclonal antibodies to 2E1 and 2B4 by comparison with purified protein standards or by comparative analysis of cytosolic and membrane fractions on the same blot. In the latter method, 4 μg of cytosolic protein (loaded in triplicate) and increasing amounts of membrane protein (loaded in duplicate) are subjected to sodium dodecyl sulfate–polyacrylamide gel electrophoresis (SDS–PAGE) in a mini-gel apparatus with a 7.5% acrylamide gel (0.75 mm thick). Electrophoretic separation is carried out at 100 V until the dye front reaches 1 mm from the bottom of the gel for 2E1 variants; the dye front is run off the gel with 2B4 variants to allow their resolution from a cytosolic protein of bacterial origin of similar molecular weight that prevents accurate immunoquantitation. The protein is transferred to nitrocellulose paper at 60 V for 1 hr, and P450 is visualized by dye precipitation[16] or enhanced chemiluminescence[8] using secondary antibodies linked to alkaline phosphatase or peroxidase following overnight blocking of nonspecific sites in 20 mM Tris–Cl, pH 7.6, containing 137 mM NaCl, 0.1% Tween 20, and 5% nonfat dry milk. Chemiluminescence is particularly useful for immunoquantitation since the band intensity on the photographic film, which is of critical importance in obtaining an appropriate range for densitometric scanning, is easily controlled by altering the time of exposure. For 2E1 and variants, the amount of P450 in membrane and cytosolic fractions is determined from a set of standards. For 2B4 and variants, the intensity of the immunoreactive band produced with 4 μg of cytosolic protein is compared to that of different amounts of membrane protein with a laser densitometer. The amount of cytosolic and membrane protein that gives an equivalent intensity and the amount of protein obtained in the two *E. coli* fractions are then used to calculate the percentage of expressed protein in each fraction. The subcellular distribution has also been determined for 2E1 (Δ3–29) by determination of the CO complex in membrane and cytosolic fractions, and found to be the same as that determined by immunoquantitation. Since 2B4 (Δ2–27) fused to GST was found to have the same subcellular distribution as 2B4 (Δ2–27), it is proposed that the spectrophotometric assay for GST activity with 1-chloro-2,4-dinitrobenzene might be used to more easily monitor the localization of P450–GST fusion proteins.

Purification Procedures

P450 cytochromes such as 2B4 (Δ2–27) that are expressed at low levels and predominantly in the cytosol present a greater challenge for purification due to the higher total protein content in this bacterial fraction. After unsuccessful attempts to obtain suitable fractions of 2B4 (Δ2–27) for study, the cDNA for the shortened 2B4 was fused to GST, and the fusion protein was expressed in *E. coli* as described earlier and purified on a glutathi-

---

[16] V. S. Fujita, D. J. Thiele, and M. J. Coon, *DNA Cell Biol.* **9**, 111 (1990).

one(GSH)–agarose column. Induced cells are washed three times in 10 mM Tris–Cl, pH 8.0, and resuspended in 30 mM potassium phosphate buffer, pH 7.4, containing 150 mM NaCl, 20% glycerol, and 0.1% 2-mercaptoethanol (PBS). Cells are treated with 1% n-octyl β-D-glucopyranoside with stirring at room temperature for 20 min and are lysed with two passes in a French pressure cell. The cell lysate is diluted with 1 volume of PBS, submitted to centrifugation at 140,000 g to remove insoluble components, and loaded onto a GSH–agarose column equilibrated with PBS containing 0.3% octylglucoside. The yield of P450 in the supernatant fraction is generally about 85% of that in the lysate; the column capacity is about 40 nmol/ml resin. The column resin is washed with 10 column volumes of 100 mM Tris–Cl, pH 7.4, containing 120 mM NaCl, 20% glycerol, and 0.3% octylglucoside (TBSO) and is then extruded under pressure. The fusion protein is eluted batchwise overnight and then several times the next day with an equal volume of TBSO containing 20 mM GSH at 4°. The preparation is dialyzed against TBS to remove GSH and detergent, and the fusion protein is cleaved at room temperature with 8 U of thrombin/nmol P450 for 36 hr. The preparation is passed over another GSH–agarose column to remove GST and a small amount of uncleaved protein. The final preparation is concentrated and dialyzed against 100 mM potassium phosphate, pH 7.4, containing 20% glycerol and 0.1 mM EDTA in a ProDicon chamber and has a final content of 6 nmol P450/mg protein. SDS–PAGE gels (10%) stained with Coomassie blue indicate a few proteins of higher molecular weight and one other protein of lower molecular weight in these preparations; comparison with standard 2B4 purified from liver microsomes of phenobarbital-treated animals demonstrates the absence of apoprotein.

Catalytic Assays

Reaction mixtures (0.5 to 1.0 ml) contain 0.05 to 0.25 μM P450 2B4 or 2E1, NADPH-cytochrome P450 reductase at a 2 : 1 to 3 : 1 molar ratio with respect to P450, 30 to 45 μg/ml 1,2-dilauroylglycero(3)-phosphorylcholine, 50 or 100 mM potassium phosphate buffer (pH 7.4 for all substrates except pH 6.8 with p-nitrophenol), and 80 mM ethanol, 100 mM N-nitrosodiethylamine, 100 μM p-nitrophenol, 1 mM d-benzphetamine, 1 mM N,N-dimethylaniline, or 100 mM 1-phenylethanol. Ascorbic acid (1 mM) is also included in the p-nitrophenol hydroxylase assay. After a 3-min equilibration at 30° for all reaction mixtures except those containing N-nitrosodiethylamine and ethanol (which are incubated at 37°), NADPH is added as the final component(1 to 2 mM). The incubation is at the same temperature in a shaking water bath for a period of time (10 to 30 min) corresponding to the linear portion of the time course. The reactions are quenched on ice by the addition of trichloroacetic acid or perchloric acid. Specific details of reconstitution and reaction conditions for each shortened P450 and substrate are provided elsewhere.[3,9] Formaldehyde formed by the N-demethylation of d-benzphetamine and N,N-dimethylaniline is determined by the method of Cochin and Axelrod,[17] nitrocatechol formation from p-nitrophenol is assayed spectrophotometrically according to the procedure of Koop,[18] acetaldehyde production from ethanol and ethylene generation from N-nitrosodimethylamine are measured by gas chromatography of head space gas,[19,20] and 1-phenylethanol oxygenation to acetophenone is determined with the use of a high-pressure liquid chromatographic assay.[21] Purified 2B4 and NADPH-cytochrome P450 reductase are from liver microsomes of adult male rabbits treated with phenobarbital,[22,23] and 2E1 is purified from liver microsomes of adult male rabbits treated with acetone.[24]

Properties of Truncated P450s

Subcellular Localization of P450 Cytochromes Modified in the N-Terminal Region

Of the full-length and variant P450 2B4 and 2E1 forms expressed in E. coli, only 2E1 (Δ3–51) is significantly degraded as judged by the presence of numerous immunoreactive bands migrating ahead of the 2E1 standard. As expected for enzymes containing a highly hydrophobic N-terminal region, full-length 2E1 and 2B4 are predominantly targeted to the membrane in E. coli as shown in Table I. Interestingly, removal of the N-terminal region (S1 and L1) of 2E1 does not alter the subcellular localization of the expressed protein, whereas deletion of the same region in 2B4 changes the P450 to a predominantly cytosolic localization. Insertion of positive charges into the N-terminal region increases the fraction of P450 2E1 (Δ3–29) that is expressed in the cytosol, and replacement of the somewhat hydrophobic N-terminal region of shortened 2E1 with the hydrophilic N-terminal region of P450$_{BM-3}$ decreases membrane targeting such that 80% of the cytochrome is expressed in the cytosol. Deletion of the first two hydrophobic segments

[17] J. Cochin and J. Axelrod, J. Pharmacol. Exp. Ther. 125, 105 (1959).
[18] D. R. Koop, Mol. Pharmacol. 29, 399 (1986).
[19] D. R. Koop, E. T. Morgan, G. E. Tarr, and M. J. Coon, J. Biol. Chem. 257, 8472 (1982).
[20] X. Ding and M. J. Coon, Drug Metab. Dispos. 16, 265 (1988).
[21] A. D. N. Vaz and M. J. Coon, Biochemistry 33, 6442 (1994).
[22] D. A. Haugen and M. J. Coon, J. Biol. Chem. 251, 7929 (1976).
[23] J. S. French and M. J. Coon, Arch. Biochem. Biophys. 195, 565 (1979).
[24] D. R. Koop, E. T. Morgan, G. E. Tarr, and M. J. Coon, J. Biol. Chem. 257, 8472 (1982).

TABLE I

CELLULAR LOCALIZATION IN *E. coli* OF P450 2E1 AND 2B4 VARIANTS MODIFIED IN THE
N-TERMINAL REGION[a]

P450 variant	Segment deleted/modified	Hydropathicity index of N-terminal region[b]	Distribution (% total P450)	
			Membrane	Cytosol
2E1	None	26.5	70 ± 5	30 ± 5
2E1 (Δ3–29)	S1, L1	5.0	65 ± 5	35 ± 5
2E1 (Δ3–29) (1+)	S1, L1	3.2	55 ± 5	45 ± 5
2E1 (Δ3–29) (2+)	S1, L1	1.8	50 ± 5	50 ± 5
BM3:2E1	S1, L1	−4.3	20 ± 5	80 ± 5
2E1 (Δ3–48)	S1, L1, S2	−4.5	45 ± 5	55 ± 5
2B4	None	25.5	73 ± 5	27 ± 5
2B4 (Δ2–20)	S1	−8.9	33 ± 6	67 ± 6
2B4 (Δ2–27)	S1, L1	2.8	32 ± 5	68 ± 5
2E1:2B4	S1, L1	5.0	41 ± 3	59 ± 3
BM3:2B4	S1, L1	−4.3	27 ± 3	73 ± 3

[a] From Larson *et al.*[2] and Pernecky *et al.*[8,9]

[b] The average hydropathicity index for the first 20 amino acid residues is determined for each form [J. Kyte and R. F. Doolittle, *J. Mol. Biol.* **157**, 105 (1982)] with the SOAP program of PC/Gene (Intelligenetics); positive values indicate a relatively hydrophobic segment whereas negative values indicate a more hydrophilic segment. The contribution of the N-terminal Met to the index is given even though this residue in 2E1 (Δ3–29) has been shown to be removed during expression in *E. coli*.

TABLE II

ACTIVITY OF FULL-LENGTH AND SHORTENED 2B4 AND 2E1[a]

P450 isozyme	Substrate	Activity (nmol product/ min/nmol P450)		Activity of shortened (% of full-length)
		Full-length	Shortened[b]	
2E1	Ethanol	35 ± 2	35 ± 1	100
	N-Nitrosodiethylamine	4.9 ± 0.2	4.7 ± 0.2	96
	p-Nitrophenol	6.7 ± 0.3	6.9 ± 0.4	103
2B4	*d*-Benzphetamine	54 ± 2	22 ± 1	41
	N,N-Dimethylaniline	27 + 0.5	21 ± 0.2	78
	1-Phenylethanol	6.3 + 0.3	2.2 ± 0.2	35

[a] From Larson *et al.*[3] and Pernecky *et al.*[9]

[b] P450 2E1 (Δ3–29) or P450 2B4 (Δ2–27).

(S1 and S2) in 2E1 to yield 2E1 (Δ3–48) that contains a hydrophilic N-terminal region also decreases the amount of P450 localized in the membrane fraction. Membrane targeting is not as affected by modifications in the hydropathy of the N-terminal region of 2B4 (Δ2–27). In no case does washing the membrane fraction with Na_2CO_3, pH 11, release significant amounts of the variant P450 2B4 or 2E1 cytochromes, thus demonstrating the integral nature of the membrane association. A ferrous carbonyl difference spectrum can be demonstrated for all P450 cytochromes following Tergitol NP-10 solubilization of membrane fractions, PEG fractionation of cytosolic fractions, or S-Sepharose column chromatography, except for 2E1 (Δ3–48). These results demonstrate that P450 cytochromes differ in the extent to which the hydrophobic N-terminal region influences membrane targeting and provide evidence for membrane targeting domains beyond the first two hydrophobic regions.

Catalytic Activities

Experiments with purified 2E1 (Δ3–29) indicate that the absence of the N-terminal region does not affect the catalytic activity with a variety of substrates as shown in Table II, has no effect on Michaelis–Menton parameters determined with *p*-nitrophenol, does not perturb interaction with the reductase, and does not alter phospholipid or cytochrome b_5 stimulation of catalytic activity.[3] However, when compared to the full-length 2B4, the rate of phenylethanol oxidation to acetophenone catalyzed by 2B4 (Δ2–27) is lowered to 35%, and the rates of benzphetamine and *N,N*-dimethylaniline N-demethylation are decreased to 41 and 78%, respectively. Taken together, these data indicate that the N-terminal region contributes to a varying extent to catalytic rates with different P450 cytochromes.

Acknowledgments

This work was supported by Grants DK-10339 and AA-06221 from the National Institutes of Health. S.J.P. was a postdoctoral trainee of the National Institute of Environmental Health Sciences, Grant ES-07062.

[4] Purification of Functional Recombinant P450s from Bacteria

By F. PETER GUENGERICH, MARTHA V. MARTIN, ZUYU GUO, and YOUNG-JIN CHUN

Introduction

A number of expression systems have been utilized to produce recombinant P450s, as discussed in this and an earlier volume.[1] Bacteria have considerable advantages because of the high levels of expression, ease of manipulation, and relatively low cost. Aspects of the choice of vectors and modification of 5' terminals for optimal expression are discussed elsewhere in this volume.[2]

While P450s can be studied within some of the other vector systems, purification is usually necessary for most studies with the enzymes produced in bacteria. In general, purification is relatively easy and considerably more efficient than from mammalian tissues. This chapter is based largely on this laboratory's experience with recombinant human P450s expressed in *Escherichia coli* and, to a lesser extent, *Salmonella typhimurium*.[3–9]

Experimental Details

Construction of Plasmids and Growth of E. coli

The N-terminal amino acid sequences of the P450s expressed in this laboratory are shown in Fig. 1. In all cases the native sequence gave little or no expression and the modified sequence used was found to be the best

---

[1] M. R. Waterman and E. F. Johnson, *Methods Enzymol.* **206** (1991).

[2] H. J. Barnes, *Methods Enzymol.* **272,** Chap. 1, 1996 (this volume).

[3] E. M. J. Gillam, T. Baba, B.-R. Kim, S. Ohmori, and F. P. Guengerich, *Arch. Biochem. Biophys.* **305,** 123 (1993).

[4] P. Sandhu, T. Baba, and F. P. Guengerich, *Arch. Biochem. Biophys.* **306,** 443 (1993).

[5] P. Sandhu, Z. Guo, T. Baba, M. V. Martin, R. H. Tukey, and F. P. Guengerich, *Arch. Biochem. Biophys.* **309,** 168 (1994).

[6] E. M. J. Gillam, Z. Guo, and F. P. Guengerich, *Arch. Biochem. Biophys.* **312,** 59 (1994).

[7] Z. Guo, E. M. J. Gillam, S. Ohmori, R. H. Tukey, and F. P.Guengerich, *Arch. Biochem. Biophys.* **312,** 436 (1994).

[8] E. M. J. Gillam, Z. Guo, Y.-F. Ueng, H. Yamazaki, I. Cock, P. E. B. Reilly, W. D. Hooper, and F. P. Guengerich, *Arch. Biochem. Biophys.* **317,** 374 (1995).

[9] E. M. J. Gillam, Z. Guo, M. V. Martin, C. M. Jenkins, and F. P. Guengerich, *Arch. Biochem. Biophys.* **319,** 540 (1995).

```
              1          10         20         30         40
1A1    MLFPISMSATEFLLASVIFCLVFWVMRASRPQVPKGLKNP...
1A1'   MAFPISMSATEFLLASVIFCLVFWVMRASRPQVPKGLKNP...

1A2    MALSQSVPFSATELLLASAIFCLVFWVLKGLRPRVPKGLK...
1A2'   MA          LLLAVFLFCLVFWVLKGLRPRVPKGLK...

2C10   MDSLVVLVLCLSCLLLLSLWRQSSGRGKLPPGPTPLPVIG...
2C10'  MA          RQSSGRGKLPPGPTPLPVIG...

2D6    MGLEALVPLAVIVAIFLLLVDLMHRRQRWAARYP...
2D6'   MA          RQVHSSWNLP...

2E1    MSALGVTVALLVWAAFLLLVSMWRQVHSSWNLPPGPFPLP...
2E1'   MA          RQVHSSWNLPPGPFPLP...

3A4    MALIPDLAMETWLLLAVSLVLLYLYGTHSHGLFKKLGIPG...
3A4'   MA          LLLAVFLVLLYLYGTHSHGLFKKLGIPG...
3A5'   MA          LLLAVFLVLLYLYGTRTHGLFKRLGIPG...
```

Fig. 1. Human P450s expressed in pCW vector. The top line in each set shows the native sequence and the lower lines (with the prime) show modified sequences used for expression.[3–9]

of a series of constructs tested. Considerations used in optimizing 5′ termini include (i) use of GCT as the second codon (coding for Ala), (ii) reduction of free energy for secondary structure formation by changing G:C pairs to A:T in the first 10 codons,[10] (iii) removal of hydrophobic segments, and (iv) alignment to insert the sequence MALLLAVFL ..., first used to express bovine P450 17A,[2,10] at a region with some similarity.[3–9] The vector pCW (tac/tac promoter) has been used in all of our own expression work.

Plasmids are stored at $-20°$ and used to prepare glycerol stocks.[3] These glycerol stocks can be stored for at least several months at $-70°$ and utilized to seed new bacterial preparations. In early studies we utilized E. coli JM109 but generally use E. coli DH5α cells in our work. Cells transformed with the plasmid of interest are grown overnight at $37°$ in Luria-Bertani (LB) medium containing 100 μg ampicillin ml^{-1}. A 10-ml aliquot is used to inoculate 1.0 liter of Terrific broth (TB) media containing 0.20% bacto-peptone (w/v) and supplemented with 100 μg ampicillin ml^{-1}, 1.0 mM thiamine, and trace elements [0.25 ml of stock preparation (liter of culture)$^{-1}$—composition: 27 g FeCl$_3$ · 6H$_2$O, 2.0 g ZnCl$_2$ · 4H$_2$O, 2.0 g CaCl$_2$·6H$_2$O, 2.0 g Na$_2$MoO$_4$, 1.0 g CaCl$_2$·2H$_2$O, 1.0 g CuCl$_2$, 0.5 g H$_3$BO$_3$, and 100 ml concentrated HCl (liter)$^{-1}$]. In the cases of P450s 1A1, 2C10, 2D6, and 2E1 we have found that the addition of 0.5 mM δ-aminolevulinic acid (ALA) and FeCl$_3$ results in a substantial increase in the amount of

[10] H. J. Barnes, M. P. Arlotto, and M. R. Waterman, Proc. Natl. Acad. Sci. U.S.A. **88,** 5597 (1991).

holo-P450 formed (three- to four-fold with 1A1, 2C10, and 2E1; with 2D6, ALA is obligatory for holoenzyme expression). The 1.0-liter cultures are shaken at 250 rpm in an Innova incubator (New Brunswick Scientific, Edison, NJ; 2.8-liter Fernbach flasks) for 4 hr at 28–32°, depending on the particular P450 construct.[3–9] Isopropyl β-D-thiogalactoside (IPTG) is added to 1.0 mM and shaking is continued for 24–42 hr, depending on the construct.[3–9]

Cells are chilled on ice and harvested by centrifugation at 4×10^3 g for 10 min.

Assays

The most direct means of monitoring P450 production in bacteria is to do direct difference spectroscopy (Fe^{2+} · CO vs Fe^{2+}).[11] This can be done directly on washed and resuspended cells for spectral analysis. A computer-driven Aminco-Chance DW-2 spectrophotometer (OLIS, Bogart, GA) is used in our own laboratory; it has the advantage of being able to utilize turbid samples.

Sodium dodecyl sulfate (SDS)–polyacrylamide gel electrophoresis is done according to the basic procedure of Laemmli[12] as modified[13]; staining can be done with ammonical silver[14] or Coomassie brilliant blue R-250 dye. Immunoblotting is done as described[13,15]; however, many polyclonal antibodies contain antibodies that recognize bacterial proteins (due to exposure of rabbits to bacteria etc.) and adsorption with bacteria may be necessary to reduce the number of extraneous bands.

General procedures for purification of NADPH-P450 reductase and cytochrome b_5 are given elsewhere.[13,16,17] Procedures for catalytic assays are given in detail or referred to elsewhere.[3–9,13]

Protein concentrations are estimated using a bicinchoninic acid BCA procedure (Pierce Chemical Co., Rockford, IL).

Preparation of Membranes

All steps are done at 0–4°. Pelleted bacterial cells are resuspended in 100 mM Tris–acetate buffer (pH 7.6) containing 500 mM sucrose and 0.5

[11] T. Omura and R. Sato, *J. Biol. Chem.* **239**, 2370 (1964).
[12] U. K. Laemmli, *Nature (London)* **227**, 680 (1970).
[13] F. P. Guengerich, *in* "Principles and Methods of Toxicology" (A. W. Hayes, ed.), p. 1259. Raven Press, New York, 1994.
[14] W. Wray, T. Boulikas, V. P. Wray, and R. Hancock, *Anal. Biochem.* **118**, 197 (1981).
[15] F. P. Guengerich, P. Wang, and N. K. Davidson, *Biochemistry* **21**, 1698 (1982).
[16] Y. Yasukochi and B. S. S. Masters, *J. Biol. Chem.* **251**, 5337 (1976).
[17] Y. Funae and S. Imaoka, *Biochim. Biophys. Acta* **842**, 119 (1985).

mM EDTA [$\sim$70 mg wet weight cells (ml)$^{-1}$]. The suspension is diluted with an equal volume of H$_2$O and 0.10 mg lysozyme ml^{-1}; the preparation is gently shaken for 30 min (to hydrolyze the outer membrane). The resulting spheroplasts are pelleted at 4×10^3 g for 10 min and resuspended ($\sim$0.5 g ml^{-1}) in 100 mM potassium phosphate buffer (pH 7.4) containing 6 mM magnesium acetate, 20% glycerol (v/v), and 0.10 mM dithiothreitol (DTT). These spheroplasts can be stored frozen at $-70°$ until further use, when they are thawed in a water bath at room temperature. Protease inhibitors are added during thawing: phenylmethylsulfonyl fluoride (PMSF, stored at $-20°$ in n-propanol), 1.0 mM; leupeptin, 2 μM; bestatin, 10 μM; and aprotinin, 0.04 U ml^{-1}. Cells are lysed with two 20-sec bursts (70% full power) of a Branson sonicator (Branson Sonic Power, Danbury, CT) while the cells are in an ice-salt bath ($\sim$$-10°$). A Rosett Cell is useful for larger scale sonications because the cooling is better (use on ice, not at $-10°$).

The resulting lysate is subjected to centrifugation at 10^4 g for 10 min and the resulting supernatant is then centrifuged at 10^5 g for 60 min (3.5 $\times 10^4$ rpm in a Beckman 45 Ti rotor) to precipitate the membranes (see modification for P450 2D6, *vide infra*). The pellet is resuspended in 50 mM Tris–acetate buffer (pH 7.6) containing 0.25 mM EDTA and 0.25 M sucrose and is stored at $-70°$.

Purification of P450s by Ion-Exchange Chromatography

Escherichia coli membranes are diluted to a protein concentration of 2.0 mg ml^{-1} in 20 mM Tris–acetate buffer (pH 7.4) containing 20% glycerol (v/v), 1.0 mM EDTA, 1.0 mM DTT, 0.625% sodium cholate (w/v), and Triton N-101 (Fig. 2). Sodium cholate is prepared from recrystallized cholic acid (50% aqueous C$_2$H$_5$OH) and is used as a 20% aqueous stock solution. The Triton N-101 is added from a 20% aqueous stock (w/v) to concentrations of 0.62% with P450s 1A1, 2C10, 3A4, and 3A5 and to 1.25% in the case of P450s 1A1 and 2E1.

We have found it necessary to add inhibitory ligands to some of the P450s to prevent denaturation.[5–7] In the case of P450 1A1 and 1A2, 30 μM α-naphthoflavone (αNF, 7,8-benzoflavone) is added during solubilization and in ion-exchange chromatography buffers.[5,7] With P450 2E1, 30 μM 4-methylpyrazole (4-MP) is added.[6] (These inhibitors show no effect when added to growth cultures.) We have also added 30 μM quinidine to P450 2D6 as a precaution.[9]

The solubilized membrane solution is stirred for 60 min and centrifuged at 10^5 g for 60 min. The clarified supernatant is applied to a column of DEAE-Sephacel (Pharmacia, Piscataway, NJ; 2.5 $\times$ 10 cm for a preparation of $\sim$1 liter) that has been equilibrated with the solubilization buffer. The

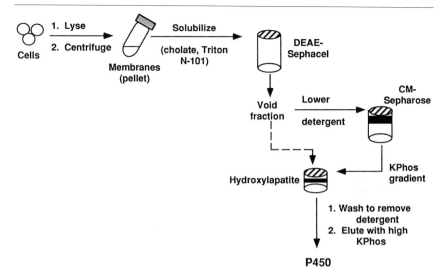

FIG. 2. General scheme of purification of recombinant P450s from *E. coli*.

column is washed with ~3 column volumes of the same buffer and the fractions containing P450 (red color) are pooled. The high concentrations of detergent favor binding of bacterial impurities to the column, and the detergent concentrations have been adjusted to favor recovery of the P450. At this point the P450 has generally been purified considerably and, for some purposes, may be treated directly to remove residual detergent.[5]

Further purification is usually done using a second ion-exchange step. The P450 fraction recovered from the DEAE step is concentrated (from ~1000 ml) to ~200 ml using an Amicon ultrafiltration device with a PM-30 membrane (Amicon, Danvers, MA) and then stirred with ~24 g of Amberlite XAD-2 beads (Sigma Chemical Co., St. Louis, MO) for 1 hr. The suspension is filtered through glass wool or Whatman No. 1 filter paper, diluted to the original volume (~1 liter) with 20% glycerol, adjusted to pH 6.5 with CH_3CO_2H, and applied to a 1.5 × 7-cm column of CM-Sepharose Fast Flow (Pharmacia) equilibrated with 10 mM potassium phosphate buffer (pH 6.5) containing 0.1 mM EDTA, 0.1 mM DTT, and 20% glycerol (v/v). The column is washed with several volumes of the equilibration buffer and then with the same buffer in which the phosphate concentration has been raised to 50 mM. A gradient of 20 to 200 mM phosphate is applied (200 ml) with the same buffer to which 0.20% Triton N-101 has been added. Fractions of 3 ml are collected and analyzed for P450 (A_{417}) and for purity by SDS–polyacrylamide gel electrophoresis.

Overall yields at this point vary from 22 to 86%.[3–8]

Purification of P450 2D6

The general ion-exchange approach used for the other P450s (Fig. 2) was not successful with the P450 2D6 we expressed, and an alternate strategy was used.[9]

The membrane fraction obtained after an initial centrifugation at 10^5 g for 45 min contains only ~5% of the P450 and is discarded. The supernatant is subjected to further centrifugation overnight (16 hr) and the resulting pellet (~1800 mg protein) is suspended in 200 ml of 10 mM Tris–HCl buffer (pH 7.4) containing 50 mM NaCl, 1.0 mM EDTA, and 30 μM quinidine (added to stabilize P450 2D6). An aqueous Triton X-114 solution [~10%, cycled three times between 4° (60 min) and 37° (overnight) to make the micelle size more uniform][18] is added to a final concentration of 0.70% (w/v). The clear solution is allowed to stand at 0° for 30 min and is then warmed to 37° (in polycarbonate tubes) for 30 min. Centrifugation of the cloudy material (10 min, 4×10^3 g, ambient temperature) yields a tight pellet, which is discarded, and a supernatant containing most of the P450.

The above supernatant (200 ml) is dialyzed twice against 50 volumes of 20 mM potassium phosphate buffer (pH 7.5) containing 0.1 mM EDTA, 20% glycerol (v/v), 0.10 mM DTT, and 30 μM quinidine (at 4°) and is applied to an *E. coli* flavodoxin affinity column,[19] which has been equilibrated with the dialysis buffer, at ~2 ml min^{-1}. (An overexpression plasmid for *e. coli* flavodoxin is available from Professor M. R. Waterman of this department.) The column is then washed with 300 ml of the equilibration buffer containing Triton N-101, and P450 2D6 is eluted with a linear gradient of 0 to 0.5 M NaCl in the same buffer. Fractions are analyzed for P450 by A_{417} and SDS–polyacrylamide gel electrophoresis. The overall yield of P450 (from membranes) is ~15%.[9]

The capacity of the affinity column depends on the conditions used for preparation. In our experience, a 1.5×6-cm column binds about 65 nmol of P450. The column may be regenerated by washing with the buffer containing Triton N-101 and 1.0 M NaCl and then reequilibration. We have also found that such a column can bind recombinant P450s 1A2 and 3A4. The general usefulness of the method in purifying these enzymes is under investigation.

Purification of Recombinant P450: NADPH-P450 Reductase
  Fusion Proteins

In addition to the human P450s, we have also expressed two P450: rat NADPH-P450 reductase fusion proteins in *E. coli* and purified these using

---

[18] C. Bordier, *J. Biol. Chem.* **256**, 1604 (1981).
[19] C. M. Jenkins and M. R. Waterman, *J. Biol. Chem.* **269**, 27401 (1994).

methods developed for NADPH-P450 reductase.[16] The P450 3A4:rat NADPH-P450 reductase construct is the same as reported by Shet *et al.*[20] and the purification procedure is modified from a report for a P450 17A fusion protein by that group.[21] We have also prepared a human P450 1A2:rat NADPH-P450 reductase construct using a similar strategy and expressed and purified and protein.[22] The expression levels for both proteins are generally 70–150 nmol (P450) liter^{-1}. Membranes are prepared as in the cases of P450 expressions.

E. coli membranes are diluted to a protein concentration of 3.0 mg ml^{-1} in 10 mM potassium phosphate buffer (pH 7.5) containing 20% glycerol (v/v), 0.5 mM EDTA, 0.10 mM DTT, 0.5 mM PMSF, and 1.0% Emulgen 911 (w/v). (All procedures are at 4°.)

The solubilized membrane is stirred for 30 min and is subjected to centrifugation at 10^5 g for 60 min. The clarified supernatant is applied to a column of DE-52 (Whatman, Fairfield,NJ; 2.5 × 9-cm column for ~800 ml of solubilized membranes) that has been equilibrated with 50 mM potassium phospate buffer (pH 7.8) containing 20% glycerol (v/v), 0.5 mM EDTA, 0.10 mM DTT, 1.0 μM FMN, and 0.20% Emulgen 911 (w/v). The column is washed with several column volumes of the same buffer. A gradient of 0 to 500 mM KCl is applied in the same buffer. Fractions of 3 ml are collected and analyzed for P450 (A_{417}) and NADPH-P450 reductase (cytochrome c reductase assay) and for SDS–polyacrylamide gel electrophoresis.

Further purification is done using an affinity column step. The fusion protein fraction recovered from the DE-52 step is applied to a 1.5 × 7-cm column of 2′,5′-ADP-Sepharose 4B (Pharmacia) equilibrated with 300 mM potassium phosphate buffer (pH 7.7) containing 20% glycerol (v/v), 0.10 mM EDTA, 0.10 mM DTT, 1.0 μM FMN, and 0.20% Emulgen 911 (w/v). The column is washed with the same buffer and then with the same buffer containing 0.15% sodium cholate instead of Emulgen 911 (about 250 ml of each). The P450:rat NADPH-P450 reductase fusion protein is eluted with a linear gradient of 0 to 10 mM AMP in the same buffer (200 ml total volum). Fractions are analyzed for P450 by A_{417} and NADPH-P450 reductase by the cytochrome c reduction assay and SDS–polyacrylamide gel electrophoresis. Residual cholate can be removed by extensive dialysis or by repetitive concentration (by ultrafiltration) and dilution. The overall yields for this process have been ~15% based on the P450 in the membrane fraction.

[20] M. S. Shet, C. W. Fisher, P. L. Holmans, and R. W. Estabrook, *Proc. Natl. Acad. Sci. U.S.A.* **90,** 11748 (1993).
[21] M. S. Shet, C. W. Fisher, M. P. Arlotto, C. H. L. Shackleton, P. L. Holmans, C. A. Martin-Wixtrom, Y. Saeki, and R. W. Estabrook, *Arch. Biochem. Biophys.* **311,** 402 (1994).
[22] Y.-J. Chun and F. P. Guengerich, unpublished results.

Removal of Detergent

The general procedure involves dialysis of a sample versus 10 mM potassium phosphate buffer (pH 6.5) containing 20% glycerol (v/v) and application to a small (1.5 × 3 cm) column of hydroxylapatite (Bio-Gel HTP, Bio-Rad Laboratories, Hercules, CA) that has been equilibrated with the dialysis buffer (the concentration of EDTA should be kept <0.1 mM with hydroxylapatite columns to avoid complexation with the matrix). The column should not be more than one-half red due to binding of P450, or a larger column is needed. The column is washed extensively with the equilibration buffer, and the detergent is monitored at 280 nm. When this reading is <0.01, the column is eluted with 500 mM potassium phosphate buffer (pH 7.5) containing 20% glycerol (v/v) and 0.1 mM DTT. This procedure not only removes detergent but also inhibitors added for stabilization in the presence of detergents (αNF, 4-MP, quinidine). If color remains bound to the column, 0.05% sodium cholate (w/v) (or more) is added to the buffer to facilitate elution. The red fractions are dialyzed against 10 mM potassium phosphate buffer (pH 7.5) containing 1.0 mM EDTA, 0.1 mM DTT, andd 20% glycerol (v/v) to lower the phosphate concentration. If cholate is needed for elution, dialysis should be more extensive (four changes over 48 hr).

P450s 1A1 and 1A2 precipitate in low salt in the absence of detergent. Therefore, these should be dialyzed against 100 mM potassium phosphate buffer prior to storage.

The proteins can be stored essentially indefinitely at -20 or $-70°$. However, repeated thawing and refreezing should be avoided, so it is practical to aliquot preparations into fractions (that would be used in typical experiments) prior to freezing.

Functional Reconstitution

The purified proteins are mixed with NADPH-P450 reductase and phospholipids prior to measurement of catalytic activity, in the same way as P450s isolated from tissues.[13] In our experience, catalytic activities are higher with the purified P450s than when NADPH-P450 reductase is added to bacterial membranes containing the P450. With P450s 1A1, 1A2, 2C10, and 2D6, typical catalytic activities are seen with L-α-dilauroyl-*sn*-glycero-3-phosphocholine.[4,5,7,9] P450 2E1 activities usually require cytochrome b_5.[6] Many (but not all) catalytic activities of P450s 3A4 and 3A5 require cytochrome b_5, glutathione, a mixture of phospholipids, cholate, and MgCl$_2$.[3,8,23]

---

[23] Y.-F. Ueng, T. Shimada, H. Yamazaki, and F. P. Guengerich, *Chem. Res. Toxicol.* **8,** 218 (1995).

Alternate Methods of Purification

Other investigators have employed other purification procedures from *E. coli*. Larson *et al.*[24] expressed rabbit P450 2E1 and partially purified the protein using S-Sepharose chromatography in the presence of ocytylglucoside. Chiang expressed rat P450 7A and purified the protein to homogeneity using *n*-octyiamino-Sepharose and hydroxylapatite chromatography.[25] Winters and Cederbaum[26] expressed human P450 2E1 and also did partial purification with S-Sepharose.

One approach is to add an oligo-His region to a P450 terminus to a protein and then utilize affinity chromatography on a Ni^{2+}–agarose column. This approch has been used with P450s 17A modified at the C terminus[19] and, more recently, with P450 2D6 modified at the N terminus.[27] It appears necessary to have nonionic detergent present during the Ni^{2+} chromatography, and these can be removed as described here. It should be emphasized that the modification of terminals with oligo-His units or alternate amino acids to enhance expression (Fig. 1) has not been shown to alter catalytic activities of proteins, although the possibility that this could happen in other cases cannot be ruled out.

Expression and Purification from S. typhimurium

In the course of investigation with P450s expressed with the pCW vector we found that human P450 1A2 could be expressed in *S. typhimurium* TA1538 at levels ~1/4 as high as *E. coli* DH5α.[28] A bulk preparation of *S. typhimurium* transfected with the vector was handled in the same way as *E. coli* to purify P450 1A2. Thus, the procedures described here can probably be generally considered for expression in other bacteria.

Other Properties of Bacterial Recombinant P450s

In principle the final specific content of P450 (nmol P450 heme/mg protein) should be ~20, since the M_rs of the P450s are ~50,000. The values reported for preparations in this laboratory range from 9 to 23.[3-9] The variation is probably not due to lack of heme (which has been an issue in

[24] J. R. Larson, M. J. Coon, and T. D. Porter, *J. Biol. Chem.* **266,** 7321 (1991).

[25] Y. C. Li and J. Y. L. Chiang, *J. Biol. Chem.* **266,** 19186 (1991).

[26] D. K. Winters and A. I. Cederbaum, *Biochim. Biophys. Acta* **1156,** 43 (1992).

[27] A. Kempf, U. M. Zanger, and U. A. Meyer, *Arch. Biochem. Biophys.* **321,** 277 (1995).

[28] P. D. Josephy, L. S. DeBruin, H. L. Lord, J. Oak, D. H. Evans, Z. Guo, M.-S. Dong, and F. P. Guengerich, *Cancer Res.* **55,** 799 (1995).

baculovirus expression)[29] but to inherent error in the accuracy of the typical protein estimations.

All P450s expressed with the N-terminal sequence MALLLAVFL ... have been found to be blocked at the N terminus.[3–5,8] This block is apparently due to the retention of *N*-formyl-Met.[30] It does not appear to influence membrane localization or catalytic activity. P450s expressed with other sequences do not show such a block and, when Ala is coded for by the second codon, the N-terminal Met is removed as predicted.[4,6,7,9]

---

[29] A. Asseffa, S. J. Smith, K. Nagata, J. Gillette, H. V. Gelboin, and F. J. Gonzalez, *Arch. Biochem. Biophys.* **274**, 481 (1989).

[30] M.-S. Dong, L. C. Bell, Z. Guo, D. R. Phillips, I. A. Blair, and F. P. Guengerich. Submitted for publication.

[5] Application of Electrochemistry for P450-Catalyzed Reactions

By RONALD W. ESTABROOK, KEVIN M. FAULKNER, MANJUNATH S. SHET, and CHARLES W. FISHER

The heme proteins called P450 catalyze the metabolism of a great variety of different chemicals. Many of these are oxidation reactions where P450s function as mixed function oxidases.[1] The unique oxygen chemistry associated with these reactions permits the introduction of an atom of molecular oxygen in a regio- and stereo-specific manner. Therefore, P450s are excellent candidates as catalysts for the synthesis of high-value speciality chemicals that are difficult to synthesize by conventional chemical oxidation techniques. In addition to molecular oxygen and the chemical substrate to be metabolized, P450s require a source of electrons (NADPH) and an electron transport protein(s) for the transfer of two electrons (in sequence) from NADPH to the hemeprotein, e.g., the microsomal flavoprotein, NADPH-P450 reductase, or the mitochondrial electron transport sequence containing a flavoprotein and an iron–sulfur protein.[2] The use of NADPH as a source of electrons serves as an economic limitation, however, when designing reaction systems that require large-scale incubations and that may operate for extended time periods.

---

[1] H. S. Mason, *Adv. Enzymol.* **19**, 79 (1957).

[2] T. Omura, R. Sato, D. Y. Cooper, O. Rosenthal, and R. W. Estabrook, *Fed. Proc., Fed. Am. Soc. Exp. Biol.* **24**, 1181 (1965).

Recently we have applied[3] the controlled potential electrolysis method for the delivery of electrons via an electromotively active mediator that interacts with the flavoprotein component of the microsomal P450 system. In this way, the requirement for NADPH is avoided. Most satisfactory results have been obtained using recombinant fusion proteins engineered to contain the heme-domain of a P450 linked to the flavoprotein-domain of NADPH-P450 reductase,[4] although a reconstituted system containing both individual components (i.e., purified P450 and purified NADPH-P450 reductase) also appears to work.[3] Guryev and Gilevich[5] have published an abstract in which they describe the use of a similar electrochemical system to drive the activity of the adrenal mitochondrial P450scc, in the presence of the iron–sulfur protein, adrenodoxin, for the conversion of cholesterol to pregnenolone. The mediator employed is a low potential macrobicyclic nitrogen-caged cobalt complex[6] called cobalt sepulchrate[3+], (S)-[(1,3,6,8,10,13,16,19-octaazabicyclo-[6.6.6]eicosane)cobalt(III)][3+]. Selection of this mediator was fortuitous since it retains its chirality during reversible oxidation–reduction and does not form a peroxy-bridge dimer following reaction with oxygen.[6] The reaction of the cobalt[2+] complex with oxygen is a second-order reaction.

We have used the electrochemical method for a number of P450-catalyzed reactions. A sampling of results are summarized in Table I which shows the applicability of the method for a variety of reactions catalyzed by different P450s, e.g., the hydroxylation of steroids, the N-demethylation of drugs, and the omega-oxidation of fatty acids. Of interest is the observation that the rates of substrate metabolism obtained by the electrochemical method are generally similar to those using NADPH as the source of electrons.

Design of the Reaction Vessel

Figure 1 shows the apparatus currently in use in the authors' laboratory. A 10- or 20-ml beaker containing a magnetic flea is suspended in a continuous flow, temperature-controlled water bath placed over a magnetic stirrer. A platinum wire working electrode, to which a 10×20-mm strip of mesh

[3] K. M. Faulkner, M. S. Shet, C. W. Fisher, and R. W. Estabrook, *Proc. Natl. Acad. Sci. U.S.A.* **92,** 7705 (1995).

[4] C. W. Fisher, M. S. Shet, and R. W. Estabrook, *Methods Enzymol.* **272,** Chap. 2, 1996 (this volume).

[5] O. Guryev and S. Gilevich, *in* "Abstracts of The 8th International Conference on Cytochrome P450," Lisbon, Portugal (M. C. Lechner, ed.), p. 172. John Libbey Eurotext Paris, 1993.

[6] I. I. Creaser, J. M. Harrowfield, A. J. Herlt, A. M. Sargeson, J. Springborg, R. J. Geue, and M. R. Snow, *J. Am. Chem. Soc.* **99,** 3181 (1977).

TABLE I
COMPARISON OF RATES OF METABOLISM CATALYZED BY P450-FUSION PROTEINS USING
ELECTROLYSIS OR NADPH

Fusion protein and substrate	Type of reaction	Rate (nmol/min/nmol P450)	
		Electrolysis	NADPH
rF450[mBov17A/mRatOR]L1			
Progesterone	17-Hydroxylation	20	28
Pregnenolone	17-Hydroxylation	5.0	5.0
rF450[mRat4A1/mRatOR]L1			
Lauric acid	Omega hydroxylation	2.0	5.0
rF450[mHum3A4/mRatOR]L1			
Testosterone	6β-Hydroxylation	2.6	10.0
Erythromycin	N-Demethylation	1.2	2.2
Benzphetamine	N-Demethylation	2.0	4.6
rF450[mHum1A2/mRatOR]L1			
Caffeine	N-Demethylation	0.75	1.5
Imipramine	N-Demethylation	1.2	2.0
CYP102-BM3			
Lauric acid	w(−) hydroxylations	110	900

platinum gauze is attached, together with a silver/silver chloride reference electrode, is connected to a Bioanalytical Systems, Inc. (West Lafayette, Indiana) CV-27 Voltammograph and the electrodes are immersed in the stirred reaction solution. A platinum wire, placed in a tube containing 3 M KCl and isolated from the reaction solution by a porous frit, serves as the auxiliary counterelectrode. A membrane-covered oxygen electrode (YSI 5331 Oxygen Probe, Yellow Springs Instrument Co., Inc., Yellow Springs, OH) is connected to a polarizing voltage unit[7] and placed in the reaction solution to monitor changes in the concentration of oxygen during the reaction. The output of the oxygen electrode is interfaced with an IBM computer by a Strawberry Tree A/D board. Gas mixtures can be introduced into the reaction solution via a gassing tube containing a fritted glass disc. The oxygen concentration in the reaction solution is maintained by the flow of gasses using a suitable regulator. In some cases, changes in the optical absorbance of the reactants are measured by continual flowing of the reaction solution through inlet and outlet tubes connected via a peristalic pump to a flow cuvette placed in the cell compartment of an Aminco DW2a

[7] R. W. Estabrook, *Methods Enzymol.* **10,** 41 (1965).

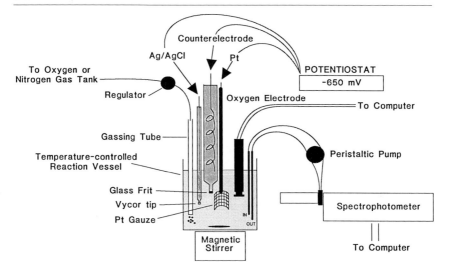

FIG. 1. Diagram of the reaction vessel used for electrochemistry experiments showing the placement of the working and counterelectrodes, the gassing tube for maintaining a fixed oxygen concentration, an oxygen electrode for monitoring the concentration of oxygen in the reaction solution, and outlet and inlet ports for the flow of the reaction solution to a spectrophotometer.

spectrophotometer. This spectrophotometer is also interfaced to an IBM computer for continuous recording of absorbance changes.

Reactants

A 5 mM stock solution of cobalt(III) sepulchrate (Aldrich Chemical Co., Milwaukee, WI) is prepared daily in 50 mM Tris–Cl buffer, pH 7.4, 10 mM MgCl$_2$, and 150 mM KCl (TMK buffer). Recombinant P450s, either as the heme protein modified with a histidine-domain at the carboxy terminus or as a fusion protein containing a P450 heme-domain connected by two or more amino acids to the flavoprotein-domains of NADPH-P450 reductase, were isolated and purified as described.[4,8,9] Radioactive substrates are obtained from Amersham Life Sciences (Arlington Heights, IL). Catalase, NADPH, and cytochrome c are obtained from Sigma Chemical Co. (St. Louis, MO). Metabolites can be resolved by either reverse-

---

[8] M. S. Shet, C. W. Fisher, P. L. Holmans, and R. W. Estabrook, *Proc. Natl. Acad. Sci. U.S.A.* **90**, 11748 (1993).

[9] C. W. Fisher, M. S. Shet, D. L. Caudle, C. A. Martin-Wixtröm, and R. W. Estabrook, *Proc. Natl. Acad. Sci. U.S.A.* **89**, 10817 (1992).

phase or normal-phase HPLC using a Waters 840 Chromatography Data Station as described.[10]

Reaction System

A typical reaction is carried out as follows: An aliquot of TMK reaction buffer mix is placed in the reaction beaker sufficient to immerse the electrodes (generally 6 to 10 ml). The cobalt(III) sepulchrate solution is added, and the mixture is stirred for about 10 min to attain temperature equilibration (generally 37°). The current is turned on at the potentiometer, set at −650 mV, and the decrease in oxygen content of the reaction mixture monitored. After about 10 min when the oxygen content of the reaction solution reaches a steady-state level of approximately 5–10 μM oxygen, an aliquot of catalase and radioactive substrate are added. Catalase destroys accumulated hydrogen peroxide formed by electrolysis causing a transient increase in the oxygen content of the reaction solution. The detection level of the oxygen electrode is increased by amplifying the sensitivity of response by 5- or 10-fold. After again attaining a steady-state level of about 10 μM oxygen, an aliquot of recombinant P450 protein is added and samples are removed for the assay of product formation at the times indicated. The steady-state level of oxygen in the reaction solution is maintained by regulating the flow of gas as described earlier. The current flow by the electrodes can also be monitored by the potentiostat. Likewise, the extent of reduction of the hemeprotein (or cytochrome c as a surrogate) can be determined spectrophotometrically by measurement of absorbance changes in the flow cuvette placed in the sample compartment of the DW2a spectrophotometer.

Results

Figure 2A shows the results of an experiment where 50 μM progesterone or 50 μM pregnenolone was converted to their 17α-hydroxy metabolites using the purified fusion protein rF450[mBov17A/mRatOR]L1[4] (a fusion protein containing the heme-domain of the bovine P450 17A, which catalyzes the 17α-hydroxylation of steroids, linked to the truncated flavoprotein-domain of NADPH-P450 reductase). The following protocol was used for this experiment: 4 ml of a buffer mixture (TMK) was added to the reaction beaker together with 1 ml of 5 mM cobalt(III) sepulchrate (prepared in the TMK buffer mixture). After temperature equilibration the electrode current was turned on, using a polarizing voltage of −650 mV, and the utilization of oxygen was monitored. After 10 min the steady-state level

[10] M. P. Arlotto, J. M. Trant, and R. W. Estabrook, *Methods Enzymol.* **206**, 454 (1991).

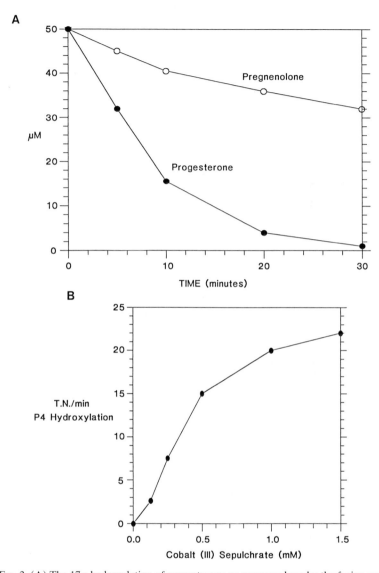

FIG. 2. (A) The 17α-hydroxylation of progesterone or pregnenolone by the fusion protein rF450[mBov17A/mRatOR]L1 (0.2 μM final concentration) using an electrochemically driven reaction system. The sequence of addition of reactants is described in the text. The reaction temperature is 37°. The percentage conversion of progesterone and pregnenolone to 17α-hydroxyprogesterone and 17α-hydroxypregnenolone was determined by reversed-phase HPLC using a Waters 380 Work Station coupled to a Radiometer Flo-1 radioactivity detector as described.[10] (B) Effect of varying concentrations of cobalt(III) sepulchrate on the initial rate of formation of 17α-hydroxyprogesterone from progesterone (P4). The rate of the reaction is expressed as turnover number (T.N.), i.e., nmol of progesterone metabolized/min/nmol of rF450[mBov17A/mRatOR]L1. Other conditions as described for A.

$(10\ \mu M)$ of oxygen in the reaction solution was established. An aliquot of catalase ($40\ \mu l$ of a 5-mg/ml solution) was added, and the release of oxygen by destruction of the hydrogen peroxide formed was seen in the oxygen electrode tracing. After about 3 min, a 25-μl aliquot of a 10 mM solution of either radiolabeled [³H]progesterone or pregnenolone (dissolved in ethanol) was added followed by the addition of 35 μl of a 29 μM solution of purified rF450[mBov17A/mRatOR]L1 (0.2 μM P450, final concentration). The oxygen concentration was maintained at about 10 μM by gassing with oxygen and monitoring the oxygen concentration by the oxygen electrode. At the times indicated, 0.5-ml samples were removed and added to 5 ml of methylene chloride and rapidly mixed. Samples were processed for HPLC analysis as described.[11] Parallel experiments (not shown) were carried out using NADPH (1 mM final concentration) plus an NADPH-regenerating system containing sodium isocitrate plus isocitrate dehydrogenase, and essentially similar rates of 17α-hydroxylation of the steroids were observed by either the electrochemistry or the NADPH-supported method. A similar protocol has been used with a number of different fusion proteins containing specific P450s as summarized in Table I.

The rate of steroid hydroxylation is dependent not only on the concentration of P450 present in the reaction solution, but also on the concentration of the mediator, cobalt(III) sepulchrate. As shown in Fig. 2B, a concentration of about 0.3 mM cobalt(III) sepulchrate is required to obtain 50% of the maximal rate of this reaction by the electrochemical method.

Comments

The ability to use electrochemical methods to support P450-catalyzed reactions now opens the possibility of developing methods for the synthesis of large amounts of chemicals with stereo- and regio-specific functional groups containing oxygen. To date, we have been able to increase the concentration of a substrate undergoing oxidation (lauric acid) to 1 mM and have been able to run the reaction for at least 2 hr with no significant destruction of the P450 catalyst. Thus it seems feasible to carry out the oxidative conversion of relatively large amounts of a chemical.

Our initial studies revealed the electrolytic reduction of molecular oxygen concomitant with the generation of hydrogen peroxide. The formation of reactive oxygen species, such as superoxide and hydrogen peroxide, is detrimental to the P450 catalyst. Inclusion of catalase to destroy the accumulated hydrogen peroxide serves as a means of protecting the P450

---

[11] M. S. Shet, C. W. Fisher, M. P. Arlotto, C. H. L. Shackleton, P. L. Holmans, C. A. Martin-Wixtröm, Y. Saeki, and R. W. Estabrook, *Arch. Biochem. Biophys.* **311,** 402 (1994).

from this destruction. Further, the reduced mediator, cobalt(II) sepulchrate, is oxidized by molecular oxygen. Use of a low oxygen tension during the electrochemical reaction greatly reduces this additional source of reduced forms of oxygen. We routinely maintain the oxygen concentration of the reaction solution at about 10 μM oxygen—a concentration sufficient to saturate the P450 with oxygen but to reduce the second-order reaction of oxygen with the reduced mediator.

We have examined other electromotively active chemicals to see if they will also serve as mediators. To date, we have tested methyl viologen, the ruthenium salt, Ru(acac)$_3$, phenosafranine, flavin mononucleotide, and methylene blue, and these do not serve as adequate mediators in our system. The only chemical, in addition to cobalt(III) sepulchrate, that we have found to function as a suitable mediator is Neutral Red, $N^8,N^8,3$-trimethyl-2,8-phenazinediamine monohydrochloride, but the violet color of this chemical interferes with spectrophotometric measurements as well as with the measurement of formaldehyde by the Nash reagent.

The use of bulk electrolysis to drive a P450-catalyzed reaction, in place of NADPH, now permits the construction of reactor systems using P450s as catalysts. The studies of Guryev and Gilevich[5] as well as our own[3] illustrate the feasibility of this method for new processes of chemical synthesis.

Acknowledgment

This work was supported in part by a grant from the National Institute of Health (GM16488).

[6] Yeast Expression of Animal and Plant P450s in Optimized Redox Environments

By DENIS POMPON, BENEDICTE LOUERAT, ALEXIS BRONINE, and PHILIPPE URBAN

Introduction

Yeast *Saccharomyces cerevisiae* offers a low-cost and efficient way to express heterologous P450s. Basic methods and advantages of this host were reviewed by F. Peter Guengerich in a previous volume of this series.[1]

---

[1] F. P. Guengerich, W. R. Brian, M. A. Sari, and J. T. Ross, *Methods Enzymol.* **206**, 130 (1991).

Nevertheless, the limiting amounts of endogenous P450-reductase (CPR) present in this organism and the limited sequence similarity of yeast redox enzymes (CPR and cytochrome b_5) with human liver or plant equivalents are severe limitations when high specific activities of expressed P450s are required.[2,3] For example, human P450 3A4 poorly recognizes the yeast CPR and does not couple at all with the endogenous cytochrome b_5 (cyt. b5).[4]

A first solution was the construction of artificial gene fusions encoding chimeric proteins composed of the heterologous P450 to express fused in frame to a yeast, rat, or human CPR.[5-10] This approach improves the turnover numbers of several P450s but does not permit one to easily include a third component such as cyt. b5 or to modulate relative enzyme stoichiometries. In addition the protein engineering required may have some unpredictable consequences on P450 functions, thus making this system questionable for toxicological predictions in human drug development. Alternatively, coexpression systems involving multiple plasmids or multiple expression cassettes on a single plasmid have been developed.[7,9,11,12] However, these approaches suffer from genetic instabilities of strains when several or large plasmids are used. In addition, endogenous yeast genes encoding CPR and cyt. b5 are still expressed simultaneously, leading to a mixed redox environment. To avoid these limitations, we developed by multiple genomic modifications yeast featuring redox environment optimized for P450 functions.[13,14]

[2] P. Urban, C. Cullin, and D. Pompon, *Biochimie* **72**, 463 (1990).

[3] H. Murakami, Y. Yabusaki, T. Sakaki, M. Shibata, and H. Ohkawa, *J. Biochem.* (*Tokyo*) **108**, 859 (1990).

[4] G. Truan, C. Cullin, P. Reisdorf, P. Urban, and D. Pompon, *Gene* **125**, 49 (1993).

[5] Y. Yabusaki and H. Ohkawa, *in* "Frontiers in Biotransformation" (K. Ruckpaul and H. Rein, eds.), Vol. 4, p. 169. Akademie-Verlag, Berlin, 1991.

[6] H. Murakami, Y. Yabusaki, T. Sakaki, M. Shibata, and H. Ohkawa, *DNA* **6**, 189 (1987).

[7] M. Shibata, T. Sakaki, Y. Yabusaki, H. Murakami, and H. Ohkawa, *DNA Cell Biol.* **9**, 27 (1990).

[8] T. Sakaki, M. Shibata, Y. Yabusaki, H. Murakami, and H. Ohkawa, *DNA Cell Biol.* **9**, 603 (1990).

[9] T. Sakaki, S. Kominami, S. Takemori, H. Ohkawa, M. Akiyoshi-Shibata, and Y. Yabusaki, *Biochemistry* **33**, 4933 (1994).

[10] N. E. Wittekindt, F. E. Würgler, and C. Sengstag, *DNA Cell Biol.* **14**, 273 (1995).

[11] H. P. Eugster, S. Bärtsch, F. E. Würgler, and C. Sengstag, *Biochem. Biophys. Res. Commun.* **185**, 641 (1992).

[12] C. Sengstag, H. P. Eugster, and F. E. Würgler, *Carcinogenesis* (*London*) **15**, 837 (1994).

[13] D. Pompon, G. Truan, A. Bellamine, and P. Urban, *in* "Assessment of the Use of Single Cytochrome P450 Enzymes in Drug Research" (M. R. Waterman and M. Hildebrand, eds.), p. 97. Springer-Verlag, Berlin, 1994.

[14] P. Urban, G. Truan, A. Bellamine, R. Lainé, J. C. Gautier, and D. Pompon, *Drug Metab. Drug Interact.* **11**, 169 (1994).

Transcriptional Control and Expression Efficiency

Increasing the plasmid copy number is the first way to enhance expression levels. This copy number depends on the balance between the positive pressure related to the selection marker(s) and the toxicity of heterologous protein overproduction. The use of an inducible promoter, which allows delay of the expression phase until the end of the growth phase, favors high copy numbers. The status is maintained during the induction phase even in the absence of selection, provided that limited or no cell division occurs. In the system we developed, the *GAL10-CYC1* hybrid promoter is fully repressed by the glucose of the medium used to support the exponential growth phase. Derepression (yielding to limited expression) is achieved at the end of the exponential phase when glucose is exhausted and when cells rely on ethanol utilization for growth. Finally, the addition of galactose triggers full induction during the subsequent, almost stationary, phase. The use of an *ADE2* complementation marker on plasmid (pYeDP60, Fig. 1) is of particular interest since rich culture media become spontaneously limiting for adenine at high cell densities.

For expression cassettes integrated into the yeast genome, the copy number is stable independently of culture conditions. Thus, expression level relies only on the promoter efficiency. The choice of inducible versus constitutive promoters is directed in this case by long-term stability requirements (i) when the gene product is not toxic (e.g., human epoxide hydrolase), the choice is not limited; (ii) when the expressed gene is essential for cell viability (e.g., cyt. b5 at low levels of CPR),[15] a constitutive promoter is required; or (iii) when the gene product is toxic (a high CPR level), an inducible promoter is required. Multiple genomic integrations can be achieved by repeated single integrations or by association of independently modified loci using classical yeast genetics.

Cloning and Formatting the cDNA to Express

Efficient expression, either from plasmid-borne or from integrated cassette, requires the deletion of the 5'- and 3'-noncoding regions originally present in the heterologous cDNA before its insertion into a yeast transcription unit.[16] Promoter and terminator sequences generally provide, in addition to the transcriptional control, a yeast-featured 5'-noncoding part (excluding the initiation codon) and 3'-processing signals for the hybrid mRNA generated. Cloning and reformatting are now easily performed in a single

---

[15] G. Truan, J. C. Epinat, C. Rougeulle, C. Cullin, and D. Pompon, *Gene* **149,** 123 (1994).
[16] D. Pompon, *Eur. J. Biochem.* **177,** 285 (1988).

step by polymerase chain reaction (PCR) amplification from total poly(A)$^+$ mRNAs or from total cDNAs. Use of a proofreading DNA polymerase (*Pyrococcus furiosus,* Stratagene, CA) is highly recommended. When present, self-complementary hairpin structures surrounding the cDNA initiation codon, which could be deleterious for expression,[17] have to be removed by the introduction of suitable silent codon changes in the PCR step.[18] Primer sequences are also used to introduce a restriction site absent from the cDNA immediately upstream of the initiation codon and downstream of the stop codon. The PCR-amplified cDNA fragment is usually cloned first into a standard *E. coli* vector for sequencing before transfer into the yeast expression vector.

Replicative and Integrative Vectors

Two types of generic vectors are presented in Fig. 1. Replicative vectors include the 2200-bp long part of the yeast 2-μm minicircle DNA (designated as the yeast origin). Such plasmids autonomously replicate in yeast at a medium (5–20) copy number. Integrative vectors cannot replicate in yeast and require genome integration by homologous recombination for propagation (single copy). Both vector types can hold similar expression cassettes based either on the glucose-repressed, galactose-inducible *GAL10-CYC1* promoter or on the constitutive promoter derived from the yeast phosphoglycerate kinase gene (*PGK*p). The main advantage of *PGK*p is that it is similarly active in glucose and in galactose-containing media, allowing its use, in combination with *GAL10-CYC1*, to vary the relative stoichiometries between components during coexpression of several heterologous genes. *PGK, ADH1,* and *CPR1* terminators are functionally equivalents. Yeast replicative and integrative vectors of the pYeDP series can be propagated in *Escherichia coli* at high copy numbers. The pYeDP series shares a common *URA3* marker (uracil auxotrophy) associated in some cases with a second marker permitting complementation of adenine (in *ade2* strain), tryptophan (in *trp1* strain), or leucine (in *leu2* strain) auxotrophies. A widely used vector is pYeDP60 (*URA3* and *ADE2* markers) which is very stable in rich medium at a high cell density.

Integrative (site-directed) vectors of the pYeDP series are composed of two functional segments delimited by *Not*I restriction sites. The first part supports self-replication in *E. coli*. The second moiety includes at both extremities the site-targeting sequences for homologous recombination, for example, at the *CPR1* locus (the yeast gene that encodes CPR, at locus

---

[17] S. B. Baim and F. Sherman, *Mol. Cell. Biol.* **8,** 1591 (1988).
[18] P. Urban, G. Truan, J. C. Gautier, and D. Pompon, *Biochem. Soc. Trans.* **21,** 1028 (1993).

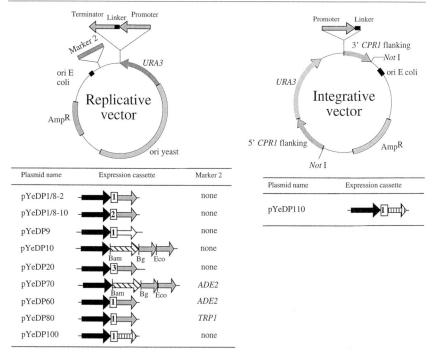

Plasmid name	Expression cassette	Marker 2
pYeDP1/8-2	[1]	none
pYeDP1/8-10	[2]	none
pYeDP9	[1]	none
pYeDP10	Bam Bg Eco	none
pYeDP20	[3]	none
pYeDP70	Bam Bg Eco	ADE2
pYeDP60	[1]	ADE2
pYeDP80	[1]	TRP1
pYeDP100	[1]	none

Plasmid name	Expression cassette
pYeDP110	[1]

FIG. 1. Yeast expression vectors. The different polylinkers (boxed numbers) used to insert the heterologous coding sequence between the yeast promoter and terminator are polylinker 1, *Bam*HI/*Sma*I/*Kpn*I/*Sac*I/*Eco*RI; polylinker 2, *Eco*RI/*Sac*I/*Kpn*I/*Sma*I/*Bam*HI; and polylinker 3, *Bam*HI/*Cla*I/*Bsu*36I/*Eco*RI. The restriction site immediately downstream of the promoter is underlined and must be as close as possible to the cDNA initiation codon in the construct. In the tabulation, solid arrows represent *GAL10-CYC1* promoters; dotted arrows, *PGK* terminators; hatched arrows, *PGK* promoters; striped arrows, *CPR1* terminators; open arrow, *ADH1* terminator. Bam, *Bam*HI; Bg, *Bgl*II; Eco, *Eco*RI. The *Bgl*II site in pYeDP70 is not unique.

YHR042w on chromosome VIII), a *URA3* selection marker, and a cloning site or a polylinker dedicated to receive standard expression cassettes. The site-directed integration results in a simultaneous clean disruption of the *CPR1* gene (Fig. 2). Integration events are selected based on *URA3* complementation. This selection marker can be recovered for further use by selecting spontaneous or induced *ura3* mutants (marker inactivation) as 5-fluoroorotate-resistant clones.[19]

[19] J. Boeke, F. Lacroute, and G. R. Fink, *Mol. Gen. Genet.* **197,** 345 (1984).

Exchange of promoter

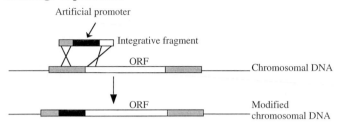

Exchange of coding sequence

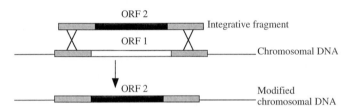

FIG. 2. Gene engineering in yeast by homologous recombination.

Combining Genetic Modifications to Humanize Strains

Integrative vectors were used to build engineered strains: W(hR) expressing the human CPR instead of the yeast CPR,[18] W(B) expressing human cyt. b5,[4] W(E) expressing the human epoxide hydrolase,[20] and W(R) overexpressing the yeast CPR[4] from a single parental strain W(N) (Fig. 3 and Table I). These strains are thus isogenic with the exception of the *CPR1* locus. The expression cassettes in engineered strains are all under the control of the *GAL10-CYC1* galactose-inducible promoter.

For W(R), the recombination event occurred within the ORF of the *CPR1* gene and led to a promoter substitution without change in the CPR ORF. W(R) cells do not express detectable levels of CPR when grown on glucose (as observed in a *CPR1*-disrupted strain), but a high expression level is achieved in galactose. CPR expression reaches a plateau in approximately 10 hr in SL(A)I or in 5–6 hr in YPL(A) medium (see below). In strains expressing, conditionally, CPR (whatever the origin), the CPR deficiency induces a significant instability of the yeast mitochondrial genome, frequently leading to slow-growing clones unable to use galactose as a carbon source in minimal medium. This can be avoided by checking transformants for growth on glycerol-containing N3 medium [yeast extract,

[20] J. C. Gautier, P. Urban, P. H. Beaune, and D. Pompon, *Eur. J. Biochem.* **211,** 63 (1993).

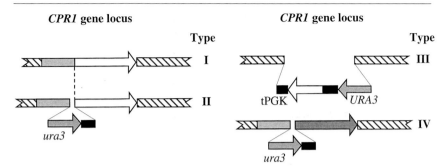

FIG. 3. Structures of the *CPR1* locus in the various strains described. Type I represents the wild-type *CPR1* locus and types II–IV represent the different type of engineered locus as found in W(R), type II; W(B) and W(E), type III; and W(hR), type IV.

10 g/liter; bactopeptone, 10 g/liter; glycerol, 2% (by vol) in 50 m*M* phosphate buffer, pH 6.2].

In the W(B) strain, the *CPR1* gene has been substituted by the expression cassette for human cyt. b5. Induced W(B) cells produce both yeast

TABLE I
CHARACTERISTICS OF ENGINEERED YEAST STRAINS

Name of strain	Type[a] of locus	Ploidy	CPR[b] type	P450 reductase activity[c]		Cytochrome b_5[b] type	Epoxide[d] hydrolase	Ref.
				In glucose	In galactose			
W(N)	I	Haploid	Y	100	100	Y	No	8
W(R)	II	Haploid	Y	≤0.1	2500–3000	Y	No	8
W(B)	III	Haploid	No	≤0.1	≤0.1	Y + H	No	8
W(E)	III	Haploid	No	≤0.1	≤0.1	Y	H	20
W(hR)	IV	Haploid	H	≤0.1	50–100	Y	No	22
W(B,R)	III + II	Diploid	Y	≤0.1	750	Y + H	No	8
W(R,N)	II + I	Diploid	Y	50	750	Y	No	8
W(B,N)	III + I	Diploid	Y	50	50	Y + H	No	8
W(E,R)	III + II	Diploid	Y	≤0.1	750	Y	H	20
WAT11	IV	Haploid	AT	N.D.[e]	200–300	Y	No	—
WAT21	IV	Haploid	AT	N.D.	300–400	Y	No	—

[a] Refers to the constructs shown in Fig. 3.
[b] Y, H, and AT stand, respectively, for yeast, human, and *A. thaliana*.
[c] Expressed as nanomoles of cytochrome *c* reduced per min per mg microsomal protein in 50 m*M* Tris–HCl buffer, pH 7.4, in the presence of 1 m*M* KCN.
[d] In both strains expressing human epoxide hydrolase, the microsomal specific activity was 2 nmol of styrene hydrolyzed per min and per mg protein in 50 m*M* Tris–HCl buffer, pH 7.4.
[e] Not determined.

and human cyt. b5 in similar amounts (each at 100–150 pmol per mg protein) but no CPR.[4] Haploid strains W(R) and W(B) of complementary mating types were crossed to give the W(B,R) diploid strain. This strain produces high microsomal levels of both yeast CPR and human cyt. b5, but at half the levels of W(R) and W(B) alone (Table I).[4] The same method can be used to build any other combination. For example, the W(E) strain producing human epoxide hydrolase was built similarly to W(B); mating of W(E) and W(R) led to a W(E,R) strain producing high levels of both yeast CPR and human epoxide hydrolase.[20]

Selecting the Right Strain for a Given Use

A fourfold reduction of human CPR expression is observed on genomic integration of the expression cassette giving W(hR), compared to the value observed in the yeast transformed by the pYeDP60-borne cassette. A similar copy number effect was observed for human epoxide hydrolase and for human cyt. b5 expression upon genomic integration.[4,20] However, the yeast and two plant CPRs (*Arabidopsis thaliana* ATR1 and ATR2) are expressed at a level similar from cassettes based on a multicopy plasmid or integrated in the genome in W(R), WAT11, and WAT21 strains (Table I).

The lower expression of P450-associated enzymes upon genomic integration is frequently not detrimental since the highest expression is not always required for saturation. As an example, the human cyt. b5 effect on P450 3A4 activity or the epoxide hydrolase level in coupled reactions with P450 1A1 is saturating in the integrated strains.[4,20] For some human P450 isoforms like 1A1, the highest turnover number is nevertheless obtained with W(R), whereas W(hR) led to an improved P450 yield, probably due to a decreased P450 destruction by CPR-produced oxygen radicals and peroxides. Coexpression of human cyt. b5 strongly improves the activity of some P450 isoenzymes (particularly CYP3A4),[4,21] but has little or no effect on others (e.g., CYP1A2). Because the potentiating effect of human cyt. b5 dramatically increases at a low CPR level, coexpression of human cyt. b5 can compensate for the lower CPR activity in W(hR).

Depending on the P450 isoform, plasmid, strain, and culture conditions, the microsomal content in the recombinant P450 ranges from 30 to 500 pmol of spectrally detectable P450 per mg protein. Engineered strains can be transformed as wild-type strains by any P450 expression pYeDP vectors, provided compatible selection markers are used. Since in our conditions

[21] M. A. Peyronneau, J. P. Renaud, G. Truan, P. Urban, D. Pompon, and D. Mansuy, *Eur. J. Biochem.* **207,** 109 (1992).

only the recombinant P450 is detectable, microsomal fractions can be used as such for spectral analysis and turnover number calculation.

Application to *in Vivo* Bioconversion

A large number of activities can be assayed using living yeast cells as biocatalysts. A few milliliters of culture are, in most cases, sufficient to achieve detectable bioconversion in a few minutes.

Conversion of Cinnamate to 4-Hydroxycinnamate by CYP73

The W(R) strain overexpressing the yeast CPR was transformed by pYeDP60 carrying a cDNA encoding *Helianthus tuberosus* cinnamate 4-hydroxylase.[22] Transformed yeast cells are grown in 100 ml YPGE (see below) until cell density reaches $7-8 \times 10^7$ cells per ml. Galactose is added to a final concentration of 20 g/liter, and the culture is continued until density reaches 2×10^8 cells per ml. A culture aliquot is diluted to 7×10^6 cells per ml with fresh YPL medium containing 0.2 mM cinnamate. At various incubation times (each 10 min) at 28°, a 1-ml aliquot of the culture is removed and microfuged to discard cells (30 sec at 10,000 rpm). A 0.5-ml aliquot of the supernatant is diluted with 0.1 ml aqueous 25% trifluoroacetic acid to quench the reaction, and 10 μl of the acidified supernatant is analyzed by reverse-phase HPLC onto a RP18 Spheri-5 Brownlee column eluted with H_2O/CH_3CN (85:15, by vol) at a 1-ml/min flow rate. The rate of 4-hydroxycinnamate accumulation is linear with time and corresponds to an *in vivo* conversion of 8×10^7 molecules per cell per min in the diluted cell suspension, corresponding to the bioconversion of 7.2 g cinnamate per day per liter. A similar experiment for ethoxyresorufin conversion to resorufin catalyzed by human P450 1A1 led to a conversion rate of 8×10^5 molecules per cell per min corresponding to the bioconversion of 0.2 g per day per liter.

Experimental Procedures

Yeast Strains and Plasmids

Basic yeast strain W(N) (W303-1B) as well as ORF-void replicative plasmids of the pYeDP series are freely available from Dr. D. Pompon. Other strains and plasmids are covered by patents, but can be made available for nonprofit research under certain acceptable conditions.

---

[22] P. Urban, D. Werck-Reichhart, H. G. Teutsch, F. Durst, S. Régnier, M. Kazmaier, and D. Pompon, *Eur. J. Biochem.* **222,** 843 (1994).

Culture Media

SGI and SLI synthetic media contain, respectively, 20 g/liter glucose (G) or 20 g/liter galactose (L) in 1 g/liter bactocasamino acids (Difco), 6.7 g/liter yeast nitrogen base without amino acid (Difco), and 40 mg/liter DL-tryptophan. SGAI and SLAI are identical to SGI and SLI except that 30 mg/liter adenine is added. YPGE complete medium contains 5 g/liter glucose, 10 g/liter yeast extract (Difco), 10 g/liter bactopeptone (Difco), and 3% (by vol) ethanol. YPG(A) and YPL(A), respectively, contain 20 g/liter glucose or 20 g/liter galactose in 10 g/liter yeast extract, 10 g/liter bactopeptone, and, when required, 30 mg/liter adenine (A).

Selection of Transformants

Yeast cells are transformed by the improved lithium acetate procedure described by Gietz.[23] Efficient cell transformation requires the use of a glucose-containing medium (YPGA). Yield is about 10^4–10^5 transformants per μg of replicative DNA vector for W(N) and is 100-fold lower for CPR-deficient strains in glucose. Transformation from cells grown on galactose is highly inefficient and should be avoided. Efficiency is also reduced about 100-fold for transformations with integrative vectors. The transformed cells are transferred onto SG(A)I plates (depending on the vector used). Positive clones are subcloned on the same selective medium and are tested for growth on minimal medium containing galactose and for growth on N3 complete medium (respiration capability). When integrative transformation is performed, the structure of the recombined genomic locus has to be checked by PCR and Southern blotting (only one to two clones out of four results from integration at the expected locus).

Cultures

Two culture methods have been developed. The "low density procedure" is best adapted when low amounts of material with highly reproducible properties (including specific contents) are required. This method is characterized by a low volume yield (0.5–5 nmol P450 per liter) and gives variable specific contents depending on the isoform produced. The "high density procedure" is best adapted for all applications requiring large amounts (up to micromoles) of the recombinant microsomal P450 at a high specific content (100–400 pmol per mg protein). Yield generally reaches 50–400 nmol per liter in laboratory conditions and up to more than 1 μmol

---

[23] D. Gietz, A. St. Jean, J. R. A. Woods, and R. H. Schiestl, *Nucleic Acids Res.* **20,** 1425 (1992).

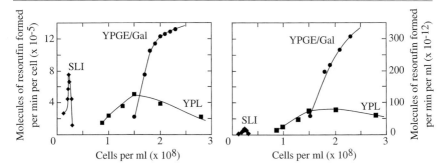

FIG. 4. Bioconversion rates measured for different cell densities and culture conditions. (Left) Specific bioconversion rate per yeast cell versus culture density. (Right) Bioconversion rate per culture volume versus culture density. SLI and YPGE/Gal refer, respectively, to the low density and high density procedure described. Culture procedures in SLI and YPL were identical.

per liter in a fermentor. Figure 4 presents the evolution of the human CYP1A1 expression in W(R) cells from bioconversion measurements at different induction times and for different culture procedures. For the sake of clarity, we have plotted the bioconversion rate (proportional to total P450) against cell density (during both the growth and the induction phases). These results clearly show that induction in a synthetic medium is only optimal over a narrow range of cell densities. Conversely, in a complete medium, the optimal bioconversion rates were measured as a plateau over a wider range of much higher cell densities.

Low Density Procedure. A clone is first transferred from the SG(A)I storage plate [for long-term storage freezing at $-80°$ in SG(A)I, 20% (by vol) glycerol is recommended] to an SL(A)I plate and grown overnight for galactose adaptation. The galactose-induced clone is toothpicked into 30 ml of SL(A)I liquid medium and grown to stationary phase. A 1:200 dilution in 250 ml of fresh SL(A)I is grown overnight at $28°$ in a 1-liter flask placed on a shaking incubator (130 rpm) until cell density reaches exactly 2×10^7 cells per ml (tolerance on density: $\pm10\%$). Cells are either immediately harvested for an on-line microsome preparation or alternatively stored for a short period (up to 3–4 hr) at $4°$ before processing.

High Density Procedure. A colony streaked onto a SG(A)I plate is toothpicked into 30 ml of SG(A)I and grown to stationary phase (overnight). A 1:50 dilution is made into 500 ml YPGE medium and cells are grown at $28°$ in a shaking incubator until cell density reaches 8×10^7 cells per ml (~24–36 hr depending on the strain). Induction is started by the addition of 10% (by vol) of a sterile aqueous solution of 200 g/liter galactose (before induction, the culture can be stored if necessary up to 24 hr at $4°$).

The induction is continued for 8–15 hr (depending on the strain) until the cell density reaches $2–5 \times 10^8$ cells per ml. Cells should be preferentially harvested and processed for cellular fractionation immediately; storage is not recommended.

Fermentation Scaleup. This method is basically an extrapolation of the laboratory-scale high density procedure with the following modifications: 15 liters of YPGE medium is inoculated with 1 liter of a culture at exponential phase in SGI medium (3×10^6 cells per ml). Culture is performed at 28° under high aeration (1 vol air per vol per min, 300 rpm) until the cell density reaches $8–10 \times 10^7$ cells per ml. A solution of sterile galactose (500 g/liter) is then added to a final concentration of 20 g/liter, and the culture is continued for 6–15 hr (see earlier) while maintaining a dioxygen saturation in the 10–20% range by adjusting the rate of agitation.

Cell Breaking and Subcellular Fractionation

Two complementary methods of cell lysis have been developed. The first is based on a mechanical disruption using glass beads whereas the second is based on enzymatic digestion of the cell wall followed by assisted hypotonic lysis. The mechanical breaking is particularly adapted for medium and high cell densities or for large-scale cultures, whereas the enzymic treatment is better for small-size (up to 500 ml), low density (up to 6×10^7 cells per ml) cultures. This last method is also obligatory when intact mitochondrial fractions have to be recovered.

Buffers

TE buffer: 50 mM Tris–HCl, pH 7.4, 1 mM EDTA
TEK buffer: 0.1 M KCl in TE
TEM buffer: 71 mM 2-mercaptoethanol in TE
TMS buffer: 1.5 M sorbitol, 20 mM Tris–MES, pH 6.3, 2 mM EDTA
TES buffer A: 1.5 M sorbitol in TE
TES buffer B: 0.6 M sorbitol in TE
TSG buffer: 0.6 M sorbitol, 20% glycerol (by vol) in TE
TEG: 20% (by vol) glycerol in TE

Mechanical Procedure. The procedure (duration: ~2 hr) is described for a 250-ml culture at a cell density of approximately $2–5 \times 10^8$ cells per ml (equivalent to a light scattering of 30–40 $A_{600\,nm}$ on a Perkin-Elmer L555 spectrophotometer). If necessary, a scaleup proportional to the cell quantity could be performed.

Step 1: Recover the cells by centrifugation at 7000 rpm for 4 min (JA17 rotor, Beckman, Fullerton, CA). Resuspend the cells in TEK to a concentration of 0.5 g wet cells per ml ($2–3 \times 10^9$ cells per ml) and leave at room temperature for 5 min. Recover cells by centrifugation and resuspend cells in the minimal volume of TES B (~1 ml per each 100 ml of initial culture).

Step 2: Gently add glass beads (diameter 0.45/0.50 mm, B. Braun Scientific glassperlen, ref. 854-170/1, Melsungen, Germany) until skimming (as exactly as possible) the top of the cell suspension. Disrupt the cell walls mechanically by hand shaking for 4–5 min in the cold room with up and down movements at two movements per sec (tube volume must at least be fourfold the suspension volume). Check the quality of the disruption with a microscope and repeat the disruption step if a significant part of the cells remains intact. For large-scale cultures (15 liters) the manual disruption step is as follows: 300 g of glass beads (diameter 0.45–0.50 mm, B. Braun) is added to the washed cells resuspended in TES B at 0.5 g wet cells per ml. The mix is passed four times through a Dynomill system equipped with a glass jacket driven at 4500 rpm and refrigerated (circulating refrigerant fluid at $-20°$) to maintain the solution in the 5–12° range.

Step 3: Now proceed on ice. Add 5 ml of TES B to the crude extracts, wash the beads, and withdraw the supernatant. Repeat this treatment twice. Pool the three 5-ml supernatants. Centrifuge at 5° for 10 min at 15,000 rpm (JA17 rotor). Discard the pellet (which contains intact cells, nuclei, and disrupted mitochondria). At this point it is sometimes advantageous to dilute the supernatant two- to threefold with TES B before precipitation in order to limit contamination by cytosoluble proteins. Precipitate microsomes by adding NaCl to the previous supernatant to a final concentration of $0.15 M$ and polyethylene glycol (PEG)-4000 to a final concentration of 0.1 g/ml. Leave at least 15 min on ice. Recover the microsomal fractions by centrifugation (10 min at 10,000 rpm in JA17 rotor). Pellets are resuspended in TEG at 20–40 mg protein per ml and can be kept frozen at $-70°$ for months without any detectable activity changes.

Enzymatic Breaking Procedure. The method (duration: ~4 hr) is described for a culture containing a total of 5×10^9 cells. Scaleup all components correspondingly to the culture size.

Step 1: Wash harvested cells with 25 ml TEK. Resuspend cells in TEM to a cell density of $2–3 \times 10^8$ cells per ml and incubate at room temperature for 5–10 min (a critical step). Recover and resuspend cells into 3 ml TMS and add 15 mg of cytohelicase (lytic enzymes from *Helix pomatia* ref. 249701 from Biosepra, Locke Drive, MA) and 5 mg of yeast lytic enzyme (from *Arthrobacter luteus* ref. 153526 from ICN Biochemicals, Costa Mesa, CA) previously dissolved in 5 ml of TMS. Incubate at 28° in a shaking incubator (130 rpm). Every 10 min, withdraw a 10-μl aliquot of the cell suspension, dilute it as quickly as possible in 1 ml of water in a disposable spectrophotometer cuvette, and measure the initial (first reading) over the final (stable reading) A_{600} ratio. Continue incubation until the OD ratio is at least

10 or until the incubation time exceeds 40 min. Recover spheroplasts by centrifugation at 5° for 5 min and 7000 rpm (JA17 rotor).

Step 2: Now proceed on ice. Gently wash spheroplasts with 30 ml of TES A. Recover spheroplasts by centrifuging for 5 min at 7000 rpm (JA17 rotor). Spheroplast lysis is carried out by resuspending washed spheroplasts in 10 ml of TES B. Vortex the suspension at high speed for 2 min and sonicate (gently) using two 15-sec bursts with a Vibra-Cell sonicator set at 100 W. Wait 5 min (on ice) and centrifuge the suspension at 3500 rpm for 4 min (JA17 rotor). The pellet corresponds to the nuclear fractions and to unbroken cells. Centrifuge the previous supernatant at 10,000 rpm for 10 min. The pellet corresponds to crude intact mitochondria.

Step 3: Add $CaCl_2$ to the supernatant to a final concentration of 16–18 mM, leave for at least 15 min on ice, and centrifuge for 10 min at 10,000 rpm (JA17 rotor). Alternatively, microsomes can be precipitated with PEG-4000/NaCl as described in the mechanical procedure. Resuspend the pellet in TEG at a final concentration of 10–15 mg protein per ml. The microsomal fractions can be stored at $-70°$ for months without any detectable activity changes. The PEG method gives microsome suspensions of lower optical diffusion than the calcium method and is best adapted for spectral analysis, but residual PEG may cause some interference in Western blotting studies. Both methods can be used for activity analysis. Alternatively, ultracentrifugation can also be used and allows a twofold improvement of P450-specific content per mg of total protein, but is more time-consuming and cannot be adapted for large-scale cultures.

Acknowledgments

Expression of the human CYP3As was performed in cooperation with Drs. D. Mansuy and P. H. Beaune. Expression of the *Helianthus tuberosus* CYP73A1 was performed in cooperation with Drs. M. Kazmaier, D. Werck-Reichhart, and F. Durst. The development work for this system was supported by the Institut National de la Recherche Médicale and by the Association de la Recherche sur le Cancer. The reader should be advised that several of the presented strains and methods are covered by issued and pending French and international patents.

[7] Generation of the Cytosolic Domain of Microsomal P450 52A3 after High-Level Expression in *Saccharomyces cerevisiae*

By ULRICH SCHELLER, THOMAS JURETZEK, and WOLF-HAGEN SCHUNCK

Introduction and Principles

Most of the presently known eukaryotic P450 forms are integral membrane proteins of the endoplasmic reticulum (ER). They are generally assumed to consist of an NH_2-terminal membrane anchor region and of a large cytosolic domain.[1,2] Using the method described here, these two parts of an ER-resident P450 protein can be separated. The procedure was established for P450 52A3 (trivial name P450Cm1) and allows one to purify its cytosolic domain as a soluble and functionally active polypeptide.[3] The basic principle is to produce the P450 protein in its native membrane-bound state and to remove the membrane anchor in a subsequent step by means of sequence-specific proteolysis at a designed cleavage site.

The main experimental steps are as follows:

1. Genetic engineering to establish the sequence Ile-Glu-Gly-Arg at the end of the putative membrane anchor region (amino acid position 63–66 in P450Cm1). This tetrapeptide serves as a recognition site for blood coagulation factor Xa, a restriction protease known to cleave immediately after this sequence.[4]

2. Production of the modified P450 [termed as (Xa)P450Cm1] as a native ER-resident protein by means of heterologous expression in *Saccharomyces cerevisiae*.

3. Purification of (Xa)P450Cm1, cleavage with factor Xa, and isolation of the cytosolic domain [$\Delta(1–66)$P450Cm1].

There are several potential applications of this method. First, it is suitable in testing the membrane topology of the P450 protein and in analyzing to what extent the removed parts are required for its enzymatic properties. Second, it can be used to study the functional role and strength of those interactions between the P450 protein and the ER membrane that remain after removing the NH_2-terminal membrane anchor. Third, it appears as a

[1] R. Nelson and H. W. Strobel, *J. Biol. Chem.* **263**, 6038 (1988).
[2] S. D. Black, *FASEB J.* **6**, 680 (1992).
[3] U. Scheller, R. Kraft, K.-L. Schröder, and W.-H. Schunck, *J. Biol. Chem.* **269**, 12779 (1994).
[4] K. Nagai and H. C. Thogersen, *Nature (London)* **309**, 810 (1984).

promising way to overcome the common difficulties in crystallizing ER-resident P450 proteins. This hope is based on the progress achieved with the soluble domains of other membrane proteins,[5] including cytochrome b_5,[6] NADH-cytochrome b_5 reductase,[7] and NADPH-P450 reductase.[8]

Procedures

This section describes in detail the protocols used in our laboratory for high-level expression in *S. cerevisiae,* purification of the P450 protein carrying the factor Xa recognition site, sequence-specific proteolysis, and isolation of the cytosolic domain. Genetic engineering methods that follow standard procedures are only briefly summarized; for data on the functional properties of the cytosolic domain of P450Cm1, we refer to our recent publication.[3] These protocols may also serve as a guide to apply the general experimental strategy to other microsomal P450 forms. Some of the problems to be expected, as well as alternative strategies, are discussed in "Conclusions and General Comments."

Yeast Strain and Expression Plasmid

The yeast strain *S. cerevisiae* GRF18 (MATα, his 3–11, his 3–15, leu 2–3, leu 2–112, can[r]) and the expression vector YEp51 used in this study were kindly provided by D. Sanglard[9] and J. R. Broach,[10] respectively. YEp51 contains elements required for replication (2-μm DNA-derived sequences), selection (LEU2 marker gene), and expression (*GAL10* promoter) in yeast and can be amplified in common *Escherichia coli* strains.

Insertion of the Factor Xa Recognition Site

The NH_2-terminal region of P450Cm1 (for the sequence, see *Schunck et al.*[11]) contains two hydrophobic segments designated HS1 (amino acid

[5] H. Michel, *in* "Crystallization of Membrane Proteins" (H. Michel, ed.), p. 73. CRC Press, Boca Raton, FL, 1991.

[6] F. S. Mathews, P. Argos, and M. Levine, *Cold Spring Harbor Symp. Quant. Biol.* **36,** 387 (1972).

[7] K. Miki, S. Kaida, N. Kasei, T. Iyanagi, K. Kobayashi, and K. Hayashi, *J. Biol. Chem.* **262,** 11801 (1987).

[8] S. Djordjevic, D. L. Roberts, M. Wang, T. Shea, M. G. Camitta, B. S. Masters, and J. J. Kim, *Proc. Natl. Acad. Sci. U.S.A.* **92,** 3214 (1995).

[9] D. Sanglard and J. C. Loper, *Gene* **76,** 121 (1989).

[10] J. R. Broach, Y.-Y. Li, L.-C. Wu, and M. Jayaram, *in* "Experimental Manipulation of Gene Expression" (M. Inoye, ed.), p. 83. Academic Press, New York, 1983.

[11] W.-H. Schunck, F. Vogel, B. Gross, E. Kärgel, S. Mauersberger, K. Köpke, C. Gengnagel, and H.-G. Müller, *Eur. J. Cell Biol.* **55,** 336 (1991).

residues 17–34) and HS2 (48–66). In order to modify the end of HS2 (residues 63–66) from Ile-Asp-Ile-Ile to the factor Xa recognition site (Ile-Glu-Gly-Arg), three codons have to be exchanged in the P450Cm1 cDNA. This is achieved by performing a recombinant polymerase chain reaction according to published procedures,[12] and we refer to our recent paper for the specific primers used.[3]

The resulting cDNA encoding the modified P450 protein [(Xa)P450Cm1] is ligated into the *Sal*I/*Bam*HI site of the plasmid YEp51, thus placing it under control of the *GAL10* promoter. After amplification in *E. coli* SURE (Stratagene), the constructed plasmid is used to transform *S. cerevisiae* GRF18 according to a method described in Keszenman-Pereyra and Hieda.[13] Transformants are selected on YMG agar plates (see below) at 28°.

Yeast Culture and P450 Expression

High-level expression of the P450 cDNAs is based on the host/vector system *S. cerevisiae* GRF18/YEp51. It takes advantage of the regulation properties of the *GAL10* promoter which is almost completely repressed by glucose and strongly induced by galactose. This allows a two-step cultivation procedure including (i) growth on glucose to produce most of the biomass and (ii) induction with galactose to produce P450. The method can be tested in shaking flasks and is easily scaled up. Cell densities and P450 levels to be reached are 3×10^8 cells/ml, 0.18 nmol P450Cm1/10^8 cells, and 0.12 nmol (Xa)P450Cm1/10^8 cells. A 20-liter bioreactor routinely yields a biomass of 300 g (yeast wet weight) containing 8000 and 6000 nmol of P450Cm1 and (Xa)P450Cm1, respectively.

Culture Medium. The yeast minimal medium YM is composed of (mg per liter) NH_4SO_4, 6000; KH_2PO_4, 2000; K_2HPO_4, 320; $MgSO_4 \times 7H_2O$, 1400; NaCl, 1000; $Ca(NO_3)_2 \times 4H_2O$, 800; H_3BO_3, 1.0; $CuSO_4$, 0.08; KJ, 0.2; $FeCl_3$, 1.0; $MnSO_4$, 0.8; $ZnSO_4$, 0.8; $NaMoO_4$, 0.4; (+)-biotin, 0.004; D(+)-pantothenic acid (Ca^{2+} salt), 0.8; folic acid, 0.004; *myo*-inositol, 4.0; aminobenzoic acid, 0.4; pyridoxine–HCl, 0.8; riboflavine, 0.4; thiamine–HCl, 0.8; nicotinic acid, 0.8; and L-histidine, 100. In addition, YMG contains 2% (w/v) D(+)-glucose.

Preculture

1. Inoculate 10 ml YMG medium with a single yeast colony and grow the culture with shaking (240 rpm) at 28° for about 16 hr to reach a cell density of approximately 1×10^8 cells/ml.

[12] R. Higushi, *in* "PCR Technology" (H. A. Erlich, ed.), p. 61. Stockton Press, New York, 1989.
[13] D. Keszenman-Pereyra and K. Hieda, *Curr. Genet.* **13**, 21 (1988).

2. Transfer 2 ml to each of five 500-ml shaking flasks containing 200 ml YMG medium, grow as described earlier.

Large-Scale Biomass Production and P450 Expression

1. Inoculate a bioreactor containing 19 liter YMG medium with approximately 1 liter of preculture to reach a cell density of 5×10^6 cells/ml.

2. Perform cultivation, maintaining pH (4.8), temperature (28°), pO_2 (at least 70%), and stirring rate (800 rpm), until the initially added glucose (estimated with teststrips or by other means) is completely consumed. This takes 15–18 hr and results in a cell density of about 1×10^8 cells/ml.

3. Add a 60% (w/v) D(+)-galactose (chemical purity >98%) solution to give a final concentration of 3% (w/v). Determine the cellular P450 content by means of CO-difference spectra: after a lag phase of 2–4 hr it should increase almost linearly over 14 hr.

4. Harvest the cells in a refrigerated continuous flow centrifuge (flow rate 300 ml/min at 30,000 g).

Note. With intact yeast cells, P450 is only slowly reduced by dithionite. Therefore, samples for CO-difference spectra are prepared as follows: pellet the cells from 20 ml culture, resuspend in 1 ml 50 mM Tris/HCl buffer, pH 7.4, containing 0.4 M sorbitol, 0.5 mM dithiothreitol (DTT), and 1 mM EDTA, disintegrate by vortexing with glass beads (0.5 mm), and dilute the sample to 5 ml with 200 mM potassium phosphate buffer, pH 7.4, containing 1 mM EDTA, 0.5 mM DTT, and 0.1% Tween 20.

Preparation of Microsomes

For rapid preparation of the P450 containing membrane fraction, yeast cells harvested from the bioreactor are disintegrated mechanically and microsomes are isolated after several centrifugation steps by Ca^{2+}-mediated sedimentation. Depending on the P450 form expressed, the microsomes obtained have a specific P450 content of 0.8 (P450Cm1) or 0.6 (XaP450Cm1) nmol per mg of protein and are usually free of P420.

Cell Disintegration

1. Wash the harvested cells twice with ice-cooled 50 mM Tris/HCl buffer, pH 7.4, and take up with the same buffer containing 40% glycerol, 5 mM DTT, and 10 mM EDTA to give a highly concentrated cell suspension (300 g yeast wet weight in a total volume of 400 ml).

2. Disrupt cells in a cooled Dyno-mill (W.A. Bachofen Maschinenfabrik, Basel, Schweiz) under the following conditions: a 600-ml grinding container filled with glass beads (0.5–0.8 mm) to leave a free volume of

about 130 ml, a temperature of 0–4°, and a breakage time of 3 min per 130-ml cell suspension at a grinding speed of 3000 rpm.

3. Dilute the cell homogenate with 10 volumes of 50 mM Tris/HCl buffer, pH 7.4, containing 0.4 M sorbitol and centrifuge at 5000 g for 5 min.

4. Centrifuge the cell-free supernatant for 15 min at 10,000 g.

5. Add a 2 M CaCl$_2$ solution to the supernatant to give a final concentration of 20 mM and centrifuge for 20 min at 20,000 g.

6. Resuspend the microsomal pellet in the Tris buffer containing 0.4 M sorbitol to give a final P450 concentration of about 15 nmol/ml.

7. Freeze aliquoted microsomes in liquid nitrogen and store them at −80°.

Purification of (Xa)P450Cm1

Starting from the microsomal fraction, (Xa)P50Cm1 and wild-type P450Cm1 can be purified by the same protocol, which includes (i) solubilization with sodium cholate, (ii) purification on ω-amino-n-octyl Sepharose 4B, and (iii) final purification and detergent exchange on hydroxyapatite. In order to handle large quantities of P450 (usually 2500–3000 nmol), the solubilization mixture is not centrifuged and a simple and time-saving batch procedure is used for the second step. Thus, it takes only about 3 hr to enrich the P450 proteins to a specific content of about 7 nmol/mg of protein in a yield of approximately 50%. After purification on hydroxyapatite, both the P450 proteins are homogenous in SDS–polyacrylamide gel electrophoresis [as shown for (Xa)P450Cm1 in Fig. 1, lane 2] and have a specific P450 content of more than 16 nmol/mg of protein.

Buffers

10 mM potassium phosphate buffer, pH 7.3, containing 0.5 mM DTT, 1 mM EDTA, and 20% (buffer A) or 35% (buffer B) glycerol, respectively

Solubilization. Microsomes (5 g of protein, 2500–3000 nmol P450) from cells expressing (Xa)P450Cm1 and P450Cm1, respectively, are thawed and diluted to a protein concentration of 10 mg/ml with buffer A (500 ml final volume). Solubilization of the membranes is initiated by a dropwise addition of a 10% (w/v) solution of sodium cholate to give a final concentration of 0.8%. After stirring the suspension for 30 min at 4°, the sodium cholate concentration is adjusted to 1.2%.

Batch Purification on ω-amino-n-octyl Sepharose 4B. The solubilization mixture is added without centrifugation to the same volume of an ω-amino-n-octyl Sepharose 4B suspension in buffer A containing 1.2% (w/v) sodium cholate. Optimal binding is achieved at a P450/gel ratio of 7 nmol/ml of gel and 1 hr of batch incubation at 4° under gentle stirring. The resin is then sedimented by centrifugation (3000 g, 2 min) and washed threefold with four gel volumes of buffer B containing (i) 0.3%, (ii) 1.8%, and (iii) 0.5% sodium cholate. P450 is eluted by repeatedly washing the gel with one gel volume of buffer B containing 0.5% cholate and 0.5% Tween 20. Again, gentle centrifugation is used to separate the eluted P450 from the gel.

Column Chromatography on Hydroxyapatite. Supernatant fractions containing the eluted P450 are pooled (about 1500 nmol in 1 liter) and applied to a hydroxyapatite (HA) column (2.6 × 40 cm, packed with fast-flow HA, Calbiochem) that has been equilibrated with buffer A and 0.3% (w/v) sodium cholate. The bound P450 protein is washed with 10 column volumes of the equilibration buffer to remove as much Tween 20 as possible and finally eluted with a linear gradient of 10–400 mM potassium phosphate in buffer A containing 0.3% sodium cholate at a flow rate of 1 ml/min. P450 elutes at a phosphate concentration of about 150 mM, and the pooled fractions have a P450 content of about 40–60 nmol/ml.

Note. Purification on ω-amino-n-octyl Sepharose 4B requires a very careful adjustment of all the parameters given earlier. In particular, ionic strength, pH, concentrations of cholate and glycerol, and P450/detergent ratios are decisive for the specifity of the binding and washing steps. Selective elution is based on the choice of appropriate concentrations of the weak nonionic detergent Tween 20 combined with cholate and a low ionic strength.

Sequence-Specific Proteolysis

Large amounts of the functionally active cytosolic domain of P450Cm1 are prepared by factor Xa-mediated proteolytic cleavage of the purified (Xa)P450Cm1 protein. Under optimized conditions (see also *Notes and Comments*), the spectrally detectable amount of P450 (as judged by CO-difference spectra) remains constant over the incubation period, and a cleavage efficiency of about 70% is achieved, resulting in liberation of a truncated protein lacking the first 66 amino acids [Δ(1–66)P450Cm1] and migrating in SDS–PAGE with an apparent M_r of 48,000 compared to 55,000 for the full-length (Xa)P450Cm1 (Fig. 1, lane 4). This does not occur when treating wild-type P450Cm1 with factor Xa (Fig. 1, lane 5). We describe here the experimental conditions for cleavage of 200-nmol aliquots of purified (Xa)P450Cm1.

Cleavage Buffer

50 mM Tris/HCl, pH 7.8, 100 mM NaCl, 1 mM $CaCl_2$, 1 mM EDTA, 0.5 mM DTT, 20% glycerol.

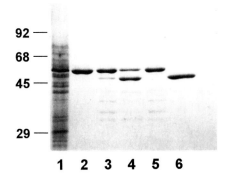

FIG. 1. Preparation and purification of the cytosolic domain of P450Cm1. Microsomal (10 μg) and purified proteins (1 μg) were analyzed on a 10% polyacrylamide–SDS gel and visualized with Coomassie blue. Lane 1, microsomes containing (Xa)P450Cm1; lane 2, purified (Xa)P450Cm1; lane 3, purified (Xa)P450Cm1 after treatment with factor Xa without the addition of CHAPS; lanes 4 and 5, purified (Xa)P450Cm1 and P450Cm1, respectively, after treatment with factor Xa in the presence of CHAPS (optimized cleavage conditions); lane 6, cytosolic domain Δ(1–66)P450Cm1 purified on *n*-octyl Sepharose and hydroxyapatite. The molecular mass standards on the left side are in kilodaltons.

Buffer Exchange and Cleavage Reaction

1. Apply the purified (Xa)P450Cm1 to a Superose 12 HR 10/30 gel filtration column (Pharmacia), previously equilibrated with cleavage buffer containing 0.1% (w/v) 3-[(cholamidopropyl)dimethylammonio]-1-propanesulfonic acid (CHAPS).

2. Elute the P450 with the same buffer and dilute the pooled fractions to 20 nmol/ml with cleavage buffer.

3. Add 1 mg/ml factor Xa solution in distilled water to give a P450/factor Xa ratio of 10 : 1 (w/w).

4. Add 10% (w/v) CHAPS solution in cleavage buffer to give a final detergent concentration of 2.0%. Thus, the cleavage mixture contains 200 nmol (12 mg) of (Xa)P450Cm1 and 1.2 mg of factor Xa in a total volume of 26 ml.

5. Perform proteolysis at 4° for 6 hr without shaking.

6. Stop cleavage by immersing in liquid nitrogen and store at −80° until further use.

Notes and Comments. This procedure is based on an extensive optimization of the cleavage conditions for (Xa)P450Cm1. We expect that the developed method can only be successfully applied to other P450 forms if a similar compromise is found in terms of the following criteria:

(i) Stability of the P450 protein: achieved here by the additions of glycerol and DTT, short incubation times, and low temperatures.

(ii) Efficiency of cleavage is largely dependent on the addition of a suitable detergent, which improves the accessibility of the cleavage site, most likely by dissolving P450 oligomers and aggregates. In the present case, CHAPS is most effective among the detergents tested (detergent kit No. 1124714 from Boehringer); compare Fig. 1, lanes 3 and 4.

(iii) Specifity of cleavage is normally a minor problem with factor Xa since the recognition site Ile-Glu-Gly-Arg does not occur in P450 proteins known so far. However, it has been reported that Arg-Lys bonds may be cleaved, but with low efficiency.[14] Specifically, in the P450Cm1 protein, the Arg(455)-Lys(456) bond is slowly attacked if the protein is handled under nonoptimized (partially denaturating) conditions. However, this unspecific reaction can be almost completely suppressed by measures that stabilize the P450, suggesting that the unspecific cleavage site is protected in the native protein.

(iv) Factor Xa activity: unfortunately, most of the conditions that improve the stability of P450 (glycerol, low temperatures, use of phosphate buffers instead of Tris, etc.) and the accessibility of its designed cleavage site decrease at the same time of the proteolytic activity of factor Xa. This can be easily tested using the synthetic substrate N-benzoyl-Ile-Glu-Gly-Arg-p-nitroanilide (Sigma): take 900 μl buffer containing the components to be tested, mix with 100 μl of a 3 mM substrate solution, add 1 μl of a 1-mg/ml factor Xa solution, and record the absorbance change at 405 nm.

Purified factor Xa is commercially available, but it can also be prepared and purified with high yield from bovine plasma by simple methods described previously.[14]

Purification of the Cytosolic Domain

Based on the large differences in hydrophobicity of full-length and shortened P450Cm1, the latter can be separated from the remaining uncleaved (Xa)P450Cm1 by hydrophobic interaction chromatography on *n*-octyl Sepharose. For final detergent removal, a chromatographic step on hydroxyapatite is performed. About 80 nmol of an electrophoretically homogenous cytosolic domain (Fig. 1, lane 6) having a specific P450 content of more than 18 nmol/mg of protein is routinely prepared from 200 nmol of (Xa)P450Cm1. Its NH$_2$-terminal amino acid sequence is found to be identical to the sequence of the wild-type P450Cm1 starting at position Lys-67,

---

[14] K. Nagai and H. C. Thogersen, *Methods Enzymol.* **153,** 461 (1987).

confirming that the (Xa)P450Cm1 was cut exactly after the inserted protease recognition site.

Chromatography on n-octyl Sepharose. After proteolysis, the cleavage mixture is loaded at a flow rate of 2 ml/min to a Pharmacia XK26/40 column, packed with 170 ml of *n*-octyl Sepharose CL-4B, and equilibrated with 500 mM potassium phosphate buffer, pH 7.4, 20% glycerol, 0.1% sodium cholate, 1 mM EDTA, 0.5 mM DTT, and 0.5 mM phenylmethylsulfonyl fluoride. The column is then washed with about 3 volumes of equilibration buffer, and a broad, red band corresponding to Δ(1–66)P450Cm1 migrates slowly along the column and elutes out. When the column is further washed with the same buffer containing 0.5% Emulgen 911, a second red band corresponding to (Xa)P450Cm1 is eluted. The eluted fractions of Δ(1–66)P450Cm1 are pooled and dialyzed for about 12 hr against 100 volumes of 10 mM potassium phosphate buffer, pH 7.3, containing 20% glycerol, 1 mM EDTA, and 0.5 mM DTT.

Removal of Detergents. For removal of CHAPS and sodium cholate still present, the dialyzed sample of Δ(1–66)P450Cm1 is loaded onto a small hydroxyapatite column (0.5 × 5 cm, packed with fast-flow HA, Calbiochem) and is extensively washed with 50 ml of buffer A, and the P450 is eluted in one step with buffer A containing 400 mM phosphate.

Notes. Δ(1–66)P450Cm1 is a soluble protein that shows the same spectral properties as wild-type P450Cm1 and is functionally active.[3] It forms dimers with an apparent M_r of about 90,000 as revealed by gel filtration and analytical ultracentrifugation studies.

Liberation of the Cytosolic Domain of P450Cm1 from Intact Microsomes

This is an additional method not used for purification. However, in a simple way it yields important information on membrane integration and topology of the P450 protein. The procedure is based on a treatment of intact microsomes containing (Xa)P450Cm1 with factor Xa. The major immunodetectable product is indistinguishable from that obtained with the purified (Xa)P450Cm1 described earlier and remains almost completely in the supernatant fraction after ultracentrifugation. The results show that the cytosolic domain of P450Cm1 is a readily soluble protein. In contrast, the microsomal (Xa)P450Cm1 (as well as the wild-type P450Cm1) is not dissociable from the membrane fractions even after treatment with 0.1 M sodium carbonate at pH 11.5, indicating its tight membrane integration.

1. Wash 200 μl of microsomes containing approximately 3 nmol of (Xa)P450Cm1 twice with cleavage buffer without the addition of CHAPS and glycerol, and adjust the protein concentration to 3 mg of total protein per ml.

2. Add factor Xa solution (1 mg/ml) to the suspension to give a protease/total protein ratio (w/w) of 1 : 50.

3. Carry out proteolysis for 12 hr at 4°.

4. Centrifuge the cleavage mixture for 1 hr at 200,000 g.

5. Dissolve the pellet in Laemmli sample buffer.[15]

6. Precipitate the supernatant with trichloracetic acid, centrifuge at 15,000 g for 15 min, rinse the pellet with acetone, and dissolve it in Laemmli sample buffer.

7. Perform SDS–polyacrylamide gel electrophoresis[15] and Western blot analysis with antibodies against P450Cm1[11] using standard protocols.

Conclusions and General Comments

The method described in this chapter allows one to produce the cytosolic domain of P450Cm1 in a scale and quality suitable for further studies on structure–function relationships. Our strategy intentionally avoids a direct formation of the truncated P450 protein in the host cells as is done by other researchers expressing P450 cDNAs deleted for the sequences that encode the putative membrane anchor region.[16–19] Instead, this region is split off only after producing the native membrane-bound P450 protein. In the case of P450Cm1, this approach has the clear advantage of maintaining both high-level expression and the spectral and enzymatic activity of the P450 protein. In comparison, only very low amounts of the cytosolic domain of P450Cm1 were available after its direct biosynthesis in *S. cerevisiae*.[20] Possible reasons include the instability and rapid degradation of the truncated protein, and even impaired folding in the absence of proper membrane integration could be discussed in analogy to recent reports on other P450 forms.[21] Therefore, the present method may be generally useful if the otherwise easier cDNA deletion approach yields only poor results.

However, to apply this strategy successfully to other microsomal P450 forms, several preconditions and possible pitfalls should be considered. Most of the problems are expected to arise from the necessity to introduce the factor Xa recognition site. As with any modifications of the amino acid

[15] U. K. Laemmli, *Nature* (*London*) **227,** 680 (1970).

[16] Y. C. Li and J. Y. L. Chiang, *J. Biol. Chem.* **266,** 19186 (1991).

[17] S. J. Pernecky, J. R. Larson, R. M. Philpot, and M. J. Coon, *Proc. Natl. Acad. Sci. U.S.A.* **90,** 2651 (1993).

[18] Y. Yabusaki, H. Murakami, T. Sakaki, M. Shibata, and H. Ohkawa, *DNA* **7,** 701 (1988).

[19] C. Cullin, *Biochem. Biophys. Res. Commun.* **184,** 1490 (1992).

[20] B. Wiedmann, P. Silver, W.-H. Schunck, and M. Wiedmann, *Biochim. Biophys. Acta* **1153,** 267 (1993).

[21] B. J. Clark and M. R. Waterman, *J. Biol. Chem.* **267,** 24568 (1992).

sequence, this may result in unpredictable changes of structure and function of the P450 protein. Therefore, the expression levels and properties of the modified P450 have to be very carefully analyzed in comparison to that of the wild-type protein. We recommend starting the experiments by testing different locations of the tetrapeptide along the P450 primary structure. Moreover, it may be helpful to change some of the culture conditions during P450 expression (decrease of temperature and pO_2 and/or addition of compounds to stimulate heme biosynthesis) if stability of the modified protein in the host organism is a problem. The next and more specific question is to find appropriate cleavage conditions. In the present case, the factor Xa site was already readily accessible with the microsomal P450 protein, indicating (i) that the tetrapeptide sequence established was located on the cytoplasmic side of the membrane and (ii) that it was not protected in the native P450 molecule. Clearly, the latter fact is an essential precondition of the whole procedure, and again it may be useful to have a number of P450 variants constructed with different positions of the cleavage site.

Finally, each of the different microsomal P450 forms will probably require some specific conditions to avoid denaturation and, at the same time, improve the efficiency of the cleavage reaction. These can be elaborated, however, by taking into consideration the general criteria discussed earlier (see *Notes and Comments* in the section on *Sequence-Specific Proteolysis*).

Acknowledgments

This work was supported by Grant 0310257A from the Bundesministerium für Bildung, Wissenschaft, Forschung und Technologie, Germany. We thank K.-L. Schröder and R. Kraft for support in large-scale cell cultivation and protein sequencing, respectively. We are also grateful to T. Zimmer and R. Menzel for helpful discussions.

[8] Use of Heterologus Expression Systems to Study Autoimmune Drug-Induced Hepatitis

By Sylvaine Lecoeur, Jean-Charles Gautier, Claire Belloc, Aline Gauffre, and Philippe H. Beaune

Some xenobiotics, including drugs, are toxic through an autoimmune mechanism. The main targets are the blood cells, the skin, the kidney, and the liver.[1,2] This chapter focuses on drugs leading to hepatic diseases and for which metabolism plays an important role. The clinical characteristics and the hypothesis put forward to explain the triggering of the hepatitis are presented and then we show how the heterologous expression systems can help prove the different steps of the scheme. The two main drugs used as examples are tienilic acid[3,4] and dihydralazine.[5,6]

Hepatitis caused by this type of drug presents very characteristic features[2,7]: delay between the beginning of the treatment and the onset of symptoms; no obvious dose–toxicity relationship; fever, rash, and eosinophilia often accompany the hepatitis; and presence of autoantibodies which are the hallmark of the disease.

The following scheme (Scheme 1) was put forward to explain the first steps leading to the hepatitis.[8]

Heterologous expression systems can be used to elucidate several of these pathways: identification of the target of the autoantibodies and even the epitopes (step 4); identification of the enzyme responsible for the production of the reactive metabolite (step 1); and identification of the target of the reactive metabolite(s) (step 2).

---

[1] M. Kammuller and N. Bloksma, in "Immunotoxicology and Immunopharmacology" (J. H. Dean, M. I. Luster, A. E. Munson, and I. Kimber, eds.), p. 573. Raven Press, New York, 1994.

[2] L. R. Pohl, Semin. Liver Dis. 10, 305 (1990).

[3] J. C. Homberg, C. Andre, and N. Abuaf, Clin. Exp. Immunol. 55, 561 (1984).

[4] P. H. Beaune, P. M. Dansette, D. Mansuy, L. Kiffel, M. Finck, M. Amar, J. P. Leroux, and J. C. Homberg, Proc. Natl. Acad. Sci. U.S.A. 84, 551 (1987).

[5] J. Nataf, J. Bernuau, D. Larrey, M. C. Guillin, B. Rueff, and J. P. Benhamou, Gastroenterology 90, 1751 (1986).

[6] M. Bourdi, D. Larrey, J. Nataf, J. Bernuau, D. Pessayre, M. Iwasaki, F. P. Guengerich, and P. H. Beaune, J. Clin. Invest. 85, 1967 (1990).

[7] D. Pessayre, in "Progress in Hepatology 93" (J. P. Miguet and D. Dhumeaux, eds.), p. 23. John Libbey Eurotex, Paris, 1993.

[8] P. Beaune, D. Pessayre, P. Dansette, D. Mansuy, and M. Manns, Adv. Pharmacol. 30, 199 (1994).

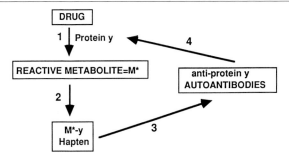

SCHEME 1. Postulated initial events in drug-induced autoimmune hepatitis. Step 1: Formation of reactive metabolites (M*) by protein y. Step 2: Covalent binding of M* on y. The complex behaves as a neoantigen. Step 3: Stimulation of the immune system. Step 4: The autoantibodies recognize complex M*-y and the native protein y.

Expression of P450 in Heterologous Systems

General Strategy

cDNAs for the different P450s were obtained and formatted, with specific restriction sites on each side of the cDNA, by polymerase chain reaction (PCR) from either a cDNA clone or a human liver mRNA. The cDNA was then inserted in a vector such as pUC19 (New England Biolabs, Beverly, MA). When the cDNA was obtained from human liver mRNA, it was sequenced by the dideoxy chain termination method[9] using a T7 sequencing kit (Pharmacia LKB, St. Quentin, France). After digestion by restriction endonuclease, cDNA was subcloned into the V60 vector as described in article [6] (Pompon et al.) or in the bacterial expression vector pGex (cf. this section). Therefore it was possible to obtain the same P450 expression in yeast (allowing one to obtain an active enzyme) or in bacteria (allowing one to very easily obtain pure antigen in high amounts) (Scheme 2).

Expression in Yeast

Human P450 1A1, 1A2, 2C8, 2C9, 2C18, 2C19, 2D6, 2E1, 3A4, and 3A5 were expressed in yeast; the yeast vector V60 and the strain of *Sacharomyces cerevisiae* W(R) fur[10] are described in detail elsewhere in this volume.

[9] F. Sanger, A. R. Coulson, G. F. Hong, D. F. Hill, and G. B. Peterson, *J. Mol. Biol.* **162,** 729 (1982).
[10] G. Truan, C. Cullin, P. Reisdorf, and D. Pompon, *Gene* **125,** 49 (1993).

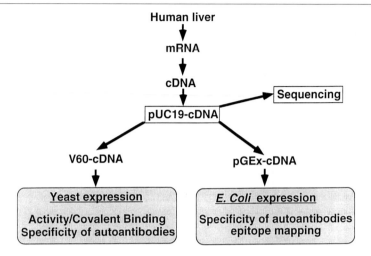

SCHEME 2. General strategy to express P450s in yeast and in bacteria.

Expression in Bacteria

Human P450 1A2 and 2C9 were expressed in a bacterial system, with vector pGEX 2T (Pharmacia, les Ulis France)[11,12] used as described by the purchaser P450 and fragments were expressed as a fusion protein with glutathione-*S*-transferase.[11,12] Full-length cDNA was obtained by RT-PCR from human liver mRNA as described earlier.[13] Specific fragments were obtained by PCR from the full-length cDNA inserted in pGEX 2T. The design of the primers allowed us to add restriction sites at the ends of the amplified fragment. The following fragments were expressed (see Fig. 4): for P450 2C9, nucleotides 1 to 883 (a) and nucleotides 884 to 1488 (b); for P450 1A2, nucleotides 1 to 635 (c), 637 to 1415 (d), and 1410 to 1564 (e). All the fragments were inserted in *Bam*HI and *Eco*RI restriction sites, except fragment e which was inserted in *Bgl*II and *Eco*RI. The amplified fragments were ligated with the plasmid digested with the same restriction enzymes and the ligation product was used to transform the *Escherichia coli* JM 101 strain according to standard protocol.[14] Colonies growing on LB medium containing 50 mg/ml ampicillin (LBA) (Sigma, St. Louis, MO)

[11] D. R. Smith and K. S. Johnson, *Gene* **67**, 31 (1988).
[12] C. Belloc, J. Cosme, S. Baird, S. Lecoeur, I. de Waziers, J. P. Flinois, and P. H. Beaune, *Toxicology* **106**, 207 (1996).
[13] J. C. Gautier, P. Urban, P. H. Beaune, and D. Pompon, *Eur. J. Biochem.* **211**, 63 (1993).
[14] J. Sambrook, E. F. Fritsch, and T. Maniatis, "Molecular Cloning: A Laboratory Manual." Cold Spring Harbor Lab. Press, Cold Spring Harbor, NY, 1989.

were used to isolate the plasmid, which was checked by restriction mapping.[14] Bacteria were then used to produce large amounts of proteins according to the following protocol: an isolated colony of bacteria was grown in 5 ml LBA overnight at 37°. The culture was then diluted 1/10 in 50 ml LBA and grown at 37° for 2 hr. Fifty microliters of 100 mM isopropyl β-thiogalactoside (Sigma) was added and the cells were further grown for 7 hr at 20°. The cells were then centrifuged for 10 min at 4000 g at 4°. The pellet was resuspended in 5 ml of chilled phosphate-buffered saline (PBS: 135 mM NaCl, 15 mM K_2HPO_4, 81 mM NaH_2PO_4, 27 mM KCl, pH 7.4); 1% Triton X-100, final concentration, was added and the cells were lysed by sonication (Branson sonifier 450; fractions of 5 ml were treated for 20 sec; foaming was avoided). The heterologous fusion protein was usually found in the pellet which was obtained after centrifugation of the lysate (30 min at 12,000 g, 4°). It was used for electrophoresis and immunoblot as described later.

Covalent-Binding Studies

The covalent binding of reactive metabolites was measured according to a previously described method.[15] Yeast microsomes corresponding to 50 pmol of each P450 studied or human liver microsomes corresponding to 200 pmol total P450 were incubated for 5 min with 0.1 mM tienilic acid (concentration corresponding to about $10 \times K_m$),[16] 0.15 mM NADP, 2.5 mM glucose 6-phosphate (G6P), and 0.2 IU glucose 6-phosphate dehydrogenase (G6PD) in a 150-μl final volume of buffer (100 mM Na_2HPO_4, 10 mM $MgCl_2$, 1 mM diethylenetriaminepentaacetic acid, pH 7.4). Aliquots (50 μl) of incubation mixture were loaded onto glass fiber filter disks previously dipped in 10% (w/v) trichloroacetic acid (TCA). Filters were washed once with 5% TCA, twice with methanol, and once with ethyl acetate. Filters were then dried and counted in a scintillation liquid (ACS II, Amersham, Buckinghamshire, UK).

The specific covalent binding was assessed by incubation of human liver microsomes (1 nmol P450) or yeast microsomes (50 pmol P450) with 1 mCi [^{14}C]tienilic acid (56 mCi/mmol), HPLC purity 98%, Amersham) 0.15 mM NADP, 2.5 mM G6P, and 2 IU G6PD in 1 ml of the same buffer described earlier for 30 min at 37°.

Proteins from the incubation mixture were precipitated by adding 1 ml 10% (w/v) TCA and washing with 2 ml 5% TCA, ethyl acetate, acetone,

[15] H. Wallin, C. Schelin, A. Tunek, and B. Jergil, *Chem.-Biol. Interact.* **38**, 109 (1981).

[16] S. Lecoeur, E. Bonierbale, D. Challine, J. C. Gautier, P. Valadon, P. M. Dansette, R. Catinot, F. Ballet, D. Mansuy, and P. H. Beaune, *Chem. Res. Toxicol.* **7**, 434 (1994).

and phosphate buffer (each washing corresponding to a precipitation and centrifugation 10 min at 5000 g).

The final pellet of proteins was submitted to electrophoresis and was transferred onto a nitrocellulose sheet (see next section). After transfer, part of the sheet corresponding to six lanes was cut every 2 mm perpendicular to the direction of migration, and each band was counted after the addition of 5 ml of scintillator (ACS II). The radioactive count of each band gave a covalent binding profile which was compared with the location of P450 by antibodies (Figs. 1 and 2).

In order to localize the P450s, the central lanes of the nitrocellulose sheet were kept for immunoblots and probed with anti-human P450 polyclonal antibodies raised in rabbit. These antibodies were prepared in our laboratory.[12,17] The immunoblotting method is described in next section.

Electrophoresis and Immunoblotting

For the characterization of the autoantibodies, 2.5 pmol of P450 in yeast microsomes, 10 μg of human liver microsomes, or 10 μl of final bacteria suspension was adjusted to 12 μl with 10 mM Tris–HCl, pH 7.4. Electrophoresis solubilizing buffer (8 μl) was added [30% (v/v) glycerol, 0.2 M Tris–HCl, pH 6.8, 3% (w/v) sodium dodecyl sulfate (SDS), 0.05% (w/v) pyronine Y]. For specific covalent binding, the final protein pellet (described earlier) was dissolved in 0.5 ml of solubilization buffer (30% glycerol, 0.2 M Tris–HCl, pH 6.8, 1% SDS, 0.01% pyronine). The whole sample was loaded into six wells. All the samples were heated for 2 min at 100° with 7.5% (v/v) mercaptoethanol and loaded on an SDS–polyacrylamide gel electrophoresis (PAGE) (4% stacking and 9% separating gels[18]). Proteins were electrotransferred (1 hr, 400 mA) to a nitrocellulose sheet.[19] The sheet was incubated in 1% (w/v) polyvinylpyrrolidone (Sigma), 0.5% (v/v) Tween 20, PBS for 30 min, then overnight with the autoantibodies diluted 1/10,000 in PBS. After washing [0.5% (v/v) Tween 20, PBS, six times for 5 min], the sheets were incubated for 30 min with peroxidase-conjugated anti-human antibodies (Dako, Copenhagen, Denmark), diluted 1/20,000 in PBS, and washed six times with PBS for 5 min. Immunoblots were developed, as described in Beaune et al.,[17] with luminol (enhanced chemoluminescence, ECL western kit, Amersham, UK) as substrate, according to the manufacturer's recommendations.

[17] P. H. Beaune, P. Kremers, F. Letawe-Goujon, and J. E. Gielen, *Biochem. Pharmacol.* **34,** 3547 (1985).
[18] U. K. Laemmli, *Nature (London)* **227,** 680 (1970).
[19] H. Towbin, T. Staehelin, and J. Gordon, *Proc. Natl. Acad. Sci. U.S.A.* **76,** 4350 (1979).

TABLE I
COVALENT BINDING OF TIENILIC ACID TO YEAST AND HUMAN MICROSOMES[a]

	Covalent binding[b] (pmol/min per nmol P450)	Human liver P450 concentration (nmol P450/mg microsomal protein)	Calculated % of covalent binding[e]
Human liver	98	0.49±0.11[c]	
Control yeast	<0.1	—	—
P450 1A1	77	0.0025[d]	0.7
P450 1A2	44	0.025[d]	3.8
P450 2C8	8±1	Not determined	—
P450 2C9	358±17	0.075[d]	92.8
P450 2C18	36±6	0.015[d]	2.2
P450 2C19	41±7	Not determined	—
P450 2D6	6	0.025[d]	0.5
P450 3A4	<0.1	0.25[d]	<0.1

[a] Conditions were described in the Covalent-Bonding Studies section, using human liver microsomes (1 nmol P450) or yeast microsomes (50 pmol P450), 0.1 mM [^{14}C]tienilic acid, and an NADPH-generating system. Control yeast are transformed with unmodified plasmid V60.

[b] Data represent the mean of duplicate experiments. For 2C8, 2C9, 2C18, and 2C19, data represent the mean of three experiments ± SD.

[c] Mean of eight experiments ± SD.

[d] Guengerich and Turvy.[20]

[e] The sum of the covalent binding by all P450 tested was measured, and the percentage corresponding to the fraction of covalent binding due to each P450 was calculated.

For some of the experiments, rabbit antibodies were used instead of the autoantibodies, at a dilution of 1/50,000.

Patients sera were obtained from patients suffering from either tienilic acid (anti-LKM2)- or dihydralazine (anti-LM)-induced hepatitis. Their characteristics have been previously described.[6,16]

Results and Discussion

The use of P450s expressed in the yeast allowed us to study the production of reactive metabolites and the covalent binding of tienilic acid metabolites (steps 1 + 2). Table I shows that P450 2C9 is the main P450 producing tienilic acid-reactive metabolites. Because it is the major form of the P450 2C subfamily expressed in the liver,[20] it is clear that P450 2C9, among all P450s in human liver microsomes, is quantitatively the most involved in

[20] F. P. Guengerich and C. G. Turvy, J. Pharmacol. Exp. Ther. 256, 1189 (1991).

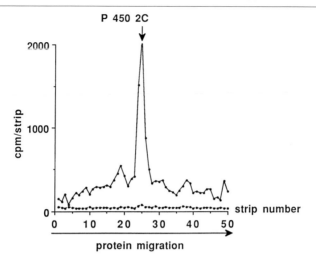

Fig. 1. Covalent binding of [^{14}C]tienilic acid metabolites to human liver microsomes. Conditions were as described in the Covalent Binding Studies section using liver microsomes (1 nmol P450), 18 μM tienilic acid, and an NADPH-generating system. Arrow indicates migration of human liver P450 2C. —o— without NADPH-generating system; —•— with NADPH-generating system.

the production of tienilic acid-reactive metabolites. This has been confirmed in yeast by Lopez-Garcia et al.[21] with a different approach. The target of the reactive metabolites (step 2) was assessed by the identification of the microsomal protein to which the reactive metabolites were bound. Figure 1[16] shows one very major radioactive peak in human liver microsomes that is well individualized and comigrates with the band recognized by rabbit anti-human P450 2C sera. This is consistent with the binding of reactive metabolites to one protein, namely the P450 which produced them. The identity of this protein was further confirmed by using yeasts expressing one single P450. The covalent-binding profile obtained showed a radioactive peak only in the case of P450 2C9. It comigrated with this P450[16] (Fig. 2). This method did not enable the detection of a significant specific covalent binding of the reactive metabolites to any of the P450 2C8, 2C18, 2C19, 1A2, 2D6, and 3A4 expressed in yeasts. It confirmed that P450 2C9 is the main target of tienilic acid-reactive metabolites. Instead of counting the nitrocellulose bands, it is possible to expose the nitrocellulose for autoradiography or to stain the neoantigen with antibodies directed against the drug.[16]

[21] M. Lopez-Garcia, P. M. Dansette, P. Valadon, C. Amar, P. H. Beaune, F. P. Guengerich, and D. Mansuy, Eur. J. Biochem. 213, 223 (1993).

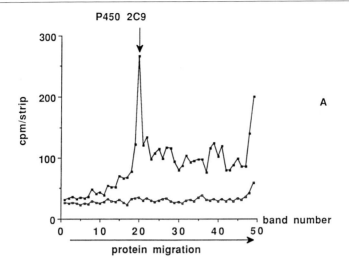

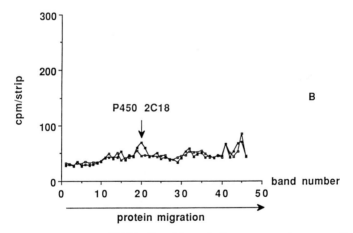

FIG. 2. Covalent binding of [^{14}C]tienilic acid metabolites to yeast microsomes. Conditions were as described in the Covalent Binding Studies section using 50 pmol of P450 2C9 (A) or P450 2C18 (B), 18 μM tienilic acid, and an NADPH-generating system. The immunoblot was developed with anti-human P450 2C. —○— without NADPH-generating system; —●— with NADPH-generating system.

    Thus, heterologous expression systems were able to help in proving that one enzyme produced the reactive metabolite and that the same enzyme was the target of this reactive metabolite. It has been identified as P450 2C9. Unfortunately, it was not possible to use such systems to study the produc-

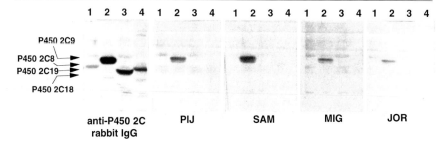

FIG. 3. Recognition of P450 2C8, 2C9, 2C18, and 2C19 expressed in yeast and in human liver by anti-LKM2 and by rabbit IgG human anti-P450 2C. Lanes 1, 2, 3, and 4 correspond to 2.5 pmol of P450 2C8, 2C9, 2C18, and 2C19, respectively. The immunoblot was performed with anti-P450 2C rabbit IgG (diluted 1/50,000) or with autoantibodies (diluted 1/10,000); PIJ, SAM, MIG, and JOR are the codes for the different patient sera. Anti-human IgG labeled with peroxidase (diluted 1/20,000) was used as the second antibody. Immunoblots were performed with luminol.

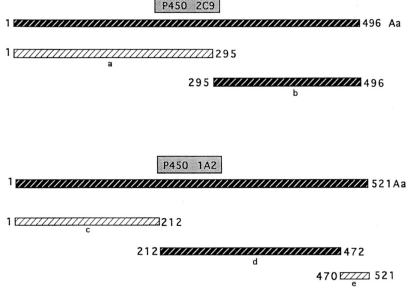

FIG. 4. Schematic representation of the fragments used to localize the binding of the autoantibodies. The cDNA fragments were amplified by PCR before cloning in pGEX bacterial vector. a, b, c, d: fragments inserted in BamHI, EcoRI restriction sites. e: fragment inserted in BglII, EcoRI restriction sites. The corresponding peptides were expressed in the bacteria as a fusion protein with glutathion-S-transferase. Positions of the fragments are indicated by amino acid numbers. Dark hatched bars represent peptides recognized by anti-LKM2 (for P450 2C9) or anti-LM (for P450 1A2) autoantibodies. Light hatched bars represent peptides not recognized by the autoantibodies.

tion and binding of dihydralazine-reactive metabolites since the background with yeasts was too high. However, it is quite reasonable to think that P450 1A2 plays the same role for dihydralazine as P450 2C9 does for tienilic acid.[22]

Heterologous systems (each expressing a single P450 were also used to test the specificity of recognition of the autoantibodies. All of the anti-LKM2 sera recognized P450 2C9 but not P450 2C8, 2C18, or 2C19 (Fig. 3). P450 1A1, 1A2, 2D6, or 3A4 inserted in the yeast was not recognized (results not shown). Anti-LM were also shown to recognize specifically P450 1A2 and not P450 1A1, both expressed in the yeast and in the bacteria.[23,24]

The target of autoantibodies was further studied by epitope mapping. The use of P450 2C9 and 1A2 fragments expressed in a bacterial system allowed one to identify the B-cell epitopes. In the two cases, the sera (anti-LKM2 and anti-LM) had a similar pattern of recognition: both were very specific for one protein and only the carboxy-terminal half of the protein was recognized (Fig. 4). These results have encouraged us to pursue the identification of the B- and T-cell epitopes, which is in progress in our laboratory.

In conclusion, heterologous systems were quite useful in proving some of the hypotheses initially put forward:

step 1: one specific P450 in each disease produces the reactive metabolite;

step 2: the reactive metabolite bound specifically to the P450 producing it, creating a neoantigen able to trigger the disease; and

step 4: the autoantibodies, a hallmark of the disease, are specific for the P450 target and the producer of the reactive metabolite. They also seem to recognize only one part of the protein.

Acknowledgment

This work was supported by the Bioavenir program (Rhône-Poulenc-Rorer and French ministère de la recherche).

[22] M. Bourdi, M. Tinel, P. H. Beaune, and D. Pessayre, *Mol. Pharmacol.* **45,** 1287 (1994).

[23] P. H. Beaune, M. Bourdi, C. Belloc, J. C. Gautier, F. P. Guengerich, and P. Valadon, *Toxicology* **82,** 53 (1993).

[24] M. Bourdi, J. C. Gautier, J. Mircheva, D. Larrey, A. Guillouzo, C. André, C. Belloc, and P. H. Beaune, *Mol. Pharmacol.* **42,** 280 (1992).

[9] Recombinant Baculovirus Strategy for Coexpression of Functional Human Cytochrome P450 and P450 Reductase

By CAROLINE A. LEE, THOMAS A. KOST, and COSETTE J. SERABJIT-SINGH

Introduction

Human cytochrome P450s can be heterologously expressed in *Escherichia coli*, yeast, mammalian, and insect cells.[1] Some isozymes of P450 have been obtained as catalytically active microsomal fractions from yeast[2] and mammalian cells,[3] but membrane-bound recombinant CYP3A4 from nonmammalian expression systems is inactive, requiring purification and reconstitution with NADPH-dependent cytochrome P450 reductase. This deficiency is overcome by coexpressing P450 with the supporting electron transport proteins in the recombinant baculovirus-insect cell system. The insect cell system allows for the high level expression of recombinant proteins targeted to the correct cellular compartment.[4]

The baculovirus-insect cell expression system has gained widespread popularity.[5,6] A multitude of biologically active cytoplasmic, secreted, and membrane-bound proteins have been successfully expressed via the strong viral polyhedrin and p10 gene promoters in recombinant viruses.[5] Baculovirus vector design, recombinant virus generation, and cell culture methods continue to improve. Vectors have been developed that incorporate a wider range of cloning sites and viral promoters, including the basic protein and p39 promoters.[7,8] Baculovirus vectors have been constructed with multiple promoters that allow the simultaneous expression of two or more recombinant proteins via a single recombinant baculovirus.[9] The time and effort involved in generating recombinant viruses have been reduced by the use

---

[1] F. J. Gonzalez and K. R. Korzekwa, *Annu. Rev. Pharmacol. Toxicol.* **35**, 369 (1995).

[2] B. J. Clark and M. R. Waterman, *Methods Enzymol.* **206**, 100 (1991).

[3] F. P. Guengerich, W. R. Brian, M.-A. Sari, and J. T. Ross, *Methods Enzymol.* **206**, 130 (1991).

[4] M. D. Summers and G. E. Smith, *Tex. Agric. Exp. Stn. [Bull.]* **1555** (1987).

[5] V. A. Luckow, *in* "Recombinant DNA Technology and Applications," p. 97. McGraw-Hill, New York, 1991.

[6] G. E. Smith, M. D. Summers, and M. J. Fraser, *Mol. Cell. Biol.* **3**, 2156 (1983).

[7] M. S. Hill-Perkins and R. D. Possee, *J. Gen. Virol.* **71**, 971 (1990).

[8] S. M. Thiem and L. K. Miller, *Gene* **91**, 87 (1990).

[9] U. Weyer and R. D. Possee, *J. Gen. Virol.* **72**, 2967 (1992).

of linearized viral DNA[10] and the bacmid system.[11] In addition to the commonly used Sf9 and Sf21 insect cell lines, new host cell lines such as *Trichoplusia ni* cells (BTI-TN-5BI-4) (established at Boyce Thompson Institute, Ithaca, NY, and commercially available as High Five from Invitrogen Corp.) have become available. These cells can be cultured in large-scale bioreactors to produce large quantities of recombinant protein. This chapter describes the use of the baculovirus-insect cell expression system as a tool to obtain functional P450s by the simultaneous expression of both P450 and P450 reductase cDNAs by a single recombinant baculovirus.

Materials

Shuttle Plasmids and Promoters

Due to the large size of the baculovirus genome (134 kbp),[12] no unique restriction sites exist for the direct subcloning of P450 cDNA into the genome. Thus, the P450 cDNA is first subcloned into a shuttle plasmid that is then transfected together with viral DNA into insect cells to generate a recombinant baculovirus. Shuttle plasmids vary in the type and number of viral promoters. The polyhedrin or p10 gene promoters are present in a wide variety of vectors. Vectors that use weaker promoters such as the 39K or basic protein promoters (pAcJP1 and pAcMP2/3) (PharMingen) have been used successfully to express proteins other than cytochrome P450's.[7,13] Shuttle vectors with multiple viral promoters, allowing the expression of two to four foreign cDNAs via a single recombinant virus, are available; pAcUW51 (PharMingen), p2Bac (Invitrogen), and pAcUW31 (Clontech) contain both the p10 gene and polyhedrin gene promoters, and pAcAB3 and pAcAB4 (PharMingen) contain three and four viral gene promoters, respectively.

Methods for Recombinant Virus Formation

The most widely used method used to generate recombinant viruses employs homologous recombination in insect cells. Shuttle plasmids with multiple promoters are compatible with linear viral DNA, marketed as

[10] P. A. Kitts and R. D. Possee, *BioTechniques* **14**, 819 (1993).
[11] V. A. Luckow, S. C. Lee, G. F. Barry, and P. O. Olins, *J. Virol.* **67**, 4566 (1993).
[12] M. D. Ayres, S. C. Howard, J. Kuzio, M. Lopez-Ferber, and R. D. Possee, *Virology* **202**, 586 (1994).
[13] N. B. Rankl, J. W. Rice, T. M. Gurganus, J. L. Barbee, and D. J. Burns, *Protein Express Purif.* **5**, 347 (1994).

Baculogold (PharMingen) or BacPAK6 (Clontech), to produce, via homologous recombination, baculoviruses carrying multiple foreign cDNAs. Baculogold or BacPAK6 generate up to 99% recombinant viruses as opposed to 0.1 to 1% using nonlinear viral DNA previously described by Gonzalez et a.[14] A newer method to generate recombinant baculoviruses in *E. coli*[11] (Bac-to-Bac kit, GIBCO-BRL) has not been evaluated as multiple viral promoter shuttle plasmids are not currently available for this system.

Cell Lines and Media

The two most common cell lines used in the baculovirus expression system are Sf9 (*Spodoptera frugiperda*) (ATCC and Invitrogen) and *T. ni* cells (Invitrogen). In our laboratory, both cell lines are propagated as suspension cultures in 500-ml sterile disposable plastic culture Erlenmeyer flasks in an orbital shaker at 135 rpm in an enclosed temperature-controlled cabinet at 27°. The maximum culture volume per flask is one-fourth to one-third the volume of the vessel in order to ensure adequate aeration and cell growth. In our experience, clumping of cells is minimal.

Sf9 cells are grown in Grace's complete supplemented medium (1X)(GIBCO-BRL) containing 10% fetal bovine serum (FBS), 0.1% pluronic F-68, and 50 μg/ml gentamycin. Pluronic F-68 is added to protect the cells from shear stress. Other media that can be used are IPL-41 (can be supplemented with 10% FBS) and Sf900II (serum-free medium). In our experience, the expression of active CYP3A4 in Sf900II medium was 5 to 150 times lower than that in Grace's medium, depending on the absence or presence of heme fortification. We have not determined whether IPL-41 medium is equivalent to Grace's supplemented (1X) medium.

T. ni cells are grown in serum-free medium, Ex-Cell 405 (JRH Biosciences), containing 50 μg/ml gentamycin. The Ex-Cell 405 medium (liquid) contains 0.1% pluronic F-68. We have observed that *T. ni* cells are more sensitive than Sf9 cells to high cell density and recommend avoiding cell densities greater than 4–5 $\times$ 10^6 cells/ml. Another medium recently developed for *T. ni* cells is GIBCO's Express Five SFM medium. We have not evaluated this medium to determine if it improves the level of P450 expression.

Heme Supplements

The overexpression of recombinant P450 in insect cells may exceed the supply of endogenous heme. The culture medium can be fortified with any

[14] F. J. Gonzalez, S. Kimua, S. Tamura, and H. V. Gelboin, *Methods Enzymol.* **206,** 93 (1991).

of several heme supplements such as hemin adducts (hemin chloride or hemin arginate) or heme precursors (5-aminolevulinic acid and iron citrate).[15] Hemin chloride can be prepared in either 0.1 N NaOH/100% ethanol[16] or in a bovine serum albumin solution.[17] Hemin arginate is commercially available from Leiras (Finland) as a 25-mg/ml solution.

Procedures

Cell Culture

Sf9 or *T. ni* cells can be grown either as a monolayer, described by Gonzalez *et al.*,[14] or in suspension cultures. To start a suspension culture from a frozen Sf9 or *T. ni* cell stock, the cells (1 ml) are rapidly thawed at 37°, diluted with 10 ml of the appropriate medium, and seeded into a 25-cm^2 T-flask. The cells are maintained at 27° in a T-flask for 2 or 3 days until the cells are actively dividing. The entire cell suspension (approximately 11 ml; pipette gently to release cells from the bottom of the T-flask) is transferred to a 100-ml sterile disposable shake flask containing 10 ml of fresh culture medium and incubated at 27° on a shaking platform at 135 rpm. The cells are typically maintained at a density of 0.5×10^5 to 3×10^6 cells per ml and are subcultured twice weekly. We have propagated our insect cells in suspension cultures for up to 9 months without any measurable decrease in expression levels.

Recombinant Virus Formation

The shuttle plasmid pAcUW51, constructed to contain both P450 reductase cDNA downstream of the p10 promoter and the P450 downstream of the polyhedrin promoter, is transfected with PharMingen's Baculogold-linearized viral DNA to generate the recombinant baculoviruses in monolayers of Sf9 cells. Although the Sf9 cells are propagated as suspension cultures, these cells adhere to a petri dish within 15 to 30 min after plating. The success of a transfection is highly dependent on the health and growth phase of the cells. Sf9 cells should be growing in log phase ($1-2 \times 10^6$ cells/ml) with greater than 98% viability as determined by trypan blue exclusion. Sf9 cells (2×10^6) in Grace's complete supplemented medium are seeded into 60-mm dishes and are allowed 15 to 30 min to attach to

[15] M. Nanji, P. Clair, E. A. Shephard, and I. R. Phillips, *Biochem. Soc. Trans.* **22,** 122S (1994).
[16] A. Asseffa, S. J. Smith, K. Nagata, J. Gillette, H. V. Gelboin, and F. J. Gonzalez, *Arch. Biochem. Biophys.* **274,** 481 (1989).
[17] J. T. M. Buters, K. R. Korzekwa, K. L. Kunze, Y. Omata, J. P. Hardwick, and F. J. Gonzalez, *Drug Metab. Dispos.* **22,** 688 (1994).

the petri dish. The insect cells are transfected with 0.5 μg of Baculogold and 3 μg of shuttle plasmid with foreign cDNA via a cationic liposomes transfection kit according to the manufacturer's protocol (Invitrogen Corp.). After 48 hr, the medium, containing infectious virus particles, is harvested and stored at 4°.

Plaque Assay

The recombinant virus is purified by a plaque assay. A plaque is a focus of cells originating from infection by a single infectious particle and represents a clonal population. The most important factor in obtaining clearly delineated plaques is that the cells are healthy, in log phase growth, and at the appropriate cell density for infection. Overconfluent cells do not support good plaque formation and with underconfluent cells, the plaques are diffuse and more difficult to visualize.[18] Sf9 cells are used in the plaque assay. The cells (2×10^6) are seeded in a 60-mm dish in 4 ml of Grace's complete supplemented medium and allowed 30 min for attachment. The medium is removed and 1 ml of diluted virus (for first round purification: 10^{-1}, 10^{-2}, and 10^{-3}, and 10^{-4} dilutions are made in Grace's complete supplemented medium) is added. The plates are incubated for 1 hr at room temperature on a level surface, rocking gently once or twice during this period. The virus is removed, and the plates are overlaid with 5 ml of 1X Grace's supplemented medium containing 50 μg/ml gentamycin, 10% FBS, and 1% low melting point agarose. The agarose overlay is prepared by autoclaving a stock solution of 2% low melting point agarose (w/v) in distilled water. The sterile agarose is liquefied in a microwave and cooled to 37° in a water bath. The 2X Grace's medium containing 50 μg/ml gentamycin and 10% FBS is preheated in a 37° water bath. Equal volumes of agarose and Grace's media are mixed and immediately added to the cells. The agarose is allowed to solidify for 10 min before transferring plates to the incubator. The plates are incubated at 27° in a humidified environment (plastic box lined with moistened paper towels) for 5 to 7 days. Polyhedrin negative recombinant virus-derived plaques refract light differently than areas of live cells or plaques arising from infection with a nonrecombinant virus, giving rise to shiny cloudy-looking areas of approximately 0.5 to 3 mm in diameter that can be detected with a stereomicroscope.[18] Three to six plaques are picked with a P-200 pipetman, and the virus-containing agarose plug is dispensed into 1 ml of Grace's complete medium. It is best to pick plaques from the higher dilution plates to ensure purity. The virus

[18] D. R. O'Reilly, L. K. Miller, and V. A. Luckow, "Baculovirus Expression Vectors: A Laboratory Manual." Freeman, New York, 1992.

is eluted from the agarose plug either by vortexing (30 sec) or by allowing the agarose plug to incubate overnight at 4°. The plaque assay is repeated with several of the plaques. The second round of plaque purification should yield a pure virus stock.

Virus Amplification

Infection of Sf9 cells with plaque-purified virus is used to obtain a high titer virus stock. In our laboratory, *T. ni* cells do not produce high virus titers, typically a log value lower than titers obtained from Sf9 cells. Amplification of the virus stock from an agarose plug to 150 ml requires two stages. In the first stage, 4×10^6 Sf9 cells are seeded in a T-25-cm^2 flask with Grace's complete supplemented medium in a total volume of 5 ml and allowed 30 min for cell attachment. The medium is removed and 1 ml of the pure virus stock (agarose plug) is added. The cells are incubated for 1 hr followed by the addition of 4 ml of Grace's complete supplemented medium. The cells are incubated for 3 days at 27°, and the supernatant is collected. The supernatant is centrifuged at 800 g for 5 min to remove cell debris. In the second stage, the entire 5 ml of virus stock from the T-25 flask is added to a 150-ml suspension culture of Sf9 cells seeded at 1×10^6 cells/ml in Grace's complete supplemented medium and incubated at 27° for 3 days. After 3 days, the supernatant is harvested and centrifuged at 800 g for 10 min. The titer of the virus stock can be determined either by plaque assay or by end-point dilution as described in the laboratory manual by O'Reilly *et al.*[18] Typically, virus titers range between 10^7 and 10^8 pfu/ml. Virus stocks are stored in the dark at 4°.

Expression

Culture Conditions

Four factors are critical to the successful expression of P450s in either Sf9 or *T. ni* cells: (1) cells should be actively growing in log phase for 2–3 days with viability $> 98\%$; (2) heme supplements; (3) multiplicity of infection (MOI) > 1; and (4) establishment of the optimal time to harvest the insect cells. Various heme supplements gave similar results with CYP3A4 coexpressed with CYPOR (CYP3A4-OR) via a single recombinant baculovirus in *T. ni* cells; testosterone 6β-hydroxylase activity was comparable to that of human liver microsomal CYP3A4 without the addition of exogenous cytochrome b_5.[19] However, for other P450s we have observed more repro-

[19] C. A. Lee, S. H. Kadwell, T. A. Kost, and C. J. Serabjit-Singh, *Arch. Biochem. Biophys.* **319**, 157 (1995).

TABLE I

COMPARISON OF CYP3A4-OR MICROSOMES FROM HEMIN CHLORIDE (2 μg/ml)-FORTIFIED *T. ni* AND Sf9 CELLS AT 72 hr POSTINFECTION

Parameter	Unit of measurement	Sf9	*T. ni*
Specific content[a]	pmol P450/mg microsomal protein	48 ± 7	105 ± 17[c]
Specific activity[a]	nmol 6β-hydroxytestosterone/min/nmol P450[b]	26 ± 2	66 ± 2[c]
CYPOR activity[a]	units/mg microsomal protein	1696 ± 806	3904 ± 1626[c]
CYPOR:CYP3A4	Molar ratio[d]	9 : 1	10 : 1

[a] Mean ± SD, n = three independent preparations.
[b] All incubations were conducted with 200 μM testosterone.
[c] *T. ni* versus Sf9 microsomes, P < 0.05; two sample student's t test.
[d] The molar amount of CYPOR was determined assuming a specific activity of 50,000 units/mg CYPOR and a molecular weight of 77,000 g/mol.

ducible heme incorporation with 100 μM iron–citrate/100 μM 5-aminolevulinic acid fortification than hemin chloride, as assessed by P450 carbon monoxide difference spectrum.[20] We have found that P450 levels are higher with a synchronous infection, achieved with a MOI of 2. We base the optimal harvest time on catalytic activity rather than on Western blots, especially when both P450 and CYPOR are coexpressed via a single recombinant baculovirus. The optimal time to harvest the infected cell cultures is 72 hr postinfection for all of the P450s expressed so far in our laboratory.

Selection of Cell Type

Sf9 cells have been commonly used for the expression of many recombinant proteins. However, the use of *T. ni* cells is increasing rapidly. These two cell lines supplemented with 2 μg/ml of hemin chloride exhibited differences in the levels and specific activity of coexpressed CYP3A4-OR (Table I). The yield of total microsomal protein (from 5 × 10^8 cells) was approximately threefold greater in *T. ni* cells than in Sf9 cells. The spectral P450 content, testosterone 6β-hydroxylase, cytochrome c reductase, and the molar ratio of CYPOR to CYP3A4 of microsomes were greater with *T. ni* vs. Sf9 cells. We use *T. ni* cells for the production of the recombinant P450 and Sf9 cells for the generation of the recombinant viruses, amplification of virus stocks, and virus titering.

[20] T. Omura and R. Sato, *J. Biol. Chem.* **239,** 2379 (1964).

Isolation of Microsomal Fraction

After 72 hr of infection, we generally observe 70 to 80% viability with virtually 100% of the cells infected, as indicated by the swollen appearance of the cells. The cells are collected by centrifugation at 800 g for 10 min. The cell pellet is washed twice with phosphate-buffered saline (PBS) (approximately 10 times the volume of the pellet) and flash frozen in an ethanol/dry ice bath until use. The frozen cell pellet is thawed in a 37° water bath and resuspended in 6 times the volume (v/v) of 100 mM potassium phosphate buffer, pH 7.4. Homogenization with either a hand-held motor-driven Teflon pestle on ice (approximately 6 to 7 full strokes) or sonication results in the lysis of 60 to 70% of the cells. The homogenate is centrifuged

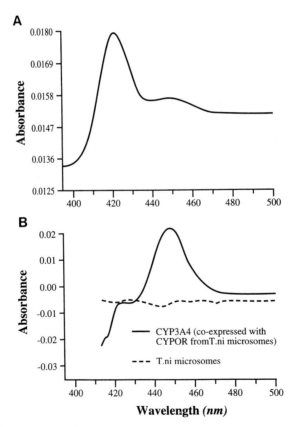

FIG. 1. Fe^{2+}–CO versus Fe^{2+} difference spectra of microsomal CYP3A4 coexpressed with CYPOR isolated from hemin chloride-fortified *T. ni* insect cells (A) with residual hemin chloride contamination and (B) without hemin chloride contamination.

at 10,000 g for 30 min, and the supernatant is transferred to a clean tube for centrifugation at 100,000 g for 70 min. The microsomal pellet is resuspended in 100 mM potasium phosphate buffer, pH 7.4.

When insect cell cultures are fortified with hemin chloride, two washes with PBS and the large volumes of buffer used in homogenization are adequate to remove the unincorporated hemin chloride. Residual hemin chloride will interfere in the carbon monoxide difference spectra measurement, resulting in a large peak at 420 nm that can obscure the peak at 450, as seen in Fig. 1A. Figure 1B shows a typical P450 CO difference spectrum where the unbound heme has been effectively removed.

Scaleup

Suspension cultures of *T. ni* cells are ideal for scaleup expression of recombinant P450s in bioreactors.[21] The conditions used in the bioreactors, such as the MOI, heme fortification, and time of harvest, are similar to small-scale culture. However, 5% FBS is added to the culture media, and the harvested cells are washed twice with PBS containing a cocktail of protease inhibitors (2 mM p-aminobenzamidine, 1 μg/ml aprotinin, 1 μg/ml pepstatin, 1 μg/ml leupeptin, and 10 μM 4-amidinophenyl-methanesulfonyl fluoride). A 36-liter bioreactor yields approximately 4 g of microsomal CYP3A4-CYPOR protein.

Conclusion

The baculovirus-insect cell combines the advantages of the high level expression of P450 achieved with yeast and *E. coli* and the high specific activity observed with mammalian cells. The ability to coexpress P450 and CYPOR by a single recombinant baculovirus ensures a fixed ratio of these proteins in the insect microsomal membrane. The insect lipid environment supports the catalytic activity, allowing the use of the microsomal fraction for catalytic studies. The catalytic activity of the recombinant CYP3A4-OR is similar to that observed for human liver microsomal CYP3A4. The turnover values for the formation of 6β-hydroxytestosterone with recombinant CYP3A4-OR and human liver microsomes were 66.6 and 16.9 nmol/min/nmol P450, respectively.[19] The baculovirus-insect cell system is readily amenable to scaleup, providing an ample supply of functional recombinant P450s.

[21] L. K. Overton and T. A. Kost, "Baculovirus Expression Systems and Biopesticides," p. 233. Wiley-Liss, New York, 1995.

Acknowledgments

We thank Sue Kadwell, Laurie Overton, and Christine Hoffman for technical assistance; Frank Gonzalez, Jeroen Buters, and Charles Crespi for materials and advice; and Douglas Rickert and Dhiren Thakker for their encouragement of this work.

Section II

Enzyme Assays

[10] *In Vitro* Assessment of Various Cytochromes P450 and Glucuronosyltransferases Using the Antiarrhythmic Propafenone as a Probe Drug

By HEYO K. KROEMER, SIGRID BOTSCH, GEORG HEINKELE, and MONIKA SCHICK

Introduction

The antiarrhythmic propafenone is widely used in the treatment of various cardiac arrhythmias.[1] The drug undergoes extensive metabolism in man (Fig. 1) by both P450 enzymes (formation of 5-hydroxypropafenone and N-desalkylpropafenone) and glucuronosyltransferases (formation of propafenone glucuronide and 5-hydroxypropafenone glucuronide). Several lines of *in vivo* and *in vitro* evidence indicate that the major metabolite in man is 5-hydroxypropafenone.[2,3] The polymorphic CYP2D6 has been identified to catalyze 5-hydroxylation, and, hence, two subsets of patients termed extensive metabolizers (EM) and poor metabolizers (PM) can be identified based on their propafenone clearance. Assignment of CYP2D6 phenotype prior to or during therapy with propafenone is of clinical relevance since PMs have a significantly higher incidence of CNS-related side effects.[2] Moreover, β-blockade is more pronounced in this subset of patients.[4] Assessment of individual phenotype during propafenone therapy is hampered by the high affinity of the drug to CYP2D6 which blocks metabolism of the commonly used probe drugs debrisoquine[2] or sparteine (M. Eichelbaum, unpublished observation). It has been suggested to use the ratio of 5-hydroxypropafenone/propafenone in plasma[5] for phenotyping. A different approach is based on the fact that the parent compound is glucuronidated and only traces of unconjugated propafenone are excreted in urine. Lack of CYP2D6 in a PM patient results in reduced clearance

[1] C. Funck-Brentano, H. K. Kroemer, J. T. Lee, and D. M. Roden, *N. Engl. J. Med.* **322,** 518 (1990).

[2] L. A. Siddoway, K. A. Thompson, B. C. McAllister, T. Wang, G. R. Wilkinson, D. M. Roden, and R. L. Woosley, *Circulation* **75,** 785 (1987).

[3] H. K. Kroemer, G. Mikus, T. Kronbach, U. A. Meyer, and M. Eichelbaum, *Clin. Pharmacol. Ther.* **45,** 28 (1989).

[4] J. T. Lee, H. K. Kroemer, D. J. Silberstein, C. Funck-Brentano, M. D. Lineberry, A. J. J. Wood, D. M. Roden, and R. L. Woosley, *N. Engl. J. Med.* **322,** 1764 (1990).

[5] R. Latini, M. Belloni, R. Bernasoni, E. Capiello, P. Giani, D. Landolina, and L. M. Castel, *Eur. J. Clin. Pharmacol.* **42,** 111 (1992).

F<small>IG</small>. 1. Chemical structure of propafenone and the major metabolites (an asterisk indicates the chiral center).

via 5-hydroxylation and in turn leads to a higher fractional excretion of propafenone glucuronides in urine. Thus, urinary excretion of propafenone glucuronides can be used for assignment of CYP2D6 phenotypes.[6] The second P450-dependent pathway is the formation of N-desalkylpropafenone. Although metabolic clearance of N-dealkylation is low compared to 5-hydroxylation and cannot compensate for the loss of 5-hydroxylation in PMs, this metabolic step contributes to the non-CYP2D6-related interaction potential of propafenone (e.g., the propafenone/theophylline interaction). Recent work from our laboratory indicated that CYP3A4 (70%) and CYP1A2 (30%) are involved in the formation of N-desalkylpropafenone.[7] In summary, the metabolism of propafenone has been studied in detail during recent years. *In vitro* methods have been developed in our laboratory for monitoring each of the metabolic pathways listed earlier. These methods can be used to characterize the microsomal fraction of human livers and/ or respective expression systems for their capacity to form 5-hydroxypropafenone, N-desalkylpropafenone, and propafenone glucuronide. The following sections describe the experimental conditions for *in vitro* assessment of each metabolite.

[6] S. Botsch, G. Heinkele, C. O. Meese, M. Eichelbaum, and H. K. Kroemer, *Eur. J. Clin. Pharmacol.* **46,** 133 (1994).
[7] S. Botsch, J. C. Gautier, P. Beaune, M. Eichelbaum, and H. K. Kroemer, *Mol. Pharmacol.* **43,** 120 (1993).

In Vitro 5-Hydroxylation of Propafenone as a Marker of
CYP2D6 Activity

5-Hydroxypropafenone formed from propafenone in the presence of
the microsomal fraction of human liver is quantified by HPLC. The mobile
phase consists of 20 mM sodium perchlorate buffer (pH 2.5; 5.1684 g of
sodium perchlorate in 2000 ml water) and acetonitrile 60:40 (v/v). Adding
0.64 ml of 60% perchloric acid results in a pH of 2.5. The mobile phase
should be filtered and sonicated for 15 min. Peaks are separated on a
Spherisorb ODS I (5 μM) column (110 × 0.5 mm) and are detected at 208
nm. The flow rate is 1.0 ml/min.

Incubation is performed in the presence of 50 μg of microsomal protein
(preparation of microsomes follows standard procedures[8]) dissolved in
30 μl phosphate buffer (0.1 M NaH$_2$PO$_4$ × H$_2$O is titrated with 0.3 M
Na$_3$PO$_4$ × 12 H$_2$O to pH 7.4; 0.01 M MgCl$_2$ is added). Ten microliters of
a solution of racemic propafenone (1 to 20 μM) dissolved in the phosphate
buffer and 1 mg NADPH are added. Phosphate buffer is then added to a
final volume of 250 μl. The mixture is vortexed and incubated at 37° for
30 min. The reaction is stopped by adding 10 μl of 60% perchloric acid.
Following a 10-min centrifugation at 10,000 g, 100 μl of the supernatant is
injected into the HPLC system.

The amount of 5-hydroxypropafenone formed is assessed by an external
standardization procedure. A calibration curve is established from 50 to
1000 pmol/250 μl. Two quality control samples of 150 and 750 pmol/250
μl are quantified with each run. Retention time is 9.3 min for 5-hydroxypro-
pafenone and 21 min for propafenone.

At substrate concentrations up to 10 μM, only 5-hydroxypropafenone
is formed. Using the incubation conditions described earlier results in a
K_m of an EM liver of about 0.5 μM. V_{max} for 5-hydroxylation is achieved at
substrate concentrations of 10 μM and is closely correlated to the CYP2D6-
mediated bufuralol hydroxylation in the same liver samples and to the
CYP2D6 expression assessed by means of Western blotting. Livers from
PM patients form only traces of 5-hydroxypropafenone, and the anti-LKM
1 antibody (directed against CYP2D6 activity) completely blunts the
5-hydroxylation activity of human liver microsomes.[3] Separate incubation
of S- and R-propafenone indicates stereoselectivity in the 5-hydroxylation
process; V_{max} of S-propafenone is significantly higher than that of R-propa-
fenone.[9] R-Propafenone is a competitive inhibitor of the 5-hydroxylation
of the S-enantiomer and vice versa (K_i values are 2.9 and 5.2 μM for the

[8] P. J. Meier, H. K. Müller, B. Dick, and U. A. Meyer, *Gastroenterology* **85**, 682 (1983).
[9] H. K. Kroemer, C. Fischer, C. O. Meese, and M. Eichelbaum, *Mol. Pharmacol.* **40**, 135 (1991).

inhibition of the 5-hydroxylation of S-propafenone by the R-enantiomer and vice versa, respectively.[9] This enantiomer/enantiomer interaction of S- and R-propafenone for CYP2D6-mediated 5-hydroxylation is observed in humans and has consequencs for both drug disposition and drug action.[10] In summary, several lines of evidence indicate that *in vitro* formation of 5-hydroxypropafenone can be used as an index of CYP2D6 activity. At higher substrate concentrations (>20 μM) N-dealkylation activity can be monitored in the same sample. We observed, however, inhibition of 5-hydroxypropafenone formation at substrate concentrations higher than 100 μM. For this reason a separate method for *in vitro* characterization of propafenone N-dealkylation has been established.

In Vitro N-Dealkylation of Propafenone as a Marker of CYP3A4 and CYP1A2 Activity

Quantification of N-desalkylpropafenone is performed by HPLC. The mobile phase consists of aqueous tetrabutylammonium sulfate (0.01 M) and methanol (56:44, v/v). Peaks are separated on a 5-μm C_{18} reverse-phase column (15 × 0.46 cm) and detected at 220 nm. The flow rate is 0.8 ml/min. Retention time is 8.4 min for N-desalkylpropafenone.

Incubations are carried out in the presence of 50 μg of microsomal protein dissolved in 20 μl of phosphate buffer (pH 7.4; described earlier). Racemic propafenone is added in a concentration range from 20 to 640 μM. One milligram of NADPH is added and the final volume is 100 μl. Incubations are carried out at 37° for 35 min. The reaction is stopped by the addition of 10 μl of 30% perchloric acid (v/v). Following centrifugation at 10,000 g for 5 min, 50 μl of the supernatant is injected into the HPLC system. Calibration is performed by external standardization. Calibration curves are linear from 50 to 1000 pmol. K_m averages 125 μM. V_{max} is correlated with expression of CYP3A and CYP1A2 as determined by Western blotting which indicates a contribution of these enzymes.[7] Antibodies directed against CYP3A and CYP1A2 inhibit the formation of N-desalkyl-propafenone by 54 ± 10 and 24 ± 16%, respectively.[7] The calcium channel blocker verapamil, which is a substrate for CYP1A2 and CYP3A4,[11] blocks N-dealkylation of propafenone in a competitive manner. N-Desalkylpropa-fenone is formed by yeast cells, which are genetically engineered for stable expression of CYP3A4 and CYP1A2.[7]

[10] H. K. Kroemer, M. F. Fromm, K. Bühl, H. Terefe, G. Blaschke, and M. Eichelbaum, *Circulation* **89**, 2396 (1994).
[11] H. K. Kroember, J. C. Gautier, P. Beaune, C. Henderson, C. R. Wolf, and M. Eichelbaum, *Naunyn-Schmiedeberg's Arch. Pharmacol.* **348**, 332 (1993).

In summary, N-dealkylation of propafenone is mediated by CYP3A4 (70%) and CYP1A2 (30%). This method is suitable for monitoring the combined activity of these enzymes in human liver microsomes or expression systems. Formation of 5-hydroxypropafenone can also be quantified in this system, thereby enabling combined assessment of CYP2D6, CYP3A4, and CYP1A2. We have, however, not validated this assay for quantification of 5-hydroxypropafenone.

In Vitro Glucuronidation of Propafenone

As described earlier, glucuronidation is a major metabolic pathway of propafenone, particularly in PM patients. To enable exact quantification in an *in vitro* system, R- and S-propafenone-O-glucuronide (sodium salt) were prepared chemically using a modification of Königs–Knorr reaction according to the method of Rudolph and Steinhart[12] and Heinkele and Meese,[13] and characterized by elemental analysis, ^{13}CNMR, and $(+/-)$-FAB-MS (optical and chemical purity: >95%). Reference substrate is available on request from one of the authors (G.H.).

R- and S-propafenone glucuronides are quantified by HPLC. Introduction of the glucuronic acid moiety leads to the formation of diastereoisomers which have different retention times (27 and 29.5 min for R- and S-propafenone glucuronide, respectively). Thus, formation of the glucuronide from the S- and R-propafenone following incubation of racemic propafenone can be readily monitored; Fig. 2). The detector is set at a wavelength of 248 nm. A C_{18} reverse-phase column (filled with ODS 5 μm, 25 $\times$ 0.46 cm i.d.) is used with a flow rate of 0.8 ml/min. The mobile phase consists of aqueous tetrabutylammonium hydrogensulfate (0.01 M) and methanol (56 : 44, v/v).

Incubations are carried out at 37° (incubation buffer: 100 mM Tris and 5 mM MgCl$_2$) for 60 min in the presence of 175 μg microsomal protein and 5 mM UDPGA. The final volume is 100 μl, and the reaction is stopped by heating for 10 min at 60°. After centrifugation at 10,000 g for 5 min, 50 μl of the supernatant is injected in the HPLC system. The formation of propafenone glucuronides is linear in a concentration range of 50–250 μg microsomal protein and in a time range from 10 to 60 min for 175 μg microsomal protein.

For kinetic experiments in the presence of the microsomal fraction of human livers, propafenone is used in concentrations of 40, 80, 120, 160, 240, 320, 480, 640, 960, 1280, and 1920 μM. HPLC chromatograms are

---

[12] M. Rudolph and H. Steinhart, *Carbohydr. Res.* **176**, 155 (1988).
[13] G. Heinkele and C. O. Meese, in preparation.

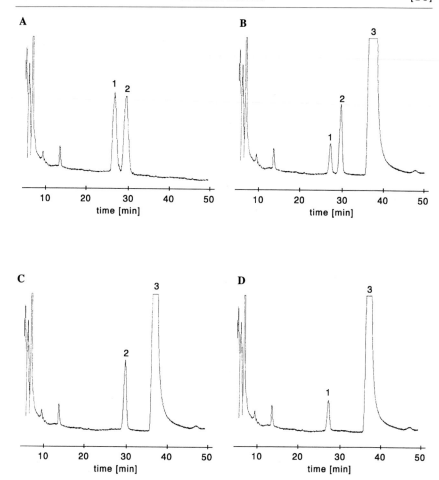

Fig. 2. HPLC chromatograms of R-propafenone glucuronide (1), S-propafenone glucuronide (2), and racemic propafenone (3). (A) Calibration point (250 pmol of R/S-propafenone glucuronide). (B) Incubation of 480 μM racemic propafenone (175 μg microsomal protein, 60 min at 37°). (C) Incubation of 240 μM R-propafenone (175 μg microsomal protein, 60 min at 37°). (D) Incubation of S-propafenone glucuronide (175 μg microsomal protein, 60 min at 37°).

displayed in Fig. 2. The formation of diastereomeric glucuronides after the incubation of propafenone follows a monophasic Michaelis–Menten kinetics. The average K_m is 360 and 505 μM for formation of R- and S-propafenone glucuronide, respectively. V_{max} is about 5 and 11.5 pmol/μg protein/hr for the formation of R- and S-propafenone glucuronide, respectively. The formation of 5-hydroxypropafenone and N-desalkylpro-pafenone can also be quantified in this system, thereby enabling combined assessment of CYP2D6, CYP3A4, CYP1A2, and the UGT involved in the glucuronidation of propafenone.

Conclusions

Both phase I and phase II metabolism of the antiarrhythmic propafenone has been subjected to detailed investigations. The P450 enzymes involved in the formation of 5-hydroxypropafenone and N-desalkylpropafenone have been identified and the *in vitro* system described earlier enable monitoring of the activity of these enzymes. Identification of the UGT which mediates formation of propafenone glucuronide is a topic of further work in our laboratory. Development and validation of a combined *in vitro* assay for both oxidative and phase II metabolism of propafenone will allow investigations of a coregulation of both enzyme systems which has been suggested for other substrates based on *in vivo* experiments.[14]

Acknowledgments

This work was supported by the Robert Bosch-Foundation, Stuttgart, Germany, and by a grant from the Deutsche Forschungsgemeinschaft (Kr 945 2-1).

[14] K. W. Bock, D. Schrenk, A. Förster, E. U. Griese, K. Mörike, D. Brockmeier, and M. Eichelbaum, *Pharmacogenetics* **4**, 209 (1994).

[11] *In Vivo* and *in Vitro* Measurement of CYP2C19 Activity

By PETER J. WEDLUND and GRANT R. WILKINSON

Introduction

The genetic polymorphism, now known to be associated with CYP2C19, was discovered[1] long before the enzyme was identified and its molecular mechanisms explained.[2] This was because of the marked difference in the 4′-hydroxylation of S-mephenytoin in extensive metabolizers (EM), who

[1] G. R. Wilkinson, F. P. Guengerich, and R. A. Branch, *Pharmacol. Ther.* **43**, 53 (1989).
[2] J. A. Goldstein and J. Blaisdell, *Methods Enzymol.* **272**, Chap. 23, 1996 (this volume).

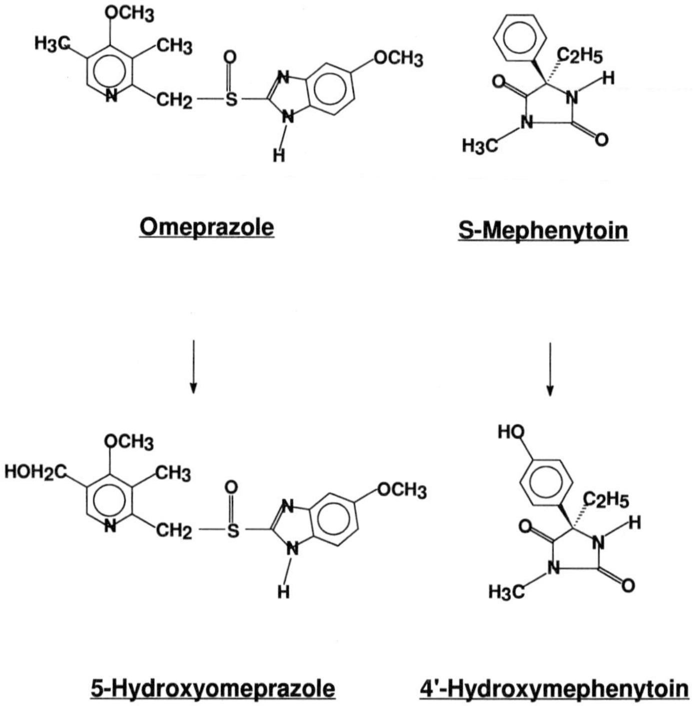

FIG. 1. CYP2C19-mediated pathways of metabolism of S-mephenytoin and omeprazole.

express CYP2C19, and poor metabolizers (PM), in whom there is a functional deficiency in this metabolic pathway. As a result, phenotypic trait measurements were developed that permitted characterization of an individual's CYP2C19 level of activity, after administration of a single dose of racemic mephenytoin as an *in vivo* probe.[3,4] This approach has been widely used throughout the world by many investigators in order to define the frequency of the PM phenotype in various types of populations. In addition, determination of the 4'-hydroxylation of S-mephenytoin provides a measurement of CYP2C19 catalytic activity in various *in vitro* preparations such as human liver microsomes and heterologous expression systems. An alternative CYP2C19 substrate, omeprazole, is beginning to be increasingly used as an *in vivo* probe because of its negligible side effects and, consequently, is a safer drug than mephenytoin (Fig. 1).

[3] A. Küpfer and R. Preisig, *Eur. J. Clin. Pharmacol.* **26,** 753 (1984).
[4] P. J. Wedlund, W. S. Aslanian, C. B. McAllister, G. R. Wilkinson, and R. A. Branch, *Clin. Pharmacol. Ther.* **36,** 773 (1984).

The methods described in this chapter primarily reflect approaches that we have found to be successful in our laboratories for over a decade. In addition, the reported experiences of other investigators are presented.

In Vivo Phenotyping with Mephenytoin

Mephenytoin is clinically available as a racemate and its metabolism is essentially stereospecific; the S-enantiomer is extensively and rapidly 4'-hydroxylated by CYP2C19 whereas R-mephenytoin is primarily and more slowly N-demethylated by another isoform(s).[5] Thus, in PMs the formation and subsequent urinary elimination of the metabolite is substantially lower than in EMs. This difference is the basis for one of the phenotyping procedures. Because deficiency in CYP2C19 also markedly alters the pharmacokinetic profile of the S-enantiomer but not that of R-mephenytoin, the enantiomeric ratio of, for example, the urinary excretion of parent drug is significantly affected.[5] This provides an alternative approach to phenotyping.

Clinical Protocol

Phenotyping using either approach involves the administration of an oral, 100-mg (460 μmol) dose of racemic mephenytoin (Mesantoin, Sandoz Pharmaceuticals, East Hanover, NJ) and collection of urine over the next approximately 8 hr. For convenience the anticonvulsant is frequently administered prior to bedtime and an overnight urine sample is collected. Some investigators also collect the second overnight urine (24–32 hr) after drug administration when measuring the $S:R$ enantiomeric ratio since this increases the interphenotypic difference.[6] It also circumvents a stability problem associated with an acid labile S-mephenytoin metabolite but, obviously, is less convenient than the single collection protocol. A blank urine collected prior to drug administration is also frequently obtained to test for the presence of interfering substances, especially when the person being studied is taking other medications; in normal volunteers such interference is rare.

The possibility of a sedative side effect from mephenytoin administration is greater in PMs than EMs,[7,8] but these are never serious or debilitating,

[5] P. J. Wedlund, W. S. Aslanian, E. Jacqz, C. B. McAllister, R. A. Branch, and G. R. Wilkinson, J. Pharmacol. Exp. Ther. **234,** 662 (1985).

[6] E. Sanz, T. Villén, C. Alm, and L. Bertilsson, Clin. Pharmacol. Ther. **45,** 495 (1989).

[7] K. Nakamura, F. Goto, W. A. Ray, C. B. McAllister, E. Jacqz, G. R. Wilkinson, and R. A. Branch, Clin. Pharmacol. Ther. **38,** 402 (1985).

[8] J. D. Balian, N. Sukhova, J. W. Harris, J. Hewett, L. Pickle, J. A. Goldstein, R. L. Woosley, and D. A. Flockhart, Clin. Pharmacol. Ther. **57,** 662 (1995).

especially if the drug is administered at bedtime. Nevertheless, in small-sized subjects and those of Asian descent, a 50-mg dose rather than a 100-mg dose may be more advisable.[9]

Measurement of Urinary 4'-Hydroxymephenytoin

After its formation, 4'-hydroxymephenytoin is glucuronidated and appears in the urine as this conjugated form. Accordingly, the aglycone must be liberated prior to quantification either by incubating with β-glucuronidase, which provides an analytically clean final extract, or by acid hydrolysis, which is more rapid and less expensive but yields a dirtier extract. Diphenylhydantoin, phenobarbital, and 5-isopropyl-5-phenylhydantoin have all been successfully used as internal standards for the HPLC-based assay. Similarly, liquid–liquid or solid-phase extraction methods are both suitable, providing the matrix is slightly acidic. For example, internal standard (20 μg phenobarbital), 1 ml 0.1 M acetate buffer (pH 4.5), and β-glucuronidase (2000 Fishman units) are added to 0.5 ml urine and the mixture is incubated at 37° for 12 hr. After extraction with 5 ml dichloromethane, the organic phase is separated, transferred to a Reactivial (Pierce, Rockville, IL), and evaporated to dryness under nitrogen at 45°. When using acid hydrolysis, 1 ml of concentrated HCl is added to 0.5 ml urine along with an internal standard (10 μg 5-isopropyl-5-phenylhydantoin) prior to incubating at 100° for 1 hr. After cooling, the mixture is transferred to a Chem Elut solid-phase extraction column (CE 1003, Varian Assoc., Harbor City, CA) and allowed to filter for 5 min. Analytes are removed from the column with 5 ml dichloromethane, which is then evaporated to dryness. Prior to analysis, the residue, regardless of the extraction procedure, is reconstituted with 100 μl 40% methanol in water and 20 μl is injected onto a C_{18} HPLC column (5-μm particle size, 4.6 × 150 mm; Alltech Assoc., Deerfield, IL). A typical mobile phase composition is 25% (v/v) acetonitrile, 75% 0.05 M phosphate buffer (pH 4.0) with 0.05 M acetic acid or 0.02 M perchloric acid added to the aqueous phase to improve column performance. An alternative mobile phase is 35% methanol, 65% 0.05 M phosphate buffer (pH 6.5). UV measurement at 211 nm (214 nm for a fixed wavelength detector) is usually used to estimate the analyte levels, and a calibration curve between 1 and 100 μg/ml is defined using authentic 4'-hydroxymephenytoin (Ultrafine Chemicals, Manchester, UK) for each sample run.

In EMs the 0- to 8-hr urinary recovery of 4'-hydroxymephenytoin is between 5 and 52% (25 to 240 μmol) of the administered dose (mean 18–20%). In contrast, from undetectable to 3% of the dose (mean 0.5–0.8%)

[9] R. Setiabudy, K. Chiba, M. Kusaka, and T. Ishizaki, *Br. J. Clin. Pharmacol.* **33,** 665 (1992).

is excreted in PMs. An alternative phenotypic trait measurement is the mephenytoin hydroxylation index:

$$\frac{\text{molar dose of } S\text{-mephenytoin } (230 \ \mu\text{mol in } 100 \ \text{mg racemic dose})}{\mu\text{mol } 4'\text{-hydroxymephenytoin excreted in } 0\text{–}8 \ \text{hr}}.$$

In individuals who express CYP2C19, this value ranges from about 0.6 to 20 whereas in PMs a much higher value (30–2500) is observed. Importantly, the 4'-hydroxymephenytoin concentration in urine samples from PMs is often close to the lower limit of sensitivity of the assay, which may vary between laboratories. Thus, an antimodal value that discriminates between the two phenotypes cannot be defined with absolute precision and exhibits interlaboratory variability. As a result, phenotypic misclassification may occur when the 0- to 8-hr recovery is in the range of about 15–25 μmol. A further concern relevant to this trait measurement is its dependence on complete urine collection over the 0- to 8-hr period since a low recovery of the 4'-hydroxy metabolite could also reflect poor subject compliance. Because of this, determination of the amount of creatinine in the sample (>50 mg in 0–8 hr) is advisable following the identification of a putative PM. Alternatively, data regarding a second *in vivo* probe drug that was coadministered with the mephenytoin dose can be used to address this issue.

Measurement of the Urinary Mephenytoin Enantiomeric Ratio

Because of extensive metabolism in EMs and a slow rate of elimination in PMs, only a small fraction of the administered dose is excreted unchanged in 8 hr and urine concentrations are well below the detection limit of the HPLC assay used to determine the 4'-hydroxy metabolite. On the other hand, the marked differences in the disposition of the *R*- and *S*-enantiomers in EMs lead to a large difference in the isomers' rates of urinary excretion that is not present in CYP2C19-deficient individuals,[5] i.e., PMs. Accordingly, an alternative phenotyping procedure is based on determining the enantiomeric ratio of mephenytoin in the 0- to 8-hr urine sample, using stereospecific analysis based on capillary gas chromatography (Fig. 2).

Liquid–liquid extraction of the enantiomers from urine (5 ml) is achieved with 6 ml dichloromethane which, after separation from the aqueous phase, is washed successively with 1 ml 0.1 N NaOH and 1 ml 0.01 N HCl. The final organic phase is then transferred to a Reactivial and evaporated to dryness under nitrogen at 45°. Alternatively, solid-phase extraction can be performed after adding 1 ml 0.01 M acetic acid to 2 ml urine and placing the mixture on a Chem Elut, CE 1003 column for 5 min. The analytes are then eluted with 6 ml dichloromethane which is then evaporated to dryness,

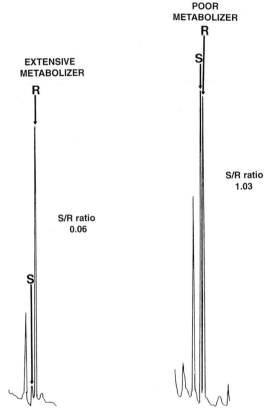

FIG. 2. Comparison of typical chromatograms of the *S*- and *R*-enantiomers of mephenytoin extracted from 0- to 8-hr urine samples obtained from an EM and PM, following oral administration of 100 mg racemic mephenytoin.

as previously described. With both procedures, the extract residue is reconstituted with 10 μl ethylacetate prior to analysis.

Chromatographic separation of the two mephenytoin enantiomers (Ultrafine Chemicals, Manchester, UK) is achieved using a Chirasil-Val capillary column (25 m in length, 0.32 mm, internal diameter; Alltech Assoc.). The injector and detector temperatures are both maintained at 225° whereas the column temperature is kept at 180°. Detection is either by a nitrogen-phosphorus or a flame ionization detector; the analytical sensitivities are comparable but the former provides a more selective and cleaner chromatogram but is considerably more expensive and requires greater maintenance than the latter. The column helium flow rate is 1.2 ml/min with a split ratio between 1:10 and 1:15 depending on column resolution and sensitivity.

At the detector, the helium makeup flow is 25 ml/min with hydrogen and air flows of 3.5 and 150 ml/min, respectively, for the nitrogen-phosphorus ionization detector. Comparable rates when using a flame ionization detector are 35 and 430 ml/min, respectively. An oxygen scrubber is placed between the helium source and the column to prevent destruction of the stationary phase when the column is heated. The peak heights or areas of the S- and R-mephenytoin peaks are measured directly from the chromatogram and are expressed as a ratio. No analytical internal standards or calibration curves are required.

The measured ratio is equivalent to the comparable ratio of the areas under the plasma concentration–time profiles of the enantiomers during the urine collection period.[10] Originally and in many population studies, the $S:R$ ratio was used, which has a reciprocal and, therefore, rectangular hyperbolic relationship with CYP2C19 activity. On the other hand, the $R:S$ ratio is linearly related to such activity so that the smaller its value the lower the 4'-hydroxylating ability. In PMs, the enantioselective disposition of mephenytoin is small so that the 0- to 8-hr urinary $S:R$ ratio is between 0.9 and 1.2. In contrast, EMs have values which range from less than 0.05 to 0.8. A ratio greater than 1.2 should be regarded with suspicion since it probably reflects an artifact associated with an acid-labile metabolite of S-mephenytoin that has been converted back to the parent enantiomer by hydrolysis during sample storage.[11] Such conversion is unpredictable even when samples are stored at $-20°$ and its probability increases with storage time. Because of this, it is advisable to repeat the analysis of a sample with an $S:R$ ratio above 0.80, following pretreatment of 1 ml urine with 0.1 ml 12 N HCl for 60 min before following the previously described extraction procedure. Since hydrolysis is rapid, any incubation period greater than 5 min is satisfactory. The ratio is significantly increased by such acid treatment in samples from EM subjects, but remains essentially unchanged in PMs.[12,13] The problem associated with the acid-labile metabolite can be minimized, but not necessarily obviated, by storing the sample at $-20°$ and analyzing within 4 weeks. Alternatively, it may be circumvented by extracting the urine immediately after collection and then storing the dried dichloromethane extract at $-20°$ until analysis. Urine samples collected between 24 and 32 hr after mephenytoin administration contain little or no acid-labile metabolite and so do not require this additional procedure. On the other

---

[10] E. Jacqz, S. D. Hall, R. A. Branch, and G. R. Wilkinson, *Clin. Pharmacol. Ther.* **39**, 646 (1986).

[11] P. J. Wedlund, B. J. Sweetman, G. R. Wilkinson, and R. A. Branch, *Drug Metab. Dispos.* **15**, 277 (1987).

[12] Y. Zhang, R. A. Blouin, P. J. McNamara, J. Steinmetz, and P. J. Wedlund, *Br. J. Clin. Pharmacol.* **31**, 350 (1991).

[13] G. Tybring and L. Bertilsson, *Pharmacogenetics* **2**, 241 (1992).

hand, the $S:R$ in EMs at this later time is smaller (0.1 to 0.5) than that determined in a 0- to 8-hr sample, whereas the value in PMs is still close to unity.

Usually either of the described phenotyping procedures permits characterization of an individual as either an EM or a PM. Occasionally, however, such discrimination is not always so clear-cut, usually because of the mentioned analytical or stability problems. In this case, measurement of the other trait measurement may clarify the situation or, alternatively, the individual can be rephenotyped following administration of another dose of mephenytoin. Despite these approaches, trait values in a very small number of individuals may still be uninterpretable based on the assumption that only two phenotypes exist. One reason for this is the very low frequency of "intermediate metabolizers" (IM) who have an $S:R$ enantiomeric ratio consistent with the PM phenotype (>0.9) but who excrete more (20–60 μmol) of the 4'-hydroxy metabolite than would be expected if this was the case but at a rate less than EMs.[14] The plasma disposition of S-mephenytoin in such individuals, who do not represent the heterozygous genotype, is also in between that of PMs and EMs. It is likely that such individuals have an allelic variant of CYP2C19 with different catalytic activity compared to that of the wild-type enzyme. It is possible that the $S:R$ enantiomeric ratio measured at 24–32 hr may also identify IMs since occasional subjects have been noted to have values between 0.2 and 0.5 whereas in most EMs this trait value is about 0.1.[6,13]

In Vivo Phenotyping with Omeprazole

The 5-hydroxylation of omeprazole is mediated by CYP2C19 and large interphenotypic differences are present in the disposition of the H^+,K^+-ATPase inhibitor and its metabolite.[15,16] In addition, the drug has a short elimination half-life (1 hr) and a far wider therapeutic margin than mephenytoin. Accordingly, omeprazole may be used as an alternative and safer *in vivo* probe of CYP2C19 activity as demonstrated by the good agreement that has been noted between the phenotyping results obtained with both drugs.[8,17,18]

[14] P. A. Arns, D. Ryder, L. White, G. R. Wilkinson, and R. A. Branch, *Pharmacologist* **32**, 140 (1990).
[15] T. Andersson, C.-G. Regårdh, Y.-C. Lou, Y. Zhang, M.-L. Dahl, and L. Bertilsson, *Pharmacogenetics* **2**, 25 (1992).
[16] D.-R. Sohn, K. Kobayashi, K. Chiba, K.-H. Lee, S.-G. Shin, and T. Ishizaki, *J. Pharmacol. Exp. Ther.* **262**, 1195 (1992).
[17] M. Chang, G. Tybring, M.-L. Dahl, E. Götharson, M. Sagar, R. Seensalu, and L. Bertilsson, *Br. J. Clin. Pharmacol.* **39**, 511 (1995).
[18] M. Chang, M.-L. Dahl, G. Tybring, E. Götharson, and L. Bertilsson, *Pharmacogenetics* **5**, 358 (1995).

The phenotyping procedure involves the administration of a single 20-mg dose of omeprazole (Prilosec, Merck & Co., Inc., West Point, PA), after an overnight fast. A peripheral venous blood sample (10 ml) is then obtained 2–3 hr after drug administration and the plasma levels of omeprazole and 5-hydroxyomeprazole are determined by an HPLC-based procedure.[19] The concentration ratio of parent drug to metabolite is then used as the phenotypic trait value.

Currently, about 250 Caucasian individuals have been phenotyped by this procedure and the results compared to those obtained with mephenytoin and, also, with the subject's genotype.[8,18] Based on these studies, an antimode has been identified at about 7, with EMs having a plasma concentration ratio below this (range 0.05 to 5.6), and a larger metabolic ratio being indicative of the PM phenotype. In fact, the presence of metabolite in the plasma is often not detectable in PMs.

In Vitro Determination of CYP2C19 Activity

Measurement of CYP2C19 catalytic activity *in vitro* has mainly been limited to human liver preparations such as microsomes. However, the developed procedures are also applicable to other preparations, e.g., heterologous recombinant systems specifically designed to express the enzyme.

Typically, a standard reaction mixture consists of microsomes (0.1 to 0.5 mg protein), 5 mM MgCl$_2$, and mephenytoin (50–200 μM racemate or S-enantiomer) in a final volume of 0.25 to 0.5 ml 100 mM phosphate buffer (pH 7.4) containing an NADPH-generating system consisting of 4 mM glucose 6-phosphate, 0.5 mM NADP$^+$, and 1 unit glucose-6-phosphate dehydrogenase. The reaction is initiated by the addition of NADP$^+$ and incubation proceeds for 30 min at 37°. The addition of an analytical internal standard solution containing 2% sodium azide followed by dichloromethane terminates the reaction and allows extraction of the 4′-hydroxymephenytoin formed during the incubation (0.01 to 0.20 nmol/min/mg protein). Subsequently, the organic layer is removed and evaporated to dryness and, after reconstitution in methanol, the analytes are determined by HPLC in the same manner as previously described for *in vivo* phenotyping.

Since the concentration of 4′-hydroxymephenytoin is far lower than in urine, analytical sensitivity is a significant problem. One approach is to detect and measure the metabolite at 204 nm rather than at 211 nm, but this requires the use of only the highest quality HPLC solvents because this wavelength is close to their UV cutoff value. A better approach to this problem is the use of a ^{14}C-labeled substrate along with radiometric

[19] D. J. Birkett, T. Andersson, and J. O. Miners, *Methods Enzymol.* **272,** Chap. 14, 1996 (this volume).

analysis.[20,21] Moreover, the recent commercial availability of [*N-methyl-*[14]C]mephenytoin (Amersham Corp., Arlington Heights, IL) now makes this a generally available approach. Also, the labeled 4'-hydroxymephenytoin can be analyzed without extraction, e.g., the microsomal incubation is terminated by the addition of an equal volume of methanol and centrifuged at 10,000 g for 10 min.[22] Subsequently, an aliquot of the supernate is injected onto the HPLC column and the radioactivity associated with the 4'-hydroxy metabolite is determined by an on-line radiochemical detector[21,22] or the corresponding eluant volume mixed with an appropriate scintillant mixture and analyzed in a liquid scintillation counter.

Conclusions

The genetic polymorphism in CYP2C19 activity has been widely studied, and the involvement of the enzyme in the metabolism of drugs other than the prototype substrate, i.e., mephenytoin, continues to be of basic and applied scientific interest. The recent discovery of the molecular genetic mechanisms responsible for the PM phenotype will, undoubtedly, result in further interest, especially in the relationship between genotype and phenotype. The described analytical methodologies have been extensively validated and widely applied, thus they provide the tools necessary for such studies.

Acknowledgment

Research leading to the development and application of the described assays was supported by USPHS Grant GM 31304.

[20] T. Shimada, J. P. Shea, and F. P. Guengerich, *Anal. Biochem.* **147,** 174 (1985).
[21] W. R. Brian, P. K. Srivastava, D. R. Umbenhauer, R. S. Lloyd, and F. P. Guengerich, *Biochemistry* **28,** 4993 (1989).
[22] J. A. Goldstein, M. B. Faletto, M. Romkes-Sparks, T. Sullivan, S. Kitareewan, J. L. Raucy, J. M. Lasker, and B. I. Ghanayem, *Biochemistry* **33,** 1743 (1994).

[12] Chlorzoxazone: An *in Vitro* and *in Vivo* Substrate Probe for Liver CYP2E1

By DANIELE LUCAS, JEAN-FRANÇOIS MENEZ,† and FRANÇOIS BERTHOU

Introduction

Cytochrome P4502E1 (CYP2E1) is involved in the oxidation of ethanol and a diverse group of suspect carcinogens, including nitrosamines, benzene, phenol, chloroform, trichloroethylene, ethylene dihalides, methylene dihalides, styrene, butadiene, vinyl halides, and urethane.[1,2] To understand the regulation of this enzyme in humans and to delineate its role in numerous human diseases, a reliable noninvasive assay was needed that could be safely used with a large number of human subjects. Furthermore, the same chemical probe would have had to be convenient for *in vitro* studies, including the determination of CYP2E1 catalytic activities in microsomal preparations or in cell cultures. Since the report of Peter *et al.*[3] demonstrating that the 6-hydroxylation of chlorzoxazone (CHZ) is catalyzed primarily by CYP2E1 in human liver microsomes (Fig. 1), chlorzoxazone, a potent muscle relaxant used in the treatment of painful muscle spasms, has been shown to be an ideal chemical probe specific for CYP2E1 in both *in vitro* and *in vivo* studies.

Selectivity of CYP2E1 in Chlorzoxazone 6-Hydroxylation

It is well documented that the major oxidative metabolite of chlorzoxazone is 6-hydroxychlorzoxazone.[4] By using different approaches, including correlation studies, immuno and chemical inhibition, and metabolism by human CYP2E1 purified from liver samples, it has been established that CYP2E1 catalyzed this reaction. This conclusion was extended by Carrière *et al.*[5] who also found that CYP1A1 was capable of catalyzing the same

---

† Deceased.

[1] C. S. Yang, J. S. H. Yoo, H. Ishizaki, and J. Hong, *Drug Metab. Rev.* **22**, 147 (1990).

[2] F. P. Guengerich, D. H. Kim, and M. Iwasaki, *Chem. Res. Toxicol.* **4**, 168 (1991).

[3] R. Peter, R. G. Böcker, P. H. Beaune, M. Iwasaki, F. P. Guengerich, and C. S. Yang, *Chem. Res. Toxicol.* **3**, 566 (1990).

[4] A. H. Conney and J. J. Burns, *J. Pharmacol. Exp. Ther.* **128**, 340 (1960).

[5] V. Carrière, T. Goasduff, D. Ratanasavanh, F. Morel, J. C. Gautier, A. Guillouzo, P. H. Beaune, and F. Berthou, *Chem. Res. Toxicol.* **6**, 852, (1993).

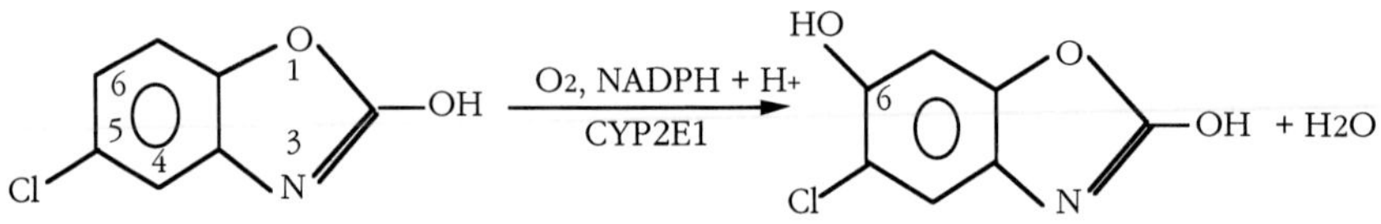

5-chloro-2(3*H*)-benzoxazolone (chlorzoxazone, CHZ)   6-hydroxychlorzoxazone (6-OH CHZ)

FIG. 1. P450 metabolism of chlorzoxazone.

reaction. The formation of 6-OH CHZ has now been used in numerous *in vitro*[6,7] studies as well as *in vivo* studies[8–10] as a specific probe of CYP2E1.[11–13] However, some recent reports have suggested that other human hepatic P450 enzymes, namely CYP1A2[14] and CYP3A4, may contribute to chlorzoxazone 6-hydroxylation. This controversy has recently been solved by two findings based on the same approach which used nine human P450 isoforms produced by bacteria[15] or yeasts[16] genetically engineered. CYP2E1 was shown to be the main enzyme able to 6-hydroxylate chlorzoxazone (Fig. 2).

In Vitro Chlorzoxazone 6-Hydroxylation Assay

Samples

CHZ has been used as a probe for CYP2E1 activity in microsomes from various sources (e.g., liver, kidney, lung cells) as well as directly in cell

[6] S. E. Clarke, S. J. Baldwin, J. L. Bloumer, A. D. Ayrton, R. S. Suzio, and R. J. Chenery, *Chem. Res. Toxicol.* **7,** 836 (1994).

[7] Y. Amet, F. Berthou, J. P. Salaün, L. Le Breton, and J. F. Ménez, *Biochem. Biophys. Res. Commun.* **203,** 1168 (1994).

[8] D. O'Shea, S. N. Davis, R. B. Kim, and G. R. Wilkinson, *Clin. Pharmacol. Ther.* **56,** 359 (1994).

[9] C. Girre, D. Lucas, E. Hispard, C. Ménez, S. Dally, and J. F. Ménez, *Biochem. Pharmacol.* **47,** 1503 (1994).

[10] R. Zand, S. P. Nelson, J. T. Slattery, K. E. Thummel, T. F. Kalhorn, S. P. Adams, and J. M. Wright, *Clin. Pharmacol. Ther.* **54,** 142 (1993).

[11] J. Mapoles, F. Berthou, A. Alexander, F. Simon, and J. F. Ménez, *Eur. J. Biochem.* **214,** 735 (1993).

[12] W. Tassaneeyakul, M. E. Veronese, D. J. Birkett, F. J. Gonzalez, and J. O. Miners, *Biochem. Pharmacol.* **46,** 1975 (1993).

[13] S. Barmada, E. Kienle, and D. R. Koop, *Biochem. Biophys. Res. Commun.* **206,** 601 (1995).

[14] S. Ono, T. Hatanaka, H. Hotta, M. Tsutsui, T. Satoh, and F. J. Gonzalez, *Pharmacogenetics* **5,** 143, (1995).

[15] H. Yamazaki, Z. Guo, and F. P. Guengerich, *Drug Metab. Dispos.* **23,** 438 (1995).

[16] V. Carrière, F. Berthou, S. Baird, C. Belloc, P. H. Beaune, and I. De Waziers, in press, 1996.

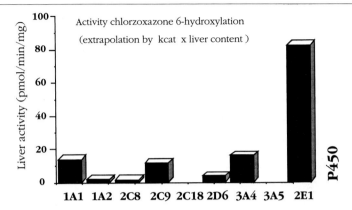

FIG. 2. Estimated contribution of nine recombinant human P450 enzymes to chlorzoxazone 6-hydroxylation. Data were extrapolated from turnover number (min^{-1}) × liver content (pmol/mg protein) in P450 isoforms.[17,18]

cultures. Almost all water-soluble organic solvents, such as dimethyl sulfoxide, methanol, ethanol, or glycerol, are known to be substrates of CYP2E1 and, accordingly, are competitive inhibitors. Therefore, their use should be avoided during incubation. Glycerol 0.8% (v/v), usually added to the microsomal samples for their storage, did not significantly inhibit the monooxygenase activity in the reaction medium.

Reagents

> Substrate: Chlorzoxazone (Sigma, St. Louis, MO), molecular weight 169.58. Chlorzoxazone stock solution: 40 mM in 60 mM freshly prepared KOH. [2-^{14}C]Chlorzoxazone (50 mCi/mmol, Amersham, UK)
> Metabolite: 6-Hydroxychlorzoxazone (Ultrafine Chemicals, Manchester, UK), molecular weight 185.58
> Reaction buffer: Potassium phosphate buffer, 0.1 M, pH 7.4
> Cofactor: NADPH (Sigma), 10 mM in reaction buffer
> Deproteinizing agent: Orthophosphoric acid 43% (v/v) (Merck)
> HPLC solvent: Acetonitrile, HPLC grade (Merck)

[17] F. P. Guengerich and C. G. Turvy, *J. Pharmacol. Exp. Ther.* **256,** 1189 (1991).
[18] T. Shimada, H. Yamazaki, M. Minura, Y. Inui, and F. P. Guengerich, *J. Pharmacol. Exp. Ther.* **270,** 414 (1994).

Extraction solvent: Chloroform/2-propanol (85:15, v/v), pro analysis (Merck)
Internal standard: Phenacetin or acetophenetidin (Sigma), molecular weight 179.2. Stock solution: 1.5 mg/ml in methanol grade HPLC

Incubation Conditions

Microsomes. The standard incubation mixture contains proteins (0.4 mg), 0.1 M potassium phosphate buffer, pH 7.4, 0.1 mM EDTA, saturating concentrations of cofactors (e.g., 1 mM NADPH or NADPH-generating system: 1 mM NADP, 10 mM glucose 6-phosphate and 2 IU glucose-6-phosphate dehydrogenase), and chlorzoxazone (400 μM) in a total volume of 1 ml. Chlorzoxazone has an apparent K_m of 39 ± 7 μM in human liver microsomes[3] and 6 ± 4 μM with pure CYP2E1 produced by genetically engineered bacteria.[15] Incubations are initiated following a 3-min preincubation at 37° by the addition of NADPH and are generally carried out for 20 min in a shaking water bath at 37°. When low enzymatic activities are expected, incubation time has to be increased provided that the linearity of the time course is established over that period. The reaction is terminated by adding 50 μl of 43% H_3PO_4. At this step, an internal standard such as 5-fluoro-2(3H)benzoxazolone[3] (5 μg) or phenacetin (7.5 μg) as previously suggested[19] can be added. The latter is more convenient as it is commercially available. The incubation medium is extracted with 2 ml of chloroform/2-propanol (85:15, v/v) in Teflon-capped vials, using a vortex. After centrifugation at 3000 g for 10 min, the organic phases are dried by filtration on Na_2SO_4 and are then evaporated to dryness under a N_2 stream at 37°. The products of chlorzoxazone oxidation are analyzed using HPLC.

Cell Culture Medium.[5,11,13] Cells (about 2.5 × 10⁶ cells/4 ml culture medium) are incubated for 24 hr in the presence of 400 μM chlorzoxazone dissolved in dimethyl sulfoxide (<0.25%; v/v). After incubation, the culture medium is removed and treated with 150 μl 43% H_3PO_4 and extracted with 6 ml of cloroform/2-propanol (85/15; v/v). The organic phase is dried by filtration on Na_2SO_4 and then evaporated to dryness under N_2.

Use of [2-¹⁴C]Chlorzoxazone. Labeled CHZ can be added to microsomes (0.05 μCi per sample, i.e., 0.125 mCi/mmol) or to culture medium (0.25 μCi, i.e., 0.15 mCi/mmol) in order to control the specificity of the reaction. As labeled CHZ is delivered in an ethanol solution, it must be evaporated to dryness before adding to unlabeled CHZ. Samples are then treated as described earlier. Chlorzoxazone and its metabolite are quantified using either HPLC with double detection (UV and radioactivity; Fig. 3) or

[19] I. L. Honigberg, J. T. Stewart, and J. W. Coldren, *J. Pharmacol. Sci.* **68,** 253 (1979).

thin-layer chromatography (TLC), which constitutes an alternative method for laboratories that do not have access to HPLC equipment.

Chromatography Analysis

HPLC. Dry residues are dissolved in 200 μl of the mobile phase consisting of 0.5% (v/v) glacial acetic acid in water/acetonitrile (75:25, v/v). CHZ and its metabolite, 6-OH CHZ, are analyzed by HPLC on a Nucleosil C18 (Machery-Nagel, Germany) column (250 × 4.6 mm, 5 μm). Twenty microliters of sample is injected onto the column. The mobile phase flows at a rate of 1 ml/min for 20 min before being modified (0.5% acetic acid in water/acetonitrile 25:75, v/v) for 10 min. The column is then reequilibrated with the initial solvent mixture (10 min) before injection of the next sample. Detection is performed at 287 nm, an intermediary wavelength between the absorption maximum of 6-OH CHZ (295 nm) and that of CHZ (280 nm). The order of elution is 6-hydroxychlorzoxazone, internal standard, and chlorzoxazone, and the retention times are approximately 7.25, 15.9, and 23.7 min, respectively, when phenacetin is used as the internal standard (see Fig. 4).

The amount of 6-OH CHZ is determined from the peak area ratio of the metabolite and the internal standard and compared to standard curves generated with known amounts of product added to water. However, the enzymatic rate of 6-OH CHZ formation can also be calculated from the percentage of the metabolite area to the total product area (metabolite + parent drug) provided that the linearity of detection is respected at high levels of chlorzoxazone (400 μM in incubates). This method requires that the substrates and products are recovered with equal efficiency. Between 0.5 and 16% of CHZ is transformed to 6-OH CHZ in microsomes from control rats and from human liver samples. As shown in Fig. 3, CHZ is exclusively transformed to 6-OH CHZ. The lowest limit of quantitation is equivalent to 50 pmol/ml of incubation.

TLC. Organic dried extracts from incubation reactions (100 μl in methanol) are separated on 0.25-mm precoated Silica gel F-254 plates (Merck) and developed with an acetone/hexane (45:55; v/v) solvent mixture. After a 45-min migration, plates are dried and then developed on a X-ray film for 48 hr to localize the radiolabeled 6-OH CHZ and CHZ spots. The radioactive spots are scraped and their radioactivity counted in a liquid scintillation counter. Unlabeled CHZ and 6-OH CHZ are cochromatographed as controls and revealed under UV light. R_f of 6-OH CHZ and CHZ are 0.54 and 0.75, respectively.

The enzymatic rates of formation products are calculated based on the percentage of labeled metabolite to the total radioactivity. Experiments

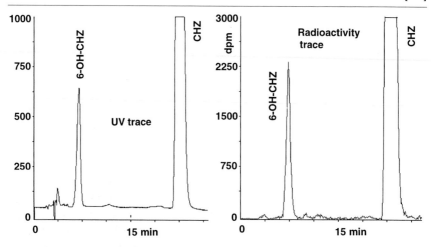

FIG. 3. HPLC profile of chlorzoxazone metabolism by a human liver microsomal sample. Detection was either by UV-287 nm or by an on-line radioactivity counter.

with labeled CHZ have shown that the efficiency of extraction is approximately 80%. Sensitivity of this method is similar to that of HPLC.

Results. The chlorzoxazone 6-hydroxylation activity presents a wide interindividual variation in human liver microsomes, i.e., between 0.25 and 9 nmol/min/mg protein. In rat liver microsomes, this value is 0.6–0.9 nmol/min/mg protein. Ethanol, acetone, pyridine, or pyrazole treatment increases this activity by 2-, 2-, 3-, and 3-fold vs control. Treatment by 3-methylcholanthrene (3-MC) also increases it by 2- to 3-fold vs control. Such an increase is due to the contribution of CYP1A1[5] which is greatly induced following this 3-MC treatment.

In Vivo Chlorzoxazone 6-hydroxylation Assay

Subjects

Parameters such as sex, age, and tobacco smoking; diseases such as cirrhosis, obesity, and diabetes; and ingestion of medications (especially chloramphenicol and isoniazid, CYP2E1 inducers, disulfiram and clomethiazole, CYP2E1 inhibitors) must be documented because of their possible influence on CYP2E1 activity. It should also be verified that subjects have no alcohol in their blood before chlorzoxazone administration because of competitive inhibition.

Drug Administration

Subjects are orally administered a 500-mg tablet of chlorzoxazone with a glass of water after a 12-hr fast. They must not drink or eat for 1 hr. A venous blood sample (10 ml) is withdrawn 2 hr later. Serums can be stored at $-20°$ for 2 months. Chlorzoxazone is a well-tolerated drug, which may occasionally produce undesirable side effects such as drowsiness, dizziness, and gastrointestinal disturbances.

Reagents

Chlorzoxazone: 500-mg tablets or pellets (generic drug commercially available in the United States)

Helix pomatia juice (IBF Biotechnics, Paris F)

Acetate buffer, 2 *M*, pH 4.5, perchloric acid 0.6 *M*, ethyl acetate (pro analysis, Merck)

Standard solutions: Stock solutions of CHZ (1 mg/ml) and 6-OH CHZ (1 mg/ml) are prepared in methanol and stored at 4°. Standard solutions are prepared from stock solution by appropriate dilution with HPLC mobile phase.

Procedure

Sample Preparation. Acetate buffer (2 *M*, 0.5 ml) and *H. pomatia* juice (20 µl) are added to 0.5 ml of serum (or plasma). Samples are hydrolyzed overnight at 37° to liberate 6-OH CHZ from their conjugates.[20] Proteins are then precipitated with 4 ml of 0.6 *M* perchloric acid. After centrifugation for 10 min at 3500 *g*, CHZ and 6-OH CHZ are twice extracted from the supernatant by shaking for 10 min with 4 ml of ethyl acetate. Following centrifugation for 10 min at 4°, the organic phases are evaporated to dryness under a stream of nitrogen at 37°. A method using solid–liquid extraction has also previously been described.[21,22]

HPLC. CHZ and 6-OH CHZ are analyzed by HPLC using the previously described method except that the mobile phase consists of 0.5% (v/v) acetic acid in water/acetonitrile (70 : 30, v/v). Under these conditions, 6-OH CHZ is eluted within 6.5 and CHZ within 17 min (Fig. 4). Peak area measurements are used for quantitation and are compared with standard solutions (0.5–20 µg/ml) of CHZ and 6-OH CHZ. Recovery from serum was 85% for 6-OH CHZ and 70% for CHZ. The lowest quantifiable concen-

[20] D. Lucas, F. Berthou, C. Girre, F. Poitrenaud, and J.-F. Ménez, *J. Chromatogr.* **622,** 79 (1993).
[21] H. Zang and J. T. Stewart, *Anal. Lett.* **26,** 675 (1993).
[22] D. Stiff, R. Frye, and R. Branch, *J. Chromatogr.* **613,** 127 (1993).

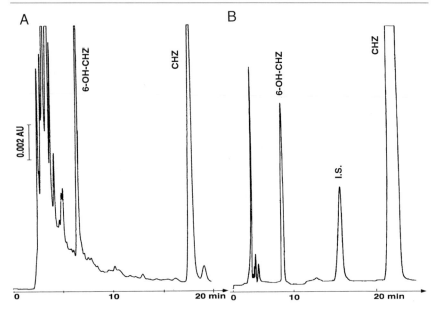

FIG. 4. Typical chromatograms of *in vivo* (A) and *in vitro* (B) 6-hydroxylation of chlorzoxazone. (A) Plasma sample 2 hr after drug intake. (B) Human hepatic microsomal sample. HPLC mobile phases as described in *in vivo* (A) and in *in vitro* (B) assays. IS, internal standard (phenacetin).

tration of these compounds in biological samples is estimated to be 0.5 μg/ml.

Results. After oral administration, CHZ is rapidly absorbed and completely metabolized in humans to 6-OH CHZ which is excreted in urine as a glucuronide conjugate. Pharmacokinetic studies showed that CHZ and 6-OH CHZ reach a C_{max} peak approximately 2 hr after administration of the drug.[9] Chlorzoxazone clearance is enhanced in alcoholics,[9] after administration of isoniazid,[10] a CYP2E1 inducer, and in obese patients.[8] In contrast, it decreased in fasted patients[8] or after administration of disulfiram,[23] a known inhibitor of CYP2E1. It does not differ between smoker and nonsmoker male controls.[9]

The blood chlorzoxazone and 6-OH chlorzoxazone concentrations were 65.7 ± 7.7 and 12.4 ± 1.1 μM[24] (mean $\pm$ SEM; $n = 32$), respectively, in control subjects after a drug intake of 500 mg. Mean values ($\pm$SEM) for

[23] E. Karash, K. Thummel, J. Mhyre, and J. Lillibridge, *Clin. Pharmacol. Ther.* **53,** 643 (1993).
[24] F. Berthou, T. Goasduff, D. Lucas, Y. Dréano, M. H. Le Bot, and J. F. Ménez, *Pharmacogenetics* **5,** 72 (1995).

the 2-h plasma chlorzoxazone metabolic ratio (CHZMR), consisting of the metabolite/drug parent ratio, was 0.34 ± 0.03[25] with a wide interindividual variation ranging from 0.05 to 0.60 in control subjects. Through an extensive study based on pharmacokinetic data, Girre et al.[9] demonstrated that the 2-hr plasma CHZMR ratio reflects the CYP2E1-mediated 6-hydroxylation. Indeed, a significant relationship was shown between this single point clearance determination and the 6-OH CHZ/CHZ AUC ratio as well as the metabolic formation clearance ($r = 0.88$; $n = 31$ subjects). This metabolic ratio is increased by 3- to 5-fold following chronic ethanol consumption and decreases rapidly during withdrawal,[25] with a half-life of CYP2E1 estimated to be 2.5 days. This CHZMR may be used as a simple and noninvasive marker for phenotyping human CYP2E1.[26]

Conclusion

In vitro studies comparing the abilities of several purified recombinant human P450 enzymes have demonstrated that CYP2E1 is the major catalyst of chlorzoxazone 6-hydroxylation in human liver. Data obtained from in vivo studies in alcoholic patients, before or after ethanol withdrawal, or in patients treated with either CYP2E1 inducers (isoniazid) or inhibitors (disulfiram) are in good agreement with data obtained from in vitro liver microsomal preparations. Therefore, chlorzoxazone should be a suitable in vitro and in vivo probe for human liver CYP2E1.

Acknowledgment

This study was supportedby the Programme Hospitalier de Recherche Clinique du CHU de Brest, France (P.H.R.C. 94, to JFM).

[25] D. Lucas, C. Ménez, C. Girre, P. Bodénez, E. Hispard, and J.-F. Ménez, Alc. Clin. Exp. Res. 19, 362 (1995).
[26] D. Lucas, C. Ménez, C. Girre, F. Berthou, P. Bodénez, I., Joannet, E. Hispard, L. G. Bardou, and J. F. Ménez, Pharmacogenetics 5, 298 (1995).

[13] Assays for CYP1A2 by Testing *in Vivo* Metabolism of Caffeine in Humans

By Bing-Kou Tang and Werner Kalow

Introduction

Caffeine (1,3,7-trimethylxanthine, 137×) undergoes *N*-demethylation, ring hydroxylation, and acetylation to give di- and monomethylxanthines, urates, and an acetylated uracil in humans (Fig. 1). The major enzyme involved in the biotransformation of caffeine and its metabolites is CYP1A2, a cytochrome P450.[1,2] Since the contribution of CYP1A2 to systemic caffeine clearance is estimated to be more than 95% of the total, clearance is the "gold standard" for an index of CYP1A2 activity for most subjects.[3] Other enzymes with contributions to caffeine metabolism are CYP2E1 (the ethanol-inducible cytochrome P450), the polymorphic *N*-acetyltransferase (NAT2), xanthine oxidase (XO), and, to a minor extent, CYP3A4 and CYP2A6. For subjects with induced CYP2E1 activity (e.g., workers in solvent factories, subjects with ethanol dependency), its contribution to systemic caffeine clearance may be more than 5%.

The urinary caffeine metabolite ratios established in this laboratory as indexes for the activity of CYP1A2, *N*-acetyltransferase, and xanthine oxidase are summarized as follows[3–5]:

In blood: CYP1A2 index = systemic caffeine clearance = Cl_{CAF}
In urine: CYP1A2 index = CMR = (AFMU + 1X + 1U)/17U
  N-Acetyltransferase index = NAT2 = AFMU/(AFMU + 1X + 1U)
  Xanthine oxidase index = XO = 1U/(1X + 1U)

A caffeine test for CYP2E1 is still being evaluated based on the shifting of the metabolite profile of caffeine 3-demethylation to the 1- and 7-demethylation pathways by CYP2E1.[2] Thus the caffeine test promises to become a single test for monitoring multiple liver enzymes identified in carcinogen activation, i.e., CYP1A2, CYP2E1, NAT2, and XO. The caf-

[1] M. A. Butler, M. Iwasaki, F. P. Guengerich, and F. F. Kadlubar, *Biochemistry* **86,** 7696 (1989).
[2] L. Gu, F. J. Gonzalez, W. Kalow, and B. K. Tang, *Pharmacogenetics* **2,** 73 (1992).
[3] W. Kalow and B. K. Tang, *Clin. Pharmacol. Ther.* **53,** 503 (1993).
[4] M. E. Campbell, S. P. Spielbery, and W. Kalow, *Clin. Pharmacol. Ther.* **42,** 157 (1987).
[5] B. K. Tang, D. Kadar, L. Qian, J. Iriah, J. Yip, and W. Kalow, *Clin. Pharmacol. Ther.* **49,** 648 (1991).

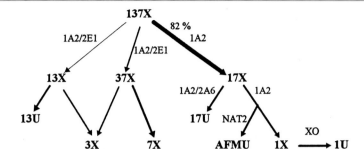

FIG. 1. Proposed caffeine metabolic pathways in humans and the major enzymes involved. 137X, caffeine = 1,3,6-trimethylxanthine; 13X, theophylline = 1,3-dimethylxanthine; 13U, 1,3-dimethylurate; 37X, theobromine = 3,7-dimethylxanthine; 17X, paraxanthine = 1,7-di-methylxanthine; 17U, 1,7-dimethylurate; 3X, 3-methylxanthine; 7X, 7-methylxanthine; AFMU, 5-acetylamino-6-formylamine-3-methyluracil; 1X, 1-methylxanthine; 1U, 1-methylurate; 1A2, CYP1A2; 2E1, CYP2E1; 2A6, CYP2A6; NAT2, the polymorphic *N*-acetyltransferase; XO, xanthine oxidase.

feine-based assays have been reviewed.[3] This chapter describes one blood and one urine test in detail that have served as indexes of CYP1A2 activity in clinical and epidemiological studies.

Blood Test for CYP1A2 Activity

Gu[6] has compared standard AUC-based measurements of systemic caffeine clearance with a two-point method measuring caffeine concentration in blood and he found a high correlation ($r = 0.98$). Thus, systemic caffeine clearance (CL_{CAF}) can be estimated by a two-point method after a single oral dose of caffeine.

$$CL_{CAF} = K_{el} \times V_d$$
$$K_{el} = -2.303 \times slope$$

$$Slope = \frac{\log C_2 - \log C_1}{t_2 - t_1}$$

$$V_d = \frac{f \times dose}{C_0 \times BW}$$

$$\log C_0 = \frac{\log C_2 + \log C_1}{2} - slope \times \frac{t_2 + t_1}{2},$$

[6] L. Gu, M.S.c. Thesis, p. 76. University of Toronto (1992).

where K_{el} is the elimination rate constant; V_d is the apparent volume of distribution; C_0, C_1, and C_2 are blood or plasma caffeine concentration at time zero, at time 1 (t_1), and at time 2 (t_2), respectively; f is the bioavailability fraction of the oral dose, which is assumed to be unity; and BW is body weight in kilograms.

A. Experimental Protocol

After abstinence from caffeine- and xanthine-containing foods or drinks (including coffee, tea, cola, chocolate products) for 2 days, the subject is asked to consume a cup of coffee containing 100 mg of caffeine (1 to 2 mg/kg body weight) in the morning. At 4 and 8 hr after coffee intake, 1 ml of blood is taken from the antecubical vein and stored in a heparinized vial (7 units/ml) at $-20°$ until analysis. Samples stored at $-20°$ are stable for at least a year. At time zero, a saliva sample (0.2 ml) may be taken for a check of compliance.

B. Analytical Determination of Caffeine Concentration in Blood, Plasma, and Saliva

To 0.1 ml of sample in a glass tube (13 × 100 mm), add 0.04 ml of 0.05% acetic acid, 0.025 ml of the internal standard solution [7-(β-hydroxypropyl)theophylline, 5 μg/ml], and 3 ml of dichloromethane : isopropanol (90 : 10, v/v). The mixture is vortexed for 30 sec and centrifuged at 2500 rpm for 5 min. The organic phase is collected and dried under a gentle stream of nitrogen at 37° for 20 to 30 min. The residue is dissolved in 0.25 ml of the HPLC mobile-phase solvent (3.0% isopropanol, 0.5% acetonitrile, 0.05% acetic acid) and 0.05 ml of the solution is injected onto an Ultrasphere ODS column (5 μm, 25 cm × 4.6 mm, Beckman Instruments, Fullerton, CA). The caffeine and metabolites are eluted at 1 ml/min and are detected by ultraviolet absorbance (AUF = 0.02) at 280 nm.[7]

All samples should be warmed to room temperature and mixed well before sampling. Saliva samples, if foaming, should be centrifuged. The standard samples were made by mixing pure chemical (concentration ranged from 1 to 40 μM) with blank plasma. A 3% solution of bovine serum albumin could be used instead of blank plasma if it is free of interfering substances.

Under these conditions, the retention times for caffeine and the internal standard [7-(β-hydroxypropyl)theophylline] were 25 and 29 min, respectively. The retention times for 17X, 37X, and 13X were 7, 11, and 13 min, respectively. The mean recovery of caffeine was over 95% at 10 μM with

[7] B. K. Tang, Y. Zhou, D. Kadar, and W. Kalow, *Pharmacogenetics* **4,** 117 (1994).

TABLE I
SYSTEMIC CAFFEINE CLEARANCE, MEAN (SD) AS CYP1A2 INDEX IN 39 HEALTHY SUBJECTS

Subject	N	CL_{CAF} (ml/min/kg)	$T_{1/2}$ (hr)	V_d (liter/kg)
Non-OC[a] users	31	1.16 (0.49)	5.76 (2.46)	0.55 (0.24)
OC users	8	0.71 (0.20)[b]	7.26 (1.61)	0.43 (0.12)
All subjects	39	1.07 (0.48)	6.20 (2.09)	0.52 (0.22)

[a] Oral contraceptives.
[b] $P < 0.05$.

a CV of about 6%. The recovery of caffeine was reduced by drying the organic phase at a temperature higher than 40° and/or for more than 30 min.

C. Results of a Population Study

The frequency distribution of systemic caffeine clearance was log normal.[7] The mean (SD) in arithmetic terms was 1.07 (0.48) ml/min/kg for 39 healthy subjects who were not cigarette smokers (Table I). The variability of the apparent volume of distribution (CV = 46%) was similar in magnitude to that of other kinetic parameters. Oral contraceptive users showed significantly decreased caffeine clearance.

Urine Test for CYP1A2 Activity

The urinary caffeine metabolism ratio (CMR) reflecting 7-demethylation of 17X, which is catalyzed by CYP1A2, showed a high correlation with systemic caffeine clearance (Cl_{CAF}) ($r = 0.92$).[4] The high correlation has been confirmed with $r = 0.77$ to 0.87.[7,8] The advantage of the urine test is that it involves collection of a single urine sample without strict control of caffeine intake.

There are some disadvantages in using CMR as an index for CYP1A2 activity. First, 17U, the denominator of CMR, was chosen because its urinary excretion is not flow dependent and because it showed only a limited person-to-person variation. However, CMR could be subject to error in persons or populations with CYP2A6 variability since this cytochrome is now known to contribute to 17U formation (Fig. 1). An association of CMR and NAT2 ratios may arise because the numerator in the CMR is the same as the denominator in the NAT2 index. The CMR values for fast and slow acetylators were reevaluated using published data from one of

[8] K. L. Rost and I. Roots, *Clin. Pharmacol. Ther.* **55,** 402 (1994).

our previous studies.[9] The mean CMR was slightly higher ($P < 0.05$) in Caucasian subjects who were fast acetylators than in slow acetylators (Table II). This trend was similar in Oriental subjects. A peculiarity of caffeine and its dimethylxanthine metabolites (e.g., theophylline) is their flow-dependent urinary excretion, a factor probably related to the diuretic effect. The CMR as the CYP1A2 index is not urine flow dependent, but some published caffeine-based ratios for CYP1A2 activity are [e.g., (17X + 17U)/137X by Butler et al.[10]]. To different degrees, 17X and 137X are urine-flow-dependent components so that the ratio may be distorted by renal factors.[7] Furthermore, the magnitude of the ratio is sensitive to the time of urine collection versus the time of caffeine intake. Therefore, this and some similar ratios are not dealt with in this chapter.

A. Experimental Protocol

After abstinence from chocolate (caffeine is allowed) and acetaminophen for 2 days and during the day of study, the subject is asked to consume a cup of coffee containing caffeine (60 to 90 mg) in the afternoon. No more than four cups should be consumed on the day of study. About 1 to 10 ml of the overnight urine (the first morning urine) is collected in a sample vial containing ascorbic acid (10 mg/ml) and is kept frozen at $-20°$ until analysis. Alternatively, the cup of coffee may be consumed in the morning and one sample of 8-hr-pooled urine will be collected.

It is necessary to abstain from chocolate and acetaminophen because chocolate may inhibit CYP1A2 activity and acetaminophen may interfere with the analytical assay. The reliability of CMR as an index for CYP1A2 declines as the collection periods decrease to below 8 hr.

B. HPLC Analysis of Urinary Caffeine Metabolites

Two separate assays are used routinely in our laboratory for the determination of CMR: one for xanthines and urates and one for AFMU. Prior to the assay, the labile AFMU is converted to the more stable AAMU (5-acetylamino-6-amino-3-methyluracil). This conversion has led to the most reliable results.[11] Measuring AFMU without this conversion will lead to an underestimate of CMR values, as is clearly illustrated in Table II.

Since the conversion is pH dependent, spontaneous transformation is difficult to avoid, and may even take place sometimes in the bladder urine prior to voiding.

[9] W. Kalow and B. K. Tang, Clin. Pharmacol. Ther. **50,** 508 (1991).
[10] M. A. Butler, N. P. Lang, J. F. Young, N. E. Caporaso, P. Vineis, R. B. Hayes, C. H. Teitel, J. P. Massengill, M. F. Lawsen, and F. F. Kadlubar, Pharmacogenetics **2,** 116 (1992).
[11] B. K. Tang, T. Zubovits, and W. Kalow, J. Chromatogr. **375,** 170 (1986).

TABLE II
BIOLOGICAL FACTORS THAT MAY AFFECT CMR AS CYP1A2 INDEX

Factor	Aspect	N	CMR mean (SEM)	Ref.
Gender	Male	30	4.8 (0.3)[b]	4
	Female	30	4.7 (0.3)[b]	4
	Male	106	6.36 (0.24)	9
	Female	54	5.85 (0.31)	9
Ethnic	White adults	42	4.7 (0.4)[b]	4
	Oriental adults	26	4.6 (0.4)[b]	4
	White adults	130	6.2 (0.2)	9
	Oriental adults	21	5.8 (0.4)	9
Age	3–11, average 7 years, white	21	7.9 (0.4)[b]	4
	6–11, average 8.9 years, white	10	8.86 (1.06)	16
	Adults 18–30	45	4.4 (0.2)[b]	4
	Adults 30–65	15	4.7 (0.4)[b]	4
	Adults, average 57 years,	94	5.9 (0.3)	17
	Adults, average 26 years,	178	5.96 (0.18)	9
	Adults, average 56 years,	125	7.3 (0.4)	17
OC user		9	3.6 (0.5)[b]	4
		18	3.90 (0.32)	9
		8	4.23 (0.79)	7
Pregnancy, epileptic		7	3.33 (0.3)	
Epileptic, nonpregnant female		7	8.08 (0.9)	18
Schistosomiasis and other factors				
12–16, average 13.5 years, black		45	3.78 (0.4)	16
Polycyclic aromatic hydrocarbons				
Smokers: 19/day		26	9.4 (0.7)[b]	4
Smokers: 10–24/day		8	8.52 (1.04)	9
Smokers: 44 pack/year		31	11.6 (1.0)	17
PBB exposed (35 ppb)		43	6.96 (0.32)[b]	19
TCDD (140 pg/g)		45	6.8 (0.7)	17
TCDD (217 pg/g) and smoker		13	10.9 (0.9)	17
Acetylation status				
Caucasian				
Slow		79	5.82 (0.26)	c
Fast		51	6.78 (0.38)	c
Oriental				c
Slow		5	5.26 (0.72)	c
Fast		16	5.93 (0.54)	c

[a] PBB, polybrominated biphenyls; TCDD, 2,3,7,8-tetrachlorodibenzo-p-dioxin; OC, oral contraceptives.
[b] CMR estimated by measurement of AFMU without convertion to AAUM.
[c] CMR values recalculated using published data from Kalow and Tang.[9]

Determination of 1X, 1U, and 17U Concentrations. This assay is similar to that of caffeine in blood described earlier. To 0.1 ml of urine in a glass tube (13 × 100 mm), 0.025 ml of the internal standard solution (1.2 mg of *N*-acetyl-*p*-aminophenol dissolved in 10 ml of 50% isopropanol water) and 3 ml of dichloromethane/isopropanol (88 : 12, v/v) are added. The mixture is vortexed for 30 sec. The organic phase is then separated and dried under a gentle stream of nitrogen at 37° for 20 to 30 min. The residue is dissolved in 0.25 ml of the HPLC mobile-phase solvent (1.3% isopropanol, 0.1% acetonitrile, 0.05% acetic acid) and 0.05 ml of the solution is injected onto an Ultrasphere IP ODS column (5 μm, 25 cm × 4.6 mm, Beckman Instruments). The caffeine and metabolites are eluted at 1 ml/min and detected by ultraviolet absorbance (AUF = 0.02) at 280 nm. Under these conditions, the retention times of 1U, 1X, and 17U and the internal standard (*N*-acetyl-*p*-aminophenol) were 8.5, 11.5, 29, and 14.5 min, respectively.

For the preparation of standard samples, 1U (1 mg) is first dissolved in about 10 ml of water of pH 9 and the pH is obtained by adding a drop of sodium hydroxide (10 *N*), which is then neutralized to pH 7 by 12 *N* HCl. This solution is mixed with other xanthines and urates and is diluted to 100 ml with blank urine (pH 3.5). The final concentrations of the metabolites are 5 to 10 mg/liter. Several 10-ml aliquots of this standard mixture can be stored at −20° until use. Precautions and pitfalls of the assay have been described.[5]

Determination of AFMU by Conversion to AAMU. To convert AFMU to AAMU, 0.05 ml of sodium hydroxide (0.25 *N*) is added to 0.05 ml of urine in a glass tube (13 × 100 mm). The pH should be at least 10. After reaction for 10 to 20 min, 0.05 ml of 0.25 *N* HCl is added to neutralize the excess sodium hydroxide. Then 0.1 ml of the internal standard solution (10 mg of benzyloxyurea dissolved in 10 ml of water) is added. An aliquot of 0.02 ml of the mixture is injected onto the TSK-20 column (G200PW, Toso Haas, Philadelphia, PA; 10-μm particle size, 300 × 7.5 mm i.d.). It is eluted with 0.1% acetic acid at a flow of 0.8 ml/min and is monitored by UV absorbance at 263 nm.[11] Under these conditions, the retention times of AAMU and the internal standard (benzyoxylurea) were 15 and 29 min, respectively.

For continuous analysis, the TSK column should be washed with 20% methanol in 10% acetic acid at a flow of 0.8 ml/min for 1 hr after 20 to 30 determinations and then conditioned for 1.5 hr before use.[5]

The use of one chromatographic system for all four metabolites has been reported.[12] The reliability of this assay needs to be defined because

---

[12] P. Dobrocky, P. N. Bennett, and L. J. Notarianni, *J. Chromatogr.* **652,** 104 (1994).

the AAMU peak overlapped by over 30% with that of an endogenous substance in the chromatogram from one urine sample of one volunteer.

Other methods using capillary electrophoresis (CE) have been published for AFMU, 1X, and 1U.[13–15] By a combination of HPLC and CE techniques, a complete baseline separation of AFMU, 1X, 1U, and 17U was achieved.[15]

C. Factors That May Affect CMR as an Index of CYP1A2 Activity

Table II[16–19] summarizes the results from 10 years of our laboratory work to detect biological factors which may have an influence on CMR as a CYP1A2 index. Significant differences were found between males and females, whites and Orientals, and of female epileptics. CYP1A2 activity was lowered by the use of oral contraceptives and was reduced in pregnancy. It was induced by cigarette smoking. Children between 3 and 11 years of age had CMR values as high as some cigarette smokers. Factory workers exposed to TCDD 20 to 30 years prior to test but with still high concentrations of TCDD in their tissues (140–217 pg/g) showed CMR values not significantly different from controls; in the smokers among these subjects, the CMR was elevated. Farmers tested not too long after exposure to PBB in Michigan showed significantly higher CMR values than controls.

[13] J. Caslavska, E. Hufschmid, R. Theurillat, C. Desiderio, H. Wolfisberg, and W. Thormann, *J. Chromatogr.* **656,** 219 (1994).

[14] R. Guo and W. Thormann, *Electrophoresis* **14,** 547 (1993).

[15] N. Rodopoulos and A. Norman, *Scand. J. Clin. Lab. Invest.* **54,** 305 (1994).

[16] C. M. Misimidrembwa, M. Beke, J. A. Hasler, B. K. Tang, and W. Kalow, *Clin. Pharmacol. Ther.* **57,** 25 (1995).

[17] W. Halperin, W. Kalow, M. H. Sweeney, B. K. Tang, M. Fingerhut, B. Timpkins, and K. Willie, *Occup. Environ. Med.* **52,** 86 (1995).

[18] M. Bologa, B. Tang, J. Klein, A. Tesoro, and G. Koren, *J. Pharmacol. Exp. Ther.* **257,** 735 (1991).

[19] G. H. Lambert, D. A. Schoeller, H. E. B. Humphrey, A. N. Kotake, H. Lietz, M. Campbell, W. Kalow, S. P. Spielberg, and M. Budd, *Environ. Health Perspect.* **89,** 175 (1990).

[14] Assays of Omeprazole Metabolism as a Substrate Probe for Human CYP Isoforms

By Donald J. Birkett, Tommy Andersson, and John O. Miners

Introduction

Omeprazole, a substituted benzimidazole inhibitor of the gastric proton pump H^+,K^+-ATPase, is essentially completely metabolized *in vivo*. The major metabolites detected in blood *in vivo* are omeprazole sulfone and hydroxyomeprazole.[1] The major urinary metabolites identified are hydroxyomeprazole and its corresponding carboxylic acid with neither omeprazole itself nor omeprazole sulfone being detected.[1,2] It is assumed that hydroxyomeprazole is converted to the carboxylic acid by cytosolic alcohol and aldehyde dehydrogenases, and that omeprazole sulfone is essentially completely further metabolized. Figure 1 shows the structure of omeprazole and its potential metabolites.

The plasma clearance of omeprazole is about 600 ml min^{-1} and the elimination half-life is very short (0.5–1 hr).[1] About half of an oral dose of omeprazole is systemically available due to first pass extraction by the liver. *In vivo,* the metabolism of omeprazole to hydroxyomeprazole, but not to omeprazole sulfone, cosegregates with the 4-hydroxylation of S-mephenytoin, indicating a role for CYP2C19 in this metabolic pathway.[3-5] *In vitro* studies of the metabolism of omeprazole in our laboratory have identified the major metabolic pathways and identified several metabolites not detected during *in vivo* studies. The CYP isoforms responsible for the various primary metabolic pathways have been identified[6,7] and the

---

[1] C. G. Regårdh, T. Andersson, P. O. Lagerström, P. Lundborg, and I. Skånberg, *Ther. Drug Monit.* **12,** 163 (1990).

[2] L. Renberg, R. Simonsson, and K. J. Hoffman, *Drug Metab. Dispos.* **17,** 69 (1989).

[3] T. Andersson, C. G. Regårdh, M. L. Dahl-Puustinen, and L. Bertilsson, *Ther. Drug Monit.* **12,** 415 (1990).

[4] T. Andersson, C. G. Regårdh, Y.-C. Lou, Y. Zhang, M.-L. Dahl-Puustinen, and L. Bertilsson, *Pharmacogenetics* **2,** 25 (1992).

[5] D. R. Sohn, K. Kobayashi, K. Chiba, K. H. Lee, S. G. Shin, and T. Ishizaki, *J. Pharmacol. Exp. Ther.* **262,** 1195 (1992).

[6] T. Andersson, P.-O. Lagerström, J. O. Miners, M. E. Veronese, L. Weidolf, and D. J. Birkett, *J. Chromatogr.* **619,** 291 (1993).

[7] T. Andersson, J. O. Miners, M. E. Veronese, Wichittra Tassaneeyakul, Wongwiwat Tassaneeyakul, U. A. Meyer, and D. J. Birkett, *Br. J. Clin. Pharmacol.* **36,** 521 (1993).

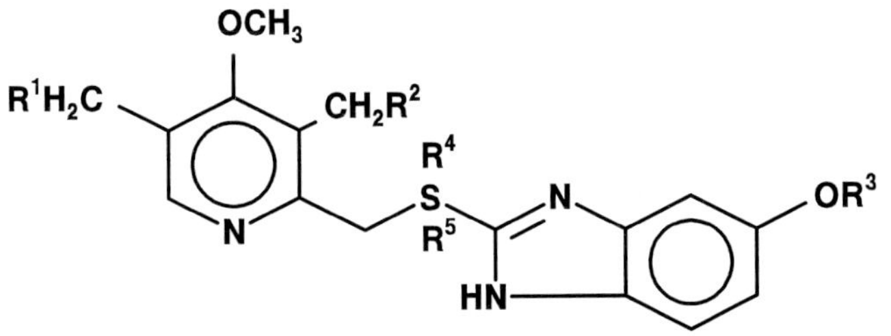

Compound	R^1	R^2	R^3	R^4	R^5
Omeprazole	H	H	CH_3	$=O$	-
Omeprazole sulfone	H	H	CH_3	$=O$	$=O$
Hydroxyomeprazole	OH	H	CH_3	$=O$	-
3-Hydroxyomeprazole	H	OH	CH_3	$=O$	-
5-O-desmethylomeprazole	H	H	H	$=O$	-
Hydroxyomeprazole sulfone	OH	H	CH_3	$=O$	$=O$
Omeprazole sulfide	H	H	CH_3	-	-

FIG. 1. The structure of omeprazole and some potential metabolites.

secondary metabolism of the two major primary metabolites similarly has been characterized.[8]

HPLC Equipment and Conditions

In this laboratory we have used a Model LC1100 solvent delivery system, a Model LC1200 variable wavelength UV-Vis detector set at 302 nm (ICI Instruments, Melbourne, Australia), a Model 7125 Rheodyne injector, and a Model SE120 BBC Goetz Metrawatt recorder (Brown-Boveri, Vienna, Austria). A Supersphere SI-60 4 μM particle size column (125 × 4 mm i.d.; E. Merck, Darmstadt, Germany) was used with an Aquapore silica 7 μM particle size guard column (15 × 3 mm i.d.; Brownlee Laboratories, CA). The mobile phase is dichloromethane : 5% NH_4OH in methanol : 2-propanol (191 : 8 : 1) run at a flow rate of 1.5 ml min^{-1}.

[8] T. Andersson, J. O. Miners, M. E. Veronese, and D. J. Birkett, *Br. J. Clin. Pharmacol.* **37,** 597 (1994).

Standards

Authentic standards for omeprazole (5-methoxy-2[[(4-methoxy-3,5-dimethyl-2-pyridinyl)methyl]sulfinyl]-1H-benzimidazole) and potential metabolites (Fig. 1) were obtained from Astra Hässle AB (Mölndal, Sweden). The internal standard (4,6-dimethyl-2-[[(4-methoxy-2-pyridinyl)methyl]sulfinyl]-1H-benzimidazole) was also obtained from Astra Hässle AB.

Incubations

Reaction mixtures contain human liver microsomes (1 mg microsomal protein), the NADPH-generating system (1 mM NADP$^+$, 10 mM glucose-6-phosphate, 2 IU glucose-6-phosphate dehydrogenase and 5 mM MgCl$_2$), and omeprazole (2.5–500 μM) in a final volume of 1 ml of 0.1 M KH$_2$PO$_4$ buffer, pH 7.4. Omeprazole is dissolved in 0.1 M KH$_2$PO$_4$ buffer, pH 7.4:methanol (10:1) and is used freshly prepared. Final concentrations of methanol of 1% or less do not inhibit the reaction. The reaction is initiated by the addition of the NADPH-generating system and is terminated after 15 min by the addition of 2 ml dichloromethane:butanol (99:1) and cooling on ice. The NADPH-generating system is omitted from blank incubations.

For studies of the further metabolism of the primary metabolites, hydroxyomeprazole and omeprazole sulfone, the same procedure is used except that the incubation time is 30 min rather than 15 min. Hydroxyomeprazole and omeprazole sulfone are dissolved in 0.1 M KH$_2$PO$_4$ buffer, pH 7.4:methanol (10:1) and added at final concentrations of 0.5–200 μM.

Sample Processing

After stopping the reaction with the addition of 2 ml dichloromethane:butanol (99:1), 100 μl of 1 M NaH$_2$PO$_4$ is added followed by the addition of internal standard (20–150 μl) dissolved in 0.1 M KH$_2$PO$_4$ buffer, pH 7.4:methanol (10:1) at a concentration of 32 μM. Extraction is performed on a vortex mixer for 1 min followed by centrifugation for 5 min at 1000 g. A 1.5-ml aliquot of the organic phase is evaporated to dryness under nitrogen, the residue is reconstituted in 150 μl of the mobile phase, and a 50-μl aliquot is injected onto the HPLC column. Sample processing is similar whether omeprazole or its primary metabolites, hydroxyomeprazole or omeprazole sulfone, are used as substrates.

Metabolite Identification

With omeprazole as the substrate, elution times of omeprazole and identified metabolites are 2 min for omeprazole sulfone, 2.5 min for omeprazole, 3.3 min for internal standard, 3.8 min for 3-hydroxyomeprazole, 11 min for 5-O-desmethylomeprazole, and 13.5 min for hydroxyomeprazole. Because no authentic standard for 3-hydroxyomeprazole is available, this initially unknown metabolite was identified by mass spectrometry.[8] A small peak corresponding to omeprazole sulfide elutes at 1.3 min, but is also seen in the blank incubations and is not therefore a metabolic product.

Hydroxyomeprazole sulfone eluting at 9 min is the major microsomal metabolite when either hydroxyomeprazole or omeprazole sulfone is used as the substrate to study the further metabolism of these primary metabolites. With omeprazole sulfone as the substrate, a further minor metabolite eluting at 4 min was identified by mass spectrometry as the pyridine-N-oxide sulfone.[8]

Assay Calibration and Validation

Omeprazole sulfone, hydroxyomeprazole, hydroxyomeprazole sulfone, and internal standard are dissolved in carbonate buffer (25 ml 0.5 M Na_2CO_3 and 65 ml of 1.0 M $NaHCO_3$ added to 910 ml H_2O): methanol (4:1) and aliquots are kept frozen at $-20°$. Standard curves are constructed in the range of 0.3 to 7 μM. Quantitation is by the peak height ratio with internal standard, and standard curves are linear over the stated range. Unknown concentrations are determined by comparison of the metabolite to the internal standard peak height ratio with those of the calibration curve.

Extraction efficiency is close to 100% for omeprazole sulfone, hydroxyomeprazole, and internal standard at concentrations in the range of 2.5–3.0 μM. Coefficients of variation for the overall assay within-day precision are in the range of 2.1–6.9% at omeprazole substrate concentrations of 5 and 500 μM. The formation of primary metabolites is linear with microsomal protein in the range of 0.25–1.5 mg and with time to a 30-min incubation time. The formation of the secondary metabolite hydroxyomeprazole sulfone is linear to a 60-min incubation time when either hydroxyomeprazole or omeprazole sulfone is used as substrate.

Assay Kinetics

Kinetic parameters for formation of primary metabolites from omeprazole are shown in Table I. In all cases, the Eadie–Hofstee plots are nonlinear, indicating the involvement of multiple CYP isoforms. Only the high-affinity activities are likely to be relevant to the *in vivo* situation. *In vitro* intrinsic clearances (V_{max}/K_m) for the high-affinity activities are about four times greater for the formation of hydroxyomeprazole than for the formation of omeprazole sulfone, consistent with the former being the major *in*

TABLE I

KINETIC PARAMETERS FOR FORMATION OF OMEPRAZOLE PRIMARY METABOLITES[a]

	Omeprazole sulfone	Hydroxy-omeprazole	5-O-Desmethyl-omeprazole	3-Hydroxy-omeprazole
K_{m1}	49 ± 15	8.6 ± 5.6	13.6 ± 13.3	52 ± 16
V_{max1}	0.19 ± 0.12	0.1 ± 0.07	—	—
K_{m2}	484 ± 190	175 ± 122	139 ± 56	321 ± 112
V_{max2}	0.4 ± 0.04	0.29 ± 0.07	—	—

[a] Values are mean (n = 4 livers) ± SD. Units for K_m are μM; for V_{max} nmol mg^{-1} min^{-1}. Based on data reported in Andersson et al.[7]

vivo metabolite.[1] High-affinity K_ms fall into two groups (about 10 μM for hydroxyomeprazole and 5-O-desmethylomeprazole and about 50 μM for omeprazole sulfone and 3-hydroxyomeprazole), indicating that different CYP isoforms are involved in the formation of these pairs of primary metabolites. Because authentic standards for the two minor metabolites were not available at the time the studies were done, V_{max} values could not be determined.

Kinetic parameters for the formation of the secondary metabolite hydroxyomeprazole sulfone are shown in Table II. The in vitro intrinsic clearances for the formation of hydroxyomeprazole sulfone are 0.8 μl min^{-1} mg^{-1} with hydroxyomeprazole as the substrate and 9.2 μl min^{-1} mg^{-1} with omeprazole sulfone as the substrate. The high in vitro intrinsic clearance for omeprazole sulfone is consistent with the extensive further metabolism of this compound in vivo.

TABLE II

KINETIC PARAMETERS FOR FORMATION
OF OMEPRAZOLE SECONDARY METABOLITES
BY HUMAN LIVER MICROSOMES[a]

	Hydroxyomeprazole sulfone formation from	
	Hydroxyomeprazole	Omeprazole sulfone
K_{m1}	131	7.6
V_{max1}	0.1	0.07
K_{m2}	—	427
V_{max2}	—	0.22

[a] Values are mean of two livers. Units are as in Table I. Based on data reported in Andersson et al.[8]

CYP Isoforms Involved in Omeprazole Metabolic Pathways

The CYP isoforms responsible for the formation of the various metabolites have been investigated by correlations with isoform selective substrates and by inhibition by isoform selective inhibitors.[7,8] Figure 2 shows the CYP isoform profiles for the formation of the various metabolites based on these studies.

Omeprazole is an effective substrate for both CYP2C19 and CYP3A4, but different metabolites are formed by the two isoforms (Fig. 2). Interestingly, hydroxyomeprazole and omeprazole sulfone show similar CYP specificity to the parent drug. Sulfone formation (whether from omeprazole or hydroxyomeprazole) is mediated by CYP3A4, and the hydroxy formation (whether from omeprazole or omeprazole sulfone) is mediated mainly by CYP2C19.

Comment

Omeprazole has a relatively complex metabolic fate, much of which has only been sorted out by application of the *in vitro* techniques described earlier. The overall metabolic pathways for omeprazole, based on *in vitro* and *in vivo* studies, are shown in Fig. 3. The microsomal assay techniques are robust and sensitive but do not detect nonmicrosomal metabolic pathways such as the conversion of hydroxyomeprazole to the carboxy compound by cytosolic dehydrogenases. *In vitro* data are consistent with *in vivo* data in indicating that the formation of hydroxyomeprazole but not omeprazole sulfone is compromised in CYP2C19-deficient subjects. As this

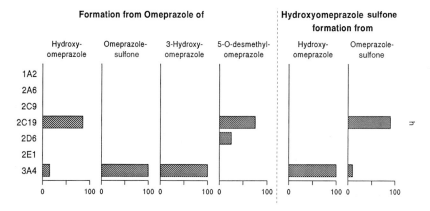

FIG. 2. CYP isoform profiles for omeprazole primary and secondary metabolism. Based on studies reported in Andersson *et al.*[7,8]

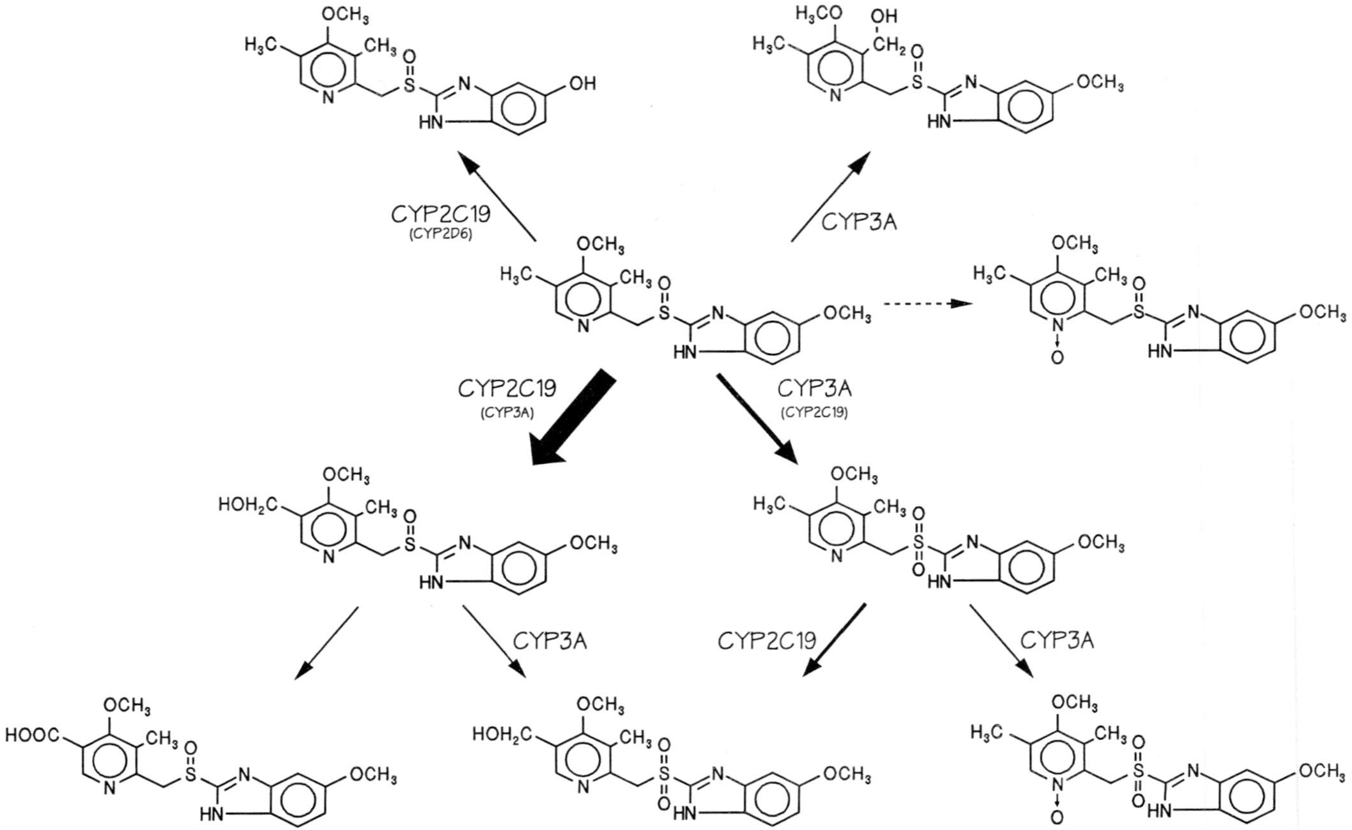

Fig. 3. Metabolic pathways for omeprazole based on *in vitro* and *in vivo* studies. The sizes of the arrows indicate the relative clearances along individual pathways. The CYP isoforms carrying out the reactions are also indicated. Adapted from Fig. 1 in Andersson *et al.*[8]

is quantitatively the major metabolic route, CYP2C19 deficiency results in a substantial decrease in the overall omeprazole clearance *in vivo*.[3–5]

Omeprazole forms a convenient *in vitro* substrate probe to assess simultaneously the activity of two CYP isoforms, CYP2C19 and CYP3A4. The formation of omeprazole sulfone seems to be exclusively carried out by CYP3A isoforms whereas hydroxyomeprazole formation is largely mediated by CYP2C19.

[15] Use of Tolbutamide as a Substrate Probe for Human Hepatic Cytochrome P450 2C9

By JOHN O. MINERS and DONALD J. BIRKETT

Introduction

Xenobiotic metabolizing isoforms of cytochrome P450 (CYP) typically exhibit distinct, but sometimes overlapping, patterns of substrate and inhibitor specificity and differ in terms of regulation. Given these characteristics, the identification of isoform-specific substrate probes has been essential for the investigation of factors which may influence the metabolism of any drug or nondrug xenobiotic in humans. Additionally, the availability of such compounds has been an important component of strategies developed to link the metabolism of newly developed drugs to an individual CYP isoform(s).[1]

In humans the elimination of the oral hypoglycemic agent tolbutamide (1-butyl-3-*p*-tolylsulfonylurea) occurs along a single pathway (Fig. 1), with the initial and rate-limiting step being methylhydroxylation to form hydroxytolbutamide.[2,3] Hydroxytolbutamide is oxidized further *in vivo* by alcohol and aldehyde dehydrogenases producing carboxytolbutamide. Overall, this pathway of metabolism accounts for up to 85% of tolbutamide clearance in humans.[3,4]

Since tolbutamide metabolism involves a single pathway, it represents a potentially useful model drug for investigating CYP activity *in vitro* and

[1] J. O. Miners, M. E. Veronese, and D. J. Birkett, *Annu. Rep. Med. Chem.* **29**, 307 (1994).
[2] E. Nelson and I. O'Reilly, *J. Pharmacol. Exp. Ther.* **132**, 103 (1961).
[3] R. C. Thomas and G. J. Ikeda, *J. Med. Chem.* **9**, 507 (1966).
[4] M. E. Veronese, J. O. Miners, D. J. Randles, and D. J. Birkett, *Clin. Pharmacol. Ther.* **47**, 403 (1990).

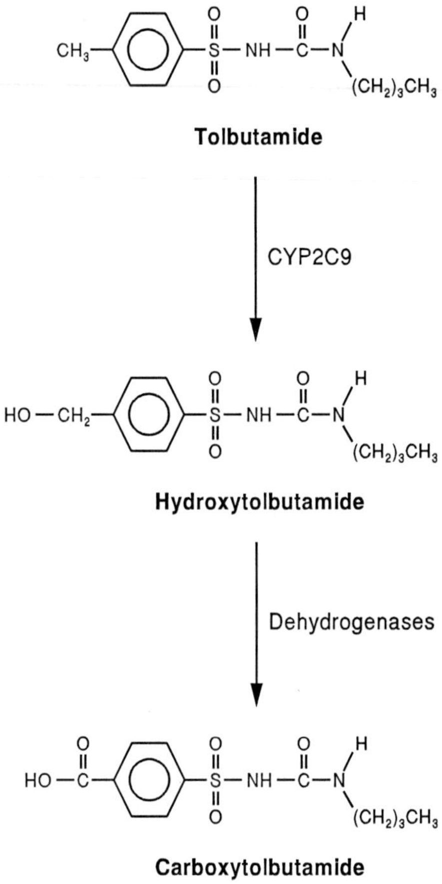

FIG. 1. The metabolic pathway for tolbutamide in humans.

in vivo. This chapter describes an established[5] high-performance liquid chromatographic (HPLC) assay for the measurement of hydroxytolbutamide formation, the rate-limiting step in tolbutamide metabolism, by human liver microsomes and CYP isoforms expressed in cell culture. The assay has been used to demonstrate that CYP2C9 is essentially solely responsible for the human hepatic hydroxylation of tolbutamide.

[5] J. O. Miners, K. J. Smith, R. A. Robson, M. E. McManus, M. E. Veronese, and D. J. Birkett, *Biochem. Pharmacol.* **37**, 1137 (1988).

Materials

Tolbutamide and sulfaphenazole were supplied by Hoechst Aust (Melbourne, Australia) and Ciba-Geigy Aust (Sydney, Australia), respectively. Chlorpropamide was purchased from the Sigma Chemical Co. (St. Louis, MO). Sources of hydroxytolbutamide include Hoechst and Ultrafine Chemicals (Manchester, UK). Aqueous stock solutions of tolbutamide (20 mM) and hydroxytolbutamide (0.1 mM) were prepared weekly by dropwise addition of NaOH (1 M) to stirred suspensions of the compounds. Aliquots of the solutions (pH approximately 9.5) were stored frozen at −20°. Working solutions of chlorpropamide, the assay internal standard, were similarly prepared weekly by dissolving chlorpropamide in 2% methanolic water to give a final concentration of 10 mg/liter.

Human liver microsomes were prepared by differential centrifugation.[6] A CYP2C9 cDNA was isolated as described previously,[7] subcloned into the expression vector pCMV5, and transfected into COS-7 cells. Cells were harvested 48 hr posttransfection and resuspended in 0.1 M phosphate buffer (pH 7.4) containing 20% (v/v) glycerol.

Methods

Incubation Conditions

The following method is employed routinely in this laboratory. Incubations are performed in borosilicate glass tubes (5 ml; 1 cm i.d.) and contain human liver microsomal (0.3 mg) or COS cell lysate (0.5 mg) protein, a NADPH-generating system (1 mM NADP, 10 mM glucose 6-phosphate, 2 IU glucose-6-phosphate dehydrogenase, and 5 mM MgCl$_2$), tolbutamide (25–2,000 μM), and phosphate buffer (0.1 M, pH 7.4; 0.5 ml) in a total volume of 1 ml. Reactions are initiated by the addition of the NADPH-generating system and are performed in air at 37° (shaking water bath) for 1.5 hr. The generating system is omitted from blanks and replaced with an equal volume of buffer. Incubations are terminated by the addition of 0.15 M phosphoric acid (1 ml) and cooling of the incubation tubes on ice. The resulting mixture is centrifuged (3000 g for 5 min), and a 1.8-ml aliquot of the supernatant fraction is transferred to a 15-ml screw-capped glass culture tube. The aqueous solution is extracted (vortex mixer for 2 min) with hexane–chloroform–isoamylalcohol (1000:250:5, 8 ml) to remove

[6] R. A. Robson, A. P. Matthews, J. O. Miners, M. E. McManus, U. A. Meyer, P. D. Hall, and D. J. Birkett, *Br. J. Clin. Pharmacol.* **24,** 293 (1987).
[7] M. E. Veronese, P. I. Mackenzie, C. J. Doecke, M. E. McManus, J. O. Miners, and D. J. Birkett, *Biochem. Biophys. Res. Commun.* **175,** 1112 (1991).

most of the unreacted tolbutamide. Tubes are then placed in a dry ice–acetone bath and the organic phase is decanted and discarded. Chlorpropamide (0.1 ml of a 10-mg/liter solution) is added to the thawed aqueous fraction which is reextracted with diethyl ether (8 ml). Phases are separated by centrifugation (3000 g for 5 min) and tubes are again placed in a dry ice–acetone bath. The ether extract is transferred to a clean conical-tip 15-ml glass tube and evaporated to dryness under a stream of N_2. Residues are reconstituted in 0.12 ml of the HPLC mobile phase, and 0.05-ml aliquots are injected onto the column.

Chromatography

The HPLC system routinely used in this laboratory comprises a Model U6K injector, M6000 solvent delivery system, Model 481 UV-VIS detector (all Waters-Millipore, Milford, MA), and a Model SE120 BBC Goetz Metrawatt chart recorder (Brown-Boveri, Vienna, Austria). The chromatograph is fitted with a Nova Pak C_{18} column (15 cm × 3.9 mm i.d., 4-μm particle size; Waters Millipore, Sydney, Australia) and operates at room temperature (18–25°). The mobile phase is acetate buffer (10 μM, pH 4.4)–acetonitrile (78:22), delivered at a flow rate of 2.0 ml/min. Peaks are monitored by ultraviolet detection at 230 nm, the absorbance maximum for hydroxytolbutamide in the mobile phase. Standard curves are constructed in the hydroxytolbutamide concentration range 0.05–5.0 μM, with standards being treated in the same manner as incubation samples. Unknown concentrations of hydroxytolbutamide are determined by comparison of hydroxytolbutamide:chlorpropamide peak height ratios with those of the standard curve.

Under the chromatographic conditions employed, retention times for hydroxytolbutamide and chlorpropamide are 3.3 and 11.3 min, respectively (Fig. 2). The chromatography of both compounds is free from interference by endogenous microsomal compounds and a range of CYP isoform-specific probes and their metabolites (see following section).

Assay Validation

Using this procedure, the recoveries of both hydroxytolbutamide and chlorpropamide from incubation mixtures are highly reproducible and almost quantitative. The mean (±SD) recovery of hydroxytolbutamide, calculated by comparing the peak height of the extracted compound with that of an equal amount injected directly onto the chromatograph, was shown to be 87.3 ± 3.7% for seven samples over the concentration range 0.05–5 μM. Chlorpropamide recovery at the single concentration used was

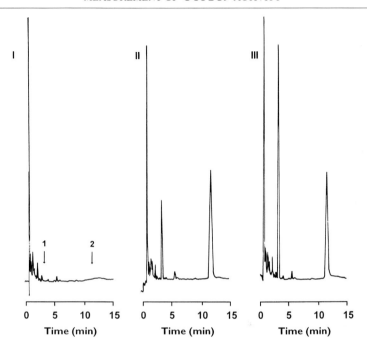

FIG. 2. Representative chromatograms of extracts of microsomal incubations. (I) Microsomal incubation performed in the absence of a NADPH-generating system and without the addition of an internal standard. (II) Standard containing 1 μM hydroxytolbutamide. (III) Incubation containing 1 mM tolbutamide. I and II show retention times of hydroxytolbutamide and chlorpropamide, respectively. Adapted from Miners et al.[5] with permission.

96.4 ± 1.1% (n = four replicates). Standard curves are linear (r^2 typically ≥0.995) and pass through the origin.

The rate of formation of hydroxytolbutamide is linear with incubation time to 3 hr and with microsomal protein concentration to at least 1.6 mg/ml for low (0.1 mM), intermediate (0.25 mM), and high (1.0 mM) substrate concentrations. Overall assay within-day precision has been determined by measuring hydroxytolbutamide formation in 12 separate incubations of the same batch of microsomes. Coefficients of variation for hydroxytolbutamide formation were 7.1, 5.0, and 3.5% at substrate concentrations of 0.1, 0.25, and 1.0 mM, respectively. Incubations performed in the presence of the CYP isoform-specific probes[1] coumarin (2A6), diethyldithiocarbamate (2E1), furafylline (1A2), S-mephenytoin (2C19), quinidine (2D6), sulfaphenazole (2C9), and troleandomycin (3A4), but without substrate (i.e., tolbutamide), indicated that none of these compounds or their metabolites interfere with the chromatography of hydroxytolbutamide or chlorpropamide.

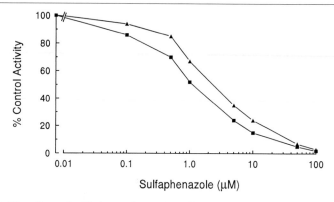

FIG. 3. The effect of sulfaphenazole on the tolbutamide hydroxylase activity of human liver microsomes (▲) and COS-7 cells transfected with a CYP2C9 cDNA (■). The tolbutamide concentration was 1 mM. Adapted from Veronese et al.[7] with permission.

The presence of dimethyl sulfoxide (1%, v/v), which is frequently necessary for the solubilization of inhibitors, has a negligible effect on the tolbutamide hydroxylase activity of human liver microsomes or cDNA-transfected COS cells.

Validation of Tolbutamide as a CYP2C9-Specific Substrate

Evidence supporting the use of tolbutamide hydroxylation as a marker for human CYP2C9 activity *in vitro* is overwhelming. Human liver microsomal tolbutamide hydroxylation exhibits Michaelis–Menten kinetics, and apparent K_m values for this reaction have been reported to range from 62 to 176 μM,[5,8,9] with a mean of 120 μM. cDNA-expressed CYP2C9 and a number of its variants catalyze the hydroxylation of tolbutamide and, importantly, apparent K_m values for the recombinant enzymes (viz. 72–132 μM) fall within the range observed with human liver microsomes.[7] Sulfaphenazole, a potent and specific inhibitor of CYP2C9, abolishes the conversion of tolbutamide to hydroxytolbutamide (Fig. 3).[5,7–9] The apparent K_i for sulfaphenazole inhibition of tolbutamide hydroxylation is approximately 0.1 μM,[5] a value typical of sulfaphenazole inhibition of human liver microsomal CYP2C9-catalyzed reactions. The tolbutamide hydroxylase activity of human liver microsomes is also inhibited competitively by other CYP2C9

---

[8] D. J. Back, J. F. Tjia, J. Karbwang, and J. Colbert, *Br. J. Clin. Pharmacol.* **26,** 23 (1988).
[9] C. J. Doecke, M. E. Veronese, S. M. Pond, J. O. Miners, D. J. Birkett, L. N. Sansom, and M. E. McManus, *Br. J. Clin. Pharmacol.* **31,** 125 (1991).

substrates, with closely matching apparent K_i and K_m values.[9,10] In contrast to the effects of sulfaphenazole and alternate CYP2C9 substrates, inhibitors of other human CYP isoforms (e.g., coumarin, diethyldithiocarbamate, furafylline, S-mephenytoin, quinidine, and troleandomycin) have little or no effect on microsomal tolbutamide hydroxylation.

As *in vitro*, there is good evidence supporting the involvement of CYP2C9 in tolbutamide hydroxylation *in vivo*. Pretreatment of human subjects with sulfaphenazole decreases the plasma clearance of tolbutamide by 80% and prolongs elimination half-life 5.3-fold.[4] Procedures for the measurement of the plasma unbound clearance of tolbutamide and the urinary tolbutamide metabolic ratio have been described elsewhere.[4] Both have been applied successfully for the measurement of tolbutamide hydroxylase (i.e., CYP2C9) activity *in vivo*.[11]

Acknowledgment

This work was supported by grants from the National Health and Medical Research Council of Australia.

[10] J. O. Miners, D. L. P. Rees, L. Valente, M. E. Veronese, and D. J. Birkett, *J. Pharmacol. Exp. Ther.* **272,** 1076 (1995).
[11] M. E. Veronese, J. O. Miners, D. L. P. Rees, and D. J. Birkett, *Pharmacogenetics* **3,** 86 (1993).

[16] Assays of CYP2C8- and CYP3A4-Mediated Metabolism of Taxol *in Vivo* and *in Vitro*

By Thomas Walle

Introduction

Taxol is one of the newer anticancer drugs, originally isolated from the Pacific yew tree, with a broad spectrum of clinical activity against solid tumors as well as acute leukemias.[1] The drug is administered by slow intravenous infusion and is cleared from the body mainly by metabolism and biliary/fecal elimination.[2,3] The principal human metabolite by far is

[1] E. K. Rowinsky and R. C. Donehower, *Semin. Oncol.* **20,** Suppl. 3, 16 (1993).
[2] T. Walle, U. K. Walle, G. N. Kumar, and K. N. Bhalla, *Drug Metab. Dispos.* **23,** 506 (1995).
[3] B. Monsarrat, P. Alvinerie, M. Wright, J. Dubois, F. Guéritte-Voegelein, D. Guénard, R. C. Donehower, and E. K. Rowinsky, *J. Natl. Cancer Inst. Monogr.* **15,** 39 (1993).

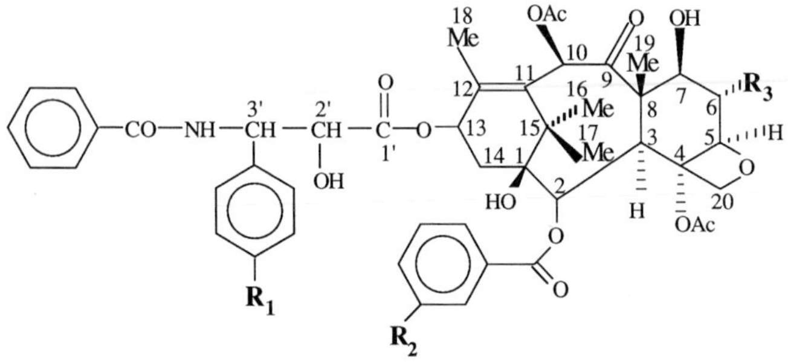

Compound	R_1	R_2	R_3
Taxol	H	H	H
6HOT	H	H	OH
3'HOT or RM1	OH	H	H
RM2	H	OH	H
DHOT	OH	H	OH

FIG. 1. The structure of taxol and its main human metabolites. Also shown are the structures of the main rat metabolites RM1 and RM2.

6α-hydroxytaxol (6HOT), identified both *in vitro,* using liver microsomes,[4] and *in vivo* in bile[3,5] and feces[2] (Fig. 1). Other human metabolic pathways include oxidation of one of the phenyl rings to 3'-(*p*-hydroxyphenyl)taxol (3'HOT), as well as oxidation in both the 6α- and 3'-*p*-phenyl positions to a dihydroxylated taxol (DHOT, Fig. 1).[2,3,6]

Because of the potential therapeutic importance of taxol metabolism, a number of investigations have been focusing on the characterization of the cytochrome P450 isoform(s) involved in its metabolism.[7–10] As taxol metabolism in animal species such as the rat is quite different from that

[4] G. N. Kumar, J. E. Oatis, Jr., K. R. Thornburg, F. J. Heldrich, E. S. Hazard, III, and T. Walle, *Drug Metab. Dispos.* **22,** 177 (1994).

[5] J. W. Harris, A. Katki, L. W. Anderson, G. N. Chmurny, J. V. Paukstelis, and J. M. Collins, *J. Med. Chem.* **37,** 706 (1994).

[6] B. Monsarrat, E. Mariel, S. Cros, M. Garès, D. Guénard, F. Guéritte-Voegelein, and M. Wright, *Drug Metab. Dispos.* **18,** 895 (1990).

[7] T. Cresteil, B. Monsarrat, P. Alvinerie, J. M. Tréluyer, I. Vieira, and M. Wright, *Cancer Res.* **54,** 386 (1994).

[8] G. N. Kumar, U. K. Walle, and T. Walle, *J. Pharmacol. Exp. Ther.* **268,** 1160 (1994).

[9] J. W. Harris, A. Rahman, B.-R. Kim, F. P. Guengerich, and J. M. Collins, *Cancer Res.* **54,** 4026 (1994).

[10] A. Rahman, K. R. Korzekwa, J. Grogan, F. J. Gonzalez, and J. W. Harris, *Cancer Res.* **54,** 5543 (1994).

of humans,[6,11] these efforts have been centered on the human enzymes. Using antibodies and, in particular, recombinant enzymes, it was recently concluded that the formation of the main metabolite, 6HOT, is catalyzed by CYP2C8[10] and one of the minor metabolites, 3′HOT, by CYP3A4,[7,9] whereas the DHOT formation is dependent on both isoforms. Thus, *in vivo* and *in vitro* measurements of taxol metabolism in humans should be a reflection of the activities of these two isoforms. This chapter describes the analytical methods used to accomplish this.

In Vivo Methodology

About 5% of the administered taxol dose is excreted in urine mainly as unchanged drug. The remainder of the dose appears as metabolites in feces[2] after biliary excretion.[3] Thus, useful measurements of taxol metabolites are limited to plasma and stool samples.

Analysis of Plasma Samples

We have used solid-phase extraction of patient plasma samples, followed by HPLC analysis with UV detection to determine unchanged taxol and its metabolites in the extract.

C_{18} Sep-Pak Vac cartridges (3 ml, Waters, Milford, MA) are conditioned with 3 ml each of methanol and deionized water. Plasma (2 ml) is applied to the cartridge, which is then washed with 3 ml each of water and 25% (v/v) methanol. After each addition, the cartridge is taken to dryness using a vacuum manifold (Waters), allowing the simultaneous processing of 12 samples. Taxol and metabolites are eluted with 3 ml of acetonitrile, which is taken to dryness under a stream of nitrogen in a 40° water bath. The samples are reconstituted in 500 μl mobile phase (35% acetonitrile in water). Four hundred microliters is subjected to HPLC analysis.

For the separation of the taxol metabolites, we use a Curosil-G, 6-μm particle size, 250 × 3.2 mm i.d., column (Phenomenex, Torrance, CA) and a Bondapak C_{18}/Corasil, 37- to 50-μm, guard column (20 × 4 mm i.d., Waters) with 35% acetonitrile at 0.6 ml/min as the mobile phase. Detection is by UV at 229 nm, which is the absorption maximum for both taxol and metabolites. A chromatogram from a plasma sample collected at 5 hr postinfusion of a 225-mg/m^2 taxol dose is shown in Fig. 2 (top). It shows all three main metabolites.

---

[11] T. Walle, G. N. Kumar, J. M. McMillan, K. R. Thornburg, and U. K. Walle, *Biochem. Pharmacol.* **46,** 1661 (1993).

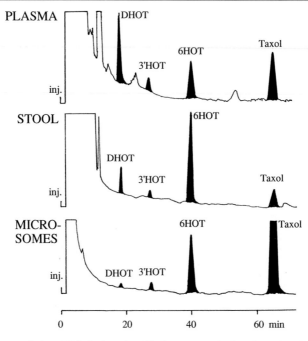

FIG. 2. Reversed-phase HPLC of taxol and its human metabolites (see Fig. 1 for structures) in extracts of biological samples. Detection was made by UV at 229 nm. (Top tracing) Patient plasma sample 5 hr after the end of a 3-hr infusion of 225 mg/m² of taxol. (Center tracing) Stool sample from the same patient 24 hr after the dose. (Bottom tracing) Human liver microsomal incubate with 10 μM taxol.

Analysis of Fecal Samples

For quantification of taxol metabolites in feces, weighed stool samples are homogenized in a blender with enough water to make a thick slurry. Aliquots (about 2 g) of the fecal homogenate (usually 12–36 hr after the taxol dose) are freeze-dried and the dry residue is extracted with 3 × 10 ml of methanol. The combined methanol extracts are evaporated to dryness, reconstituted in 0.5 ml of methanol and 1.5 ml of HPLC mobile phase (35% acetonitrile), and filtered through a hydrophilic Durapore (Millipore, Bedford, MA) syringe filter. Aliquots (100 μl) are analyzed using the same HPLC system and wavelength (229 nm) as for plasma samples. A chromatogram of a stool sample collected at 24 hr postinfusion of a taxol dose of 225 mg/m² is shown in Fig. 2 (center). 6HOT is by far the major taxol metabolite in feces.[2]

Comments

Quantitation was based on peak area measurements as compared to samples spiked with taxol. The three major metabolites have the same UV spectra. High recoveries and precision for all metabolites were confirmed by liquid scintillation spectrometry in a protocol using radioactive doses of taxol.[2] Good reproducibility was obtained without the use of internal standard, although this has been the practice in other studies.[9,12,13] As in other studies,[13,14] solid-phase extraction was preferred over solvent extraction for plasma samples because of more complete recovery of the polar DHOT metabolite and less interference from the biological matrix. The addition of acetonitrile to plasma and direct HPLC of the supernatant after centrifugation[9] gives high recoveries but more interferences. The most critical step for stool samples is thorough methanol extraction. Because of the high concentration of taxol and metabolites, there is very little interference. To decrease the retention time and thereby increase the speed of analysis, the acetonitrile content in the mobile phase can be increased, e.g., an increase from 35 to 40% decreases the retention time of taxol by one-half. Although no interferences in the assay by concomitantly used drugs have been noted so far, this may occur when combination therapies with taxol become more common. To ensure the identity of measured HPLC peaks, it is then advisable to use photodiode array detection. As taxol and its metabolites have virtually identical UV spectra, interferences by other drugs can easily be detected.

In Vitro Methodology

The metabolism of taxol appears to depend entirely on P450 metabolism in the liver in humans.[7-10] For the hepatic microsomal metabolism to its major metabolite 6HOT, K_m values of 4–20 μM and V_{max} values of 15–880 pmol min^{-1} mg microsomal protein^{-1} have been found.[7-10]

Materials

Normal human liver tissue is obtained from the Liver Tissue Procurement and Distribution System (University of Minnesota, Minneapolis, MN)

[12] L. Gianni, C. M. Kearns, A. Giani, G. Capri, L. Viganó, A. Locatelli, G. Bonadonna, and M. J. Egorin, *J. Clin. Oncol.* **13,** 180 (1995).
[13] D. S. Sonnichsen, C. A. Hurwitz, C. B. Pratt, J. J. Shuster, and M. V. Relling, *J. Clin. Oncol.* **12,** 532 (1994).
[14] M. T. Huizing, A. C. F. Keung, H. Rosing, W. van der Kuij, W. W. ten Bakkel Huinink, I. M. Mandjes, A. C. Dubbelman, H. M. Pinedo, and J. H. Beijnen, *J. Clin. Oncol.* **11,** 2127 (1993).

or other sources.[7,9,10] Flash-frozen 10-g pieces wrapped in aluminum foil have been stored at $-80°$ for several years without significant loss of P450 activity. Taxol (paclitaxel) can be obtained from Calbiochem (La Jolla, CA) or Sigma (St. Louis, MO), with a high purity compound (>99%) available from Calbiochem. Tritium-labeled taxol (10–30 Ci/mmol) in ethanol, with the majority of the tritium in the m- and p-positions of the aromatic rings, is available from Moravek Biochemicals (Brea, CA).

Incubations

Human liver microsomes are prepared by differential centrifugation of 20% (w/v) liver homogenates in isotonic KCl (1.15%), resuspended in 100 mM potassium phosphate buffer (pH 7.4) containing 1 mM EDTA and 20% glycerol, and stored at $-80°$ in small aliquots for up to 2 months. Typically, incubates contain 1 mg microsomal protein/1-ml incubate in 50 mM HEPES buffer (pH 7.4), with 2 mM $MgCl_2$, 0.4 mM NADP, 4 mM glucose 6-phosphate, and 0.2 units of glucose-6-phosphate dehydrogenase. The reaction is started by adding taxol stock solution in dimethyl sulfoxide (final concentration 10 μM) with [^{3}H]taxol (0.5 μCi/incubate). The final concentration of dimethyl sulfoxide should be kept at 0.3% or less. Incubations are carried out at 37° in a shaking water bath for 20 min. For studies of inhibition of taxol metabolism, inhibitors can be added as described.[7–10]

Assays

Determination of taxol metabolites formed during the incubation is done by applying the incubate to a C_{18} Sep-Pak Vac cartridge, as for plasma samples described earlier. Alternatively, we have extracted the incubate with 5 volumes of ethyl acetate, transferred the organic layer to a clean tube, and evaporated the solvent to dryness under a stream of nitrogen before reconstitution in mobile phase and HPLC analysis. For detection we use either UV at 229 nm as for plasma and fecal extracts described previously or radiometric detection with [^{3}H]taxol as the substrate. The latter uses either on-line detection with a Flo-One monitor (Packard Instrument Co.) or fraction collection and liquid scintillation spectrometry. A typical chromatogram from one human liver preparation, using 229-nm detection, is shown in Fig. 2 (bottom). 6HOT is the main product, followed by 3′HOT. DHOT is barely detectable.

Comments

The same considerations apply as for plasma samples. The use of a radiolabeled substrate gives higher sensitivity, particularly for 3′HOT and

DHOT, and facilitates recovery determinations. It also permits inhibition studies, where inhibitors can interfere with the UV absorbance in the taxol metabolism assay.[7] If 6HOT is the only metabolite measured, retention times of 10 and 15 min can be obtained for 6HOT and taxol, respectively, by increasing the acetonitrile content of the mobile phase to 46%.

General Comments

Measurement of 6HOT vs 3'HOT in plasma may give a reasonable account of CYP2C8 vs CYP3A4 metabolism of taxol in the whole body. However, as previously indicated,[2] only a small portion of taxol metabolites formed in the liver may appear in plasma, with the main fraction promptly excreted directly from the liver into feces via the bile. Rapid further metabolism of systemically available 6HOT and 3'HOT to DHOT further complicates the interpretation of data from plasma measurements. Fecal elimination of the metabolites, although more difficult to deal with experimentally, should give a more clear account of the total body activity of CYP2C8 and CYP3A4.

Human liver microsomal metabolism should give minimal secondary metabolism, particularly if the reaction time is kept reasonably short, thus avoiding any further metabolism of 6HOT and 3'HOT, therefore well representing activities of CYP2C8[10] and CYP3A4,[7,9] respectively. Further evaluation, particularly of the CYP2C8 activity in the presence of other drugs, will be important. Quercetin and retinoic acid have already been shown to be potent competitive inhibitors of CYP2C8-mediated 6HOT formation.[10] Other inhibitors of 6HOT formation interestingly include drugs generally known to be substrates for CYP3A4, e.g., verapamil, testosterone, felodipine, nifedipine, α-naphthoflavone, ketoconazole, miconazole, and kaempferol.[8,9] Midazolam and 17α-ethinyl estradiol are potent competitive inhibitors of 6HOT formation.[8] Inhibitors of 3'-HOT formation include drugs that are typical inhibitors of CYP3A4.[9]

Acknowledgments

This work was supported by the National Cancer Institute (grant CA63386) and by the Hollings Cancer Center.

[17] Tamoxifen Metabolism by Microsomal Cytochrome P450 and Flavin-Containing Monooxygenase

By DAVID KUPFER and SHANGARA S. DEHAL

Introduction

The observations that tamoxifen (tam) is an antiestrogen prompted its development as a therapeutic agent for hormone responsive (estrogen receptor-*positive*) breast tumors.[1,2] Surprisingly and contrary to expectations, a considerable number of patients with estrogen receptor-*negative* tumors respond to tamoxifen therapy.[3] The exact mechanism of its anticancer activity and of the evolvement of tumor resistance in patients on long-term tamoxifen therapy is not understood. A large-scale clinical trial was recently initiated to determine the potential for the prophylactic use of tamoxifen in women considered at risk of breast cancer.[4,5] However, of considerable concern are the observations that tamoxifen treatment increases the incidence of endometrial and possibly liver cancer in humans and causes hepatocellular carcinoma in rats.[6–9]

These enigmas concerning the mechanism of the beneficial or potentially deleterious action of tamoxifen and of the development of tumor resistance suggested the possibility that tamoxifen metabolism may play a role in those activities and prompted intensive investigations in several laboratories on the hepatic metabolism of tamoxifen.

Tamoxifen is metabolized by liver microsomes from animals and humans into a variety of compounds, most notably into tamoxifen *N*-oxide (tam-*N*-oxide), *N*-desmethyl (*N*-desmethyl-tam), and 4-hydroxy (4-OH-tam)

[1] V. C. Jordan, *Breast Cancer Res. Treat.* **11,** 197 (1988).
[2] V. C. Jordan, *Br. J. Pharmacol.* **110,** 507 (1993).
[3] B. J. A. Furr and V. C. Jordan, *Pharmacol. Ther.* **25,** 127 (1984).
[4] B. Fisher and C. Redmond, *J. Natl. Cancer Inst.* **83,** 1278 (1991).
[5] M. D. Wold and V. C. Jordan, *in* "Antihormones in Health and Disease" (M. K. Agarwal, ed.), Vol. 19, p. 87. Karger, Basel, 1991.
[6] M. A. Killackey, T. B. Hakes, and V. K. Price, *Cancer Treat. Rep.* **69,** 237 (1985).
[7] T. Fornander, B. Cedermark, A. Matteson, L. Skoog, T. Theve, J. Askergren, L. E. Rutqvist, U. Glas, C. Silversward, A. Somell, N. Wilking, and M. J. Hjolmar, *Lancet* **1,** 117 (1989).
[8] K. C. Fendel and S. J. Zimniski, *Cancer Res.* **52,** 235 (1992).
[9] G. M. Williams, M. J. Iatropoulos, M. W. Djordjevic, and O. P. Kaltenberg, *Carcinogenesis (London)* **14,** 315 (1993).

derivatives (see Fig. 1).[10–14] Also, smaller amounts of 3,4-dihydroxy-tam, 3′,4′-dihydroxy-tam, α-hydroxy-tam, α-hydroxy-tam-N-oxide, α-hydroxy-N-desmethyl-tam, 4-hydroxy-tam-N-oxide, and tam-epoxide have been detected.[12,15–18] Hepatic cytochrome P450 (CYP)[19] enzymes were found to catalyze the 4-hydroxylation and N-demethylation[14,20] and the formation of 3,4-catechol via 3-hydroxylation of 4-OH-tam (S. Dehal and D. Kupfer, manuscript in preparation). The flavin-containing monooxygenase (FMO) was found to catalyze N-oxidation.[21] In addition to these routes of metabolism, tamoxifen undergoes metabolic activation that is catalyzed by hepatic CYP, forming a reactive intermediate that binds covalently to proteins[22] and DNA.[23–25] Microsomal CYP3A catalyzes covalent binding of tamoxifen to proteins[26] and tam-N-demethylation.[14,20] Since tamoxifen treatment induces CYP3A and CYP2B in rats,[27,28] induction of CYPs in humans may

[10] S. D. Lyman and V. C. Jordan, in "Metabolism of Non-steroidal Antiestrogens in Estrogen/Antiestrogen Action and Breast Cancer Therapy" (V. C. Jordan, ed.), p. 191. Univ. of Wisconsin Press, Madison, 1986.

[11] A. B. Foster, L. J. Griggs, M. Jarman, M. S. van Maanen, and H.-R. Schulten, *Biochem. Pharmacol.* **29,** 1977 (1980).

[12] P. C. Reunitz, J. R. Bagley, and C. W. Pape, *Drug Metab. Dispos.* **12,** 478 (1984).

[13] R. McCague and A. Seago, *Biochem. Pharmacol.* **35,** 827 (1986).

[14] C. Mani, H. V. Gelboin, S. S. Park, R. Pierce, A. Parkinson, and D. Kupfer, *Drug Metab. Dispos.* **21,** 645 (1993).

[15] C. K. Lim, Z.-X. Yuan, J. H. Lamb, I. N. H. White, F. De Matteis, and L. L. Smith, *Carcinogenesis (London)* **15,** 589 (1994).

[16] G. K. Poon, B. Walter, P. E. Lonning, M. N. Holton, and R. McCague, *Drug Metab. Dispos.* **23,** 377 (1995).

[17] D. H. Phillips, P. L. Carmichael, A. Hewer, K. J. Cole, and G. K. Poon, *Cancer Res.* **54,** 5518 (1994).

[18] G. K. Poon, Y. C. Chui, R. McCague, P. E. Lonning, R. Feng, M. G. Rowlands, and M. Jarman, *Drug Metab. Dispos.* **21,** 1119 (1993).

[19] D. R. Nelson, T. Kamataki, D. J. Waxman, F. P. Guengerich, R. W. Estabrook, R. Feyereisen, F. J. Gonzalez, M. J. Coon, I. C. Gunsalus, O. Gotoh, K. Okuda, and D. W. Nebert, *DNA Cell Biol.* **12,** 1 (1993).

[20] F. Jacolot, I. Simon, P. Y. Dreano, P. Beaune, C. Riche, and F. Berthou, *Biochem. Pharmacol.* **41,** 1911 (1991).

[21] C. Mani, E. Hodgson, and D. Kupfer, *Drug Metab. Dispos.* **21,** 657 (1993).

[22] C. Mani and D. Kupfer, *Cancer Res.* **51,** 6052 (1991).

[23] X.-L. Han and J. G. Liehr, *Cancer Res.* **52,** 1360 (1992).

[24] I. N. H. White, F. de Matteis, A. Davies, L. L. Smith, C. Crofton-Sleigh, S. Venitt, A. Hewer, and D. H. Phillips, *Carcinogenesis (London)* **13,** 2197 (1992).

[25] D. N. Pathak, K. Pongracz, and W. J. Bodell, *Carcinogensis (London)* **16,** 11 (1995).

[26] C. Mani, R. Pierce, A. Parkinson, and D. Kupfer, *Carcinogenesis (London)* **15,** 2715 (1994).

[27] I. N. H. White, A. Davis, L. L. Smith, S. Dawson, and F. de Matteis, *Biochem. Pharmacol.* **45,** 21 (1993).

[28] E. F. Nuwaysir, Y. P. Dragon, C. R. Jefcoate, V. C. Jordan, and H. C. Pitot, *Cancer Res.* **55,** 1780 (1995).

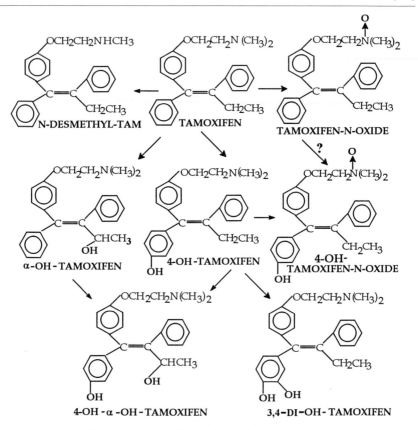

FIG. 1. Proposed pathway of tamoxifen metabolism.

have bearing for the failure of tamoxifen therapy and for its toxic side effects in humans. These aspects of tamoxifen metabolism, however, are beyond the scope of this chapter and hence will not be discussed.

This chapter describes the radiometric methods developed and used in our laboratory for detection, identification, and quantitation of several of the major tamoxifen metabolites. Also, the hepatic enzymes forming these tamoxifen metabolites are described. For analytical methods used in other laboratories, primarily involving HPLC, the reader is referred to published work.[20,29]

[29] C. K. Lim, Z.-X. Yuan, J. H. Lamb, I. N. H. White, F. de Matteis, and L. L. Smith, *Carcinogenesis (London)* **15,** 589 (1994).

Materials

NADPH, glucose 6-phosphate, glucose-6-phosphate dehydrogenase, pregnenolone-16α-carbonitrile (PCN), ethylenediaminetetraacetic acid disodium salt (EDTA), catechol-O-methyltransferase (COMT), and S-adenosyl-L-methionine iodide salt (SAM) were purchased from Sigma Chemical Co. (St. Louis, MO). ^{14}C ring-labeled tamoxifen citrate (21.1 mCi/mmol; currently available only through custom synthesis), N-[$methyl$-^{3}H]tamoxifen (85.6 Ci/mmol), and [$methyl$-^{3}H]-S-adenosyl-L-methionine (^{3}H-SAM, 15 Ci/mmol) were obtained from DuPont-NEN (Boston, MA). Phenobarbital sodium salt (PB) was from Mallinckrodt (St. Louis, MO). 3-Methylcholanthrene (3-MC) was obtained from Eastman Kodak (Rochester, NY). Corn oil (USP grade) was from Matheson, Coleman, and Bell (Cincinnati, OH). 2,3,7,8-Tetrachlorodibenzo-p-dioxin (TCDD) was obtained from NCI Chemical Carcinogen Repository (Kansas City, MO). Ultima-Gold, biodegradable scintillation fluid, was obtained from Packard (Downers Grove, IL). β-Naphthoflavone (βNF), sodium cyanoborohydride, and titanium(III) chloride were purchased from Aldrich Chemical Co. (Milwaukee, WI). Normal phase thin-layer chromatography (TLC) plates, containing fluorescent indicator and *pre*adsorbent strip, were purchased from Whatman, Inc. (Clifton, NJ). All other chemicals were of reagent grade quality and were used without further purification.

Methods

Animals and Treatment

Male and female Sprague–Dawley CD rats (90–100 g) were purchased from Charles River Breeding Laboratories (Wilmington, MA) and housed under controlled temperature (22°) and light (12-hr light/dark cycle; lights off at 7:00 PM EDT). PB treatment (37.5 mg/kg ip in 0.2 ml water, twice daily) is for 4 days. Liver microsomes (PB microsomes) are prepared 12 hr after the last dose. MC treatment (25 mg/kg ip in 0.4 ml of corn oil once daily) is for 3 days, and liver microsomes (MC microsomes) are prepared 48 hr after the last injection. PCN is injected (50 mg/kg ip as a suspension in 0.4 ml of corn oil daily for 3 days), and liver microsomes (PCN microsomes) are prepared 24 hr after the last dose. Control animals from each treatment group receive the same injection regimen of the respective vehicle only.

Preparation of Microsomes

Microsomes are prepared by homogenization of the livers in 0.25 M sucrose (5 ml/1 g liver) followed by differential centrifugation, as previously described,[30] and, unless stated otherwise, represent a pool of four to eight livers. The resulting microsomal pellet is resuspended in a 1.15% aqueous KCl solution with a Potter–Elvehjem Teflon plunger/glass homogenizer and centrifuged at 105,000 g for 1 hr. The supernatant is discarded and the microsomal pellet is covered with ~2 ml of KCl solution and stored at $-70°$ until use.

Incubations

Incubations are carried out as previously described.[22,26] After thawing, the pellet is suspended in fresh KCl solution using a Potter–Elvehjem homogenizer. An aliquot is used to determine the protein concentration by a modified procedure of Lowry et al.[31,32] Incubations are conducted in 20-ml glass scintillation vials (open to air atmosphere) containing the following constituents: 0.6 ml of sodium phosphate buffer (pH 7.4, 60 μmol) containing EDTA (0.1 μmol), 0.1 ml of MgCl$_2$ (10 μmol), 1 mg of microsomal protein suspension, [^{3}H]tamoxifen (250,000 dpm, 100 nmol) in 10 μl ethanol, a NADPH-regenerating system (glucose 6-phosphate, 10 μmol; NADPH 0.5 μmol; glucose-6-phosphate dehydrogenase, 2 IU) in 0.1 ml of sodium phosphate buffer (pH 7.4, 10 μmol), and water to a final volume of 1 ml. After preincubation at 37° for 2 min,[33] the reaction is initiated by the addition of the NADPH-regenerating system and incubated 37° for 60 min in a water bath shaker. The enzyme-catalyzed reaction is terminated by adding 10 ml of ethanol.

The aqueous ethanolic solution is filtered through a 2.4-cm Whatman GF/C glass microfiber filter (Whatman, Ltd. Maidstone, Kent, England) in a filter holder (Schleicher and Schuell, Inc., Keene, NH) attached to a vacuum filter flask. The trapped protein precipitate on the filter is washed with ethanol (20 ml) and methanol (10 ml) to elute the loosely bound tamoxifen metabolites, as previously described.[14,21] If determination of the covalently bound tamoxifen metabolite is desired, the protein remaining

---

[30] S. H. Burstein and D. Kupfer, Ann. N.Y. Acad. Sci. **191,** 61 (1971).
[31] O. H. Lowry, N. J. Rosebrough, A. L. Farr, and R. J. Randall, J. Biol. Chem. **193,** 265 (1951).
[32] C. E. Stauffer, Anal. Biochem. **69,** 646 (1975).
[33] Because of heat lability of FMO, preincubation in the absence of NADPH should be avoided when FMO catalysis is desired.

on the filter can be eluted with 3 ml of 2% aqueous sodium dodecyl sulfate and processed as previously described.[22,34,35]

Analysis of Tamoxifen Metabolites

The combined ethanol and methanol filtrate from the previously described step is evaporated to dryness under a stream of nitrogen at room temperature. The residue is taken up in 2 ml of ethanol, and the radioactivity of an aliquot (10 μl) is determined in duplicate in a Packard Tri-Carb 460CD liquid scintillation spectrometer using an automatic quench correction curve previously generated with a series of quenched ^{14}C and ^{3}H standards. Routinely, 10–20% of the ethanolic sample is used for chromatographic separation and quantification of metabolites on TLC, and the rest of the sample is stored at 0–4° under argon for the subsequent isolation and identification of the metabolites. Chromatographic separation is performed on Whatman silica gel TLC plates and is developed in $CHCl_3 : CH_3OH : NH_4OH$ (80 : 20 : 0.5; v/v/v), slightly modified from the chromatographic system described by Reunitz et al.[12] To localize and identify the metabolites, radioinert tamoxifen derivatives (see below for sources) are used as chromatographic standards and visualized under UV light. Alternatively, isolated radiolabeled metabolites are used as standards (for their preparations, see below). Radiolabeled metabolites on TLC are quantified (Figs. 2A and 2B) with a System 2000 imaging scanner (Bioscan, Inc., Washington, DC).

Source of Authentic Compounds Used as Chromatographic Standards for Identification of Tamoxifen Metabolites

Synthetic Sources

a. TAMOXIFEN, TAMOXIFEN CITRATE, AND 4-HYDROXYTAMOXIFEN. These compounds were obtained as a gift from ICI Pharmaceuticals Group (Wilmington, DE).

b. N-DESMETHYLTAMOXIFEN. This compound was kindly provided by Dr. John F. Stobaugh (University of Kansas, Lawrence, KS).

c. TAMOXIFEN-N-OXIDE. Tam-N-oxide was synthesized essentially as previously reported.[11,21,22] A 1-ml methanolic (HPLC grade) solution of tamoxifen (15 mg) is added to 0.5 ml of 30% H_2O_2, and the resulting solution is kept in the dark for 24 hr. The solution is evaporated under a stream of nitrogen at room temperature. To remove the residual H_2O_2, ethanol (1.5 ml) is added and the solution is evaporated to dryness. The

---

[34] M. J. Juedes, W. H. Bulger, and D. Kupfer, *Drug Metab. Dispos.* **15,** 786 (1987).
[35] W. H. Bulger, J. E. Temple, and D. Kupfer, *Toxicol. Appl. Pharmacol.* **68,** 367 (1983).

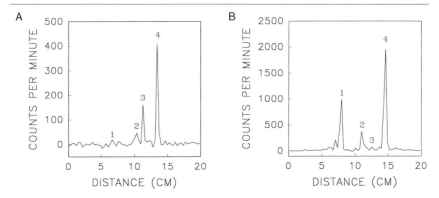

FIG. 2. TLC analysis of radiolabeled tamoxifen and its metabolites. A normal phase TLC plate was developed in $CHCl_3 : CH_3OH : NH_4OH$ (80:20:0.5; v/v/v) and metabolites were detected and quantified using a Bioscan 2000 imaging scanner. (A) Incubation of [^{3}H]tamoxifen with chicken liver microsomes (1 mg/ml) in the presence of NADPH for 60 min. (B) Incubation of [^{3}H]tamoxifen with liver microsomes (1 mg/ml) from phenobarbital-treated rats in the presence of NADPH for 60 min. Peak 1, tamoxifen-N-oxide; 2, N-desmethyltamoxifen; 3, 4-hydroxytamoxifen; 4, tamoxifen.

residue is taken up in fresh ethanol and stored at 4°. Spectral and chromatographic data of the synthesized product are identical to that previously reported for tam-N-oxide.[11,21] [^{3}H]Tam-N-oxide is prepared identically as the radioinert compound; however, [^{3}H]tamoxifen is used instead. To further confirm the identity of the compound as being tam-N-oxide, an aliquot of the compound is deoxygenated back to tamoxifen. Tam-N-oxide (0.5 mg) is dissolved in HPLC grade methanol (0.5 ml) and titanium(III) chloride (5.15 mg in 30 μl H_2O) is added.[13,21] The reaction is allowed to proceed for 1 hr at room temperature and is subsequently quenched with 0.5 ml of H_2O. The resulting solution is made alkaline with 0.1 N NaOH and extracted with diethyl ether (3 × 1.5 ml). The combined ether phase is backwashed with 2 ml of H_2O. The organic phase is evaporated under a stream of nitrogen, and the residue is taken up in ethanol (2 ml) and stored at 4° until further analysis. The product exhibits a single spot on a TLC plate with an R_f that is identical to tamoxifen.

Biosynthesis

a. [^{3}H]-4-HYDROXYTAMOXIFEN ([^{3}H]-4-OH-Tam). [^{3}H]Tamoxifen (250,000 dpm, 100 nmol) is incubated with liver microsomes from untreated adult chicken or from TCDD- or βNF-treated chick embryos in the presence of the NADPH-generating system in a final volume of 1 ml for 60 or 120 min as described under Incubations.[36] The radiolabeled 4-OH-tam,

[36] D. Kupfer, C. Mani, C. A. Lee, and A. B. Rifkind, *Cancer Res.* **54,** 3140 (1994).

representing the major metabolite formed by chicken liver microsomes corresponding chromatographically to authentic radioinert 4-OH-tam (Fig. 2A), is eluted off a TLC plate with ethanol and purified further on TLC using the previously described solvent system. Because of the lability of 4-OH-tam in light, chromatography is conducted under subdued illumination.

b. [^{14}C]- OR [^{3}H]-N-DESMETHYLTAMOXIFEN ([^{14}C]- OR [^{3}H]-N-DES-METHYL-TAM). [^{14}C]Tamoxifen (100,000 dpm, 100 nmol) or [^{3}H]tamoxifen (250,000 dpm, 100 nmol) is incubated with PB or PCN microsomes in the presence of the NADPH-generating system in 1 ml for 60 min.[14] The formation of this monodemethylated metabolite is *markedly* higher in liver microsomes from PB- or PCN-treated rats as compared with microsomes from control rats. N-Desmethyl-tam is chromatographically more polar than 4-OH-tam and less polar than tam-N-oxide (Fig. 2B). To confirm the identity of the product as N-desmethyl-tam, the putative radiolabeled N-desmethyl-tam (eluted off a TLC plate with ethanol) is treated with formaldehyde followed by reduction of the Schiff base with sodium cyanoborohydride (NaCNBH$_3$).[14] This results in a less polar radiolabeled compound that is chromatographically indistinguishable from tamoxifen.

c. [^{3}H]TAMOXIFEN-N-OXIDE ([^{3}H]TAM-N-OXIDE). [^{3}H]Tamoxifen (250,000 dpm, 100 nmol) is incubated with untreated rat liver microsomes in the presence of the NADPH-generating system in 1 ml.[21,37] A major chromatographically *highly* polar metabolite corresponding to authentic synthetic tam-N-oxide is formed (Fig. 2B). This metabolite is formed in extremely low amounts in incubations of [^{3}H]tam with chicken liver microsomes.[38] Also, little or no N-oxide is formed when heat-treated (50° for 90 sec) rat liver microsomes are used. To establish its identity, the putative tam-N-oxide eluted off a TLC plate with ethanol is deoxygenated with titanium(III) chloride into tamoxifen.[13,21] Alternatively, tam-N-oxide is enzymatically reduced by incubation with rat liver microsomes and NADPH in the presence of methimazole (an FMO inhibitor).[21] Additionally, tam-N-oxide is formed as a single metabolite by incubation of [^{3}H]tam with purified FMO, in the presence of NADPH, as previously reported.[21]

d. 3,4-DIHYDROXYTAMOXIFEN (3,4-DIHYDROXY-TAM). Incubation of tamoxifen with liver microsomes from several species indicated that 3,4-dihydroxy-tam, previously identified by mass spectrometry,[18] is minimally

---

[37] Although an ethanolic solution of tamoxifen stored at 0–4° is essentially stable, a small amount of N-oxide is generated with time because of air exposure. It is, therefore, advisable that the stock of tamoxifen be monitored and, if necessary, purified on TLC.

[38] The low level of N-oxidation of tamoxifen by chicken liver microsomes, despite the presence of considerable amount of FMO, is attributed to its restrictive active site (Dr. Daniel Ziegler, personal communication).

detected, presumably because of its slow rate of formation and/or because of its low accumulation due to its rapid conversion into other products. To circumvent this difficulty, the 3,4-dihydroxy-tam metabolite was not isolated, but was intentionally converted *in situ* into monomethylated derivative: Incubations are conducted in 20-ml glass scintillation vials open to air atmosphere with minor modifications of a previously described procedure for quantitation of catechol estrogens.[39–42] Microsomal suspension (1 mg of protein in KCl solution) is added to 0.6 ml of sodium phosphate buffer (pH 7.4; 60 μmol) containing EDTA (1 μmol), 0.1 ml of MgCl$_2$ (10 μmol), radioinert tamoxifen or 4-OH-tam (25 or 100 nmol in 10 μl of ethanol), dithiothreitol (50 nmol in 10 μl), [^{3}H]SAM (1 μCi, 200 nmol in 12 μl H$_2$O), the NADPH-regenerating system (as described earlier) in 0.1 ml of sodium phosphate buffer (pH 7.4; 10 μmol), and water to a final volume of 1 ml. After a preincubation at 37° for 2 min, the reaction is initiated by adding the NADPH-regenerating system, and the vials are incubated at 37° in a water bath shaker for 30 min. To terminate the reaction, the incubation mixture is placed on ice and 3 ml of ice-cold hexane is added. After a thorough mixing with a vortex, the resulting mixture is centrifuged and the hexane phase is removed and saved. This step is repeated with fresh hexane. The combined hexane phase is backwashed with 2 ml of water and, after centrifugation, the aqueous phase is discarded and an aliquot of the hexane phase containing the monomethylated catechol is taken for radioactivity determination. Maximal amounts of radiolabeled product(s) are obtained from 15- to 30-min incubations. A longer incubation time results in decreased product levels, possibly due to demethylation of the monomethylated catechol and/or due to its conversion into polar metabolites that are not extractable by hexane.

Discussion

Our laboratory utilized TLC for the separation, identification, and quantitation of tamoxifen metabolites formed by incubation of tamoxifen with liver microsomes of several species. Other laboratories have employed HPLC and various modes of mass spectrometry for tamoxifen metabolite identification and quantitation.

The formation of N-desmethyl-tam[14,20] and 4-OH-tam[14] is catalyzed by

---

[39] Initially, 25 IU of COMT was included in incubations; however, we later observed that there is no necessity for exogenous COMT since there is sufficient COMT in the microsomes.
[40] S. A. Li, R. H. Purdy, and J. J. Li, *Carcinogenesis* (*London*) **10,** 63 (1989).
[41] A. R. Hoffman, S. M. Paul, and J. Axelrod, *Biochem. Pharmacol.* **29,** 83 (1980).
[42] D. Kupfer, W. H. Bulger, and A. D. Theoharides, *Chem. Res. Toxicol.* **3,** 8 (1990).

cytochrome P450 monooxygenase. Various inhibitors of P450 (metyrapone, benzylimidazole, octylamine) and carbon monoxide diminish the formation of these metabolites. Interestingly, whereas male rat liver microsomes exhibited higher N-demethylation of tamoxifen than females, 4-hydroxylation was higher in females. An increase in tamoxifen N-demethylation by PB and PCN treatment appears to be due to the induction of CYP3A. This observation, coupled with the fact that monoclonal antibodies anti-P450 2B1/2 did not inhibit N-demethylation in PB microsomes, indicates the involvement of CYP3A in this reaction. Support for CYP3A involvement was obtained from observations that alternate substrates of the CYP3A subfamily inhibit N-demethylation. Additionally, troleandomycin (TAO), a mechanism-based inhibitor and a substrate of CYP3A isozymes, dramatically inhibited *in vitro* tamoxifen N-demethylation in PCN microsomes. Further, the ferricyanide-mediated dissociation of the TAO–P450 complex in liver microsomes of TAO-treated rats resulted in an enhanced N-demethylation of tamoxifen.[43,44] Inhibition of N-demethylation of tamoxifen by anti-P4503A1 antibodies in PCN- and PB-treated rat liver microsomes corroborates that CYP3A1 catalyzes the tamoxifen N-demethylation. However, attempts to demonstrate N-demethylation of tamoxifen by reconstituted CYP3A have been unsuccessful.[46] There is no significant difference in the observed rate of N-demethylation of ^{14}C ring or [*methyl*-^3H]tamoxifen by rat liver microsomes. Surprisingly, and contrary to expectations, the loss of one methyl group from tamoxifen does not result in an appreciable loss of radiolabel. This is presumably due to a primary isotope effect (C-H vs C-^3H) so that the methyl group containing hydrogen is lost preferentially over the tritiated methyl group.

4-OH-tam exhibits a much greater affinity for the estrogen receptor than tamoxifen.[3] Also, 4-OH-tam is between 50- and 100-fold more potent against proliferation of MCF-7 cells[48,49] and normal human breast cancer

[43] S. A. Wrighton, P. Maurel, E. G. Schuetz, P. B. Watkins, B. Young, and P. S. Guzelian, *Biochemistry* **24,** 2171 (1985).

[44] TAO treatment of rats induces CYP3A; however, the induced enzymatic activity is not detected because of the formation of a TAO–CYP3A complex. This complex can be dissociated by ferricyanide, restoring the CYP3A enzymatic activity.[45]

[45] A. J. Sonderfan, M. P. Arlotto, D. R. Sutton, S. K. McMillen, and A. Parkinson, *Arch. Biochem. Biophys.* **255,** 27 (1987).

[46] Several other investigators have also experienced difficulties in CYP3A reconstitution studies.[14,47]

[47] M. Halvorson, D. Greenway, D. Eberhart, K. Fitzgerald, and A. Parkinson, *Arch. Biochem. Biophys.* **277,** 166 (1990).

[48] F. Vignon, M.-M. Bouton, and H. Rochefort, *Biochem. Biophys. Res. Commun.* **146,** 1502 (1987).

[49] E. Coezy, J. L. Borgna, and H. Rochefort, *Cancer Res.* **42,** 317 (1982).

cells in culture.[50] These findings invite speculations that tamoxifen is a prodrug and that 4-OH-tam may be the active species. Although liver microsomes from several species metabolize tamoxifen, adult or embryonic chicken liver microsomes yield the highest rate of tamoxifen 4-hydroxylation.[36] Tamoxifen 4-hydroxylation was between 2.1- and 3.3-fold higher in liver microsomes of PB-, TCDD-, and βNF-treated chick embryos. Benzo [a]pyrene and α-naphthoflavone, which are substrates and inhibitors of CYP1A1 and CYP1A2, inhibited tamoxifen 4-hydroxylation in βNF-treated chick embryos. These observations suggested that the increased formation of 4-OH-tam in TCDD- and βNF-induced embryos was due to the induction of CYP1A-like isoforms. Indeed, reconstituted TCDD$_{AA}$, one of the two major hepatic P450 isoforms induced by TCDD and βNF in chick embryos that catalyzes effectively estradiol 2-hydroxylation, exhibits high tamoxifen 4-hydroxylation activity. The second P450 isoform, TCDD$_{AHH}$, was found to be inactive in 4-hydroxylation.[36,51] However, the rat or the human P450s catalyzing the 4-hydroxylation have not yet been identified. Recent correlation studies between P450 levels and 4-hydroxylase activity in human liver microsomes suggest that CYP2C8, CYP2C9, and CYP2D6 may catalyze that reaction.[52]

Metabolism of tamoxifen by liver microsomes from various species, in the presence of NADPH, results in the formation of tam-N-oxide, N-desmethyl-tam, and 4-OH-tam.[10–12,14] Classical inhibitors of P450s had no effect on N-oxide formation, indicating the participation of a non-P450 enzyme. The observation that mild heat treatment of liver microsomes (50° for 90 sec) or the addition of a low concentration of methimazole (0.2 mM) to rat liver microsomes markedly inhibited N-oxide formation strongly suggests catalysis by FMO in tamoxifen N-oxidation. Indeed, incubations of tamoxifen with purified mouse liver FMO result only in tam-N-oxide formation.[21] Organic N-oxides can undergo reduction to the tertiary amines by rat liver microsomes,[53,54] and we observed such a reduction of tam-N-oxide to tamoxifen.[21] Therefore, the amount of N-oxide observed in incubations of tamoxifen with liver microsomes represents the net result

[50] C. Malet, A. Compel, P. Spritzer, N. Bricout, H. Yaneva, I. Mowszowicz, F. Kuttenn, and P. Mayvais-Jarvis, *Cancer Res.* **48,** 7193 (1988).
[51] A. B. Rifkind, A. Kanetoshi, J. Orlinick, J. H. Capdevila, and C. A. Lee, *J. Biol. Chem.* **269,** 3387 (1994).
[52] I. N. H. White, F. de Matteis, A. H. Gibbs, C. K. Lim, C. R. Wolf, C. Henderson, and L. L. Smith, *Biochem. Pharmacol.* **49,** 1035 (1995).
[53] K. Iwasaki, H. Noguchi, R. Kato, and R. Sato, *Biochem. Biophys. Res. Commun.* **77,** 1143 (1977).
[54] M. Sugiura, K. Iwasaki, and R. Kato, *Biochem. Pharmacol.* **26,** 489 (1977).

of the FMO-catalyzed N-oxide formation (forward reaction) and the N-oxide reduction (reverse reaction).

Acknowledgments

This study was supported by U.S. Public Health Service Grant ES 00834 from the National Institute for Environmental Health Sciences. We thank ICI Pharmaceuticals Group (Wilmington, DE) for a sample of tamoxifen, tamoxifen citrate, and 4-hydroxytamoxifen, and Dr. John F. Stobaugh (University of Kansas, Lawerence, KS) for N-desmethyltamoxifen.

[18] Trimethadione: Metabolism and Assessment of Hepatic Drug-Oxidizing Capacity

By EINOSUKE TANAKA and YOSHIHIKO FUNAE

Introduction

Measurement of the activity of drug-metabolizing enzymes of the liver, which largely depend on cytochrome P450, is essential in evaluating the capacity of oxidative drug metabolism in liver disease. Several methods have recently been developed to determine the hepatic drug-oxidizing capacity *in vivo* using probe drugs, such as antipyrine, aminopyrine, and caffeine.[1,2]

Trimethadione (TMO), an antiepileptic drug, may be a more suitable candidate for estimating hepatic drug-oxidizing activity. It is rapidly absorbed from the gastrointestinal tract, distributed into total body fluids, and is extensively N-demethylated to dimethadione (DMO) by P450-dependent monooxygenases in liver microsomes. The chemical formulas of the compounds are shown in Fig. 1. Neither TMO nor DMO is bound to blood proteins or any other macromolecules in biological materials. The elimination of TMO from blood follows first-order kinetics according to a simple one-compartment model.[3,4] These pharmacokinetic properties of TMO enable the determination of hepatic microsomal function with a single blood sample. Previous studies have shown that the ratio of the serum (or plasma) concentration of DMO to TMO at 2 hr in rats and at 4 hr in humans after

---

[1] B. K. Park, *Biochem. Pharmacol.* **14,** 631 (1982).

[2] R. A. Branch, *Hepatology* **2,** 97 (1982).

[3] E. Tanaka, H. Kinoshita, T. Yamamoto, Y. Kuroiwa, and E. Takabatake, *J. Pharmaco-Dyn.* **4,** 576 (1981).

[4] E. Tanaka, A. Ishikawa, S. Kobayashi, H. Yasuhara, S. Misawa, and Y. Kuroiwa, *Comp. Biochem. Physiol.* **104C,** 205 (1993).

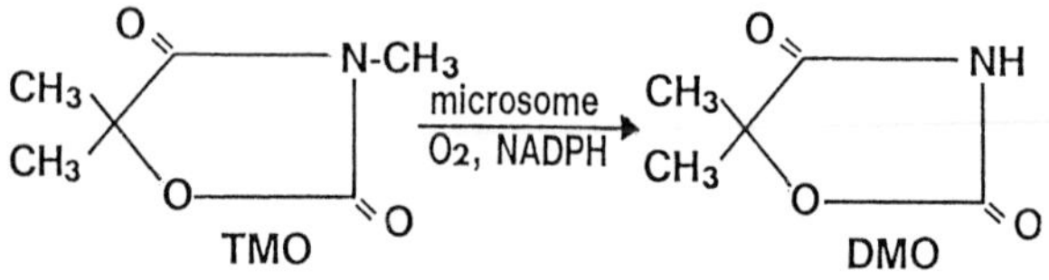

FIG. 1. Metabolism of trimethadione. TMO, trimethadione (3,5,5-trimethyloxazolidine-2,4-dione); DMO, dimethadione (5,5-dimethyloxazolidine-2,4-dione).

oral administration of TMO (designated the DMO/TMO ratio) is a good quantitative indicator of liver function and can predict the quantity of functional remnant hepatic parenchyma after hepatectomy in patients with liver disease (TMO tolerance test).[5,6]

In Vitro Studies

We attempted to identify the specific forms of P450 involved in the metabolism of TMO in hepatic microsomes or in a reconstituted system using purified P450 isozymes from rats and humans.[7]

Reagents

TMO was supplied by the Dainippon Pharmaceutical Co. Ltd (Osaka, Japan). DMO and maleimide were obtained from Tokyo Kasei (Tokyo, Japan). TMO and DMO are highly soluble in water or alcohol. Dilauroylphosphatidylcholine (DLPC) and NADPH were obtained from Sigma Chemical Co. (St. Louis, MO). Other reagents and organic solvents were obtained from Wako Pure Chemical Industries (Tokyo, Japan).

TMO Metabolism by Rat Hepatic Microsomes

Rat hepatic microsomes (200 μg) and TMO (0.5 mM) are incubated with NADPH (0.2 mM) at 37° for 30 min in 0.1 M potassium phosphate buffer, pH 7.4 (a final volume of 1.0 ml). The reaction is stopped by the addition of 15% $ZnSO_4$ (0.25 ml) and saturated $Ba(OH)_2$ (0.25 ml). The resulting solutions are centrifuged at 1800 g for 10 min. The supernatant (1 ml) is transferred to a fresh test tube. Formaldehyde produced by P450

---

[5] E. Tanaka, A. Ishikawa, K. Fukao, K. Tsuji, A. Osada, S. Yamamoto, Y. Adachi, Y. Takase, M. Abei, and Y. Iwasaki, *Int. J. Clin. Pharmacol. Ther. Toxicol.* **29**, 333 (1991).

[6] M. Abei, E. Tanaka, N. Tanaka, Y. Matsuzaki, T. Ikegami, A. Ishikawa, and T. Osuga, *J. Gastroenterol.* **30**, 478 (1995).

[7] M. Nakamura, E. Tanaka, S. Misawa, T. Shimada, S. Imaoka, and Y. Funae, *Biochem. Pharmacol.* **47**, 247 (1994).

is measured by the calorimetric method of Nash.[8] The Nash reagent (0.4 ml) is added to the tubes. They are incubated at 60° for 30 min and then the absorbance of each solution is measured at 415 nm. Nash's reagent is prepared by dissolving ammonium acetate (30 g) and acetyl acetone (0.4 ml) in distilled water (final volume 100 ml).

TMO Metabolism in a Reconstituted System

Purified P450 (30 pmol), NADPH-P450 reductase (0.3 unit), cytochrome b_5 (30 pmol), and DLPC (5 μg) are added to a test tube in this order and mixed well. DLPC is suspended in distilled water by sonication (1 mg/ml) before use. Reaction buffer, TMO, and NADPH are added to this tube and the reaction is carried out using the same method as with hepatic microsomes.

Inhibition Study Using Antibody

Immunoglobulin G (20–200 μg) is preincubated with hepatic microsomes (200 μg) at 37° for 10 min. The incubation mixture is added and the reaction is carried out with the same method as for the TMO metabolism by hepatic microsomes.

TMO Metabolism by Human Hepatic Microsomes

Human hepatic microsomes (200 μg) and TMO (0.5 mM) are incubated with NADPH (0.2 mM) at 37° for 30 min in 0.1 M potassium phosphate buffer, pH 7.4 (final volume 1.0 ml). The reaction is stopped by the addition of 2 ml 5 M NaH$_2$PO$_4$. The amount of DMO produced (nmol of DMO/min/nmol P450) is measured by gas chromatography because DMO production is less in humans than in rats.[7]

Measurement of DMO by Gas Chromatography

Small amounts of Na$_2$SO$_4$ and MgSO$_4$ (approximately each 50 mg) and 100 μl of ethyl acetate containing maleimide (internal standard; 5 μg/ml) are added to a 2.5-ml tube containing 300 μl of a reaction mixture of human hepatic microsomes. After vortex mixing for 1 min, the tubes are centrifuged at 1800 g for 5 min. A 2-μl aliquot of the organic phase is directly injected into the gas chromatograph.[9] Analysis of TMO and DMO in the reaction mixture is carried out by a Shimadzu-9A instrument (Kyoto, Japan) with flame thermionic detector (FTD). Separation is carried out on a CBP 10-

[8] T. Nash, *Biochem. J.* **55,** 416 (1953).
[9] E. Tanaka and S. Misawa, *J. Chromatogr.* **584,** 267 (1992).

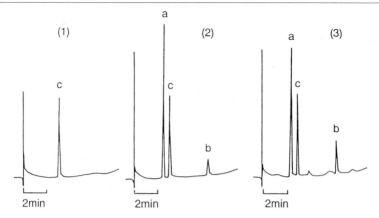

Fɪɢ. 2. Representative gas chromatograms of (1) blank human serum, (2) human serum spiked with trimethadione (TMO) and dimethadione (DMO) (5 μg/ml), and (3) a human serum sample obtained 4 hr after oral administration of TMO (4 mg/kg). Peaks: a, TMO; b, DMO; c, internal standard (maleimide).

W12-100 column (12 m $\times$ 0.35 mm i.d., film thickness 1 μm) from Shimadzu (Kyoto, Japan). The column oven temperature is held at 70° for 2 min, raised to 120° at 10°/min, and finally raised to 220° at 20°/min. The injection port temperature is 300°. The helium, hydrogen, and air flow rates are 30, 30, and 145 ml/min, respectively. A chromatogram of TMO and DMO and the internal standard is shown in Fig. 2. The retention times of TMO, the internal standard, and DMO are 2.5, 3.1, and 6.3 min, respectively. The detection limit is 10 and 50 ng/ml for TMO and DMO, respectively.

Comments

    TMO metabolism using purified P450s is shown in Table I. CYP2Cll, a major form of P450 in rat hepatic microsomes, and 2Bl exhibited high TMO *N*-demethylation activity in rats. Antibodies raised to CYP2Cll and 2Bl/2 inhibited effectively TMO *N*-demethylation in hepatic microsomes of untreated and phenobarbital-treated rats. In human liver, CYP3A and CYP2C effectively metabolized TMO to DMO. The content of the CYP3A and CYP2C subfamily makes up over 60% of the P450s present in human hepatic microsomes.[10] Therefore, measurement of TMO *N*-demethylation activity is considered to reflect human liver function.

---

[10] T. Shimada, H. Yamazaki, M. Miura, Y. Inui, and F. P. Guengerich, *J. Pharmacol. Exp. Ther.* **270**, 414 (1994).

TABLE I
TMO N-DEMETHYLATION ACTIVITIES OF PURIFIED RAT AND HUMAN P450s[7,a]

P450 isozymes	TMO N-demethylation activity	P450 isozymes	TMO N-demethylation activity
Rat			
1A1	7.4	2C7	9.0
1A2	9.0	2C11	17.0
2A1	8.4	2C12	15.4
2A2	14.8	2C13	8.0
2B1	17.2	2E1	9.6
2B2	8.4	3A2	6.2
2C6	8.0	4A2	5.0
Human			
3A4 (P450$_{NF}$)	4.4		
2C9 (P450$_{MP}$)	2.4		
1A2 (P450$_{PA}$)	0.6		

[a] Values are nmol product/min/nmol P450.

In Vivo Studies

Rat[4]

Twenty hours before the experiment, male Sprague–Dawley rats weighing 200–250 g are cannulated in the right jugular vein under anesthesia for blood sampling. The polyethylene canula (i.d. 0.58 mm, o.d. 0.96 mm) is inserted and, to avoid destruction by the rat during the experiment, is pulled subcutaneously to the nape of the neck, thus allowing free movement of the rats during the experiment.

After overnight fasting, rats are given TMO (4 mg/kg) orally. A 0.2% solution of TMO in water is administered at a dose of 2 ml/kg body weight. Blood samples (0.3 ml) are obtained from the jugular vein at 0.5, 1, 4, 6, 8, 12, 24, and 48 hr after TMO administration. Two hundred microliters of 5 M NaH$_2$PO$_4$, small amounts of Na$_2$SO$_4$ and MgSO$_4$ (approximately each 50 mg), and 100 μl of ethyl acetate containing maleimide (internal standard; 5 μg/ml) are added to a 2.5-ml tube containing 100 μl serum (or plasma). After vortex mixing for 1 min, the tubes are centrifuged at 1800 g for 5 min. The concentration of serum (or plasma) TMO and DMO is determined by gas chromatography. A 2-μl aliquot of the organic phase is directly injected into the gas chromatograph (see In Vitro Studies).

Human

The human study[5,6] is performed the morning after a fast of approximately 12 hr. TMO (4 mg/kg: commercially available as a 66.7% powder;

Mino-Aleviatin, Dainippon Pharmaceutical Co. Ltd) is given orally with 100 ml water to subjects who had given their consent and they are allowed breakfast 2 hr later. The subjects remain seated as much as possible. Blood (0.5 ml) is collected before TMO administration and after 0.5, 1, 2, 4, 6, 8, 12, and 24 hr later. The concentration of serum TMO and DMO is measured as described earlier.

Calculation of Pharmacokinetic Parameter

The half-life $(t_{1/2})$ of TMO was estimated by linear regression analysis. The apparent volume of distribution (V_d) was calculated from the TMO concentration at zero time. The apparent total body clearance (CL) of TMO was calculated from the equation $CL = dose/AUC_0^\infty$, where AUC is the area under the concentration–time curve. Because the absolute bioavailability of orally administered TMO is not known, values of CL and V_d are termed "apparent."

The possible correlation of the serum DMO/TMO ratio with CL and $t_{1/2}$ was studied at 1, 2, 4, 6, and 12 hr after TMO administration. The correlation coefficient between the serum DMO/TMO ratio and both CL and $t_{1/2}$ was the highest at 2 hr for rat[4] or 4 hr for humans.[11] These results suggest that hepatic drug-oxidizing capacity can best be estimated from the serum DMO/TMO ratio obtained at 2 and 4 hr for rats and humans, respectively, after oral administration of TMO.

Application

The serum (or plasma) DMO/TMO ratio in a single blood sample calculated 2 and 4 hr for rats and humans, respectively, after oral administration of TMO provides a useful indicator for the assessment of the hepatic drug-oxidizing capacity. The TMO tolerance test was applied in a hepatic disease[5,6] and hepatectomy studies[12] in humans. These results indicate that the serum DMO/TMO ratio reflects the severity of liver damage.

Hepatic Disease Study.[6] The histologic characteristics of patients with liver disease were classified as chronic persistent hepatitis (CPH), chronic active hepatitis (CAH), chronic active hepatitis with bridging (CAHB), and chronic active hepatitis with cirrhosis (CAHC) on the basis of standard histologic criteria. The extent of the abnormality of the serum DMO/TMO ratio was also related to the histological diagnosis; patients with CPH

[11] E. Tanaka, S. Kobayashi, K. Nakamura, E. Uchida, and H. Yasuhara, *Biopharm. Drug Dispos.* **10**, 617 (1989).

[12] A. Ishikawa, K. Fukao, K. Tsuji, A. Osada, Y. Yamamoto, M. Ohtsuka, and E. Tanaka, *J. Gastroenterol. Hepatol.* **8**, 426 (1993).

(0.56 ± 0.12; mean ± SD) or CAH (0.48 ± 0.13) exhibited no change, whereas patients with CAHB (0.37 ± 0.12, $P < 0.01$) or CAHC (0.25 ± 0.14, $P < 0.01$) showed significantly lower values compared with controls (0.59 ± 0.10).

Conclusion

These findings show that TMO may be used as a probe drug for the rapid determination of the hepatic drug-oxidizing capacity as well as the functional reserve mass of the liver.

[19] Antipyrine, Theophylline, and Hexobarbital as *in Vivo* P450 Probe Drugs

By KEES GROEN and DOUWE D. BREIMER

Introduction

This chapter focuses on antipyrine (AP), hexobarbital (HB), and theophylline (TH) as probe drugs for *in vivo* characterization of specific P450 enzymes. What these three probe drugs have in common is that they can be considered multifunctional substrates since they provide information on more than one specific P450 enzyme. The use of these drugs has recently been reviewed by Pelkonen and Breimer.[1] AP is undoubtedly the most commonly used substrate in estimating the influence of all kinds of endogenous and exogenous factors on P450 enzyme activities. Whereas initially AP elimination half-life and clearance were used as the pharmacokinetic variables to characterize total P450 enzyme activity, a more isozyme specific assay *in vivo* was developed by measuring the rate of formation (or clearance of formation) of the main AP metabolites norantipyrine or desmethylantipyrine (NORA), 3-hydroxymethylantipyrine (HMA), and 4-hydroxyantipyrine (OHA). There is strong evidence from studies in Wistar rats and humans that different P450 enzymes are involved in the formation of the three main metabolites: 3-hydroxymethylantipyrine (HMA), norantipyrine (NORA), and 4-hydroxyantipyrine (OHA).[2–8] Engel *et al.*[9] concluded from

[1] O. Pelkonen and D. D. Breimer, *Handb. Exp. Pharmacol.* **110,** 289 (1994).

[2] M. Danhof, D. P. Krom, and D. D. Breimer, *Xenobiotica* **9,** 695 (1979).

[3] M. Danhof, R. M. A. Verbeek, C. J. van Boxtel, J. K. Boeijinga, and D. D. Breimer, *Br. J. Clin. Pharmacol.* **13,** 379 (1982).

[4] M. van der Graaff, N. P. E. Vermeulen, R. P. Joeres, T. Vlietstra, and D. D. Breimer, *J. Pharmacol. Exp. Ther.* **227,** 459 (1983).

in vitro experiments with human liver microsomes that NORA is mainly formed by CYP2C. Both CYP1A2 and CYP2C were involved in the formation of HMA. CYP3A3/4 and CYP1A2 were found to be responsible for the formation of OHA. However, the latter could not be confirmed using stable expressed enzymes. As a result, assessment of clearances of formation of the different metabolites takes into account intrasubstrate selectivity in enzyme activity and thus overcomes some of the problems of previously applied methodologies.[10]

Antipyrine shows no differences in pharmacokinetics between oral and intravenous administration. This is due to the complete absorption after oral administration of this drug and the low clearance characteristics.[11] Antipyrine shows linear kinetics for doses up to 1000 mg.[12] It has also been reported that antipyrine kinetics based on saliva concentrations are a good reflection of plasma kinetics, and thus blood sampling can be avoided.[13-15] A further development has been that the measurement of AP concentrations in one sample (plasma or saliva) seems to give a reliable estimate of AP kinetics.[16]

TH is a compound that is also used as substrate for the assessment of cytochrome P450 enzyme activity and, more specifically, for the 3-methylcholanthrene-inducible enzyme forms, P450 IA1 and IA2,[5,10,17,18] Although TH is metabolized to different metabolites, *in vitro* studies led to the conclusion that at least the main metabolites are probably formed via

---

[5] M. W. E. Teunissen, L. G. J. de Leede, J. K. Boeijinga, and D. D. Breimer, *J. Pharmacol. Exp. Ther.* **233**, 770 (1985).

[6] D. D. Breimer, J. H. M. Schellens, and P. A. Soons, *Microsomes Drug Oxid. Proc. Int. Symp., 7th, 1987,* p. 232 (1988).

[7] J. H. M. Schellens, J. H. F. van der Wart, M. Danhof, E. A. van der Velde, and D. D. Breimer, *Br. J. Clin. Pharmacol.* **26**, 373 (1988).

[8] D. D. Breimer and J. H. M. Schellens, *Trends Pharmacol. Sci.* **11**, 223 (1990).

[9] G. Engel, U. Hofmann, H. Heidemann, and M. Eichelbaum, *Naunyn-Schmiedeberg's Arch. Pharmacol.* **349** (Suppl.), R 133 (1994).

[10] D. D. Breimer, *Clin. Pharmacokinet.* **8**, 371 (1983).

[11] M. Danhof, A. van Zuilen, J. K. Boeijinga, and D. D. Breimer, *Eur. J. Clin. Pharmacol.* **21**, 433 (1982).

[12] M. Danhof and D. D. Breimer, *Br. J. Clin. Pharmacol.* **8**, 529 (1979).

[13] E. S. Vesell, G. T. Passananti, P. A. Glenwright, and B. H. Dvorchik, *Clin. Pharmacol. Ther.* **18**, 259 (1975).

[14] H. S. Fraser, J. C. Mucklow, S. Murray, and D. S. Davies, *Br. J. Clin. Pharmacol.* **3**, 321 (1976).

[15] M. Danhof and D. D. Breimer, *Clin. Pharmacol.* **3**, 39 (1978).

[16] S. Loft and H. E. Poulsen, *Pharmacol. Toxicol.* **67**, 101 (1990).

[17] A. Kappas, K. E. Andersen, A. H. Conney, and A. P. Alvares, *Clin. Pharmacol. Ther.* **20**, 643 (1976).

[18] J. W. Jenne, *Chest* **81**, 529 (1982).

closely linked CYP1A enzymes.[5,19–22] The use of TH is extensively reviewed by Tröger and Meyer.[23]

From studies in rats, HB is assumed to be a good marker substrate for the phenobarbital-inducible P450 enzymes.[4] The use of HB and its characteristics have been reviewed extensively by Van der Graaff et al.[24] Recent experiments in extensive and poor metabolizers of mephenytoin demonstrated that the metabolism of both $R(-)$- and $S(+)$-HB in humans is primarily determined by CYP2C19.[25]

AP, HB, and TH can be used in experimental animals as well as in humans without side effects at the dose levels required for determination of P450 enzyme activity. Another advantage is that the absorption of AP, TH, and HB from an oral solution is rapid and complete. They have been used alone, but also in combination with each other or other probe drugs (the "cocktail" approach).[1,8]

Methods

Use in Human Volunteers

When applying AP, HB, and/or TH to human volunteers, the volunteers should have abstained from food for a period of at least 10 hr before and 4 hr after administration of the solutions since the presence of food might influence the rate and extent of absorption of the compounds. When TH is applied, subjects have to abstain from methylxanthine-containing food and beverages, starting 1 day before and continuing until completion of the study.

In healthy humans, the elimination half-life of AP ranges from 8 to 20 hr, of HB from 3 to 5 hr, and of TH from 6 to 9 hr. In some patients, these half-lives may be significantly prolonged.

[19] R. Dahlqvist, L. Bertilsson, D. J. Birkett, M. Eichelbaum, J. Säwe, and F. Sjöqvist, *Clin. Pharmacol. Ther.* **35**, 815 (1984).
[20] R. A. Robson, A. P. Matthews, J. O. Miners, M. E. McManus, U. A. Meyer, P. Hall, and D. J. Birkett, *Br. J. Clin. Pharmacol.* **24**, 293 (1987).
[21] D. J. Birkett, J. O. Miners, M. E. McManus, I. Stupans, M. Veronese, and R. A. Robson, *Microsomes Drug Oxid. Proc. Int. Symp., 7th, 1987*, p. 241 (1988).
[22] U. Fuhr, J. Doehmer, N. Battula, C. Wölfel, C. Kudla, Y. Keita, and A. H. Staib, *Biochem. Pharmacol.* **43**, 225 (1992).
[23] U. Tröger and F. P. Meyer, *Clin. Pharmacokinet.* **28**, 287 (1995).
[24] M. Van der Graaff, N. P. E. Vermeulen, and D. D. Breimer, *Drug Metab. Rev.* **19**, 109 (1988).
[25] A. Adedoyin, C. Prakash, D. O'Shea, I. A. Blair, and G. R. Wilkinson, *Pharmacogenetics* **4**, 27 (1994).

Preparation of Solutions for Oral Administration

Doses of 300 mg AP, anhydrous TH, equal to 100 mg TH, or sodium HB, equal to 500 mg racemic HB, are dissolved in 100 ml tap water, either simultaneously or separately. After administration of the solution, rinse the beaker with another 50 ml of tap water and administer as well.

Blood Sampling Schedules

When TH and/or HB are administered, blood samples (5 ml) are preferably collected in heparinized tubes just before and 0.5, 1, 2, 4, 6, 8, 10, 12, 24, and 30 hr after administration of the compounds. In the case of AP, additional samples are collected 36 and 48 hr after administration. Immediately after collection, the samples are centrifuged for 15 min at 2000 g. Plasma samples should be stored at $-18°$ until analysis.

Urine Sampling Schedule

When AP or TH is used, urine should be collected over a period of at least 48 hr. Since 5–10% of the AP dose and about 10% of the administered TH dose are excreted unchanged into urine, a correction for this fraction excreted unchanged should be applied in order to avoid overestimation of enzymatic activity.

Since HB is hardly excreted unchanged and metabolite measurement does not provide useful information on enzyme selectivities, urine collection is not required using this compound.

Use in Rats

Rats should have been fasted during a period of at least 12 hr prior to and 4 hr after oral administration of the probe drugs.

In uninduced rats, the elimination half-life of AP is 1–3 hr and of TH about 3 to 5 hr, depending on strain and age. The elimination half-lives of both R- and S-HB are between 20 and 120 min.

Preparation of Solutions and Administration

In rats, both AP and TH can be administered either orally or parenterally. AP is administered at a dose level of 15 mg/kg and in a concentration of 3 mg/ml tap water (oral) or saline (parenteral), whereas TH is administered at a dose level of 5 mg/kg and in a concentration of 1 mg/ml tap water (oral) or saline (parenteral).

HB should preferably not be given intravenously since HB has an extraction ratio of about 0.7 in rats.[24] Therefore, the best reflection of changes in hepatic enzyme activity is obtained after oral administration of HB (assessment of intrinsic clearance).

The solution for oral administration is prepared by dissolving racemic sodium HB in water. The final concentration should be about 6 mg HB/ml. Four milliliters of this solution is administered per kilogram of body weight.

Blood Sampling Scheme

After administration of TH, serial blood samples (250 μl) are collected in heparinized tubes just before and 15, 30, 45, 60, 90, 120, 150, 180, 240, 360, and 480 min after administration to obtain a reliable estimate of the pharmacokinetic variables. Using AP, an additional sample at 540 min post dose is preferred.

Using HB, blood samples (250 μl) are collected in heparinized tubes just before and 15, 30, 40, 50, 60, 75, 90, 105, 120, 150, 180, and 240 min after administration. Immediately after sampling, the blood is centrifuged for 15 min at 2000 g. Plasma should be stored at $-18°$ until analysis.

Urine Sampling

Since a significant part of the TH dose is excreted unchanged into the urine, in some rat strains over 20%, a correction should be made for this way of elimination of TH to reliably estimate the metabolic elimination of the compound. Urine should be collected during a period of at least 24 hr.

Because renal clearance of unchanged HB is practically absent, urine collection is not indicated in studies with HB as a probe drug.

Assay of AP and TH in Plasma

To 500 μl human plasma, 50 μl of an internal standard solution (8-chloroTH 4 μg/ml methanol) and 100 μl 1.0 M phosphate buffer (pH 7) are added. After 10 sec of mixing on a whirlmixer, 5 ml chloroform is added and again mixed, but now for 30 sec. After 10 min of centrifugation at 1500 g, the organic layer is transferred into another test tube and evaporated to dryness. The residue is redissolved in 200 μl eluent, consisting of 12% acetonitrile and 88% 0.05 M acetate buffer, pH 4.6. Fifty microliters is injected onto the HPLC system. The eluent flow is 1.0 ml/min, and a Nova Pack C_{18} column (100 $\times$ 8 mm i.d.) was used. TH is measured at 280 nm.[26] This analytical method can also be used when AP and TH are administered simultaneously. However, AP should be measured at 254 nm.

[26] K. Groen, M. A. Horan, N. A. Roberts, R. S. Gulati, B. Miljkovic, E. J. Jansen, V. Paramsothy, D. D. Breimer, and C. F. A. van Bezooijen, *Clin. Pharmacokinet.* **25,** 136 (1993).

The analysis of rat plasma for AP and TH can be performed using the method described by Hartley *et al.*[27] However, extraction of TH from the plasma can easiest be obtained by adding 75 μl methanol to 25 μl of the plasma sample to precipitate the proteins. After mixing and centrifugation, 25 μl of the supernatant is injected onto the HPLC system. This HPLC system consists of a Nova Pack C_{18} column (100 × 8 mm i.d.) with a Guard-Pack CN cartridge. The eluent consists of 1% glacial acetic acid and methanol (83 : 17), and the flow is 2.7 ml/min.

Assay of HB in Plasma

With respect to analysis of HB in plasma, a stereoselective method is preferred[28] since it has been shown that the metabolism of HB can be affected stereoselectively.[24,29] Although this method has been developed for the analysis of HB enantiomers in rat blood, it can also be applied for analysis in rat and human plasma. For this purpose, we slightly modified the procedure. Plasma samples of 50 μl (rat) or 200 μl (humans) are acidified with 0.5 ml of 0.37 *M* acetate buffer, pH 4.6, instead of phosphate buffer and are extracted with 5 ml of methylene chloride. Cyclobarbital is used as an internal standard. The eluent consists of 73% water and 27% methanol, and the flow is 1.0 ml/min. Absorption is measured at 214 nm (the description of this method is adapted from Chandler *et al.*).[28] When phenobarbital (PB) is present, this compound has the same retention time as R-HB using this method. Therefore, for samples from PB-induced rats, the eluent flow is 0.7 ml/min and consists of 85% water and 15% methanol. Using this method, AP can also be measured. It elutes just before S-HB. However, in that case the detection wavelength should be 254 nm until AP has been eluted and then be switched to 214 nm to measure the barbiturates.

Assay of AP and Its Metabolites in Urine

The method described here focuses on the determination of the phase I metabolites of AP. Therefore, a deconjugation step is incorporated. For deconjugation, 50 μl of *Helix pomatia* suspension containing 5.5 U/ml β-glucoronidase and 2.5 U/ml arylsulfatase (at 38°) and 500 μl of 0.05 *M* acetate buffer, pH 4.5, also containing 80 mg/ml freshly added $Na_2S_2O_5$, is added to 500 μl of urine. This mixture is then incubated for at least 3 hr at 37°.

[27] R. Hartley, I. J. Smith, and J. R. Cookman, *J. Chromatogr.* **342,** 105 (1985).
[28] M. H. H. Chandler, R. J. Guttendorf, R. A. Blouin, and P. J. Wedlund, *J. Chromatogr.* **419,** 426 (1987).
[29] M. H. H. Chandler, S. R. Scott, and R. A. Blouin, *Clin. Pharmacol. Ther.* **43,** 436 (1988).

After incubation, this sample is mixed with 50 μl of a phenacetin solution (40 μg/ml), 1 ml of a NaCl solution (200 mg/ml), and 5 ml of a chloroform : ethanol mixture (9 : 1). Avoid allowing the mixture to emulsify. Then centrifuge at 1500–2000 g for 15 min. The organic layer is then transferred to a clean test tube and 250 μl of methanol is added to stabilize NORA. This sample is then evaporated until semidryness in a stream of nitrogen at room temperature. The residue is redissolved in 200 μl of 0.01 M acetate buffer, pH 4.5, which also contains 15 mg/ml $Na_2S_2O_5$. Fifty microliters is injected into the HPLC system. A PLRP-S column is used in combination with an eluent consisting of 0.05 M phosphate buffer and acetonitrile (82 : 18). The flow rate is 1.0 ml/min and the detection wavelength is 254 nm.

Assay of Unchanged TH in Urine

For this assay, a 50-μl aliquot of internal standard solution (caffeine in methanol) is added to 500 μl urine. TH and the internal standard are extracted using 5 ml of the chloroform : ethanol (9 : 1) mixture. The organic layer is transferred to a clean test tube and evaporated. The residue is redissolved in 300 μl eluent, consisting of 20% methanol and 80% of a 1% acetic acid in water solution. A Nova pack C_{18} column (100 $\times$ 8 mm i.d.) was used. The eluent flow is 2.0 ml/min.

Data Analysis

The elimination half-lives ($t_{1/2}$) of TH and HB are calculated from the log-linear part of the plasma concentration vs time curve. The area under the plasma concentration vs time curve (AUC) is usually calculated according to the linear trapezoidal rule including extrapolation to infinity. Total body or systemic clearance (Cl_S) is calculated by dividing the dose by the AUC. The metabolic clearance (Cl_m) of TH and AP is calculated by correcting for their urinary excretion. Rate or clearance of formation of metabolites is calculated by multiplying total metabolic clearance by the formation of a metabolite excreted into urine.

To illustrate the application of the described methods, Figs. 1 and 2 give typical plasma concentration–time curves of AP, TH, $R(-)$-HB, and $S(+)$-HB in humans and rats, respectively.

Conclusion

The methods described in this chapter have proved to be reliable in estimating the pharmacokinetics of AP, TH, and HB. These model substrates have been applied very extensively in studying the influence of environmental and disease conditions on P450 enzyme activities both in

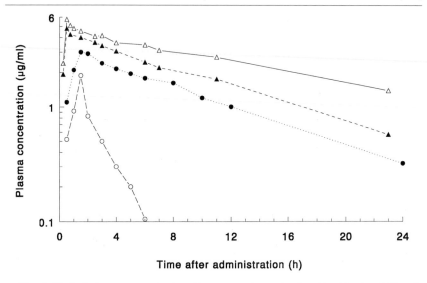

FIG. 1. Typical plasma concentration–time curves for antipyrine ($\triangle$; AP, dose 250 mg), theophylline (▲; TH, 150 mg), R-($-$)-hexobarbital ($\bigcirc$; R-HB, dose 250 mg), and S-($+$)-hexobarbital (●; S-HB, dose 250 mg) after oral administration in humans.

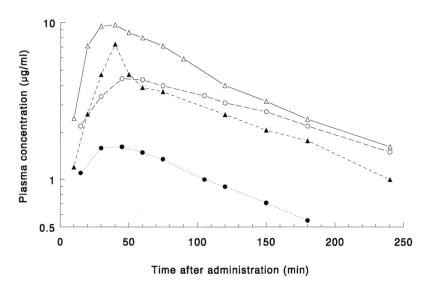

FIG. 2. Typical plasma concentration–time curves for antipyrine ($\triangle$; AP, dose 15 mg/kg), theophylline (▲; TH, dose 4.5 mg/kg), R-($-$)-hexobarbital ($\bigcirc$; R-HB, dose 12.5 mg/kg), and S-($+$)-hexobarbital (●; S-HB, dose 12.5 mg/kg) after oral administration in female BN/BiRij rats.

humans and in experimental animals. Thereby, much insight has been obtained in the factors governing overall oxidative drug-metabolizing enzyme activity *in vivo*. Studies with other probe drugs, which are more specific for single P450 enzymes, are needed to elucidate selectivity in altered enzyme activity.

[20] Imipramine: A Model Drug for P450 Research

By KIM BRØSEN, ERIK SKJELBO, and KARIN KRAMER NIELSEN

In 1955 the Swiss physician Kuhn discovered the powerful antidepressant properties of imipramine. Even today imipramine is a very important drug for the management of depression. The drug is also useful in the treatment of a variety of other clinical conditions such as panic disorder, diabetic neuropathy, and enuresis nucturna.

Imipramine is the prototype of the class of tricyclic antidepressants (Fig. 1). The drug is predominantly eliminated by oxidation catalyzed by the cytochrome P450 system in the liver.[1] The major oxidative pathways are N-demethylation in the side chain to the active metabolite desipramine (Fig. 1) and aromatic hydroxylation to 2-hydroxyimipramine. More than 90% of 2-hydroxyimipramine is excreted in the form of glucuronides in the urine. 10-Hydroxyimipramine and imipramine-N-oxide are minor metabolites. Desipramine is mainly eliminated by 2-hydroxylation (Fig. 1) and, to a much lesser extent, by further N-demethylation to didesmethylimipramine and by 10-hydroxylation. About 75% of 2-hydroxydesipramine is excreted in the urine in the form of glucuronides.[1]

With the development of an applicable thin-layer chromatography (TLC) method for the quantitative assessment of imipramine and desipramine in plasma it became possible to establish the basic pharmacokinetics of imipramine.[2] Imipramine has a high systemic clearance (0.8–1.5 liter/min, hence a pronounced first-pass metabolism (25–75%), and a large volume of distribution (v_d: 650–1100 liter) so the elimination half-life is therefore of intermediate length: 12 hr.

[1] L. F. Gram, *Dan. Med. Bull.* **21,** 218 (1974).
[2] L. F. Gram and J. Christiansen, *Clin. Pharmacol. Ther.* **17,** 555 (1975).

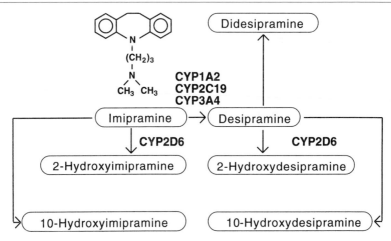

FIG. 1. The biotransformation of imipramine in humans (imipramine-*N*-oxide not included).

In a subsequent clinical study,[3,4] the TLC assay was used to determine imipramine and desipramine in plasma of patients with a fixed imipramine dose. The study showed a 30- to 60-fold variability in imipramine and desipramine levels; this pharmacokinetic variability is a characteristic for tricyclic antidepressants. The study showed that the optimal effect was achieved when the plasma concentrations ranged from 700 to 1400 nM. Accordingly, the interindividual difference in the oxidation of imipramine and desipramine is a major factor controlling the clinical outcome on a fixed dose.

Another important observation of the study was that there seemed to be a poor intraindividual correlation between the plasma levels of imipramine and of desipramine. It was therefore hypothesized that the *N*-demethylation of imipramine and the 2-hydroxylation of desipramine (and presumably also of imipramine) are catalyzed by different cytochrome P450 enzymes. With a subsequent improvement of the TLC method[5] it was discovered that there was a distinct subgroup of patients that had low or immeasurable 2-hydroxymetabolite levels but high imipramine and especially very high desipramine levels. This further supported the notion that

[3] L. F. Gram, I. Søndergaard, J. Christiansen, G. O. Petersen, P. Bech, N. Reisby, I. Ibsen, J. Ortmann, A. Nagy, S. J. Dencker, O. Jacobsen, and O. Krautwald, *Psychopharmacology* **54,** 255 (1977).

[4] N. Reisby, L. F. Gram, P. Bech, A. Nagy, G. O. Petersen, J. Ortmann, I. Ibsen, S. J. Dencker, O. J. Jacobsen, O. Krautwald, I. Søndergaard, and J. Christiansen, *Psychopharmacology* **54,** 263 (1977).

[5] L. F. Gram, M. Bjerre, P. Kragh-Sørensen, B. Kvinesdal, J. Molin, O. L. Pedersen, and N. Reisby, *Clin. Pharmacol. Ther.* **33,** 353 (1983).

the 2-hydroxylation and the N-demethylation are catalyzed by different P450s, and imipramine was one of the first drugs for which a regioselective oxidation by different P450s was proposed. This hypothesis has formed the basis for numerous subsequent studies and it has been of paramount importance for the present status of imipramine as a model drug for P450 research.

Studies in imipramine-treated patients[6] and in panels of healthy subjects phenotyped with regard to sparteine oxidation[7] showed that the sparteine oxidation phenotype has a major impact on the 2-hydroxylation of imipramine and desipramine but not on the N-demethylation. In agreement with this, subsequent in vitro studies.[8-10] confirmed that CYP2D6, the source of the sparteine/debrisoquine oxidation polymorphism, is a major enzyme in catalyzing the 2-hydroxylation of imipramine, but that the enzyme does not contribute quantitatively to the N-demethylation.

A panel study including six poor metabolizers of S-mephenytoin[11] showed that the N-demethylation clearance of imipramine was only about 50% of the value in extensive metabolizers. The importance of the S-mephenytoin oxidation phenotype, alias CYP2C19, has been confirmed in two subsequent population studies.[12,13] An in vitro study showed that CYP1A2 and CYP3A4 are two other important P450 enzymes catalyzing the N-demethylation of imipramine.[10]

The aim of this chapter is to give an account of the HPLC methods currently used in our laboratory for the assay of imipramine and its metabolites in plasma, urine, and human liver microsome preparations. Each of the two methods has been published in detail previously.[14,15]

Experimental

Chemicals and Reagents

Imipramine hydrochloride, desipramine hydrochloride, didesmethylimipramine hydrochloride, 2-hydroxyimipramine hydrochloride, 10-hy-

[6] K. Brøsen, R. Klysner, L. F. Gram, S. V. Otton, P. Bech, and L. Bertilsson, Eur. J. Clin. Pharmacol. 30, 679 (1986).
[7] K. Brøsen, S. V. Otton, and L. F. Gram, Clin. Pharmacol. Ther. 40, 543 (1986).
[8] K. Brøsen, T. Zeugin, and U. A. Meyer, Clin. Pharmacol. Ther. 49, 609 (1991).
[9] E. Skjelbo and K. Brøsen, Br. J. Clin. Pharmacol. 34, 256 (1992).
[10] A. Lemoine, J. C. Gautier, D. Azoualy, L. Kiffel, F. P. Guengerich, P. Beaune, P. Maurel, and J. P. Leroux, Mol. Pharmacol. 43, 827 (1993).
[11] E. Skjelbo, K. Brøsen, J. Hallas, and L. F. Gram, Clin. Pharmacol. Ther. 49, 18 (1991).
[12] E. Skjelbo, L. F. Gram, and K. Brøsen, Br. J. Clin. Pharmacol. 35, 331 (1993).
[13] H. Madsen, K. Kramer Nielsen, and K. Brøsen, Br. J. Clin. Pharmacol. 39, 433 (1995).
[14] K. Kramer Nielsen and K. Brøsen, J. Chromatogr. 612, 87, (1993).
[15] T. Zeugin, K. Brøsen, and U. A. Meyer, Anal. Biochem. 189, 99 (1990).

droxyimipramine free base, 2-hydroxydesipramine oxalate, 10-hydroxydesipramine fumarate, and 2-hydroxydesmethylclomipramine free base were kindly supplied by Ciba-Geigy (Basel, Switzerland). Stock solutions were prepared in ethanol (96%) and stored at −20°. Chemicals were of analytical grade and purchased from Merck (Darmstadt, Germany). β-Glucuronidase/arylsulfatase was purchased from Boehringer (Mannheim, Germany). Acetonitrile, dichlormethane, tert-butylmethyl ether, and n-butanol were of HPLC grade and obtained from Merck. Heptane was of HPLC grade and purchased from Rathburn (Walkerburn, UK). Water was purified by osmosis and distillation. The samples were eluted with a mixture of 30% acetonitrile and 70% aqueous sodium perchlorate solution, pH 2.5; the aqueous sodium perchlorate solution was prepared by the addition of 14.05 g of sodium perchlorate and 1.6 ml of 60% perchloric acid to 5000 ml of water. The aqueous solution was filtered through a Milipore filter (0.45 μm) and the eluent was degassed prior to use.

Apparatus

The HPLC system consists of Hitachi instruments (Hitachi, Tokyo, Japan): an AS-2000 autosampler with a 100-μl injector loop, a T-6300 column thermostat, a L-6200 intelligent pump, and a L-4250 UV-VIS detector with variable wavelength. The system is controlled through a D-6000 HPLC interface module and a personal computer (IBM). The RP-phenyl column (Nucleosil, 5 μm, 100 Å, 250 × 4 mm i.d., Machery-Nagel, Düren, Germany) is equipped with a guard column (Nucleosil, 7 μm, 120 Å, 20 × 4 mm i.d., Machery-Nagel). Elution is carried out at a flow rate of 1.0 ml/min and a column temperature of 30°. The column effluent is quantified at the wavelength of 220 nm.

In the following sample, pretreatment and results are discussed separately for urine/plasma and the microsomal assay.

Pretreatment of Urine and Plasma Samples

Deconjugation. In order to determine total concentrations in urine, including glucuronide conjugates, the urine is enzyme treated with β-glucuronidase/arylsulfatase prior to extraction: 470 μ of 0.2 M potassium dihydrogenphosphate plus 3% L-(+)-ascorbic acid and 30 μl of β-glucuronidase/arylsulfatase are added to 0.5 ml of urine in a 10-ml glass test tube. The mixture is vortex mixed for 2 sec. Incubation is performed in capped tubes at 37° for 16 hr in a water bath. Fifty microliters of a 2 M sodium hydroxide solution is added to the mixture. The mixture is then extracted according to the following procedure.

Extraction Procedure. To 1.0 ml of plasma or urine (or the enzyme-treated urine mixture) in a 10-ml glass test tube, 1 ml of a 0.6 M potassium carbonate solution (pH 11), 100 μl of a 2-hydroxydesmethylclomipramine solution in ethanol (5 and 20 μM for plasma and urine samples, respectively) as an internal standard, and 5 ml of heptane-*tert*-butylmethyl ether (1:1, v/v) plus 5% *n*-butanol are added. The mixture is vortex mixed for 1 min and is centrifuged for 10 min at 1400 g. The test tube is maintained at −50° (ethanol bath) until the aqueous layer is frozen. The organic layer is transferred into a test tube containing 1 ml of a 0.02 M hydrochloric acid solution. The mixture is vortex mixed for 1 min, centrifuged for 10 min at 1400 g, and frozen. The organic layer is discarded, and the aqueous layer is thawed and made alkaline (pH 11) after addition of 0.5 ml of a 0.6 M potassium carbonate solution (pH 11.3). Subsequently, 3 ml of heptane-*tert*-butyl ether (1:1, v/v) plus 5% *n*-butanol is added. The mixture is vortex mixed for 1 min, centrifuged for 10 min at 1400 g, and frozen. THe organic layer is transferred to a conical glass test tube and evaporated to dryness at 50° under a stream of nitrogen. The residue is dissolved in 100 μl of eluent, vortex mixed for 5 sec, and centrifuged for 1 min at 1400 g. A 20-μl aliquot is injected onto the column.

Selectivity. As shown in Fig. 2B, baseline separation of imipramine, the metabolites, and the internal standard is achieved with the applied conditions. About 18 min are required for the analysis. The retention times are 5.93, 6.55, 7.37, 8.25, 10.02, 12.01, 14.32, and 16.62 for 10-hydroxy-desipramine, 10-hydroxyimipramine, 2-hydroxydesipramine, 2-hydroxy-imipramine, 2-hydroxydesmethylclomipramine, didesmethylimipramine, desipramine, and imipramine, respectively. No interference from impurities produced by the plasma, urine, or the additives from the sample preparation is detected at the detection wavelength (220 nm).

Imipramine-N-oxide is lost during the sample pretreatment and is there-fore not included as a metabolite. However, in the chromatograms obtained from enzyme-treated, spiked urine, there are peaks, due to artifacts, that interfere with the integration of didesmethylimipramine and 10-hydroxy-desipramine peaks. These artifacts may be produced during the deconjuga-tion procedure. Figure 2 shows typical chromatograms for enzyme-treated, blank urine (A); enzyme-treated, spiked urine (B); patient plasma (C); urine without enzyme treatment from a healthy volunteer (D); and enzyme-treated urine from the same volunteer (E).

The detector response is linear with concentrations that range from 0.05 to 2.0 μM in plasma and with concentrations that range from 0.5 to 25 μM in urine (with and without enzyme treatment) for all seven compounds. Prior to analysis of unknown samples, calibration curves were prepared of three standard levels covering the expected concentration range. The linear

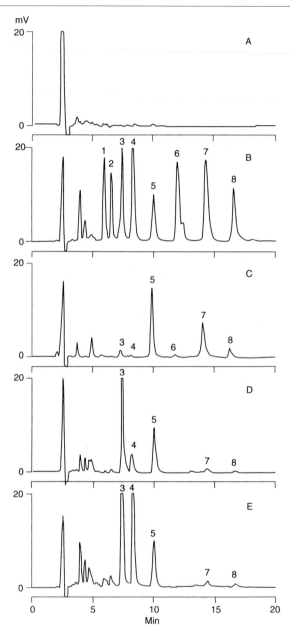

FIG. 2. (A) Chromatogram of blank, enzyme-treated urine. (B) Chromatogram of enzyme-treated urine, spiked to 10 μM with imipramine and metabolites; 100 μl of 20 μM **5** was added to the sample before extraction. (C) Chromatogram of a 1-ml plasma extract from a

calibration curves were fitted through the data point by linear regression. The quantitative analysis of an unknown sample was derived with reference to the internal standard.

The limit of detection, based on a signal-to-noise ratio of 3 : 1, is 5 nM for each compound in plasma and 10 nM in urine without enzyme treatment. In enzyme-treated urine, the detection limit for the hydroxy metabolites is 100 nM due to interfering peaks. For imipramine, desipramine, and didesmethylimipramine, the limit is 30 nM. The limit of determination based on a signal-to-noise ratio of 10 : 1 is 15 nM for plasma samples and 50 nM for untreated urine. In enzyme-treated urine, the limit is 1.0 μM for the hydroxymetabolites and 100 nM for didesmethylimipramine, desipramine, and imipramine.

Details regarding recovery, linearity, repeatability, reproducibility, and accuracy have been published previously.[14]

Assay of Imipramine Metabolism in Human Liver Microsomes

Microsomes are prepared from whole human livers obtained from kidney donor patients before circulatory arrest. The livers are kept at $-80°$ without loss of activity. There are several published methods for the preparation of microsomes. The one we use is described in detail elsewhere.[16]

Stock solutions of 10 mM of imipramine (or 5 mM of desipramine) are prepared in distilled water. (Imipramine should not be dissolved in ethanol because ethanol inhibits the microsomal oxidation.) Microsomes are incubated in a final incubation volume of 200 μl in a disodium phosphate buffer (100 mM, pH 7.4) using 50–100 μg of protein, depending on the activity of the microsomes and of the substrate concentration. The microsomal suspension is preincubated for 5 min at room temperature with 20 μl imipramine solution. The reactions are started by adding 20 μl of a

---

[16] P. J. Meier, H. K. Mueller, B. Dick, and U. A. Meyer, *Gastroenterology* **85**, 682 (1983).

---

patient receiving an oral dose of 175 mg of imipramine per day containing 60 nM **3**, 15 nM **4**, 30 nM **6**, 640 nM **7**, and 180 nM **8**; 100 μl of 5 μM **5** was added to the sample before extraction. (D) Chromatogram of a 1-ml urine extract, without enzyme treatment, from a healthy volunteer receiving an oral dose of 25 mg imipramine containing 5.7 μM **3**, 0.6 μM **4**, 0.3 μM **7**, and 0.3 μM **8**; 100 μl of 20 μM **5** was added to the sample before extraction. (E) Chromatogram of a 0.5-ml enzyme-treated urine extract from the same volunteer as in D containing 14.0 μM **3**, 12.3 μM **4**, and below 1 μM **7** and **8**; 100 μl of 20 μM **5** was added to the sample before extraction. Peaks: **1**, 10-hydroxydesipramine; **2**, 10-hydroxyimipramine; **3**, 2-hydroxydesipramine; **4**, 2-hydroxyimipramine; **5**, 2-hydroxydesmethylclomipramine; **6**, didesmethylimipramine; **7**, desipramine; and **8**, imipramine. Reprinted from Kramer Nielsen and Brøsen,[14] with permission from Elsevier Science B.V., Amsterdam, The Netherlands.

NADPH-generating system (concentration in microsomal suspension: isocitrate dehydrogenase, 1 unit/ml; NADP-Na$_2$, 1 mM; isocitrate, 5 mM; MgCl$_2$, 5 mM).

Incubations are carried out at 37° in a shaking water bath in air and the reactions are stopped after 20 min by adding 800 μl of 12.5% ammonia.

Calibration is carried out by adding standard solutions to the microsomal incubation system, inactivated by adding ammonia. Ammonia is added before the NADPH-generating system to avoid metabolism of the standards.

A 20-μl sample of the internal standard (2-hydroxydesipramine) working solution (20 μM) is added, followed by 5.0 ml dichlormethane. The tube is mixed at maximum speed for 30 sec on a vortex mixer (dual press-to-mix, Snijders, Holland). (With vortexing for more than 30 sec, broad irregular peaks appeared on the chromatograms; this is probably due to impurities arising from homogenization of either microsomes and/or the regenerating system.) After centrifugation at 1000 g for 10 min, the aqueous phase is discarded, and the organic phase is transferred to a conical tube and evaporated to dryness at 50° under a stream of nitrogen. The residue is dissolved in 100 μl of HPLC solvent, and 40 μl of this solution is injected onto the HPLC column. Unknown metabolite concentrations in incubations are quantitated by comparing the peak area ratios of the internal standards by means of a calibration curve. The calibration curves obtained passed through or nearby the origin and are linear within the concentration range of 50–1000 pmol per tube for all compounds. The limit of determination, based on a signal-to-noise ratio of 6:1, is 20 pmol.

2- and 10-hydroxyimipramine and desipramine are the only metabolites formed after incubation with imipramine. With desipramine as a substrate, 2- and 10-hydroxydesipramine and didesmethylimipramine are the only metabolites formed.[15] The fact that 2-hydroxydesipramine is not formed after incubation with imipramine makes it possible to use the compound as an internal standard. The use of 2-hydroxydesipramine offers another advantage: to serve as an antioxidant. The problem is that some batches of either dichlormethane or acetonitrile probably contain oxidants leading to further oxidation and hence disappearance of 2-hydroxyimipramine. This is especially true during the summer time when the temperature is high in the laboratory.

The formation rate of 2- and 10-hydroxyimipramine and desipramine is linear with time for up to 30 min and with a protein concentration from 25 to 100 μg protein per tube.

Details regarding recovery and intraday reproducibility have been published elsewhere.[15]

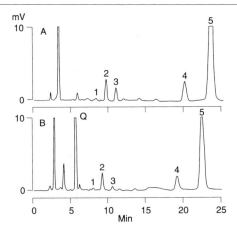

FIG. 3. Chromatograms of imipramine and its metabolites. Microsomal incubation with 32 μM imipramine (A) without and (B) with 1 μM of the CYP2D6 inhibitor quinidine (Q). **1,** 10-hydroxyimipramine; **2,** 2-hydroxydesipramine (internal standard); **3,** 2-hydroxyimipramine; **4,** desipramine; **5,** imipramine; **Q,** quinidine.

Discussion

This chapter describes isocratic, reversed-phased HPLC methods for the simultaneous assay of imipramine and six of its metabolites in plasma, urine, and human liver microsomes.

The plasma and urine method has been applied in several published studies. In a panel study[17] we showed for the first time that the minor metabolite 10-hydroxydesipramine, which is formed in detectable amounts only after desipramine intake, is not catalyzed by CYP2D6. In a population study[13] we examined whether imipramine could replace sparteine as a model drug for the determination of the oxidation phenotype. Indeed, assessment of the 2-hydroxylation of imipramine and desipramine could not separate extensive and poor metabolizers of sparteine completely. The major reason is that the 2-hydroxylation, in addition to CYP2D6, is also catalyzed by alternative, low-affinity P450s.

The microsomal HPLC method was initially developed in order to use the 2-hydroxylation of imipramine as a marker reaction for CYP2D6 *in vitro.* Thus several inhibitors of CYP2D6 have been identified or confirmed using imipramine as a probe drug, e.g., quinidine (Fig. 3) and selective serotonin reuptake inhibitors (SSRI).[8,9] The possible effects of these agents

---

[17] K. Brøsen, M. G. J. Hansen, K. Kramer Nielsen, S. H. Sindrup, and L. F. Gram, *Eur. J. Clin. Pharmacol.* **44,** 349 (1993).

on the P450s catalyzing the formation of desipramine were initially a secondary consideration. However, it was shown that the SSRI fluvoxamine is a very potent inhibitor of one of the P450s that catalyzes the N-demethylation of imipramine.[9] The P450 turned out to be CYP1A2.[18] This is the most important discovery made by means of the HPLC method.

[18] K. Brøsen, E. Skjelbo, B. B. Rasmussen, H. E. Poulsen, and S. Loft, *Biochem. Pharmacol.* **45,** 1211 (1993).

[21] Measurement of Human Liver Microsomal Cytochrome P450 2D6 Activity Using [O-Methyl-14C]Dextromethorphan as Substrate

By A. DAVID RODRIGUES

Introduction

Of the human liver microsomal cytochromes P450 (CYP), one of the best documented and polymorphically expressed is cytochrome P450 2D6 (CYP2D6, P450IID6, P450db1 or $P450_{bufI}$).[1-3] The *CYP2D6* gene is inherited as an autosomal recessive trait and separates 90 and 10% of the white European and North American population into "extensive" and "poor" metabolizer phenotypes, respectively.[2] Poor metabolizer (PM) subjects not only lack functional CYP2D6 protein, but also have impaired metabolism of more than 30 drugs. In addition, if two or more drugs are metabolized by CYP2D6 in extensive metabolizer (EM) subjects, then the likelihood of a potentially hazardous drug–drug interaction is increased. This is of particular importance when dealing with CYP2D6 substrates exhibiting a relatively narrow therapeutic index (e.g., desipramine, amitriptyline, or imipramine).

Therefore, a number of authors have reported methods for studying CYP2D6 monooxygenase activity *in vitro* as a means of screening for potential clinically relevant drug–drug interactions.[3-6] These methods

[1] U. A. Meyer, J. Gut, T. Kronbach, C. Skoda, U. T. Meier, and T. Catin, *Xenobiotica* **16,** 449 (1986).
[2] U. A. Meyer, R. C. Skoda, and U. M. Zanger, *Pharmacol. Ther.* **46,** 297 (1990).
[3] T. Kronbach, D. Mathys, J. Gut, T. Catin, and U. A. Meyer, *Anal. Biochem.* **162,** 24 (1987).
[4] S. V. Otton, T. Inaba, and W. Kalow, *Life Sci.* **32,** 795 (1982).
[5] S. V. Otton, T. Inaba, and W. Kalow, *Life Sci.* **34,** 73 (1983).
[6] T. Inaba, M. Jurima, W. A. Mahon, and W. Kalow, *Drug Metab. Dispos.* **13,** 443 (1985).

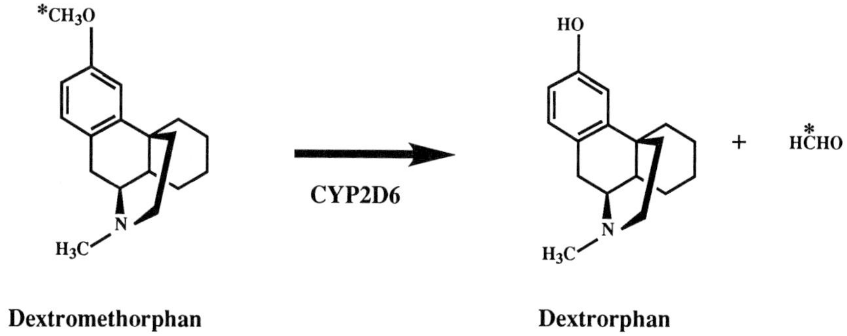

FIG. 1. The *O*-demethylation of [*O-methyl*-14C]dextromethorphan. An asterisk indicates the position of the carbon-14 label.

usually employ substrates such as debrisoquine, bufuralol, or sparteine, and involve high-performance liquid chromatographic (HPLC) analysis of the parent drug and its metabolites. Similarly, dextromethorphan has been increasingly used as a probe for monitoring CYP2D6 activity both *in vitro* and *in vivo*; the enzyme has been shown to selectively catalyze the *O*-demethylation of the parent drug (Fig. 1), yielding dextrorphan as the demethylated product.[3,7–10] Invariably, the assay involves the separation of dextrorphan from other metabolites and parent drug by HPLC with fluorescence detection.

This chapter describes a relatively simple, rapid, and sensitive assay procedure for measuring CYP2D6-dependent monooxygenase activity in human liver microsomes employing [*O-methyl*-14C]dextromethorphan as substrate.[11] The assay involves the radiometric measurement of [14C]formaldehyde after a single-step extraction procedure and does not require the use of HPLC, an internal standard, or metabolite standards. Furthermore, the method circumvents problems with interfering (coeluting and/or fluorescent) compounds and may prove useful for the bulk screening of potential CYP2D6 cosubstrates and/or inhibitors. Although only human liver

[7] P. Dayer, T. Leemann, and R. Striberni, *Clin. Pharmacol. Ther.* **45,** 34 (1989).

[8] R. J. Guttendorf, P. J. Wedlund, J. Blake, and S.-L. Chang, *Ther. Drug Monit.* **10,** 490 (1988).

[9] Z.-Y. Hou, L. W. Pickle, P. S. Meyer, and R. L. Woosley, *Clin. Pharmacol. Ther.* **49,** 410 (1991).

[10] A. Kupfer, B. Schmid, and G. Pfaff, *Xenobiotica* **16,** 421 (1986).

[11] A. D. Rodrigues, M. J. Kukulka, B. W. Surber, S. B. Thomas, J. T. Uchic, G. A. Rotert, G. Michel, B. Thome-Kromer, and J. M. Machinist, *Anal. Biochem.* **219,** 309 (1994).

microsomal CYP2D6 is discussed, the method may also prove useful for measuring CYP2D activity in nonhuman species such as rat and monkey.[12,13]

Materials

All reagents and solvents required for the assay can be obtained from various commercial sources and it is only a matter of time before [O-methyl-^{14}C]dextromethorphan is also commercially available. However, a brief description of its synthesis is warranted.

Synthesis of [O-Methyl-^{14}C]Dextromethorphan

Originally, the material was prepared from dextromethorphan (Sigma Chemical Co.) and [^{14}C]methylamine hydrochloride (Amersham International, 55 mCi/mmol).[11] The latter was used to prepare [^{14}C]diazald or N-[^{14}C]methyl-N-nitroso-p-toluenesulfonamide, while dextromethorphan had to be O-demethylated to yield dextrorphan. However, dextrorphan is now commercially available (Research Biochemicals International, Natick, MA) and [^{14}C]diazomethane can be prepared from commercially available [^{14}C]diazald (Sigma Chem. Co.). Synthesis of the final product involves reacting dextrorphan with [^{14}C]diazomethane. A solution of [^{14}C]diazomethane (16 mCi, 0.31 mmol) in ether (8 ml) is transferred via pipette to a solution of dextrorphan (111 mg, 0.43 mmol) in methanol (1 ml) contained in a thick-walled tube with a narrow neck suitable for sealing with a flame. The system is cooled with liquid nitrogen, evacuated, sealed, and heated to 40° for 22 hr.[11] After cooling to −78°, the tube is opened and a few drops of glacial acetic acid are added to quench the unreacted [^{14}C]diazomethane. The volatiles are evaporated *in vacuo*, chasing with methanol to give 3.6 mCi in the condensate. The residue (12.4 mCi) is applied to a column of silica gel (3 × 30 cm), which is eluted with hexane:acetone:NH$_4$OH (50:50:1, v/v/v), and fractions containing [O-methyl-^{14}C] dextromethorphan are combined and concentrated, chasing with ethanol. The final residue (62% yield, 10 mCi, 0.20 mmol) is dissolved in ethanol (4 ml). The identity, purity (>99%), and specific activity (51 mCi/mmol) of the final product are confirmed by radio-HPLC and mass spectrometric analysis.[11] The material can be stored in ethanol (−20°) for long periods of time (>12 months) without loss of purity. However, continual monitoring by radio-HPLC is advised since impurities can give rise to high assay backgrounds. Typically, the radiolabeled dextromethorphan is used for

[12] D. Larrey, L. M. Distlerath, G. A. Dannan, G. R. Wilkinson, and F. P. Guengerich, *Biochemistry* **23**, 2787 (1984).
[13] D. Wu, S. V. Otton, P. Morrow, T. Inaba, W. Kalow, and E. M. Sellers, *J. Pharmacol. Exp. Ther.* **266**, 715 (1993).

microsome incubations without dilution with unlabeled drug. However, when higher concentrations of [*O-methyl-*[14]C]dextromethorphan ($>40 \mu M$) are needed, it may be wise to decrease the specific activity with unlabeled drug.

Source of Microsomes

Human liver tissue can be obtained from various sources (e.g., International Institute for the Advancement of Medicine, IIAM, Exton, PA), and microsomes are prepared using standard differential centrifugation procedures.[11] The final microsome pellet is resuspended in 0.1 M potassium phosphate buffer (pH 7.4) containing 0.1 mM EDTA and glycerol (20%, v/v) so that the final concentration of protein is 10–25 mg/ml. All samples are stored at $-70°$. Alternatively, human liver microsomes can be purchased directly from various sources such as IIAM, Human Biologics Inc. (Phoenix, AZ), or *In Vitro* Technologies (Baltimore, MD). Microsomes prepared from human B-lymphoblastoid cells transfected with cDNAs for the major human CYP proteins can be obtained from Gentest Corp. (Woburn, MA).

Reagents

> Assay buffer: 0.1 M potassium phosphate (pH 7.4), containing 0.1 mM EDTA
> 179 mM semicarbazide HCl dissolved in distilled water, pH to 7.0, with NaOH
> 150 mM MgCl$_2$ in distilled water
> 0.6 M NaOH in distilled water
> Methylene chloride
> [*O-Methyl-*[14]C]dextromethorphan (51 mCi/mmol), 4.0 mM (0.2 mCi/ml), in ethanol
> [^{14}C]HCHO (Amersham International, Arlington Heights, IL; 30.1 mCi/mmol), 0.05 mM, in distilled water
> NADPH-generating system, in assay buffer: glucose 6-phosphate (0.1 M), NADP$^+$ (40 mM), glucose-6-phosphate dehydrogenase (Sigma Type VII, from baker's yeast, 20 U/ml)
> Hionic-Fluor scintillation cocktail (Packard Instrument Co., Meriden, CT)

Assay

Incubation of [O-Methyl-^{14}C]Dextromethorphan with Microsomes*

The stock solution of [*O-methyl-*[14]C]dextromethorphan (0.2 mCi/ml) can be diluted with ethanol in order to give the required working (final) concentrations of substrate. Incubations are carried out in open tubes (16 × 125-mm borosilicate glass diSPo screw-cap culture tubes, Fisher Scientific) in a shaking water bath at 37° (under air). A typical assay mixture consists of assay buffer, $MgCl_2$ (0.1 ml), human liver or B-lymphoblastoid microsomes (0.1–1.0 mg protein/ml), semicarbazide HCl (0.1 ml), and [O-methyl-^{14}C]dextromethorphan (1–40 μM, added in 10 μl of ethanol) in a final volume of 0.9 ml. After a preincubation period of 3 to 5 min, the reaction is initiated with the NADPH-generating system (0.1 ml). The reaction is allowed to proceed for the required length of time and is terminated with 0.6 M NaOH (1.0 ml). After thorough mixing using a vortex device, the samples are subsequently extracted with methylene chloride as described below. A number of incubations are carried out in the absence of the NADPH-generating system in order to determine the levels of unmetabolized [O-methyl-^{14}C]dextromethorphan remaining in the aqueous phase after extraction.

Extraction of [^{14}C]HCHO

Methylene chloride (7.0 ml) is added to the basified incubates and the tubes are closed with caps containing Teflon-faced cap liners. After mixing, using a vortex device (20 sec), the tubes are placed in a horizontal position and mechanically shaken for 1 hr at room temperature (e.g., Eberbach 6000 variable speed shaker, set at 260 oscillations/min). The phases can be separated by centrifugation (2000 g for 10 min; e.g., Sorvall Model RT6000B centrifuge), and the top portion (1.0 ml) of the aqueous phase is removed and mixed with 15 ml of Hionic-Fluor scintillation cocktail in appropriately sized liquid scintillation vials (Research Products International Corp.). The samples are analyzed for [^{14}C]formaldehyde by radioassay (e.g., Tri-Carb Packard Model 2500 liquid scintillation spectrometer) and corrected for quenching with Automatic external standardization. Standards of [^{14}C]formaldehyde (0.05–4.0 nmol/ml, 30.1 mCi/mmol) are routinely run in parallel with the microsomal assays (minus radiolabeled dextromethorphan) in order to determine the recovery of labeled formaldehyde. This extraction procedure allows for almost complete (>99%) recovery of unmetabolized [O-methyl-^{14}C]dextromethorphan in the organic phase.[11]

The quantity of formaldehyde formed is calculated from the net disintegrations per minute observed (sample minus no NADPH blank disintegrations per minute), corrected to the total volume of aqueous phase (1.0-ml aliquot/2.0 ml aqueous phase), and corrected for formaldehyde recovery (94.4 ± 1.6%). The total corrected net disintegrations per minute is then converted to nanomoles of product from the specific activity of the substrate (113,220 disintegrations/min/nmol). In turn, data are normalized with respect to the concentration of microsomal protein in the assay and the time of incubation. A standard 20-min incubation, in the presence of the NADPH-generating system and microsomal protein (0.5 mg/ml), can yield as much as 80,000 disintegrations/min/ml aqueous phase, whereas radioactivity in the absence of NADPH is considerably lower (~2000 disintegrations/min/ml aqueous phase).

Sample Data

The O-demethylation of [O-methyl-^{14}C]dextromethorphan (1–40 μM) has been studied with a panel of microsomes prepared from a series of 11 different human livers and with human B-lymphoblastoid microsomes containing various individual cDNA-expressed CYP proteins.[11] In the presence of human liver microsomes, selectivity for CYP2D6 has been verified using kinetic and immunological approaches, and with the use of known CYP-selective inhibitors.[11]

cDNA-Expressed CYP2D6

Upon incubation of [O-methyl-^{14}C]dextromethorphan with human liver microsomes or cDNA-expressed CYP2D6, the formation of [^{14}C]HCHO is linear with respect to microsomal protein concentration (0.1–1.0 mg/ml) and time of incubation up to 30 min (Fig. 2A).[11] Data obtained with various cDNA-expressed CYP proteins indicate that CYP2D6 is able to catalyze the O-demethylation of [O-methyl-^{14}C]dextromethorphan at a rate (80 ± 1.8 pmol HCHO/hr/pmol CYP) that is nearly 50-fold higher than that of other CYP forms (Fig. 2B). However, one must not lose sight of the fact that CYP2D6 represents only a minor fraction (≤5.0%) of the total CYP pool in human liver microsomes when compared to other CYP forms such as CYP3A (15–60%) and CYP2E1 (3–18%).[14] However, the use of low [O-methyl-^{14}C]dextromethorphan concentrations in the assay (≤40 μM) prevents other, non-CYP2D6, CYP forms from contributing to the O-demethylation reaction in human liver microsomes. Under these assay conditions, CYP2D6 represents the "high-affinity component" ($K_m \leq 20$ μM) in subjects expressing CYP2D6.

Enzyme Kinetic Studies with Human Liver Microsomes

Under the presently described assay conditions, [O-methyl-^{14}C]dextromethorphan O-demethylase activity varies considerably among microsome

[14] T. Shimada, H. Yamazaki, M. Mimura, Y. Inui, and F. P. Guengerich, *J. Pharmacol. Exp. Ther.* **270,** 414 (1994).

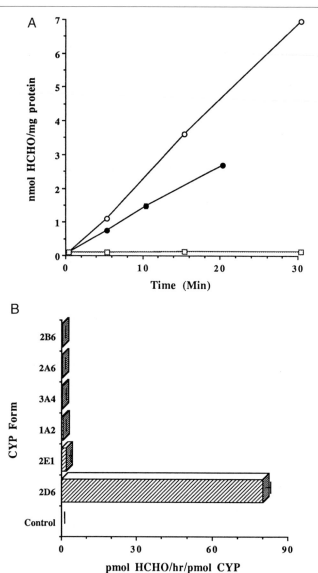

Fig. 2. *O*-Demethylation of [*O-methyl*-^{14}C]dextromethorphan in the presence of cDNA-expressed CYP2D6. (A) Time-dependent formation of [^{14}C]HCHO in the presence of control B-lymphoblastoid microsomes (□), B-lymphoblastoid microsomes containing cDNA-expressed CYP2D6 (○), or native human liver microsomes containing CYP2D6 (●). (B) *O*-Demethylase activity measured in the presence of B-lymphoblastoid microsomes containing various cDNA-expressed human CYP forms. All incubations were carried in the presence of microsomal protein (0.5 mg/ml) and [*O-methyl*-^{14}C]dextromethorphan (20 μM).

samples of different human livers (0.09–13 nmol HCHO/hr/mg; 144-fold). Despite the marked variability, the activity is significantly correlated ($r >$ 0.808, $P < 0.01$; $n \geq 10$) with the immunologically determined levels of CYP2D6 protein in a panel of human liver microsomes.[11] The work of others has shown that human liver samples can be phenotyped *in vitro* by determining the microsomal [*O-methyl*-^{14}C]dextromethorphan *O*-demethylase V_{max}/K_m ratio (Fig. 3).[7] In the case of individuals phenotyped *in vivo* as EM subjects, the *O*-demethylation of dextromethorphan is usually

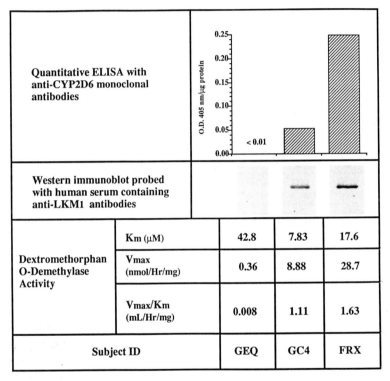

Quantitative ELISA with anti-CYP2D6 monoclonal antibodies			
Western immunoblot probed with human serum containing anti-LKM1 antibodies			
Dextromethorphan O-Demethylase Activity	Km (μM)	42.8	7.83	17.6
	Vmax (nmol/Hr/mg)	0.36	8.88	28.7
	Vmax/Km (mL/Hr/mg)	0.008	1.11	1.63
Subject ID		GEQ	GC4	FRX

FIG. 3. Measurement of CYP2D6 protein level and activity in human liver microsomes. Samples of microsomes, from the liver tissue of three organ donor subjects (GC4, FRX, and GEQ), were analyzed for CYP2D6 protein content using qualitative (Western immunoblot) and quantitative (enzyme-linked immunosorbent assay, ELISA) immunoassays.[11] The *O*-demethylation of [*O-methyl*-^{14}C]dextromethorphan (1–40 μ*M*) was measured in the presence of microsomal protein (0.5 mg/ml) to yield estimates of apparent K_m and V_{max}. Subjects GC4 and FRX ($V_{max}/K_m > 1.0$ ml/hr/mg) were characterized as "extensive metabolizers." CYP2D6 was not detected in microsomes of subject GEQ ($V_{max}/K_m < 0.05$ ml/hr/mg) who classified as a "poor metabolizer." Erythromycin *N*-demethylase (CYP3A) activity in microsomes of subject GEQ, GC4, and FRX was 0.78 ± 0.06, 0.58 ± 0.04, and 0.81 ± 0.08 nmol HCHO/min/mg, respectively.[11]

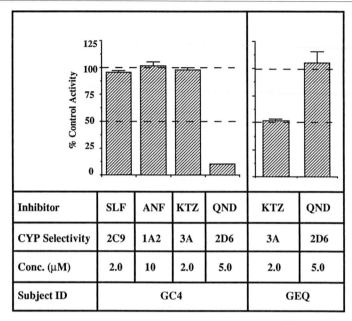

Inhibitor	SLF	ANF	KTZ	QND	KTZ	QND
CYP Selectivity	2C9	1A2	3A	2D6	3A	2D6
Conc. (μM)	2.0	10	2.0	5.0	2.0	5.0
Subject ID	GC4				GEQ	

Fig. 4. Effect of various cytochrome P450 inhibitors on [O-methyl-^{14}C]dextromethorphan O-demethylase activity in human liver microsomes. All inhibitors were dissolved in ethanol, and data are expressed as percentage activity remaining relative to a control incubation containing ethanol only ($\leq$1.0%, v/v). Control activity was 2.08 ± 0.16 and 0.21 ± 0.01 nmol HCHO/hr/mg for subjects GC4 and GEQ, respectively. Reactions were performed in the presence of microsomes (0.5 mg protein/ml), at an [O-methyl-^{14}C]dextromethorphan concentration which approximated apparent K_m (subject GEQ, 40 μM; subject GC4, 10 μM). QND, quinidine; KTZ, ketoconazole; SLF, sulfaphenazole; ANF, α-naphthoflavone. The concentrations chosen for each inhibitor ensured that the associated CYP form was inhibited >70% (CYP2C9, SLF; CYP3A, KTZ; CYP1A2, ANF).[19]

characterized by an apparent K_m and V_{max} of 0.5–10 μM and 2.5–19.5 nmol/ hr/mg, respectively. On the other hand, the rates of liver microsomal O-demethylase activity in subjects phenotyped *in vivo* as PM are lower and can vary from 0.09 to 3.0 nmol/hr/mg ($K_m > 20$ μM).[3,7,15–18] In accordance, we have found that the O-demethylation of [O-methyl-^{14}C]dextro-

[15] S. V. Otton, D. Wu, R. T. Joffe, S. W. Cheung, and E. M. Sellers, *Clin. Pharmacol. Ther.* **53**, 401 (1993).

[16] V. Fischer, B. Vogels, G. Maurer, and R. E. Tynes, *J. Pharmacol. Exp. Ther.* **260**, 1355 (1992).

[17] O. Mortimer, K. Persson, M. G. Ladona, D. Spalding, U. M. Zanger, U. A. Meyer, and A. Rane, *Clin. Pharmacol. Ther.* **47**, 27 (1990).

[18] D. Wu, S. V. Otton, P. Morrow, T. Inaba, W. Kalow, and E. M. Sellers, *J. Pharmacol. Exp. Ther.* **266**, 715 (1993).

methorphan (1–40 μM), in the presence of human liver microsomes containing CYP2D6 (Fig. 3), conforms to monophasic Michaelis–Menten kinetics ($V_{max}/K_m \geq 1.0$ ml/hr/mg). These data suggest that one enzyme is involved in the reaction. In the case of human liver microsomes devoid of CYP2D6, the V_{max}/K_m ratio is considerably lower (≤ 0.05 ml/hr/mg). To date, we have not detected any livers exhibiting the intermediary metabolizer (IM) phenotype described by others ($V_{max}/K_m = 0.05$–0.9 ml/hr/mg).[7] In subjects characterized *in vitro* as PM, it is advisable to determine if the low level of CYP2D6 is the result of poor tissue quality. This can be easily done by measuring a number of monooxygenase activities associated with other CYP forms, e.g., CYP3A, CYP2E1, etc. This is extremely important since most tissue procurement centers are unable to obtain information concerning the *in vivo* phenotypic status of individual organ donors.

Studies with CYP Form Selective Inhibitors

In the presence of human liver microsomes containing CYP2D6, the *O*-demethylation of [*O-methyl*-14C]dextromethorphan has been shown to be markedly inhibited by well-known CYP2D6 inhibitors such as quinidine and lobeline (Fig. 4).[11] Potent inhibitors of other CYP forms, e.g., ketoconazole (CYP3A), sulfaphenazole (CYP2C9), and α-naphthoflavone (CYP1A2), are ineffective.[11,19] However, [*O-methyl*-14C]dextromethorphan *O*-demethylase activity catalyzed by microsomes lacking CYP2D6 (e.g., subject GEQ) is largely insensitive to quinidine and is partially inhibited by ketoconazole (Fig. 4). These data suggest that CYP3A may partly contribute to the *O*-demethylation of [*O-methyl*-14C]dextromethorphan in the absence of CYP2D6 and may represent the "low-affinity component" ($K_m \geq 40$ μM) in microsomes of EM subjects.[3]

Acknowledgments

The author thanks the following personnel of Abbott Laboratories: Dr. Bruce Surber and Mr. Samuel Thomas (synthesis of radiolabeled dextromethorphan) and Mr. Michael Kukulka (technical assistance).

[19] A. D. Rodrigues, *Biochem. Pharmacol.* **48**, 2147 (1994).

Section III

Determination of Genotype

[22] *CYP2D6* Multiallelism

By ANN K. DALY, VIDAR M. STEEN, KAREN S. FAIRBROTHER,
and JEFFREY R. IDLE

Introduction

The cytochrome P450 enzyme debrisoquine 4-hydroxylase, which is encoded by the *CYP2D6* gene on chromosome 22, metabolizes many different classes of commonly used drugs. A variety of *CYP2D6* mutations and polymorphisms which result in an absence of active enzyme, an enzyme with reduced activity, or have apparently no effect on enzyme activity have now been described (Table I).[1-17] In addition, alleles with amplification of the active *CYP2D6* gene have been demonstrated.[16] The autosomal reces-

[1] M. Kagimoto, M. Heim, M. K. Kagimoto, T. Zeugin, and U. A. Meyer, *J. Biol. Chem.* **265,** 17209 (1990).

[2] N. Hanioka, S. Kimura, U. A. Meyer, and F. J. Gonzalez, *Am. J. Hum. Genet.* **47,** 994 (1990).

[3] A. C. Gough, J. S. Miles, N. K. Spurr, J. E. Moss, A. Gaedigk, M. Eichelbaum, and C. R. Wolf, *Nature (London)* **347,** 773 (1990).

[4] A. Gaedigk, M. Blum, R. Gaedigk, M. Eichelbaum, and U. A. Meyer, *Am. J. Hum. Genet.* **48,** 943 (1991).

[5] V. M. Steen, O. A. Andreassen, A. K. Daly, T. Tefre, A.-L. Borresen, J. R. Idle, and A.-K. Gulbrandsen, *Pharmacogenetics* **5,** 215 (1995).

[6] B. Evert, E.-U. Griese, and M. Eichelbaum, *Naunyn-Schmideberg's Arch. Pharmacol.* **350,** 434 (1994).

[7] D. Marez, N. Sabbagh, M. Legrand, J. M. Lo-Guidice, P. Boone, and F. Broly, *Pharmacogenetics* **5,** 305 (1995).

[8] F. Broly, D. Marez, N. Sabbagh, M. Legrand, P. Boone, and U. A. Meyer, *Pharmacogenetics* **5,** 373 (1995).

[9] R. Saxena, G. L. Shaw, M. V. Relling, J. N. Frame, D. T. Moir, W. E. Evans, N. Caporaso, and B. Weiffenbach, *Hum. Mol. Genet.* **3,** 923 (1994).

[10] B. Evert, E.-U. Griese, and M. Eichelbaum, *Pharmacogenetics* **4,** 271 (1994).

[11] A. K. Daly, J. B. S. Leathart, S. J. London, and J. R. Idle, *Hum. Genet.* **95,** 337 (1995).

[12] H. Yokota, S. Tamura, H. Furuya, S. Kimura, M. Watanabe, I. Kanazawa, I. Kondo, and F. J. Gonzalez, *Pharmacogenetics* **3,** 256 (1993).

[13] I. Johansson, M. Oscarson, Q.-Y. Yue, L. Bertilsson, F. Sjöqvist, and M. Ingelman-Sundberg, *Mol. Pharmacol.* **46,** 452 (1994).

[14] R. Tyndale, T. Aoyama, F. Broly, T. Matsunaga, T. Inaba, W. Kalow, H. V. Gelboin, U. A. Meyer, and F. J. Gonzalez, *Pharmacogenetics* **1,** 26 (1991).

[15] F. Broly and U. A. Meyer, *Pharmacogenetics* **3,** 123 (1993).

[16] I. Johansson, E. Lunqvist, L. Bertilsson, M.-L. Dahl, F. Sjöqvist, and M. Ingelman-Sundberg, *Proc. Natl. Acad. Sci. U.S.A.* **90,** 11825 (1993).

[17] S. Panserat, C. Mura, N. Gerard, M. Vincent-Viry, M. M. Galteau, E. Jacqz-Aigrain, and R. Krishnamorthy, *Hum. Genet.* **94,** 401 (1994).

sive poor metabolizer (PM) phenotype is due to the possession of two alleles encoding an inactive enzyme, whereas the ultrarapid metabolizer (UM) phenotype is related to an inherited amplification of functional *CYP2D6* genes. The most common allele associated with the absence of enzyme activity among European Caucasians is the *CYP2D6B* allele (frequency 0.19), followed by *CYP2D6D* (frequency 0.04).[2,18] *CYP2D6A*, *CYP2D6T*, and *CYP2D6E* each occur at frequencies of approximately 0.015 in European.[6,10,11,18] The frequency of the *CYP2D6F* and *CYP2D6G* alleles is still unclear, but preliminary indications are that both occur at frequencies of less than 0.01.[7,8] Ethnic variation occurs in the spectrum and frequency of the different sales. Moreover, it should be noted that not all poor metabolizers of debrisoquine are identified by genotyping for the inactive mutations known at present.

This chapter describes methods for the detection of the main *CYP2D6* alleles associated with the PM phenotype and discusses the use of single-strand conformational polymorphism (SSCP) analysis and DNA sequencing for the identification of new polymorphisms. The availability of long polymerase chain reaction (PCR) techniques has enabled development of improved assay methods and has also been useful in improving protocols for SSCP analysis and DNA sequencing. Gene amplification associated with the UM phenotype are currently detected by restriction fragment length polymorphism (RFLP) analysis using the restriction enzymes *Xba*I and *Eco*RI,[16] but these should be detectable by long PCR assays which are currently under development in several laboratories, including ours.

DNA Preparation

There is a wide range of methods for DNA preparation. Choice of method will depend on a number of factors, particularly the amount of material available and the type of analysis to be carried out. If a blood sample is to be genotyped only for the more common inactivating alleles (*CYP2D6A*, *CYP2D6B*, *CYP2D6T*, and *CYP2D6E*) with the aim of determining whether the individual is a poor metabolizer, there are a variety of rapid, small-scale methods available.[19–21] However, if long PCR or RFLP analysis for detection of the *CYP2D6D* allele or for identification of ultrarapid metabolizers is to be carried out or a DNA sample for long-

[18] A. K. Daly and J. R. Idle, unpublished results (1995).
[19] P. S. Walsh, D. A. Metzger, and R. Higuchi, *BioTechniques* **10,** 506 (1991).
[20] N. K. Spurr, A. C. Gough, C. A. D. Smith, and C. R. Wolf, *Methods Enzymol.* **206,** 149 (1991).
[21] J. Grimberg, S. Nawoschik, L. Belluscio, R. McKee, and A. Turck, *Nucleic Acids Res.* **17,** 8390 (1989).

term storage is required, many of the published methods are cumbersome, requiring multiple phenol extractions or purchase or relatively expensive commercial kits. We routinely prepare DNA by a method involving deproteinization with perchloric acid which is rapid and inexpensive and produces DNA which has an average size of at least 30 kb and is suitable for long-term storage at 4°.

DNA is prepared from 5 ml whole blood. Nuclei are prepared by addition of the blood to 45 ml lysis buffer (320 mM sucrose, 5 mM MgCl$_2$, 1% Triton X-100, 10 mM Tris–HCl, pH 7.4) and, after gentle mixing, centrifugation at 2000 g for 5 min at 4°. The pellet is resuspended in 2 ml 150 mM NaCl, 60 mM EDTA, 1% sodium dodecyl sulfate, 400 mM Tris–HCl, pH 7.4, and 0.5 ml 5 M sodium perchlorate added. The suspension is rotary mixed at room temperature for 15 min and is then incubated at 65° for 30 min. Two milliliters of chloroform precooled to −20° is added and the mixture is rotary mixed at room temperature for 10 min followed by centrifugation at 1400 g for 10 min. The aqueous DNA-containing upper phase is transferred to a fresh tube and 2 volumes of ethanol cooled to 4° is added. The tube is inverted gently to precipitate DNA, and the DNA is spooled onto a disposable plastic loop. The spooled DNA is washed briefly in 70% ethanol and allowed to dry at room temperature for 20 min. The DNA is then resuspended in 200 μl 10 mM Tris–HCl, 1 mM EDTA, pH 7.4, by incubation at 60° for 8 to 16 h and stored at 4° until required. We typically obtain in the region of 100 μg DNA from 5 ml blood.

This method is suitable for blood samples, lymphocytes, and a variety of cultured cells, but for liver samples, we have found that it is necessary to carry out proteinase K digestion and phenol extraction[22] to obtain good quality DNA.

Detection of PM-Associated Alleles

For routine screening of PM-associated alleles, the use of several different PCR methods is recommended. We have developed methods that are rapid, sensitive, and relatively inexpensive and involve restriction enzyme digestion of the primary PCR product, allele-specific PCR or long PCR, or combinations of these methods. Due to the presence of two or more pseudogenes with >90% homology to *CYP2D6* within the *CYP2D* gene cluster, it is esssential to ensure that only *CYP2D6*-specific sequences are amplified to avoid interference from the pseudogenes. This may be accomplished by choosing primers specific for *CYP2D6* at the 3′ end and using

---

[22] N. Blin and D. W. Stafford, *Nucleic Acids Res.* **3**, 2303 (1986).

the maximum annealing temperature compatible with adequate amplification. Alternatively, the use of nested PCR methods is particularly helpful, especially for regions where homology between *CYP2D6* and the pseudogenes is high.

With respect to restriction enzyme digestion, it is important to have an internal control for the enzyme activity in each tube. This can be an additional restriction site in the PCR product that is not subject to polymorphism or an additional PCR product with a site for the enzyme that will yield products of different noninterfering size.

As summarized in Table I, increasing numbers of *CYP2D6* alleles, particularly those associated with the PM phenotype, are being described. If the prediction of the PM phenotype only is required, the vast majority of European PMs should be identified by screening for the *CYP2D6A*, *CYP2D6B*, *CYP2D6E*, and *CYP2D6T* alleles. It is not essential to screen

TABLE I
CYP2D6 ALLELES

Allele	Mutation	Effect[a]	Ref.
No activity			
CYP2D6A	A_{2637}del	Frameshift	1
CYP2D6B	G_{1934}A; C_{188}T; C_{1062}A; A_{1072}G; C_{1085}G; G_{1749}C; G_{1934}A; G_{4268}C	Splicing defect/frameshift	1–3
CYP2D6D	CYP2D6 deleted	No enzyme	4, 5
CYP2D6E	A_{3023}C	His$_{324}$Pro (no activity *in vivo*)	6
CYP2D6F	G_{971}C; C_{1062}A; A_{1072}G; C_{1085}G	Splicing defect	7
CYP2D6G	G_{1846}T; G_{1749}C; C_{2938}T	Premature termination	8
CYP2D6T	T_{1795} del; G_{2064}A	Premature termination	9–11
Reduced activity			
CYP2D6J	C_{188}T; C_{1749}G; G_{4268}C	Pro$_{34}$Ser	12
CYP2D6Ch$_1$	C_{188}T; C_{1127}T; G_{1749}C; G_{4268}C	Pro$_{34}$Ser	13
CYP2D6Ch$_2$	C_{188}T; *CYP2D7* sequence in exon 9	Unstable enzyme	13
Effect unclear			
CYP2D6C	A_{2701}, G_{2702}, A_{2703} del	Lys$_{281}$ del	14, 15
CYP2D6L	G_{1749}C; C_{2938}T; G_{4268}C	Reduced activity *in vivo*	16, 17
Amplifications			
CYP2D6Lx2	Two genes each with G_{1749}C; C_{2938}T; G_{4268}C	Increased activity *in vivo*	16
CYP2D6Lx13	Thirteen genes each with G_{1749}C; C_{2938}T; G_{4268}C	Increased activity *in vivo*	16

[a] Effects on activity are based on current evidence but may be substrate dependent in some cases.

for the *CYP2D6D* allele because no amplification of this allele will take place and the genotype will relate to the other allele only. However, if no PCR product is obtained, the subject may be homozygous for the *CYP2D6D* allele; this should be confirmed by use of the long PCR assay for this allele described below. Identification of subjects heterozygous for mutant *CYP2D6* alleles will also be of importance in many studies since these individuals will show slower metabolism than homozygous wild-type subjects. In this case, it is important that screening for the *CYP2D6D* allele should be carried out in addition to the alleles discussed earlier. If this is not done, up to 20% of heterozygous individuals may be misclassified as homozygous wild type.

CYP2D6A and *CYP2D6B* Alleles

CYP2D6A and *CYP2D6B* alleles can be detected in a single assay. The advantage of this assay over previously described combined assays is that a simpler band pattern is obtained and that the restriction enzymes used are relatively inexpensive compared with those used in other similar assays.

Genomic DNA (0.5–1 μg) is amplified by the polymerase chain reaction using primers G1 (bp 1827–1846; 5'-TGCCGCCTTCGCCAACCACT-3') and B1 (bp 2638–2657; 5'-GGCTGGGTCCCAGGTCATAC-3'). The reaction is carried out in a total volume of 50 μl containing 10 mM Tris–HCl, pH 8.3, 1.5 mM MgCl$_2$, 5% dimethyl sulfoxide (DMSO), 50 mM KCl, 0.1% (v/v) Triton X-100 with 200 μM nucleotides, 0.25 μM primers, and 2 units Tbr polymerase (NBL Gene Sciences, Cramlington, UK). For amplification, 30 cycles of 1 min at 95°, 1.5 min at 63°, and 3 min at 70° in a PHC3 thermocycler (Techne, Cambridge, UK) are carried out.

The product (20 μl) is digested with 15 units of *Bsa*AI and 15 units of *Bst*NI at 50° for 3 hr. Longer digestion times may result in nonspecific degradation due to "star activity" from the restriction enzymes. To control for digestion by *Bsa*AI, a sample of part of the glyceraldehyde-3-phosphate dehydrogenase (GAPDH) gene amplified by separate PCR reaction and containing a *Bsa*AI site is added to the digest. A similar control for *Bst*NI digestion is not necessary since there are several sites for this enzyme in the amplified *CYP2D6* fragment in addition to the polymorphic site. The GAPDH PCR is carried out using the primers R4 (5'-AGAAACAGGAG-GTCCCTACT-3') and R5 (5'-GTCGGGTCAACGCTAGGCTG-3') and similar conditions to those for the *CYP2D6* amplification. The digest is analyzed on a 10% polyacrylamide gel using TBE as the buffer and the gel is stained with ethidium bromide. The various band patterns are represented in Fig. 1.

A further PCR assay is carried out on samples positive for both the *CYP2D6A* and *CYP2D6B* alleles to determine whether the mutations are

1 2 3 4 5

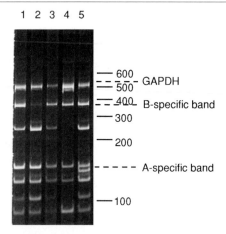

FIG. 1. Combined PCR-RFLP assay for the *CYP2D6A* and *CYP2D6B* alleles. Lane 1, CYP2D6B/wild type; lane 2, homozygous wild type; lane 3, CYP2D6B/wild type; lane 4, homozygous CYP2D6B; and lane 5, CYP2D6A/CYP2D6B.

on separate chromosomes and the subject is therefore a poor metabolizer. It should be noted that it is a theoretical possibility that the mutations associated with both alleles could occur on a single chromosome, but we have not found this in any subject analyzed in our laboratory up to the present. However, for completeness we recommend that the additional assay be performed. PCR is carried out as described earlier using the primers G1 (see above) and S1 (5'-GCTGGGTCCCAGGTCATCCG-3') at an annealing temperature of 64° for 2 min. Under these conditions, the allele-specific primer S1 allows the amplification of the *CYP2D6A* allele only. The PCR product is digested with *Bst*NI to determine whether the *CYP2D6B*-associated mutation is also present on this allele. The digestion products are analyzed by polyacrylamide gel electrophoresis (10% gel), and the allele is negative for the *CYP2D6B* mutation if a band of 279 bp is observed but is positive if a band of 390 bp is present.

CYP2D6D Allele

We have recently developed a rapid assay which, for the first time, detects the *CYP2D6D* allele by the use of long PCR technology.[5] In short, we have focused on the presence of two 0.6- and 2.8-kb direct repeats downstream of both *CYP2D7* and *CYP2D6*. The existence of these *CYP2D6*-flanking repeats will seriously complicate and interfere with the ability to design PCR primers in the area of the deletion break points to selectively amplify a deletion-specific fragment. To overcome this problem, we have located a forward primer in the nonrepeated specific sequence downstream of *CYP2D7* and a reverse primer 3′ to the repeats downstream of *CYP2D6*.[5]

Several different kits are available for long PCR. We use r*Tth* DNA polymerase XL (Perkin Elmer, Warrington, UK) which is a mixture of the enzymes r*Tth* DNA polymerase and Vent$_R$ DNA polymerase together with XL buffer (Perkin Elmer) which contains tricine, potassium acetate, glycerol, and DMSO (exact concentrations are not given by the manufacturer). Satisfactory results have also been obtained using the Expand PCR system (Boehringer-Mannheim, Mannheim, Germany). The following primers are used (20 μM stock solutions): forward primer 5′-ACCGGG-CACCTGTACTCCTCA-3′ (CYP-13; position 7020–7040 in Sequence Accession No. X58467) and reverse primer 5′-GCATGAGCTAAGGCACC-CAGAC-3′ (CYP-24; position 9374–9353 in Sequence Accession No. M33388).

All experiments are performed with standard thick-walled PCR tubes (GeneAmp PCR Reaction Tubes; Perkin Elmer) on a Perkin Elmer DNA Thermal Cycler Model TC1. To improve the PCR performance, the hot start procedure is used for all reactions, with formation of a separating wax layer between the lower and the upper reagent mixes. The long PCR assay is carried out in a total volume of 100 μl containing 1× XL buffer, 1.1 mM magnesium acetate, 200 μM of each dNTP, 0.3 μM of primers CYP-13 and CYP-24, 2 units r*Tth* DNA polymerase XL, and 0.4 μg genomic DNA. The PCR parameters are as follows: hot start at 93° for 1 min, followed by 35 cycles of denaturation at 93° for 1 min, annealing at 65° for 30 sec, and synthesis at 68° for 5 min. The program ends with a final extension for 10 min at 72°. Ten to 15 μl of each PCR reaction is then analyzed on a 0.8 to 1.0% agarose gel.

The primers CYP-13/CYP-24 amplify a 3.5-kb PCR product from the *CYP2D6D* allele (Fig. 2, lanes 1 and 4). We recommend always performing the assay in duplicate, with inclusion of control samples, to avoid false-negative results from technical errors, etc. In our hands, the method has a 100% sensitivity to detect DNA samples heterozygous or homozygous for the *CYP2D6D* allele. It should be noted that samples with a *CYP2D6B*/*CYP2D6D* genotype may amplify the 3.5-kb product with a markedly lower yield as compared to DNAs with the *CYP2D6D*/wild-type genotype, but the reason for this difference is not known. The assay does not detect the rate 11-kb *Xba*I allele.[5] We have not experienced any false-positive amplifications from DNA samples without the *CYP2D6D* allele, thereby indicating a 100% specificity of the assay. By increasing the time of the polymerization step above 5–10 min, a 15-kb PCR product amplified from the wild-type allele may be synthesized in visible amounts. Annealing at

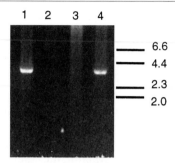

FIG. 2. Detection of the *CYP2D6D* allele by long PCR. Lanes 1 and 4 show subjects heterozygous for the *CYP2D6D* allele, and lanes 2 and 3 show homozygous wild-type subjects.

63 or 68° results in the appearance of nonspecific products or a reduced yield, respectively. It is possible to reduce the volume of the PCR reaction to 25 μl, if required.

CYP2D6T Allele

Two allele-specific PCR reactions are carried out in parallel with the primers 2G (5'-CTCGGTCTCTCGCTCCGCAC-3'; bp 2096 to 2115) and 9G (5'-CAAGAAGTCGCTGGAGCTGT-3'; bp 1776 to 1795) and 2G and 10G (5'-CAAGAAGTCGCTGGAGCTGG-3'; bp 1776 to 1795), respectively, using similar reaction mixes to that described for the *CYP2D6A* and *CYP2D6B* allele assay. The temperature conditions are 30 cycles of 1 min at 95°, 1.5 min at 54°, and 3 min at 70° in a Techne PHC-3 heating block. Primers 9G and 10G are allele specific with an additional mismatch introduced in position 18 to give improved specificity. Primer 2G is specific at the 3' end for *CYP2D6*, and these primer combinations do not amplify the corresponding sequences of *CYP2D7* and *CYP2D8* under the assay conditions.

Detection of Other Mutant Alleles

To maximize the detection rate of poor metabolizers, various recently described rare alleles can be screened for. Using published methods from other workers,[6,8] studies in our laboratories have detected the *CYP2D6E* and *CYP2D6G* alleles. In Oriental populations, the alleles referred to as *CYP2D6J* and *CYP2D6Ch* are likely to be common causes of impaired metabolism. An allele-specific PCR method suitable for unambiguous detection of these alleles has been described.[13]

Identification of Novel Alleles Using SSCP Analysis and
  DNA Sequencing

A number of methods for the detection of new polymorphisms have been described. For studies on *CYP2D6* and other cytochrome P450 genes, we have found that SSCP analysis is a relatively simple techique capable of detecting the majority of point mutations. However, with the advent of long PCR methodology and the availability of rapid techniques for sequencing these products, sequencing of exons and exon–intron boundaries directly is an alternative, especially where only a small number of samples require screening.

For SSCP analysis, all nine exons and exon–intron boundaries (including at least 20 bp of the intron) are amplified separately. Although it is possible by use of appropriate primers and PCR conditions to amplify only *CYP2D6* exons and not the corresponding homologous exons from *CYP2D7* and *CYP2D8* from genomic DNA for all exons except exon 8, we now find it is more effective to specifically amplify the entire *CYP2D6* gene using specific primers and to then reamplify the various exons separately using the primers listed in Table II.

For initial amplification of a *CYP2D6*-specific 4.4-kb product by long PCR, the reaction is carried out in a 50 μl total volume containing 0.5 μg DNA, 0.25 μM primers JM2 and JM3, 200 μM dNTPs, 1.1 mM magnesium acetate, and 2 units r*Tth* DNA polymerase (Perkin Elmer, Warrington, UK) in XL buffer (Perkin Elmer) which is composed of tricine, potassium acetate, glycerol, and DMSO. After a 1-min hot start at 93°, 30 cycles of 1 min at 93°, 2 min at 60°, and 5 min at 68° are carried out in a PHC-3 thermocycler with all ramp rates set at maximum, except for the 93 to 60° step which is carried out at a setting of 30.

Reamplification of the various exons is achieved using the primers listed in Table II.[23] Reactions are carried out in a total volume of 50 μl containing 10 mM Tris–HCl, pH 8.3, 1.5 mM MgCl$_2$, 50 mM KCl, 0.1% (v/v) Triton X-100 with 200 μM nucleotides, 0.25 μM primers, 2.5 units *Taq* polymerase, and 1 μl primary PCR product. For all the exons, satisfactory results are obtained using 30 cycles of 1 min at 93°, 1 min at 57°, and 2 min at 70° in a PHC-3 thermocycler.

We obtain good quality SSCP analysis data with standard polyacrylamide gel electrophoresis equipment using modifications of previously described methods.[24,25] In general, use of SSCP with at least two different

---

[23] S. Kimura, M. Umeno, R. C. Skoda, U. A. Meyer, and F. J. Gonzalez, *Am. J. Hum. Genet.* **45,** 889 (1989).

[24] T. Hongyo, G. S. Buzard, R. J. Calvert, and C. M. Weghorst, *Nucleic Acids Res.* **21,** 3637 (1993).

[25] P. Chaubert, D. Bautista, and J. Benhatter, *Biotechniques* **15,** 586 (1993).

TABLE II
PRIMERS FOR SPECIFIC AMPLIFICATION OF *CYP2D6* BY LONG PCR AND FOR
REAMPLIFICATION OF THE VARIOUS EXONS[a]

Code	Orientation	Sequence (in 5′ to 3′ orientation)	Location (bp)
CYP2D6			
JM2	Forward	GTGTGTCCAGAGGAGCCCAT	52–71
JM3	Reverse	TGCTCAGCCTCAACGTACCCC	+105–+124
Exon 1			
JM2	Forward	GTGTGTCCAGAGGAGCCCAT	52–71
L1	Reverse	GCTTCTGGTAGGGGAGCCTC	305–324
Exon 2			
E5	Forward	TCTGCAGTTGCGGCGCCGCT	964–983
E6	Reverse	CTGTCCCCACCGCTGCTTGC	1144–1163
Exon 3			
L2	Forward	AATGCCTTCATGGCCACGCG	1652–1671
G4	Reverse	TTGGGGCGAAAGGGGCGTCC	1933–1952
Exon 4			
G1	Forward	TGCCGCCTTCGCCAACCACT	1824–1843
G2	Reverse	CTCGGTCTCTCGCTCCGCAC	2096–2115
Exon 5			
E1	Forward	TGAGACTTGTCCAGGTGAAC	2466–2485
E2	Reverse	ATTCCTCCTGGGACGCTCAA	2754–2773
Exon 6			
L3	Forward	CCCGTTCTGTCCCGAGTATG	2859–2878
L4	Reverse	CCCCTGCACTGTTTCCCAGA	3047–3066
Exon 7			
G6	Forward	TTGACCCATTGTGGGGACGCA	3213–3232
L5	Reverse	ATCACCAGGTGCTGGTGCTG	3462–3481
Exon 8			
L6	Forward	GTGGCAGGGGTCCCAGCATC	3825–3844
L7	Reverse	GACAGGGAGCCGGGCTCCCC	4037–4056
Exon 9			
L8	Forward	GCCAGGCTCACTGACGCCC	4093–4112
L9	Reverse	GGCTAGGGAGCAGGCTGGGG	4321–4340

[a] The sequence numbering is that of Kimura *et al.*[23]

temperatures has been demonstrated to detect up to 80% of polymorphisms for a number of different genes.[26] PCR products (2.5–4.5 μl) are denatured by the addition of 25 μl formamide followed by 7 μl of gel-loading buffer (0.05% bromophenol blue, 0.05% xylene cyanol, 20 mM EDTA, 95% formamide) and are incubated at 99° for 5 min. The samples are cooled rapidly on ice and rapidly loaded on a 10% polyacrylamide (49:1 acrylamide/N,N''-methylenebisacrylamide) gel (170 × 150 × 1.5 mm) containing 0.5× TBE.

[26] A. Vidal-Puig and D. E. Moller, *BioTechniques* **17**, 490 (1994).

Each sample is analyzed on a gel run at 4° (cold room) and at room temperature (22–25°). Gels run at room temperature also contain 10% glycerol. The gel is run at 300 V until the samples have entered the gel to minimize interstrand renaturation and the voltage is then reduced to 160 V. Gels are run for approximately 15 hr.

DNA is visualized by ethidium bromide staining. Although this is not a sensitive method for staining single-stranded DNA, we have found it preferable to silver-straining techniques that are more expensive, time-consuming, and sensitive to background. The gels are stained in a solution of ethidium bromide (3 μg/ml) for 30 min and are destained in water for 10 min. The single-stranded DNA bands are not always clearly visible when viewed directly on a UV transilluminator, but careful photography can improve contrast and sensitivity. Only half the gel is photographed at one time using a Polaroid MP-4 camera, and excess light is eliminated by covering the surrounding area of the gel and the transilluminator with aluminum foil. An exposure time of 1 sec is generally required for good results. A typical result [for the polymorphism $C_{2938}T$ (Arg$_{296}$Cys) in exon 6[16]] is shown in Fig. 3.

Where differences in mobility are detected by SSCP analysis, the polymorphism is identified by DNA sequencing. Satisfactory results have been obtained by using appropriate sequencing primers with the primary *CYP2D6* long PCR product as template. A variety of methods are available for sample preparation. We have found that either purification by electrophoresis of the product onto a NA45 membrane (Schleicher and Schuell,

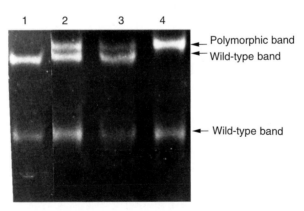

FIG. 3. Detection of the $C_{2938}T$ polymorphism in exon 6 of *CYP2D6* by SSCP analysis. Lane 1 shows a homozygous wild-type subject, lanes 2 and 3 show heterozygotes, and lane 4 shows a subject homozygous for the variant allele.

Dassel, Germany) followed by elution with 3 M NaCl or treatment of the PCR product with exonuclease I and shrimp alkaline phosphatase (PCR sequencing kit, Amersham, Bucks, UK) results in a template that gives good quality sequencing data with Sequenase II and ^{35}S-dATP.

[23] Genetic Tests Which Identify the Principal Defects in *CYP2C19* Responsible for the Polymorphism in Mephenytoin Metabolism

By Joyce A. Goldstein and Joyce Blaisdell

Introduction

A genetic polymorphism in the metabolism of the anticonvulsant drug *S*-mephenytoin has been studied extensively in humans.[1,2] In population studies, individuals can be divided into two phenotypes, extensive metabolizers (EMs) and poor metabolizers (PMs) of mephenytoin. There are large interracial differences in the frequency of the PM phenotype, with Oriental populations having a five-fold greater frequency (13–23%) compared to Caucasians (3–5%). PMs are deficient in the ability to 4'-hydroxylate the *S*-enantiomer of mephenytoin, and this results in a slower rate of metabolism of the racemic mixture. This polymorphism also affects the metabolism of a number of other clinically used drugs, including omeprazole, proguanil, citalopram, barbiturates, and, to a somewhat smaller extent, that of propranolol, certain tricyclic antidepressants, and diazepam.

The enzyme responsible for the metabolism of mephenytoin has been identified as *CYP2C19*.[3,4] Recently, we have identified the two principal genetic defects in *CYP2C19* that are responsible for the PM phenotype in humans.[5,6] The primary defect producing the PM phenotype is a single

[1] G. R. Wilkinson, F. P. Guengerich, and R. A. Branch, *Pharmacol. Ther.* **43,** 53 (1989).

[2] J. A. Goldstein and S. M. F. de Morais, *Pharmacogenetics* **4,** 285 (1994).

[3] J. A. Goldstein, M. B. Faletto, M. Romkes-Sparks, T. Sullivan, S. Kitareewan, J. L. Raucy, J. M. Lasker, and B. I. Ghanayem, *Biochemistry* **33,** 1743 (1994).

[4] S. A. Wrighton, J. C. Stevens, G. W. Becker, and M. VandenBranden, *Arch. Biochem. Biophys.* **306,** 240 (1993).

[5] S. M. F. de Morais, G. R. Wilkinson, J. Blaisdell, K. Nakamura, U. A. Meyer, and J. A. Goldstein, *J. Biol. Chem.* **269,** 15419 (1994).

[6] S. M. F. de Morais, G. R. Wilkinson, J. Blaisdell, U. A. Meyer, K. Nakamura, and J. A. Goldstein, *Mol. Pharmacol.* **46,** 595 (1994).

G → A base pair mutation in exon 5 of *CYP2C19* (*CYP2C19$_{m1}$*) which produces an aberrant splice site. This splice site appears to be used exclusively in PMs metabolizers carrying this mutation. The aberrantly spliced mRNA lacks the first 40 bp of exon 5, producing a premature stop codon and a truncated 234 amino acid protein lacking the heme-binding region. This mutation also destroys a *Sma*I site. Polymerase chain reaction (PCR)–restriction-based genotyping tests have shown that *CYP2C19$_{m1}$* accounts for ~75–83% of the defective alleles in both Oriental and Caucasian PMs. A second major mutation (*CYP2C19$_{m2}$*) was subsequently identified in a Japanese PM consisting of a G → A mutation at position 636 of exon 4 of *CYP2C19* which creates a premature stop codon. This mutation also destroys a *Bam*HI site. This defect appears to account for the remaining defective alleles in Oriental poor metabolizers but is rare in Caucasians. This chapter describes the PCR–restriction tests for the detection of *CYP2C19$_{m1}$* and *CYP2C19$_{m2}$*.

General Method

We recently described PCR-based restriction enzyme tests for the analysis of both *CYP2C19$_{m1}$* and *CYP2C19$_{m2}$* defects.[5–7] In this chapter, we have modified the original procedures slightly by selecting more specific *CYP2C19* primers (given below) for the amplification of both exon 5 (*CYP2C19$_{m1}$*) and exon 4 (*CYP2C19$_{m2}$*). The present primers decrease the possibility of amplifying closely related *CYP2C* sequences which complicate the interpretation of results. The method consists of two steps (Fig. 1). In the first step, two *CYP2C19*-specific pairs of primers are used to amplify exons 4 and 5 of *CYP2C19*, respectively. In the second step, the PCR fragments from the amplification of exon 5 are digested with *Sma*I which digests the normal or wild-type (*CYP2C19$_{wt}$*) sequence but does not digest the *CYP2C19$_{m1}$* allele. Similarly, the PCR fragments from the amplification of exon 4 are digested with *Bam*HI which digests the normal sequence, but does not digest the *CYP2C19$_{m2}$* allele. It should be noted that the digestion of the PCR products of exon 5 by *Sma*I does not distinguish between the normal and the mutant *CYP2C19$_{m2}$* alleles since this portion of the sequence is normal in *CYP2C19$_{m2}$*. Similarly, the digestion of exon 4 PCR products by *Bam*HI does not distinguish between the wild-type and the *CYP2C19$_{m1}$* alleles. The final assignment of genotype requires the comparison of digestion patterns for both sets of PCR products.

[7] J. A. Goldstein and S. M. F. de Morais, U.S. Patent Appl. USSN08/238,821 (1994).

Nomenclature

We have referred to the normal *CYP2C19* allele as *CYP2C19$_{wt}$*, the allele containing a base pair mutation that creates an aberrant splice site in exon 5 as *CYP2C19$_{m1}$*, and the allele containing a G → A base mutation at position 636 producing a stop codon in exon 4 as *CYP2C19$_{m2}$*. If additional defective alleles are discovered, these can be numbered consecutively beginning with *CYP2C19$_{m3}$*.

Oligonucleotide Primers

Primers were synthesized on an Applied Biosystems (Foster City, CA) synthesizer. Primer positions in introns are stated relative to the exons, and positions in *CYP2C19* exons are stated relative to the translational start site of the cDNA.

Primers for m1 Defect (Exon 5) Reaction A

Primer 1 (forward) 5'-CAGAGCTTGGCATATTGTATC-3' from 71 to 51 bases upstream of exon 5 in intron 4
Primer 2 (reverse) 5'-GTAAACACACAACTAGTCAATG-3' from 52 to 73 bases downstream of exon 5 in intron 5

Primers for m2 Defect (Exon 4) Reaction B

Primer 3 (forward) 5'-AAATTGTTTCCAATCATTTAGCT-3' from 21 bases upstream of exon 4 in intron 3 to base 2 in exon 4 (position 483 in the cDNA)
Primer 4 (reverse) 5'-ACTTCAGGGCTTGGTCAATA-3' from 70 to 89 bases downstream of exon 4 in intron 4

Stock Solutions

10× PCR Buffer. 0.67 *M* Tris–HCl, pH 8.8, 0.166 *M* ammonium sulfate, 0.05 *M* 2-mercaptoethanol, 67 μM EDTA, and 0.8 mg/ml bovine serum albumin. This is aliquoted in sterile microfuge tubes and stored at −20°.

dNTP Mix. 2.5 m*M* dATP, 2.5 m*M* dCTP, 2.5 m*M* dGTP, and 2.5 m*M* dTTP at pH 7.0 (prepared from equal volumes of 10 m*M* of each dNTP from USB, Cleveland, OH, and stored frozen at −20°).

Enzymes. The following enzymes may be purchased from several sources. We have used *Taq* polymerase from USB (Cat. No. 71085) and *Sma*I(10 units/μl) and *Bam*HI(20 units/μl) from New England Biolabs (Beverly, MA).

TABLE I
FINAL CONCENTRATION OF REAGENTS IN PCR REACTIONS

Reagent	Final concentration or amount
PCR buffer	1×
dNTP mix	100 μM each
MgCl$_2$	2 mM
Forward primer	0.25 μM
Reverse primer	0.25 μM
Sterile water to volume	
Taq polymerase	0.25 units/μl

6× *Stop Buffer.* 30% glycerol, 0.75% sodium dodecyl sulfate, 0.75% bromophenol blue.

DNA. Genomic DNA can be prepared by different methods. We have used DNA isolated from peripheral leukocytes by standard methods.[8] However, for simplicity, we routinely purify genomic DNA for PCR from 200 μl of whole blood using the Qiamp blood kits (Cat. No. 29104, 29106) (Qiagen Inc., Chatsworth, CA) according to the manufacturer's recommendations. This yields DNA at a predicted concentration of ~30 ng/μl, and we use 2–5 μl directly in a 20-μl PCR reaction.

Procedure

Two sets of sterile microfuge tubes are set up for each sample of DNAs to be tested: one for reaction A (*CYP2C19$_{m1}$* or *m1*) and one for reaction B (*CYP2C19$_{m2}$* or *m2*). To each tube the following is added:

2 μl 10× PCR buffer
0.8 μl dNTP mix (2.5 mM each)
1.6 μl 25 mM MgCl$_2$
0.5 μl forward primer (10 pmol/μl)
0.5 μl reverse primer (10 pmol/μl)
40 ng (20–150 ng) genomic DNA
Sterile water to a final volume of 19.9 μl
0.1 μl *Taq* polymerase (5 U/μl)
Final volume of 20 μl

The appropriate concentration of reagents is also indicated in Table I for ease in use with PCR machines requiring larger volumes.

[8] J. Sambrook, E. F. Fritsch, and T. Maniatis, "Molecular Cloning: A Laboratory Manual," 2nd Ed., p. 9.16. Cold Spring Harbor Lab., Cold Spring Harbor, NY, 1989.

Tubes are then placed in the thermocycler. All conditions given are for 20 μl volumes using a GenAmp PCR System 9600 (Perkin Elmer Corp., Norwalk, CT) and it may be necessary to modify the times for thermocyclers using larger volumes. For both the *m1* and *m2* PCR reactions, the initial denaturation is allowed to proceed at 94° for 5 min, then 37 cycles are performed with denaturation at 94° for 20 sec, annealing at 53° for 10 sec, and extension at 72° for 10 sec. The products are then extended for an additional 5 min at 72°.

Digestion Conditions. To test for the *m1* defect, add 1 μl *Sma*I (10 units/ μl) to the entire 20-μl PCR reaction A and incubate at room temperature for 3 hr (or overnight). To test for the *m2* defect, add 1 μl *Bam*HI (20 units/μl) directly to the 20-μl PCR reaction B and incubate at 37° for 3 hr (or overnight). Add stop buffer (4.2 μl) to both reactions, and analyze by gel electrophoresis on a 4% agarose gel containing ethidium bromide.

Interpretation of Results

The products of the PCR reaction A for exon 5 (*m1* defect) should yield a single band of 321 bp which will be digested completely to two smaller bands at 109 and 212 bp by *Sma*I in individuals who are homozygous for the normal (*wild-type*)(*wt*) sequence as shown in Fig. 1. The 321-bp fragment will not be digested in individuals homozygous for the *m1* defect. In contrast, DNA from individuals heterozygous for this allele will be partially digested to produce three bands of 321, 212, and 109 bp.

The products of the PCR reaction B for exon 4 (*m2*) will yield a single 271-bp fragment. This product will be completely digested to two bands of 175 and 96 bp in individuals homozygous for the *wt* sequence. The 271-bp fragment will not be digested in individuals homozygous for the *m2* defect, but will be partially digested to produce three fragments of 96, 175, and 271 bp in individuals who are heterozygous for the *m2* defect.

The results of both the A and B amplification/digestions must be taken into account to determine the identity of the two alleles. Figure 2 shows an example of the restriction patterns expected for the six known genotypes. Individuals homozygous for *m1/m1* should show only the undigested 321-bp fragment in reaction A. The presence of three bands of 321, 212, and 109 bp indicates that an individual carries one *m1* allele, but does not indicate whether the remaining allele is normal or *m2*. Similarly, the complete digestion of the 321-bp fragment to 212 and 109 bp means that the individual does not carry the *m1* mutation, but the results of the second reaction are required to determine whether the individual is *wt/wt*, *wt/m2*, or *m2/m2*. In reaction B, individuals who are homozygous for *m2* will show only one 271-bp band after digestion with *Bam*HI. However, PCR products

A

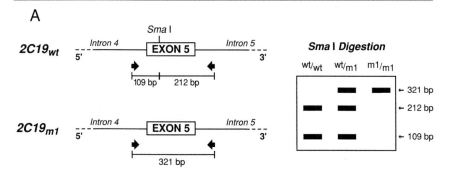

B

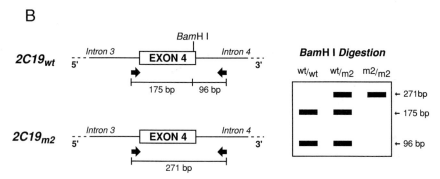

FIG. 1. Strategy used to genotype human genomic DNA utilizing PCR amplification of exon 5 followed by SmaI digestion for $CYP2C19_{m1}$ (A) and PCR amplification of exon 4 followed by BamHI digestion for $CYP2C19_{m2}$ (B). The predicted sizes of the digested DNA fragments for the various genotypes are shown. $CYP2C19_{wt}$ indicates the PCR fragment from the normal (wild-type) gene.

from individuals who have one copy of this defective allele will show three bands of 271, 175, and 96 bp. PCR products from individuals who are *wt/wt* will yield completely digested products in both reactions, but those from *m1/m1* or *m1/wt* individuals will also yield completely digested products in this reaction due to the fact that the BamHI site is present in the *m1* allele.

Troubleshooting Suggestions

A blank containing no DNA should always be run with each PCR reaction. The appearance of PCR products in the blank indicates probable contamination of one of the reagents with DNA. Standard reactions con-

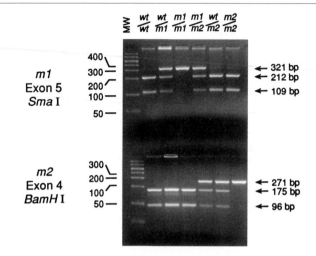

FIG. 2. Ethidium bromide-stained agarose gel showing PCR–restriction enzyme fragmentation patterns for $CYP2C19_{m1}$ (top) and $CYP2C19_{m2}$ (bottom) from individuals representing the six possible genotypes (wt/wt, $wt/m1$, $m1/m1$, $m1/m2$, $m2/m2$). The size of the molecular weight markers (MW) is shown on the left, and the sizes of the PCR–restriction fragments on the right.

taining five samples of known genotype, wt/wt, $wt/m1$, $m1/m1$, $m2/wt$, and $m2/m2$, are recommended with each PCR reaction. The wild-type genotypes should be digested completely. Incomplete digestion of the wild-type allele may result from excess DNA, insufficient enzyme/digestion time, or inappropriate salt concentrations in the restriction reaction. In heterozygotes, the intensity of the upper band should be brighter than that of the lower two bands. A higher intensity of the lower two bands in certain samples may suggest incomplete digestion and can occur if the concentration of DNA in the sample was too high or insufficient time was given for complete digestion. If incomplete digestion is suspected, the analysis should be repeated with a longer digestion time. Inappropriate annealing temperatures may lead to contamination of PCR products with other CYP2C fragments which could lead to the apparent partial digestion of $m1/m1$ or $m2/m2$ genotypes. BamHI can give star activity (cleavage of sequences similar but not identical to its recognition sequence) at high concentrations of enzyme. This generally results in digestion of the 175-bp fragment to two pieces of 32 and 143 bp. Therefore, we recommend using low amounts of this enzyme and a longer 3-hr or overnight digestion time to avoid the appearance of spurious PCR products.

The absence of PCR products in a sample may suggest inappropriate amounts of DNA or impurity of the sample. In this case, it is suggested that the DNA concentration in the PCR reaction can be increased or the cycle number can be increased to 40. Finally, we have found that treatment of the DNA with GeneReleaser (BioVentures, Inc.) enhances the amplification of impure DNA samples. The PCR reaction can be performed on whole blood treated with GeneReleaser. We have treated 1 μl blood or impure DNA with 10 μl of GeneReleaser according to the manufacturer's instructions and scaled up the PCR reactions to 50 μl. When the PCR reactions are finished, we spin down the pellet and remove 20 μl before digestion. However, we generally use DNA purified using a Qiamp blood kit or other standard purification methods.

Sensitivity and Specificity of the Tests

In our experience to date, PCR tests for $CYP2C19_{m1}$ and $CYP2C19_{m2}$ correctly predict 100% of the defective alleles in 65 Oriental PMs.[5,6,9,10] In Caucasians, $CYP2C19_{m1}$ accounts for ~87% of defective alleles in 30 unrelated PMs while the $CYP2C19_{m2}$ has been found in a single Caucasian PM (1/60 alleles.)[5,6,10,11] All 128 Caucasian and 158 Oriental EMs contain at least one normal copy of the gene. Moreover, all individuals identified as PMs genotypically were also phenotypic PMs, indicating that the specificity of the tests for identification of PMs is 100%. However, the sensitivity of the tests in identifying PMs is ~100% in Orientals, but somewhat lower (~80%) in Caucasians. This suggests that other as yet unidentified rare mutations may contribute to the PM phenotype in Caucasian individuals.

It should be noted that the interpretation of phenotyping tests is not always unequivocal. PMs have been characterized phenotypically by a high urinary S/R ratio of >1.0 or by the excretion of low amounts of 4'-hydroxy-mephenytoin in the urine after a 100-mg dose (e.g., hydroxylation indices).[1,12] However, the prediction of phenotype is complicated by the presence of an acid-labile metabolite of S-mephenytoin in the urine of some EMs, which can hydrolyze on storage to produce a high S/R ratio and the improper designation of phenotype in some EMs. Therefore, the urinary S/R ratios of any outliers from genotype/phenotype comparisons should be repeated before and after acidification of the urine if possible. Lack of compliance with some individuals failing to take the dose of mephenytoin or incomplete urine collection can also lead to problems of interpretation.

[9] S. M. F. de Morais, J. A. Goldstein, H.-G. Xie, S.-L. Huang, Y. P Lu, H. Xia, Z.-S. Xiao, N. Ile, and H. H. Zhou, *Clin. Pharmacol. Ther.* **58,** 404 (1995).

[10] S. M. F. de Morais, D. A. Price Evans, P. Krahn, K. Chiba, T. Ishizaki, and J. A. Goldstein, unpublished data.

[11] K. Brøsen, S. M. F. de Morais, U. A. Meyer, and J. A. Goldstein, *Pharmacogenetics* **5,** 312 (1995).

[12] P. J. Wedlund and G. R. Wilkinson, *Methods Enzymol.* **272,** Chap. 11, 1996 (this volume).

Acknowledgments

The authors acknowledge the work of Dr. Sonia M. F. de Morais in this laboratory which led to the identification of the defects responsible for this polymorphism. The authors also thank Dr. Douglas A. Bell and Dr. Masahiko Negishi, NIEHS, for reviewing this manuscript and for helpful suggestions.

[24] Genetic Polymorphism of Human Cytochrome P450 2E1

By VESSELA NEDELCHEVA, IRENE PERSSON, and
MAGNUS INGELMAN-SUNDBERG

Introduction

Ethanol-inducible cytochrome P450 (CYP2E1) is constitutively expressed in liver and in many other tissues. The substrate specificity is broad and includes at present at least 80 different characterized substrates, mainly small and hydrophobic substances. Among them are ethanol and other aliphatic alcohols, aldehydes like acetaldehyde, alkanes, most organic solvents, putative carcinogenic N-nitrosoamines, aromatic heterocyclic compounds, halogenated anesthetics, ethers, aromatic hydrocarbons, fatty acids, acetone, and caffeine.[1] The enzyme is induced by a variety of different chemicals, mainly substrates, and the induction in these cases is to a major extent mediated at the posttranslational level.[1]

It is evident that CYP2E1 plays an important toxicological role. It activates precarcinogens, organic solvents, and drugs to cytotoxic or carcinogenic products that might be harmful and of importance for the synergistic effect of ethanol on many types of human cancer as well as the potentiation of solvent and drug toxicity.[1,2] Furthermore, it has the ability to effectively reduce dioxygen with the subsequent formation of reactive oxy radicals that can initiate lipid peroxidation processes.[3] This reaction is implicated as being of importance in the etiology of alcohol liver disease.[4] In addition,

[1] M. J. J. Ronis, K. O. Lindros, and M. Ingelman-Sundberg, in "Cytochromes P450: Pharmacological and Toxicological Aspects" (I. Ioannides, ed.). CRC Press, Boca Raton, FL, 1996 (in press).
[2] Y. Terelius, K. O. Lindros, K. O. E. Albano, and M. Ingelman-Sundberg, in "Frontiers of Biotransformation" (H. Rein and K. Ruckpaul, eds.), Vol. 8, p. 187. Akademie-Verlag, Berlin, 1992.
[3] G. Ekström and M. Ingelman-Sundberg, Biochem. Pharmacol. 38, 1313 (1989).
[4] M. Ingelman-Sundberg, I. Johansson, H. Yin, Y. Terelius, E. Eliasson, P. Clot, and E. Albano, Alcohol 10, 447 (1993).

the enzyme metabolizes ethanol by a radical mechanism causing the formation of an α-hydroxyethyl radical that covalently modifies proteins. Autoantibodies raised against such adducts, mediated by the action of CYP2E1, are seen both in rats[5] and in humans.[6] Because of the important toxicological role of CYP2E1, a lot of research has been conducted aimed at elucidating a genetic polymorphism of the human *CYP2E1* gene and the linkage of various allelic forms to different types of cancers as well as to alcohol liver disease.

Genetic Polymorphism of the Human *CYP2E1* Gene

Restriction fragment-length polymorphism analysis (RFLP) of the human *CYP2E1* gene initially revealed polymorphism detectable with the restriction endonucleases *Taq*I,[7] *Dra*I,[8] *Rsa*I,[8] *Xmn*I[9] and *Msp*I.[10] The frequencies of the rare alleles detected with *Dra*I and *Rsa*I have been found to be about two- to fourfold higher in Oriental populations. The polymorphic sites are all present in noncoding regions of the gene. The open reading frame of the human *CYP2E1* gene has been found to be well conserved, and functional mutations are very rare.

Two major polymorphic sites have been studied in relation to disease. The first is located in the 5'-flanking region at about 1020 bp upstream, where RFLP analysis revealed the c1 and c2 alleles designating the presence or the absence of a *Rsa*I site, respectively.[11] The *Rsa*I site was found to be accompanied by the absence of a nearby *Pst*I site and vice versa.[12] Interest in this site originates from observations regarding the possible linkage to inducibility of the gene. Subcloning of this region upstream of a SV40 promoter using a chloramphenicol *O*-acetyltransferase activity as a marker of induction in HepG2 cells revealed higher expression caused by the c2 allele, but less amount of protein bound to the motif, as evident from

[5] E. Albano, P. Clot, M. Morimoto, A. Tomasi, M. Ingelman-Sundberg, and S. French, *Hepatology* **21,** 1910 (1995).

[6] P. Clot, M. Ingelman-Sundberg, G. Bellomo, M. Tabone, S. Aricò, Y. Israel, C. Moncada, and E. Albano, *Hepatology* **20,** 317 (1994).

[7] O. W. McBride, M. Umeno, H. V. Gelboin, and F. J. Gonzalez, *Nucleic Acids Res.* **15,** 10071 (1987).

[8] F. Uematsu, H. Kikuchi, T. Ohmachi, I. Sagami, M. Motomiya, T. Kamataki M. Komori, and M. Watanabe, *Nucleic Acids Res.* **19,** 2803 (1991).

[9] D. P. Kelsell, C. R. Wolf, and N. K. Spurr, *Nucleic Acids Res.* **18,** 3111 (1990).

[10] F. Uematsu, H. Kikuchi, T. Abe, M. Motomiya, T. Ohmachi, I. Sagami, and M. Watanabe, *Nucleic Acids Res.* **19,** 5797 (1991).

[11] S.-I. Hayashi, J. Watanabe, and K. Kawajiri, *J. Biochem. (Tokyo)* **110,** 559 (1991).

[12] J. Watanabe, S.-I. Hayashi, K. Nakachi, K. Imai, Y. Suda, T. Sekine, and K. Kawajiri, *Nucleic Acids Res.* **18,** 7194 (1990).

DNase footprinting analysis.[11] The frequency of the rare c2 allele is about 4% in Caucasian and 20% in Japanese populations.[1]

The other site, recognized with DraI RFLP in intron 6,[13] received interest after the description of its relationship to the incidence of lung cancer in a Japanese population.[14] The frequency of the rare allele (C) is about 10% in Caucasian and 25% in Japanese populations.[1] This allele has been sequenced and found to carry mutations near the promoter and in the 3'-flanking region, but not in the open reading frame.[4]

The polymorphic CYP2E1 alleles have been linked to various forms of cancer and to alcoholic liver disease, in particular liver cirrhosis (see Table I).[15-22] The data are, however, hitherto not conclusive. A major point for future research concerns the role of polymorphically distributed alleles in relation to the inducibility of the enzyme as monitored in vivo by measuring the ratio between chlozoxazone and the 6-hydroxylated product in blood[7] and their possible consequences. However, the important toxicological role of this enzyme still requires further investigations in the area of molecular epidemiology with respect to cancer, drug, and alcohol toxicity of value. It is, however, considered of importance to focus the research on mutations of putative functional importance and to use sufficient numbers of subjects in each group.

PCR-Based Detection of the Polymorphic Sites in the 5'-Upstream Region and in Intron 6

The CYP2E1 RsaI polymorphism is due to a C → T mutation in a restriction site for RsaI (GTAC) at −1019 bp, yielding GTAT.[11] The linked PstI polymorphism is caused by a substitution of G by C at position −1259,

[13] I. Persson, I. Johansson, H. Bergling, M.-L. Dahl, J. Seidegård, R. Rylander, A. Rannug, J. Högberg, and M. Ingelman-Sundberg, FEBS Lett. 319, 207 (1993).
[14] F. Uematsu, H. Kikuchi, M. Motomiya, T. Abe, I. Sagami, T. Ohmachi, A. Wakui, R. Kanamaru, and M. Watanabe, Jpn. J. Cancer Res. 82, 254 (1991).
[15] A. Hirvonen, K. Husgafvel-Pursiainen, S. Anttila, A. Karjalainen, and H. Vainio, Carcinogenesis (London) 14, 85 (1993).
[16] J. Watanabe, J.-P. Yang, H. Eguchi, S. Hayashi, K. Imai, K. Nakachi, and K. Kawajiri, Jpn. J. Cancer Res. 86, 245 (1995).
[17] G. S. Hamada, H. Sigimura, I. Suzuki, K. Nagura, E. Kiyokawa, T. Iwase, M. Tanaka, T. Takahashi, S. Watanabe, and I. Kino, Cancer Epidemiol. Biomarkers Prev. 4, 63 (1995).
[18] M. Tsutsumi, S. Takase, and A. Takada, Alcohol Alcohol., Suppl. 1B, 2 (1994).
[19] Y. Maezawa, M. Yamauchi, and G. Toda, Am. J. Gastroenterol. 89, 561 (1994).
[20] D. Lucas, C. Menez, C. Girre, F. Berthou, P. Bodenez, I. Joannet, E. Hispard, L.-G. Bardou, and J.-F. Menez, Pharmacogenetics 5, 298 (1995).
[21] S. Kato, P. G. Shields, N. E. Caporaso, H. Sugimura, G. E. Trivers, M. A. Tucker, B. F. Trump, A. Weston, and C. C. Harris, Cancer Epidemiol. Biomarkers Prev. 63, 45 (1995).
[22] F. Uematsu, S. Ikawa, H. Kikuchi, I. Sagami, R. Kanamaru, T. Abe, K. Satoh, M. Motomiya, and M. Watanabe, Pharmacogenetics 4, 58 (1994).

TABLE I
GENETIC POLYMORPHISM OF *CYP2E1*: LINKAGE TO INDUCIBILITY AND DISEASE

Polymorphism	Disease/property	No. of subjects	Results compared to control	Comments	Ref.
c1/c2 (5′-flanking)					
	Lung cancer	195 (Swe)	Less c2[a]	$P < 0.05$	13
		101 (Fin)	—	c2 very rare	15
		316 (Jap)	Negative	NS[d]	16
		123 (Jap)	Negative	NS	17
	Alcohol liver disease	51 (Ita)	Negative	NS	4
		50 (Jap)	More c2[b]	$P < 0.05$	18
		62 (Jap)	Less c2	$P < 0.05$	19
	CYP2E1 inducibility	74 (Fra)	Less c2	3.1-fold,[e] $P < 0.01$	20
C/D (intron 6)					
	Lung cancer	101 (Fin)	Negative	NS	15
		58 (US)	Negative	OR[f] 1.57	21
		91 (Jap)	OR 2.1 for CD[c]	$P < 0.05$	22
	Alcohol liver disease	51 (Ita)	Less C	$P < 0.01$	4
	CYP2E1 inducibility	74 (Fra)	More D	2.9-fold,[e] $P < 0.01$	20

[a] Lower prevalance of the c2 allele in the cancer group.
[b] Higher prevalance of the c2 allele among alcoholics.
[c] Odds ratio for the CD genotype.
[d] Nonsignificant.
[e] Fold induction of CYP2E1 in individuals homozygous for the D or the c1 allele as compared to heterozygotes.
[f] Odds ratio.

creating a restriction site for *Pst*I (CTGCAG).[11] The polymorphic *Dra*I site has been shown to be situated in intron 6[13] of the gene, where a T → A mutation in position 7668 liquidates the *Dra*I site (TTTAAA to TTAAAA).

Equipment and Materials

Reagents

Phosphate-buffered saline (PBS) buffer: 140 mM NaCl, 2.7 mM KCl, 1.5 mM KH$_2$PO$_4$, and 8 mM Na$_2$HPO$_4$

NLS (nuclei lysis) buffer: 10 mM, Tris–HCl, pH 8.2, 400 mM NaCl, and 2 mM Na$_2$EDTA

Proteinase K solution: 1% sodium dodecyl sulfate (SDS), 2 mM Na$_2$EDTA, and 2 mg/ml proteinase K (Boehringer-Mannheim)

10% SDS

6 M NaCl

TE buffer: 10 mM, Tris–HCl, pH 7.5, and 0.2 mM Na$_2$EDTA

10× PCR buffer: 0.1 M, Tris–HCl, pH 8.3, and 0.5 M KCl

10× TBE buffer: 400 mM Tris, adjusted with boric acid to pH 8.3; 200 mM EDTA, and 90 mM boric acid

10× high salt restriction buffer: 0.1 M Tris–HCl, pH 7.4, 0.1 M MgCl$_2$, 10 μM dithiotreitol, 1 M NaCl, and 1 mg/ml albumin (added just before use)

Ethanol 95%

Ethanol 70%

Ethidium bromide: stock 10 mg/ml

Mineral oil (Sigma)

Materials and Instruments

Falcon tubes

Swing out rotor centrifuge

Microcentrifuge tubes (0.5 ml)

Thermal cycler for PCR (mineral oil, Sigma Chemical Co., if necessary)

Horizontal gel electrophoresis device

Water bath (37°)

dNTPs: 1.25 mM each, Ultrapure dNTP set (Pharmacia Biotech)

Oligomer solutions, diluted with water to 10 μM

Thermostable DNA polymerase, 5 U/μl

SeaKem LE agarose (FMC Bioproducts)

DNA marker VIII: dilution 1/40 containing 1× loading buffer (Boehringer-Mannheim)

*Rsa*I, *Dra*I, *Pst*I (Boehringer-Mannheim), 10 U/μl

L restriction buffer (Boehringer-Mannheim)

6× loading buffer: 30% glycerol, 0.25% Orange G (BDH Laboratory Supplies, England)

500-μl tubes for PCR

Speed Vac

Procedure

Step 1: DNA Purification[23]

Two and a half milliliters of blood is diluted with 1 volume of PBS buffer and is centrifuged for 20 min at 3000 rpm in a swing out rotor

---

[23] S. A. Miller, D. D. Dykes, and H. F. Polesky, *Nucleic Acids Res.* **16**, 215 (1988).

centrifuge. The pellet is washed several times with PBS until no traces of red blood hemoglobin are left. The pellet is subsequently resuspended in 1.5 ml NLS buffer, 0.1 ml 10% SDS, and 0.25 ml proteinase K solution and is incubated at 37° overnight. For precipitation of proteins, 0.5 ml 6 M NaCl is added, the tubes are shaken vigorously for 15 sec, and the samples are centrifuged at 2500 rpm for 15 min. The supernatant is separated into clean sterile Falcon tubes. DNA is precipitated with 2 volumes (5 ml) of 95% ethanol and is washed twice with 70% ethanol. The ethanol is air dried and the DNA is dissolved in 250 μl TE buffer for 2 hr at 37°.

RsaI Polymorphism in the 5'-Flanking Region (c1/c2)

Step 2: PCR Amplification

Primers. Two primers designed to limit a 510-kb fragment containing both the *Rsa*I and the *Pst*I polymorphic sites in the 5'-flanking area are 5'-CCCGTGAGCCAGTCGAGT-3' (-1380 to -1363) and 5'-ATACA-GACCCTCTTCCAC-3' (-870 to -887). The numbering of the bases is according to Hayashi *et al.*[11] and the primers are from Persson *et al.*[13]

Procedure. Reactions are performed using 0.5-ml microcentrifuge tubes in a final volume of 25 μl. The PCR mixture contains 1$\times$ PCR buffer, 0.2 mM dNTPs, 0.25 μM of each primer, 3.15 U heat-stable DNA polymerase, and about 1.2 mM MgCl$_2$, depending on the conditions. The components are mixed and drawn to the bottom of the tube by brief centrifugation. Aliquots of 23.5 μl of the PCR mixture are pipetted to each tube, and 0.1–1 μg (1.5 μl) of DNA sample is added. The samples are overlayered with mineral oil (Sigma), tightly capped, and placed on a Cetus Perkin-Elmer DNA thermal cycler. After initial denaturation at 94° for 1.5 min, the PCR reaction is programmed in 35 cycles, with each cycle including denaturation at 94° for 1 min, an annealing temperature of 52° for 1 min, an extension period of 72° for 1 min, and a final elongation period of 7 min at 72°.

Step 3: Electrophoresis on Agarose Gels

Five microliters of the product is mixed with 1 μl of 6$\times$ loading buffer and subjected to electrophoresis on a 1.8% agarose gel containing ethidium bromide (0.125 μl/ml from a stock solution of 10 mg/ml) in 1$\times$ TBE buffer, using a horizontal gel electrophoresis apparatus. DNA marker VIII (19–1114 bp) is used for estimation of the size of the products.

Step 4: Restriction Analysis

Digestion is performed on 15 μl of the PCR product with 75 U of *Rsa*I (Boehringer-Mannheim) and 1$\times$ L buffer (recommended by the manufacturer) in a final volume of 50 μl. Digestion is carried out at 37° overnight. Reduce the volume to 20 μl in a Speed Vac and add 4 μl of 6× loading buffer. Results are analyzed after electrophoresis on agarose gels as described earlier. Two fragments of 360 and 150 bp in length are seen in subjects homozygous for the c1 allele. Incomplete digestion, which causes an additional band of 510 bp in size, indicates a subject heterozygous for the c2 allele (Fig. 1). The absence of any digestion with RsaI (only a 510-bp band) indicates that an individual is homozygous for the c2 allele. The results may be confirmed by subsequent digestion with PstI, since this polymorphic site is situated on the same fragment. Digestion with PstI is accomplished under the same reaction conditions, using 1× H buffer (Boehringer-Mannheim). These polymorphic sites are almost always in complete linkage. Thus, in order to confirm the homozygous c2 genotype, cleavage with PstI should result in the formation of two bands of 126 and 384 bp fragment size, respectively.

DraI Polymorphism in Intron 6 (C/D)

Step 2: PCR Amplification

   Primers. Primers for a 370-bp-long fragment from intron 6,[16] containing

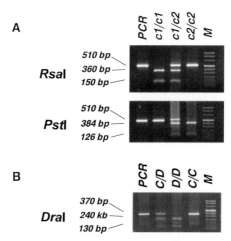

FIG. 1. Results of PCR and restriction enzyme analysis of polymorphic sites of the human *CYP2E1* gene in the 5'-upstream region (A) and in intron 6 (B). Samples for examination of mutations in the 5'-flanking region (A) are evaluated using RsaI. In case of no or incomplete cleavage with RsaI, the absence of a restriction site has to be verified using PstI. For details, see text.

the *Dra*I polymorphic site are, as follows:

primer 1: 5'-AGTCGACATGTGATGGATCCA-3' (7421 to 7441)
primer 2: 5' GACAGGGTTTCA-TCATGTTGG-3' (7773 to 7793)

Procedure. Reactions are carried out as described earlier in a final volume of 25 μl. The same PCR mixture, which is prepared for all samples, contains 1× PCR buffer, 0.2 mM dNTPs, 0.25 μM of each primer, 1.2 mM MgCl$_2$, and 3.15 U per sample heat-stable DNA polymerase. The components are mixed and centrifuged; 0.1–0.3 μg (1.5 μl) of the DNA sample is added to each tube. The samples are overlayered with mineral oil (Sigma), tightly capped, and placed on a Cetus Perkin-Elmer DNA thermal cycler. After initial denaturation at 94° for 1.5 min, the PCR reaction is programmed in 35 cycles, with each cycle including denaturation at 94° for 1 min; an annealing temperature of 60° for 1 min, an extension period of 72° for 1 min, and a final elongation period of 7 min at 72°.

Step 3: Electrophoresis on Agarose Gels

This step is carried out as described earlier. A 370-bp-long fragment is detected on 1.8% agarose gels.

Step 4: Restriction Analysis

Digestion with *Dra*I is carried out using 5 U of the enzyme per sample in a final volume of 50 μl. A digestion mixture with 1× H buffer (containing albumin) and restriction enzyme can be prepared for all samples and aliquoted into the tubes containing the PCR products. Digestion is carried out at 37° overnight. Reduce the volume to 20 μl and add 4 μl of 6× loading buffer. Results are analyzed after electrophoresis on agarose gels as described earlier. Two fragments of 240 and 130 bp size are characteristic for the homozygous wild-type genotype (DD). A band in the 370-bp site on the gel, in addition to the previous two, is evidence for incomplete digestion and indicates a heterozygous genotype (CD). The presence of a single 370-bp band after treatment with *Dra*I is due to the absence of a *Dra*I site on both alleles (CC) (see Fig. 1).

Acknowledgments

Vessela Nedelcheva was granted a short-term visiting fellowship by the European Science Foundation. The research in the authors' laboratory is supported by The Swedish Alcohol Research Fund and from The Swedish Medical Research Council.

[25] Identification of Allelic Variants of the Human *CYP1A1* Gene

By Kaname Kawajiri, Junko Watanabe, and Shin-ichi Hayashi

Introduction

Individual differences in susceptibility to chemically induced cancers are ascribed partly to genetic differences in metabolic balance in the activation and detoxification of environmental procarcinogens.[1] Human lung cancer, especially squamous cell carcinoma, requires exposure to the procarcinogens contained mainly in cigarette smoke. CYP1A1 is expressed in the lung and metabolizes polycyclic aromatic hydrocarbons, such as benzo[*a*]pyrene, in cigarette smoke.[2] Furthermore, high aryl hydrocarbon hydroxylase (AHH) inducibility, which shows a genetically determined variation among individuals, has been reported to be associated with high susceptibility to lung cancer among smokers.[3-5]

In such comparative metabolic phenotype studies, however, more attention to developing methods of determining phenotypes and to examining the possible effects of medications on both patients and controls might be needed. Obviously, if some of the genetic polymorphisms of the *CYP1A1* gene correlate to the lung cancer incidence, this might be helpful in predicting the individual risk of cancer. We found a close association of smoking-associated lung cancer with the *Msp*I and *Ile-Val* polymorphisms of the *CYP1A1* gene in a Japanese population in terms of genotype frequency comparison[6-8] and cigarette dose response.[9,10] The purpose of this chapter is to describe the identification of variant alleles of the human *CYP1A1* gene.

[1] N. Caporaso, M. T. Landi, and P. Vineis, *Pharmacogenetics* **1,** 4 (1991).
[2] T. Shimada, C.-H. Yun, H. Yamazaki, J.-C. Gautier, P. H. Beaune, and F. P. Guengerich, *Mol. Pharmacol.* **41,** 856 (1992).
[3] G. Kellermann, C. R. Shaw, and M. Luyten-Kellermann, *N. Engl. J. Med.* **289,** 934 (1973).
[4] R. E. Kouri, C. E. McKinney, D. J. Slomiany, D. R. Snodgrass, N. P. Wray, and T. L. McLemore, *Cancer Res.* **42,** 5030 (1982).
[5] R. Korsgaard, E. Trell, B. G. Simonsson, G. Stiksa, L. Janzon, B. Hood, and J. Oldbring, *J. Cancer Res. Clin. Oncol.* **108,** 286 (1984).
[6] K. Kawajiri, K. Nakachi, K. Imai, A. Yoshii, N. Shinoda, and J. Watanabe, *FEBS Lett.* **263,** 131 (1990).
[7] S.-I. Hayashi, J. Watanabe, and K. Kawajiri, *Jpn. J. Cancer Res.* **82,** 1325 (1992).
[8] K. Kawajiri, K. Nakachi, K. Imai, J. Watanabe, and S.-I. Hayashi, *Crit. Rev. Oncol. Hematol.* **14,** 77 (1993).
[9] K. Nakachi, K. Imai, S.-I. Hayashi, J. Watanabe, and K. Kawajiri, *Cancer Res.* **51,** 5177 (1991).
[10] K. Nakachi, K. Imai, S.-I. Hayashi, and K. Kawajiri, *Cancer Res.* **53,** 2994 (1993).

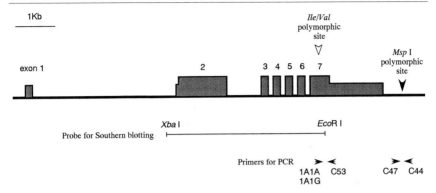

FIG. 1. Structural organization of the human *CYP1A1* gene and polymorphic sites. The DNA fragment (*Xba*I–*Eco*RI) was used as a probe for Southern blot analysis. The location of the primers used for the PCR reaction is also shown (solid arrows).

Identification of the *Msp*I Polymorphism in the Human *CYP1A1* Gene

The human *CYP1A1* gene was localized to chromosome 15 near the *MPI* locus at 15q22–24.[11] Figure 1 shows the structural organization of the human *CYP1A1* gene,[12,13] including the two genetically associated polymorphisms, *Msp*I and *Ile-Val*.

Southern Blot Analysis

Human lymphocyte DNA (8 μg) is digested with restriction nuclease *Msp*I for 3 hr at 37°, and the products are subjected to electrophoresis in 0.8% agarose for Southern blot analysis.[14] The DNA fragments are transferred to a nitrocellulose or nylon membrane filter. The filter is hybridized to the ^{32}P-labeled *Xba*I–*Eco*RI fragment of the cloned *CYP1A1* gene in a hybridization solution at 65° overnight and is washed twice with 0.1× SSC (sodium chloride + sodium citrate) containing 0.1% sodium dodecyl sulfate at 65° for 30 min followed by autoradiography against a Kodak XAR-5 film at −80° with an intensifying screen. An individual with genotype *A* (*m1/m1*) is a predominant homozygote,[6] where the *Msp*I site [264th downstream from the poly(A) additional signal] is absent. An individual homozygous for the rare allele is genotype *C* (*m2/m2*), derived from a 1-base

---

[11] C. E. Hildebrand, F. J. Gonzalez, O. W. McBride, and D. W. Nebert, *Nucleic Acids Res.* **13,** 2009 (1985).

[12] A. K. Jaiswal, F. J. Gonzalez, and D. W. Nebert, *Nucleic Acids Res.* **13,** 4503 (1985).

[13] K. Kawajiri, J. Watanabe, O. Gotoh, Y. Tagashira, K. Sogawa, and Y. Fujii-Kuriyama, *Eur. J. Biochem.* **159,** 219 (1986).

[14] E. M. Southern, *J. Mol. Biol.* **98,** 503 (1975).

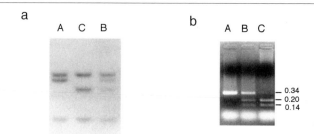

FIG. 2. The *Msp*I polymorphism of the human *CYP1A1* gene detected by Southern blot (a) or PCR–restriction nuclease digestion (b) analysis.

substitution of thymine with cytosine to form the *Msp*I site, which was confirmed by polymerase chain reaction (PCR)-direct sequencing.[15] An individual with genotype B (*m1/m2*) is heterozygous for the alleles (Fig. 2a).

PCR–Restriction Nuclease Digestion Analysis

Two synthetic oligonucleotide primers of 21 bases [C47:5′-CAGTGAA-GAGGTGTAGCCGCT-3′ and C44:5′-TAGGAGTCTTGTCTCATG-CCT-3′ from the 130th to 150th and from the 449th to 469th bases, respectively, counting from the poly(A) additional signal] were used.[15] PCR is carried out by 25 cycles under the following conditions: 1 min at 95° for denaturation, 1 min at 68°, and 1 min at 72° for primer annealing and primer extension. Other conditions used were as described previously.[16] The amplified fragments, including the *Msp*I site, are digested with *Msp*I for 2 hr at 37°, and the products are subjected to electrophoresis in a 1.8% agarose gel. Genotype A is characterized by a 0.34-kb fragment; genotype B by 0.14, 0.20, and 0.34 kb; and genotype C by 0.14 and 0.20 kb (Fig. 2b). The genotypes of the *CYP1A1* gene ascribed to the *Msp*I site were identified as restriction fragment-length polymorphisms (RFLPs) by the PCR and are in complete agreement with the results of Southern blot analysis.

Identification of the *Ile-Val* Polymorphism in the Human *CYP1A1* Gene

By PCR direct sequencing, we found a difference in one base at position 4889 in the seventh exon of the *CYP1A1* gene.[15] As shown in Fig. 3a,

[15] S.-I. Hayashi, J. Watanabe, K. Nakachi, and K. Kawajiri, *J. Biochem. (Tokyo)* **110,** 407 (1991).
[16] R. K. Saiki, D. H. Gelfand, S. Stoffel, S. J. Scharf, R. Higuchi, G. T. Horn, K. B. Mullis, and H. A. Ehrlich, *Science* **239,** 487 (1988).

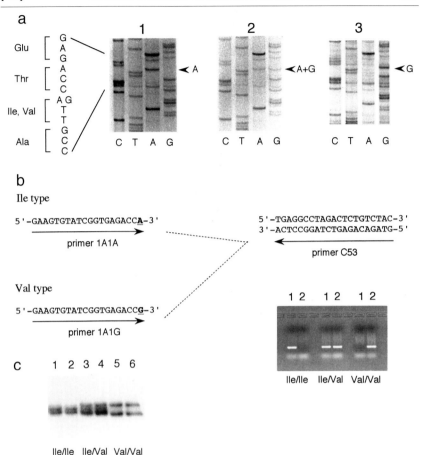

FIG. 3. Detection of the *Ile-Val* polymorphism of the human *CYP1A1* gene. (a) A novel polymorphic mutation in the coding region is shown. DNAs from individuals were examined by PCR direct sequencing. Individuals 1, 2, and 3 show the homozygous *Ile/Ile* genotype, the heterozygous *Ile/Val* genotype, and the homozygous *Val/Val* genotype, respectively. The point mutation is indicated by an arrow. (b) Detection by the allele-specific amplification method. Primers 1A1A and C53 were used for lane 1 and primers 1A1G and C53 for lane 2. (c) The SSCP profiles of the *Ile-Val* polymorphism. Lanes 1 and 2 show genotype *Ile/Ile*, whereas lanes 5 and 6 show a rare homozygote of *Val/Val*. Lanes 3 and 4 show the heterozygous *Ile/Val* genotype.

adenine in individual 1 was replaced by guanine in individual 3. This novel point mutation resulted in the replacement of Ile by Val at residue 462 in the HR2 region,[17] which was well conserved among the P450 families. The RFLP method cannot be used to detect this polymorphism because there is no suitable restriction site. Therefore, for use in screening, we developed new detection methods as follows.

Allele-Specific PCR Amplification

Two oligonucleotides of 20-mer (primer 1A1A; 5′-GAAGTGTATCG-GTGAGACCA-3′ and primer 1A1G; 5′-GAAGTGTATCGGTGA-GACCG-3′), both of which contain the polymorphic site at the 3′ end, were synthesized and each was used as a primer for allele-specific PCR amplification,[18] together with another strand of 21-mer primer (C53; 5′-GTAGACAGAGTCTAGGCCTCA-3′), which is located 200 bp downstream of a polymorphic site detected by sequencing. PCR is performed by 30 cycles under the following conditions: 1 min at 95° for denaturation, annealing at 65° for 1 min, and extension at 72° for 1 min. The other conditions are as described by Saiki *et al.*[16] The PCR products are then subjected to electrophoresis in a 1.8% agarose gel. A representative result of PCR to identify *Ile-Val* polymorphism is shown in Fig. 3b.

Single-Strand Conformational Polymorphism (SSCP)

A pair of primers (5′-GAACTGCCACTTCAGCTGTCT-3′ and 5′-GTAGACAGAGTCTAGGCCTCA-3′) was used for the screening of the *Ile-Val* polymorphism by SSCP analysis.[19] Genomic DNAs (50 ng) are used in a 5-μl PCR reaction mixture containing 10 mM Tris–HCl (pH 8.3), 1.5 mM MgCl$_2$, 50 mM KCl, 0.01% gelatin (w/v), 1.25 mM each of four deoxynucleotide triphosphates except for dCTP, which has a concentration of 0.125 mM, 1 μM of each primer, and 0.2 μl of [α-^{32}P]dCTP (3000 Ci mmol^{-1}) and *Taq* DNA polymerase. The PCR is programmed as follows: initial denaturation, 1 min at 95°; amplification, 20 sec at 95°, 2 min at 60° for 30 cycles; and elongation, 1 min at 72°. After completion of the PCR, the product is diluted 1:100 in loading buffer (95% formamide, 20 mM EDTA, 0.05% bromphenol blue, and 0.05% xylene cyanol). The DNA

[17] O. Gotoh, Y. Tagashira, T. Iizuka, and Y. Fujii-Kuriyama, *J. Biochem.* (*Tokyo*) **93,** 807 (1983).

[18] S.-I. Hayashi, J. Watanabe, K. Nakachi, and K. Kawajiri, *Nucleic Acids Res.* **19,** 4797 (1991).

[19] M. Orita, H. Iwahana, H. Kanazawa, K. Hayashi, and T. Sekiya, *Proc. Natl. Acad. Sci. U.S.A.* **86,** 2766 (1989).

fragments are subjected to electrophoresis at 35 W for approximately 3 hr in a 5% nondenaturing polyacrylamide gel with 5% glycerol at room temperature. Upon complete migration, the gels are dried and subjected to autoradiography against a Kodak XAR-5 film at −80° with an intensifying screen. A representative analysis of the *Ile-Val* polymorphism by SSCP is shown in Fig. 3c. The results obtained by these two methods are fully consistent with those obtained by direct sequencing.

Remarks

Following the procedures mentioned earlier, we compared the genotype frequencies of the two polymorphisms between healthy controls and Japa-

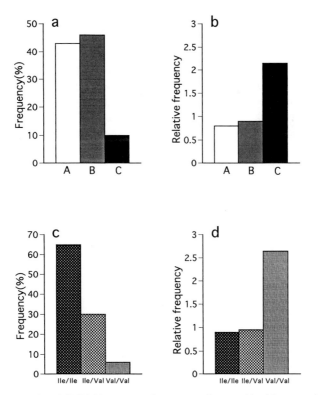

FIG. 4. Frequencies of *CYP1A1* genotypes in cancer patients and healthy controls. Distribution of the three genotypes of the *Msp*I (a) and *Ile-Val* (c) polymorphisms in healthy controls is shown. Also shown are the relative genotype frequencies of the *Msp*I (b) and *Ile-Val* (d) polymorphisms in patients with squamous cell carcinoma of the lung compared with those of healthy controls.

nese cancer patients. We isolated the lymphocyte DNA of 2500 persons from a cohort of the general population[9,10] and analyzed the DNA polymorphisms of 375 randomly selected subjects as healthy controls. The allelic frequencies of the MspI or Ile-Val polymorphism in controls were calculated to be in Hardy–Weinberg linkage equilibrium: 0.67 and 0.33 for the m1 and m2 alleles, respectively; and 0.80 and 0.20 for the Ile and Val alleles, respectively.[8] The linkage coefficient R was described by Hill and Robertson[20] to be 0.78, showing that these two loci are closely associated.[15] Namely, 99% linkage of the m1 allele with the Ile allele and 97% linkage of the Val allele with the m2 allele were demonstrated.

We have found that these two polymorphisms are associated with a high susceptibility to squamous cell carcinoma of the lung[6–8]: genotype C in the MspI polymorphism and genotype Val/Val in the Ile-Val polymorphism among the patients were more than twice as frequent as they were among the controls (Fig. 4). The patients with susceptible genotypes of the CYP1A1 gene contracted carcinoma after smoking fewer cigarettes than those with other genotypes.[9,10] A case control study also revealed that individuals with the susceptible MspI or Ile-Val genotype of CYP1A1 were at a remarkably high risk at a low dose level of cigarette smoking.[9,10] Our results on the identification of genetically susceptible individuals in terms of individual genetic predispositions and life-style may thus offer a new approach to cancer prevention. These results were reconfirmed by a subsequent study in a different Japanese population.[21] However, a lack of association between the CYP1A1 genotypes and lung cancer incidence has been reported in Caucasian populations.[22,23] This discrepancy may be ascribed largely to an ethnic difference in the allelic frequency of the polymorphism.

[20] W. G. Hill and A. Robertson, *Theor. Appl. Genet.* **38,** 226 (1968).

[21] T. Okada, K. Kawashima, S. Fukushi, T. Minakuchi, and S. Nishimura, *Pharmacogenetics* **4,** 333 (1994).

[22] T. Tefre, D. Ryberg, A. Haugen, D. W. Nebert, V. Skaug, A. Brøgger, and A-L. Børresen, *Pharmacogenetics* **1,** 20 (1991).

[23] A. Hirvonen, K. Husgafvel-Pursiainen, A. Karjalainen, S. Anttila, and H. Vainio, *Cancer Epidemiol. Biomarkers Prev.* **1,** 485 (1992).

Section IV

Plant P450s

[26] Analysis of Herbicide Metabolism by Monocot Microsomal Cytochrome P450

By SZE-MEI CINDY LAU and DANIEL P. O'KEEFE

Plant cytochromes P450 serve critical functions in numerous biosynthetic pathways, especially those involved in the production of a diverse array of secondary metabolites.[1] An economically significant function, and one which is only partially understood, is the role of P450s in the detoxification of herbicides.[2,3] This function is one which provides the first committed step in the detoxification process responsible for herbicide selectivity in a variety of plant–herbicide combinations. The role of P450s in selectivity is simple: plants that are able to rapidly metabolize the active compound to one that is nonphytotoxic survive the treatment, those that cannot (preferably most weed species) are killed. This is the basis for the use of many compounds on monocot crop species. Wheat (*Triticum aestivum*) and many of the cereal grains (e.g., oats, barley) have nearly identical resistance to many herbicides because these plants have a very similar capacity to metabolize these compounds.[4–6] The two other major monocot crop species, rice (*Oryza sativa*) and corn (*Zea mays*), have slightly different xenobiotic metabolizing capabilities, but also rely on cytochrome P450-dependent mechanisms for resistance to a variety of herbicidal compounds.[7–9]

While the P450-mediated hydroxylation of herbicides is an important step in metabolism-based resistance, it often does not alter the phytotoxicity of the compound and is not directly observable *in vivo*. Figure 1 demonstrates this with chlorsulfuron in wheat. Hydroxylation of chlorsulfuron leads to a negligible alteration in the ability of the compound to inhibit acetolactate synthase (ALS), the target enzyme for sulfonylurea and imida-

[1] F. Durst and D. P. O'Keefe, *Drug Metab. Drug Interact.* **12,** 171 (1995).
[2] D. P. O'Keefe, R. Lenstra, and C. A. Omer, *in* "Plant Protein Engineering" (P. R. Shewry and S. Gutteridge), p. 281. Cambridge Univ. Press, Cambridge and New York, 1992.
[3] D. P. O'Keefe, J. M. Tepperman, C. Dean, K. J. Leto, D. L. Erbes, and J. T. Odell, *Plant Physiol.* **105,** 473 (1994).
[4] P. B. Sweetser, G. S. Schow, and J. M. Hutchison, *Pestic. Biochem. Physiol.* **17,** 18 (1982).
[5] D. S. Frear, H. R. Swanson, and F. W. Thalacker, *Pestic. Biochem. Physiol.* **41,** 274 (1991).
[6] J. J. Anderson, T. M. Priester, and L. M. Shalaby, *J. Agric. Food Chem.* **37,** 1429 (1989).
[7] S. Takeda, D. L. Erbes, P. B. Sweetser, J. V. Hay, and T. Yuyama, *Weed Res. (Tokyo)* **31,** 67 (1986).
[8] D. E. Moreland, F. T. Corbin, and J. E. McFarland, *Pestic. Biochem. Physiol.* **47,** 206 (1993).
[9] K. E. Diehl, E. W. Stoller, and M. Barrett, *Pestic. Biochem. Physiol.* **51,** 137 (1995).

METHODS IN ENZYMOLOGY, VOL. 272

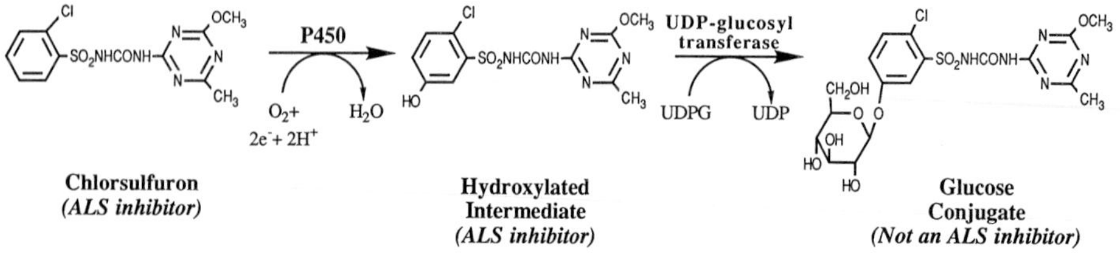

Chlorsulfuron
(ALS inhibitor)

Hydroxylated
Intermediate
(ALS inhibitor)

Glucose
Conjugate
(Not an ALS inhibitor)

P450

$O_2 +$ $2e^- + 2H^+$ H_2O

UDP-glucosyl transferase

UDPG UDP

FIG. 1. The metabolic detoxification of the herbicide chlorsulfuron in wheat. The initial hydroxylation step is carried out by microsomal cytochrome P450. Subsequent conjugation is catalyzed by a soluble UDPglucosyltransferase.

zolinone herbicides. The hydroxylated compound is not observed in intact wheat plants either; the compound appears to be transformed immediately to a glucose conjugate, which has a greatly decreased ability to inhibit ALS.[2,4] This is because hydroxylation is the overall rate-limiting step in this process, and the UDPglucosyltransferase which catalyzes the next step in metabolism acts quickly enough that an intermediate never accumulates. It is clear, however, that hydroxylation is key to the overall detoxification process since the inability to carry out this step determines sensitivity to the herbicide. The hydroxylation reaction is best observed in the absence of any conjugating enzymes *in vitro* with microsomal P450 isolated from wheat.[5]

Similar reactions carried out by many monocot species on a variety of herbicidal compounds can be observed *in vitro*. With wheat, in addition to the chlorsulfuron-5-hydroxylase, hydroxylation of other sulfonylurea herbicides and diclofop has been commonly investigated.[5,6,10–12] Corn is capable of hydroxylating nicosulfuron and similarly producing a 6-hydroxy metabolite from bentazon.[8,9,13–15] Rice can hydroxylate bensulfuron methyl in a reaction that is directly responsible for the loss of phytotoxicity (i.e., subsequent conjugation is not required).[7] All of these reactions are ones that are thought to be the primary step responsible for selectivity of these compounds. Cytochrome P450-mediated metabolism in weeds is also possible, and indeed there are problem weeds that are able to withstand treatment with some herbicides for the same reasons as the crop species survive.[4,7]

[10] J. J. McFadden, D. S. Frear, and E. R. Mansager, *Pestic. Biochem. Physiol.* **34**, 92 (1989).
[11] A. Zimmerlin and F. Durst, *Phytochemistry* **29**, 1729 (1990).
[12] A. Zimmerlin and F. Durst, *Plant Physiol.* **100**, 874 (1992).
[13] J. J. McFadden, J. W. Gronwald, and C. V. Eberlein, *Biochem. Biophys. Res. Commun.* **168**, 206 (1990).
[14] J. W. Gronwald and J. A. Connelly, *Pestic. Biochem. Physiol.* **40**, 284 (1991).
[15] A. E. Haack and N. E. Balke, *Pestic. Biochem. Physiol.* **50**, 92 (1994).

This chapter illustrates the essential steps for preparing microsomes from a variety of monocot plants and shows the assay of this material for chlorsulfuron-5-hydroxylase activity. The level of activity is high enough and the HPLC-based assay is sensitive enough that radiolabeled material is unnecessary.

Preparation of Plant Material

Microsomes are prepared from young etiolated tissues induced by naphthalic anhydride and ethanol. The aerial shoots of these plants, consisting of a coleoptile surrounding an unexpanded leaf, are most commonly used for the preparation of microsomes. Some workers have found that activity in the coleoptile is higher than the leaf,[16] however, this dissection is a very tedious first step in the preparation and is unsuitable for some species. In the whole shoot, herbicide-metabolizing activities peak at 3–4 days of growth (1 to 2 cm long) and gradually decline after that. Treatment of germinating seedlings with inducers can stimulate enzyme activity over 100-fold. Greening of the tissue is undesirable because chlorophyll and other pigments interfere with the analysis and also because the inducible herbicide hydroxylase activity in the microsomal fraction drops precipitously during greening, for unknown reasons. Any spring wheat variety may be suitable; we have obtained similar results with Era, Polk, and Olaf.

Seeds (200 g) are soaked 2–4 hr prior to planting and then towel dried. The seeds are placed in a plastic bag and 0.5% naphthalic anhydride is added. The mixture is shaken until the seeds are coated evenly with naphthalic anhydride. Seeds are then sown on a bed of wet vermiculite, covered with a thin layer of vermiculite, and germinated in the dark at 28°. The seedlings are subirrigated with 0.5 mM CaSO$_4$ for 2 days and then with 10% ethanol in 0.5 mM CaSO$_4$ for 24 hr. After 3 days, etiolated shoots are excised and used for the preparation of microsomes.

Preparation of Microsomes

Microsomal isolation is carried out according to Frear *et al.*[5] Buffers are saturated with N$_2$ for 30 min prior to use. Etiolated shoots (10–20 g) are harvested and immediately immersed in 10 mM Na$_2$S$_2$O$_5$ to reduce the aerobic formation of endogenous inhibitors following excision. The tissue is homogenized in a chilled mortar with 1.5 volumes of 0.1 M potassium phosphate buffer, pH 7.5, containing 1 mM LiCO$_3$, 5 mM dithiothreitol, 10% glycerol (v/v), and 0.1% bovine serum albumin (BSA). The homoge-

---

[16] F. W. Thalacker, H. R. Swanson, and D. S. Frear, *Pestic. Biochem. Physiol.* **49,** 209 (1994).

nate is strained through several layers of well-rinsed, chilled cheesecloth. The filtrate is then centrifuged at 1000 g for 5 min, discarding the pellet, and at 6000 g for 20 min, discarding the pellet again, and finally at 100,000 g for 90 min. The pellet from this final centrifugation is homogenized in ~1 ml of 0.1 M potassium phosphate buffer, pH 7.5, containing 1.5 mM mercaptoethanol and 10% glycerol. Finally, the microsomal preparation is passed through a Sephadex G-25M (Pharmacia PD-10) gel filtration column with 0.1 M potassium phosphate buffer, pH 7.5, containing 1.5 mM mercaptoethanol. The passage of the microsomes through this column is easily monitored visually by the yellow color and by the appearance of cloudy drops at the outlet. Gel filtration is believed to remove some endogenous inhibitors of the hydroxylase activity; however, in our system it is more important to remove low molecular weight impurities that comigrate with the chlorsulfuron product on the reverse-phase column and limit the sensitivity of the assay. Before storage, the microsomal preparation is made 10% (v/v) in glycerol and is drip-frozen as beads in liquid nitrogen and stored under liquid nitrogen (long-term storage at $-80°$ is also suitable).

Chlorsulfuron-5-hydroxylase Assay

Chlorsulfuron-5-hydroxylase activity is assayed by reverse-phase HPLC. The assay has been streamlined to eliminate any unnecessary steps (extraction especially) and to carefully control the volume so that quantitation is accurate. A typical assay contains

25–50 μg microsomal protein
400 μM chlorsulfuron in 0.1 M MOPS, pH 7
2 mg/ml BSA
400 μM NADPH
20 mM glucose 6-phosphate
0.1 U glucose-6-phosphate dehydrogenase (from *Leuconostoc mesenteroides*)
25 μl final 0.1 M MOPS, pH 7

The reaction is started by the addition of NADPH. Glucose 6-phosphate and the dehydrogenase are added to regenerate NADPH for prolonged incubation, although for times less than 30 min they are unnecessary with wheat microsomes. The addition of BSA increases the activity by 50–100%. After up to 1 hr incubation at room temperature, the reaction is terminated by adding 75 μl acetonitrile. The mixture is microfuged for 15 min to remove most of the precipitated proteins. The supernatant is transferred to another microfuge tube containing 150 μl of water and is then analyzed by reverse-phase HPLC. Two hundred microliters is run on a Zorbax ODS column (3 μm, 6.2 mm $\times$ 8 cm) using a combination of linear gradients (flow = 1.5 ml/min): starting with 20% of solvent B in solvent A, progressing from 20 to 75% of solvent B between 1 and 10 min, from 75 to 95% of solvent B between 10 and 12 min, and holding at 95% of solvent B between 12 and 15 min. Solvent A is 1% phosphoric acid in water and solvent B is 1% phosphoric acid in acetonitrile. Singly hydroxylated herbicide metabolites, which are more polar than the parent, will elute about 1 min before the herbicide peak, using these solvents and gradient. For chlorsulfuron, the signal is monitored at 225 nm, using a diode-array detector. Selection of an optimum wavelength for monitoring is a balance between sensitivity and selectivity. Figure 2 shows a typical assay of chlorsulfuron metabolism by wheat microsomes. In addition to the single major metabolite formed, it is worth noting that there are two minor unidentified potential metabolites at 5.2 and 6.6 min in this chromatogram. Formic acid (0.1%) or trifluoroacetic acid (0.1%) can be substituted for phosphoric acid in the HPLC solvents with little effect on retention or resolution. While neither of these modifiers

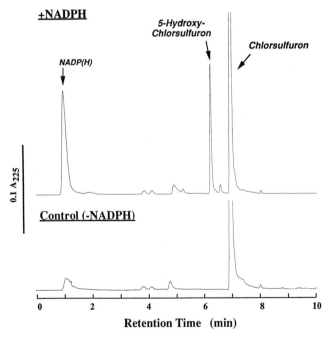

FIG. 2. Chromatogram showing the separation of herbicide metabolites by the procedure described in the text. The total height of the chlorsulfuron peak is 0.84 A in the upper chromatogram and 1.11 in the lower one. The peak at ~1.1 min in the upper chromatogram is derived from NADPH used in the assay.

is as transparent as phosphoric acid at lower wavelengths, they are particularly useful when it is necessary to separate or isolate the metabolite in a volatile buffer (e.g., for subsequent LC/MS).

Routinely, a control without NADPH is included in the assay to ensure that the metabolite is indeed a NADPH-dependent product. With every new microsomal preparation, it is also advised to run another control that includes NADPH but no substrate in order to verify that no endogenous compound coelutes with the metabolite. Sensitivity of the assay as described here is high; without using radioisotopes it is possible to measure down to about 2 pmol of metabolite (80 nM in a 25-μl reaction). An assay run on a preparation as described here should yield a rate of at least 0.4 nmol mg^{-1} min^{-1} for wheat microsomes induced with naphthalic anhydride and ethanol. This is generally 2 to 4 times the rate observed with naphthalic anhydride induction alone and >100 times the rate with no induction treatment. The amounts measured are a function of many variables; the quality of the preparation is probably the most important. The length of incubation time in the assay also helps determine the overall rate, and excessively long incubations should be avoided; at room temperature the rate is only linear for about 20 min.

General Considerations

The procedures outlined here have been successfully used with only minor variations to prepare enzymatically active microsomal preparations from wheat, corn, and rice. Weed species can be used as sources of enzyme as well, and we have prepared active fractions from wild oats (*Avena fatua*), jointed goatgrass (*Aegilops cylindrica*), green foxtail (*Setaria viridis*), downy brome (*Bromus tectorum*), and annual ryegrass (*Lolium multiflorum*). The herbicide undergoing metabolism is not limited to chlorsulfuron, and we have measured the metabolism of metsulfuron methyl, nicosulfuron, bensulfuron methyl, diclofop, and bentazon under nearly identical conditions. In some cases the chromatographic conditions may need to be varied to optimize separation of parent herbicide, metabolite, and endogenous compounds.

At the microsomal level of purity, the most useful comparative parameter between substrates and different microsomal preparations is the Michaelis constant (K_m) of the P450-catalyzed reaction. This value, which is related to the affinity of the enzyme for substrate, is one of the few ways to distinguish between different enzymes in crude plant microsomal fractions. At the extreme, it is possible to distinguish multiple forms of an enzyme that have substantially different K_m values for the same substrate. At a minimum, the K_m provides an indication of the amount of substrate necessary for maximum rate. The wheat microsomal chlorsulfuron hydroxylase has a K_m for chlorsulfuron of 46 μM. As Fig. 3 shows, this value is pH dependent and increases markedly above pH 7. Note that the V_{max} of this reaction exhibits a maximum at pH 8; however, the specificity factor (V_{max}/K_m), which defines the rate at low substrate concentrations, is maximal at pH 7. This is an important distinction since the maximum rate is desirable for *in vitro* studies but the rate at limiting substrate concentration is more significant physiologically. The value of a substrate concentration/rate analysis on a crude microsomal preparation can be further illustrated with a comparison of wheat and wild oats, a weed that is tolerant to chlorsulfuron. Wild oats has identical activity (an NADPH-dependent chlorsulfuron-5-hydroxylase), when it is induced the same way, and is present in these microsomes at nearly the same specific activity (i.e., the apparent V_{max} in the two preparations are roughly equal). However, the wild oat enzyme has a K_m for chlorsulfuron of 600 μM, demonstrating a markedly lower affinity for this substrate. This in turn shows that the specificity factor is >10-fold higher in wheat, although a simple assay run at a saturating substrate concentration would give the same rate for each.

We have thus far not mentioned the spectroscopically determined P450 content of the microsomal fractions in this work, primarily because it may be one of the most misleading descriptions of the quality of the preparation. Microsomes prepared from etiolated shoots of most grasses contain a large amount of a cytochrome P450, allene oxide synthase (AOS), that is enzy-

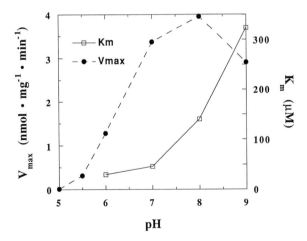

FIG. 3. The pH dependence of apparent steady-state enzyme kinetic parameters of wheat microsomal chlorsulfuron-5-hydroxylase.

matically and physically very distinct from most monooxygenase P450s.[17] The AOS P450 is catalytically self-sufficient, catalyzing the formation of an allene oxide from a fatty acid hydroperoxide.[18] The AOS P450 is also distinguishable from other P450s in the markedly low affinity of the ferrous form for CO (reduced tulip bulb AOS has a CO affinity of $2.9 \times 10^3 \ M^{-1}$ compared to $1.8 \times 10^5 \ M^{-1}$ for avocado CYP71A1).[17] This property can be used to assess the relative content of AOS in microsomal fractions, and preparations from noninduced plants typically contain predominantly AOS P450 by this criteria. Induction treatments, such as those described here, result in a very little increase[5] in spectroscopically detectable P450, but the CO affinity increases in a heterogeneous fashion. This suggests that these treatments lead to some loss in AOS P450 and an increase in other reductase-dependent forms. No evidence to date shows that the AOS P450 itself carries out any reductant-dependent monooxygenase reaction. However, some investigators have observed xenobiotic metabolizing activity in P450s from tulip bulb, a tissue which appears to contain almost entirely AOS P450.[17,19,20] AOS presence in a microsomal fraction is an important factor to account for in the determination of total P450 (from CO difference spectra) and in attempts to interpret substrate-binding spectra.

Finally, we must evaluate the basic assumptions that are made about the enzyme, or enzymes, that are under study when microsomes are prepared from plants that have been treated with one or more inducers. There is very little evidence that addresses whether the uninduced, naphthalic anhydride induced, or the naphthalic anhydride plus ethanol induced activities represent expression of multiple distinct enzymes or different expression levels of the same enzyme. It is possible to treat the data from these preparations as if from a single enzyme; however, that is simply a functional expedient and is not meant to imply a conclusion. Evidence from inhibitor studies of these preparations is consistent with two or more distinct enzymes accounting for the activity. This possibility needs to remain foremost during any interpretation of data from crude microsomal fractions.

Acknowledgments

The authors acknowledge the excellent technical assistance of Roberta Perkins and are grateful to D. Stuart Frear for his many useful suggestions and advice.

[17] S. C. Lau, P. A. Harder, and D. P. O'Keefe, *Biochemistry* **32**, 1945 (1993).
[18] W. Song and A. R. Brash, *Science* **253**, 781 (1991).
[19] A. Topal, N. Adams, W. C. Dauterman, E. Hodgson, and S. L. Kelly, *Pestic. Sci.* **38**, 9 (1993).
[20] H. Haniskova, E. Frei, H. H. Schmeiser, P. Anzenbacher, and M. Stiborová, *Plant Sci.* **110**, 53 (1995).

[27] Microsome Preparation from Woody Plant Tissues

By Jerry Hefner and Rodney Croteau

Introduction

Woody tissues carry out many of the same cytochrome P450-dependent primary metabolic reactions as their herbaceous counterparts (e.g., in the biosynthesis of phytosterols and gibberellins) and additionally are often rich in secondary metabolites (terpenoids, alkaloids, phenylpropanoids), the biosynthesis of which commonly involves cytochrome P450-catalyzed steps.[1] In many cases, metabolic activity in such tissues is restricted to a relatively thin annular layer of living cells (cambium, phloem, xylem parenchyma) between the nonliving phellem (cork) and the interior heartwood and sapwood derived from old xylem.[2] The study of biochemical processes in general, and cytochrome P450 reactions in particular, in woody tissues has been slow to develop, in large part because of two major problems. Such tissues are often fibrous, at best, and too dense to homogenize by common methods,[3] at worst, and they usually contain very low levels of protein per unit mass, even when extraction is efficient. Woody tissues are commonly pigmented, especially the bark, and often accumulate considerable quantities of oils, resins, and phenolics that can inactivate microsomal enzymes, complicate product isolation and analysis, and, in the case of P450 proteins, interfere with spectroscopic evaluation.

A range of approaches has been developed to deal with these problems. These include very rigorous tissue disruption techniques which, however, should not be so harsh as to inactivate the target enzyme or to unnecessarily generate fines from nonliving structural materials which serve only to occlude microsomes upon centrifugal separation. Polymeric adsorbents (polyvinylpyrrolidone, polystyrene, anion-exchange resin) are often employed to remove hydrophobic materials and phenolics.[4–6] Reducing agents,[7,8] such

[1] T. Higuchi, ed., "Biosynthesis and Biodegradation of Wood Components." Academic Press, New York, 1985.

[2] J. D. Mauseth, "Plant Anatomy," p. 297. Benjamin/Cummings, Menlo Park, CA, 1988.

[3] J. L. Hall and A. L. Moore, eds., "Isolation of Membranes and Organelles from Plant Cells." Academic Press, New York, 1983.

[4] W. D. Loomis, Methods Enzymol. 31, 528 (1974).

[5] T. H. Lam and M. Shaw, Biochem. Biophys. Res. Commun. 39, 965 (1970).

[6] W. D. Loomis, J. D. Lile, R. P. Sandström, and A. J. Burbott, Phytochemistry 18, 1049 (1979).

[7] J. W. Anderson and K. S. Rowan, Phytochemistry 6, 1047 (1967).

[8] B. E. Haissig and A. L. Schipper, Jr., Phytochemistry 14, 345 (1975).

as $Na_2S_2O_5$, as well as inhibitors of phenoloxidase may also be of use in dealing with the phenolic problem.[9-11] Such approaches, however, are not without complications arising from significant alteration in buffer viscosity or suspended solids content, and from the inadvertent inhibition of the target enzyme.

An extensive search of the literature revealed that microsomes have been rarely prepared from tissue displaying fully lignified, secondary growth. Of over 500 literature references to microsome preparation from higher plants during the period 1974 to 1995, fewer than 2% dealt with woody tissues. Protocols designed to overcome the problems associated with microsome isolation from woody tissue have been described by Yoshida[12] and Sutinen.[13,14] The homogenization methods used in these procedures necessitate the separation of the bark tissue[15] from the wood, either by peeling in the case of stem material or by grinding frozen tissue sections with a mortar and pestle. Once the bark has been removed from the wood, small pieces are either ground wet with an abrasive using a motor-driven mortar and pestle or homogenized using a polytron. These disruption procedures are routinely performed at 0 to 4° in the presence of a medium containing reducing agents, protease inhibitors, and polymeric adsorbents.

This chapter describes a series of methods developed from our studies on cytochrome P450 terpenoid hydroxylases[16] involved in resin acid and taxol biosynthesis in gymnosperms, including pine, fir, and yew species.[17-20] These techniques range from protocols for small-scale survey work, in which tissue can be completely pulverized, to intermediate-scale preparations involving shearing of tissue and hammer milling, to large-scale isolation requiring a substantial chopping mill (Wiley mill). All of these operations

[9] W. D. Loomis and J. Battaile, *Phytochemistry* **5**, 423 (1966).
[10] J. W. Anderson, *Phytochemistry* **7**, 1973 (1968).
[11] W. D. Loomis, *Methods Enzymol.* **13**, 555 (1969).
[12] S. Yoshida, *Plant Physiol.* **57**, 710 (1976).
[13] M.-L. K. Sutinen, *Acta Univ. Ouluensis, Ser. A* **240**, doctoral thesis (1992).
[14] A. Ryyppö, E. M. Vapaavuori, R. Rikala, and M.-L. Sutinen, *J. Exp. Bot.* **45**, 1533 (1994).
[15] In this chapter, we use the word bark in an operational sense to describe all tissue from the cambium outward that can be separated from the nonliving, woody core by peeling, abrasion, or other physical means.
[16] C. Mihaliak, F. Karp, and R. Croteau, *Methods Plant Biochem.* **9**, 261 (1993).
[17] C. Funk and R. Croteau, *Arch. Biochem. Biophys.* **308**, 258 (1994).
[18] C. Funk, E. Lewinsohn, B. Stofer-Vogel, C. L. Steele, and R. Croteau, *Plant Physiol.* **106**, 999 (1994).
[19] R. Croteau, M. Hezari, J. Hefner, A. Koepp, and N. G. Lewis, *ACS Symp. Ser.* **583**, 72 (1995).
[20] R. Croteau, J. Hefner, M. Hezari, and N. G. Lewis, *in* "Phytochemicals and Health" (H. E. Flores and D. L. Gustine, eds.), p. 94. American Society of Plant Physiologists, Rockville, MD, 1995.

are performed at subambient temperatures, employ buffers designed to minimize the influence of deleterious coextractives, and maintain membrane integrity, and are adaptable to a wide spectrum of woody tissues.

Buffers for Isolation of Microsomes from Woody Tissue

High quality microsomes may be extracted from woody tissues using a variety of buffers; however, these media should all contain agents which protect cytochrome P450 from inactivation and promote efficient recovery of this enzyme. The pH of the buffer should be between 6.6 and 7.6 to protonate phenolics, minimize their oxidation, and maximize the effectiveness of polymeric adsorbents.[4,9] A sodium phosphate or HEPES buffer at a concentration of 0.1 to 0.2 M is recommended. The antioxidants sodium ascorbate and $Na_2S_2O_5$ are generally added to the extraction buffer at a concentration of 10 mM each to maintain phenolics in the reduced state, as quinones are particularly deleterious. The vinylpyrrolidone polymers, PVP-40[21] and PVPP,[22] are also added at a concentration of 1% (w/v) to adsorb phenolics and prevent their binding to proteins. In our experience, these precautions completely eliminate problems with phenolics during extraction of both soluble and microsomal enzymes, and present no harm to either.[17,23,24] Many of the hydrophobic metabolites and pigments that are coextracted from woody tissues may be removed from the homogenate by adding washed Amberlite XAD-4[25] to the extraction buffer at a concentration of 10% (w/v). The extraction medium should also contain reagents which are necessary to preserve cytochrome P450 in active form. These include 15–30% glycerol or 0.3 M sorbitol, 10 mM dithiothreitol (DTT), and 1 mM EDTA.

In some woody species, such as yew, endogenous phospholipase A_2 activity can cause disruption of microsomal membranes and partial solubilization of cytochrome P450 by generating powerful detergents such as lysolecithin. Phospholipase A_2 activity can be minimized by reducing the level of free calcium through the addition of 2 mM EGTA[26] to the extraction buffer and by adding inhibitors of this enzyme, including 10 mM procaine, 10 mM lidocaine, and 1 mM p-bromoacetophenone. In our experience, proteolysis has not been a major problem during microsome preparation

---

[21] Soluble polyvinylpyrrolidone with an average molecular weight of 40,000.
[22] Insoluble polyvinylpolypyrrolidone or Polyclar AT.
[23] E. Lewinsohn, M. Gijzen, T. J. Savage, and R. Croteau, *Plant Physiol.* **96,** 38 (1991).
[24] A. E. Koepp, M. Hezari, J. Zajicek, B. Stofer Vogel, R. E. LaFever, N. G. Lewis, and R. Croteau, *J. Biol. Chem.* **270,** 8686 (1995).
[25] This polystyrene resin should be prepared according to the procedure outlined in Loomis.[4]
[26] EGTA replaces EDTA.

from woody plant species. However, the serine/thiol protease inhibitor leupeptin is generally added to the extraction buffer, as a precaution, at a concentration of 1 μg/ml. The addition of other class-specific protease inhibitors, such as phenylmethylsulfonyl fluoride, antipain, and pepstatin, may slightly increase cytochrome P450 yield; however, they usually are not required. The extraction buffer should be used shortly after its preparation, as some of the buffer components are not indefinitely stable under ambient atmosphere (e.g., ascorbic acid). Typically, 4 to 5 ml of extraction buffer is used per gram of tissue in the following homogenization procedures.

Tissue Selection

The first two methods presented in this chapter are sufficient to homogenize and extract young woody stem tissue (stems less than 0.5 cm in diameter). The hammer mill and Wiley mill are capable of processing stems up to 1.0 cm in diameter. If older tissue is used, the bark must first be separated from the wood at the vascular cambium before homogenization.

Small-Scale Tissue Extraction for Activity Surveys

A quick and efficient method for woody tissue disruption, which requires no special equipment, is a pulverization technique developed by Lewinsohn et al.[23] This method is capable of producing a good quality homogenate; however, only small amounts of tissue (1 to 3 g) can be processed per batch. Young stems or sections of bark are collected and frozen in liquid nitrogen. Needles may be removed from the stems by hand while wearing protective gloves. The tissue is then cut to fit inside a 6 × 11-cm or 8 × 14-cm manila envelope. An appropriate amount of well-frozen tissue is added to the envelope, which is then placed on a steel block or plate (anvil) and struck repeatedly with the flat end of a heavy ball peen hammer. The pulverized, still frozen tissue is then quickly added to the required amount of extraction buffer.

Intermediate-Scale Tissue Extraction by Shearing

The Waring blender has found much use in the preparation of microsomes from herbaceous tissues.[27,28] This method of homogenization can be used successfully with woody stems provided that young tissue is used.[17,18]

[27] K. Higashi, K. Ikeuchi, M. Obara, Y. Karasaki, H. Hirano, S. Gotoh, and Y. Koga, *Agric. Biol. Chem.* **8,** 2399 (1985).
[28] D. P. O'Keefe and K. J. Leto, *Plant Physiol.* **89,** 1141 (1989).

However, in our experience with microsome preparation from yew species, the extraction efficiency can be increased if the bark is first stripped from the stems. Typically, fresh stems are collected and the needles removed by hand. The tissue is weighed (10 to 100 g can be processed per batch) and cut into 1- to 2-cm sections which are then added to the prechilled chamber with the requisite amount of cold extraction buffer. The mixture is normally processed at the highest speed setting for 10 sec.

Unfortunately, there are two major disadvantages to this technique when compared to the other methods presented here. The first drawback is the oxidation of phenolic compounds, with subsequent protein binding, that may be promoted by whipping air into the homogenate. If phenolic oxidation is a problem, working in an inert atmosphere glove box may be necessary.[4] Alternatively, foam arrestors and blender chambers which allow replacement of the ambient atmosphere with inert gas (nitrogen or argon) are available.[29] The second problem concerns protein yield. Microsomes from yew have been routinely prepared using both the hammer mill (see below) and the Waring blender. Comparison of membrane-bound protein yield obtained with each method, using similar amounts of stem tissue, consistently showed that microsomes prepared using the hammer mill contained considerably higher levels of protein than those produced using the Waring blender. The main advantage to the use of a blender in the production of woody tissue homogenates is that these devices are generally available.

Intermediate-Scale Tissue Extraction Using a Hammer Mill

The most efficient and least troublesome intermediate-scale homogenization method is hammer milling. The hammer mill has been used extensively in this laboratory to process stem tissue from fir, pine, and yew.[24,30] Quantities of stem tissue ranging from 5 to 200 g can be processed rapidly using this technique. Moreover, this milling procedure is designed such that tissue homogenization occurs with minimal exposure to oxygen. Another advantage to using the hammer mill for woody tissue extraction is that the force of compression can be adjusted so that stems are either completely pulverized or only debarked (i.e., the bark and cambium are powdered while the majority of the woody stems remain intact).[31]

---

[29] Such attachments are available from a number of suppliers, including the Cole-Palmer Instrument Co.

[30] R. E. LaFever, B. Stofer Vogel, and R. Croteau, *Arch. Biochem. Biophys.* **313**, 139 (1994).

[31] Selective homogenization of stem tissue is sometimes desirable to avoid releasing compounds located within the wood. The milling of stems with sufficient force to homogenize only the bark tissue alleviates the need for preparative bark stripping.

The lack of commercial availability is the only disadvantage associated with this device. However, such a mill of any size can be constructed from hardened nickel–copper alloy at a local machine shop. The design and dimensions of the hammer mill used in this laboratory have been described in detail.[30] The mill consists of a large chamber where the stems are pulverized and a matching piston, which is grooved at the end that contacts the tissue and machined with a striking head at the other end.

A routine preparation of yew microsomes using the hammer mill employs about 60 g of young stems. The stems are cut to lengths of 5 to 8 cm and are held in a cooler containing liquid nitrogen. The needles are removed by agitation and the frozen stems are then quickly weighed. Once the tissue has been prepared, the mill chamber and piston are cooled with liquid nitrogen. When most of the liquid nitrogen has evaporated from the chamber, about 20 g of frozen stems is added.[32] Next, the piston is lowered into the chamber and struck repeatedly with a 16-pound sledge hammer. The piston is periodically rotated 90° during this pulverizing process. The frozen tissue powder can be easily removed by inverting the chamber over a large piece of weighing paper, at which time the tissue is added to chilled extraction buffer. Milling is repeated until all the tissue is processed.[33]

Large-Scale Tissue Extraction Using a Chopping Mill

It is often necessary to process large amounts of tissue in order to obtain sufficient quantities of enzyme for purification. Hammer milling, although efficient, is time-consuming when the amount of tissue to be processed approaches a kilogram. The use of a Wiley mill for this purpose greatly decreases processing time (approximately 20 min per kg tissue) and provides a homogenate of good quality for microsome isolation.[34] It is recommended that a Wiley mill model No. 1 (Thomas Scientific, Swedesboro, N.J.), or equivalent, containing a sieve plate with 1-cm-diameter circular openings be used for such tasks.

Both microsomes and soluble enzymes from fir and pine have been routinely prepared in this laboratory using the Wiley mill.[17,35] A large-scale enzyme preparation from grand fir stems or lodgepole pine bark begins with about 500 g of fresh tissue. It is desirable to remove any lateral branches

---

[32] Milling of 15- to 20-g portions gives optimum homogenization in a mill of this size.

[33] It may be necessary to periodically cool the chamber and piston between loads to maintain an appropriate temperature.

[34] It is worth noting at this point that the limiting factor in microsome preparation at this scale is usually not the homogenization method but the time and capacity needed for centrifugation.

[35] M. Gijzen, E. Lewinsohn, and R. Croteau, *Arch. Biochem. Biophys.* **289**, 267 (1991).

from stem material and to cut all tissue into 7- to 10-cm sections. The tissue sections are then frozen in a cooler containing liquid nitrogen. Many of the brittle needles can be removed from the stems by agitation or stirring while submerged in liquid nitrogen. Any remaining needles must be removed by hand while wearing protective gloves. Once the tissue has been prepared, the Wiley mill should be chilled with liquid nitrogen. In order to adequately chill all necessary parts of the mill, the release plate is opened slightly and 100 to 200 ml of liquid nitrogen is slowly poured into the hopper while the machine is running. Not only does this process serve to chill all the components of the mill which contact the tissue (i.e., the chopping chamber, cutting head and blades, sieve plate, and collection box) but it freezes and expels any residual material left in the mill from the previous run. This cooling process should be repeated until the collection box is very cold to the touch and all debris from previous runs has been removed.[36] With the mill operating and the release plate fully closed, the hopper is filled with tissue. A flat board is then placed over the mouth of the hopper to prevent tissue from being hurled from the machine when the release plate is fully opened. Once this amount of tissue is fully ground, it is desirable to slowly add a few hundred milliliters of liquid nitrogen through the hopper to maintain the mill at low temperature. This batchwise grinding process is repeated until all of the tissue is pulverized.[37]

Extraction, Filtration, and Centrifugation

Regardless of which homogenization method is used, the tissue should be allowed to equilibrate with the extraction buffer at 4° with gentle stirring or agitation for 5 to 15 min. The extract is then filtered through eight layers of cheesecloth (or, in the case of small volumes, one layer of Miracloth) in order to remove polystyrene beads and large tissue debris. When filtration is complete, the extract should be centrifuged at 3000 g for 20 min to remove cellular debris and the insoluble PVPP[22] (that was added to the extraction buffer). The supernatant is then centrifuged at 20,000 g for 20 min to remove dense organelles. If necessary, the 20,000-g supernatant can be passed through a 50-μm nylon mesh to remove any floating debris (such as small wood fibers). Finally, this supernatant is centrifuged at 105,000 g for 90 min.

Microsomal pellets may be stored at −80° under argon without resuspension.[16] Alternatively, pellets may be resuspended with the aid of a glass

---

[36] The rapid addition of large amounts of liquid nitrogen in order to speed up the cooling process may result in damage to the mill.

[37] For large preparations, it may be necessary to periodically empty the collection box as the grinding chamber will become clogged if the collection box is overfilled.

homogenizer in a buffer consisting of 25 mM HEPES, pH 7.0–7.5, 1 mM EGTA, 10 mM MgCl$_2$, 25 mM KCl, 5 mM DTT, 1 μg/ml leupeptin, and 20% glycerol. Generally, the suspension can be stored at −80° for at least 2 weeks without significant loss of enzymatic activity.

Acknowledgments

This research was supported by Grant CA55254 from the National Institutes of Health and Grant 94-37302-0614 from the U.S. Department of Agriculture. We thank Christopher Steele for helpful discussions.

[28] Detection, Assay, and Isolation of Allene Oxide Synthase

By ALAN R. BRASH and WENCHAO SONG

Introduction

The enzyme(s) now commonly referred to as allene oxide synthase (AOS), *CYP74*,[1] was reported in 1966 as an activity in flaxseed that metabolized linoleic acid hydroperoxide to an α-ketol derivative.[2] Early publications established that other associated products from unsaturated fatty acid hydroperoxides are γ-ketols and cyclopentenone fatty acids.[3,4] These derivatives are now recognized to arise by hydrolysis and rearrangements of the initial enzymic product, the very unstable allene oxide (Fig. 1).[5,6] The early literature refers to AOS as "hydroperoxide isomerase"[2,3] or "hydroperoxide cyclase,"[4] and "hydroperoxide dehydrase" has also been used.[7] In plants, the 12,13-epoxyallene oxides are substrates for a specific cyclase that promotes almost complete conversion to a chiral cyclopentenone, 12-oxophytodienoic acid, a metabolic precursor of jasmonic acid.[8]

[1] W.-C. Song, C. D. Funk, and A. R. Brash, *Proc. Natl. Acad. Sci. U.S.A.* **90,** 8519 (1993).
[2] D. C. Zimmerman, *Biochem. Biophys. Res. Commun.* **23,** 398 (1966).
[3] H. W. Gardner, *J. Lipid Res.* **11,** 311 (1970).
[4] D. C. Zimmerman and P. Feng, *Lipids* **13,** 602 (1975).
[5] M. Hamberg, *Biochim. Biophys. Acta* **920,** 76 (1987).
[6] A. R. Brash, S. W. Baertschi, C. D. Ingram, and T. M. Harris, *Proc. Natl. Acad. Sci. U.S.A.* **85,** 3382 (1988).
[7] M. Hamberg and H. W. Gardner, *Biochim. Biophys. Acta* **1165,** 1 (1992).
[8] M. Hamberg, *Biochem. Biophys. Res. Commun.* **156,** 543 (1988).

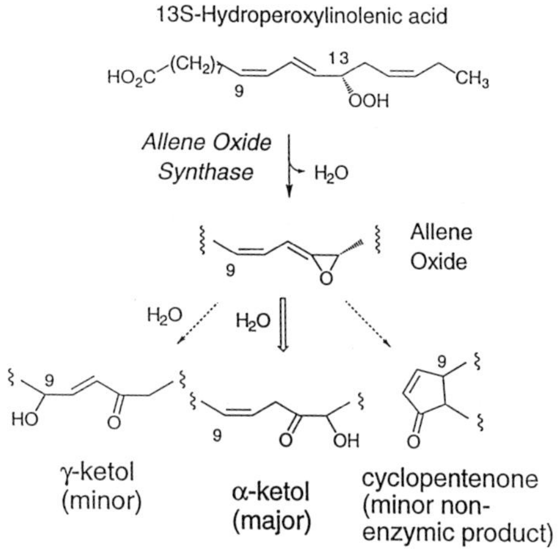

13S-Hydroperoxylinolenic acid

FIG. 1. Pathways of allene oxide synthesis and degradation. The major allene oxide-derived products starting from 13S-hydroperoxylinolenic acid are a γ-ketol (12-keto-9-hydroxyoctadec-10E,15Z-dienoic acid), an α-ketol (12-keto-13-hydroxyoctadec-9Z,15Z-dienoic acid), and a cyclopentenone (12-oxophyto-10,15Z-dienoic acid).

Preparation of Fatty Acid Hydroperoxide Substrate for AOS Assay

13-Hydroperoxylinoleic acid, 13-hydroperoxylinolenic acid, or 15-hydroperoxyeicosatetraenoic acid (15-HPETE) (15-hydroperoxyarachidonic acid) can be purchased (e.g., Cayman Chemical, Ann Arbor), but repetitive assays require multimilligram supplies of substrate that can be prepared for a small fraction of the commercial cost.

Materials

Soybean Lipoxygenase (Sigma Type V, Cat. No. L 6632; Sigma Uses the Old Name "Lipoxidase"). The "type V" is the most convenient form of the enzyme because it is an ammonium sulfate milky slurry that can be pipetted (instead of having to weigh out small amounts of the solid enzyme). This preparation is stable for years at 4°. Five million Sigma units is sufficient to prepare >1 g of hydroperoxide.

Linolenic Acid [e.g., Sigma Cat. No. L 2376, or Nu Check Prep, Inc. (Elysian, MN) Cat. No. U-62-A]. Linoleic acid or arachidonic acid is equally acceptable. Prepare a stock solution of 10 or 20 mg/ml in ethanol and store at −20° or below under nitrogen or argon.

Buffer. The soybean lipoxygenase L-1 (the isozyme in the Sigma "type V" preparation) catalyzes specific reactions at high pH, and typically pH 8.5–10.5 buffers are used. We routinely use 0.1 M K_2HPO_4 (pH not adjusted and usually observed at pH 8.5–9).

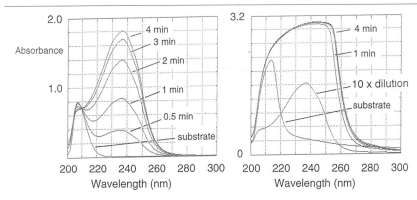

FIG. 2. Preparation of 13S-hydroperoxylinolenic acid using soybean lipoxygenase. (Left) Analytical scale reaction. UV recordings of substrate (25 μg α-linolenic acid in 1 ml 0.1 M K$_2$HPO$_4$) and product formation on addition of 0.25 μg/ml soybean lipoxygenase (type V, Sigma). Reaction is at room temperature. (Right) Preparative reaction. UV recordings of substrate (10 mg α-linolenic acid added in 1 ml EtOH to 50 ml of oxygen-saturated 0.1 M K$_2$HPO$_4$) and the formation of a hydroperoxide product following the addition of 2 μg/ml lipoxygenase. Although the UV is saturated, product formation can be quantified accurately in diluted aliquots. The absence of distinct chromophores in the 280-nm region is a reflection of the specific oxygenation and minimal formation of oxodiene by-products.

Soybean Lipoxygenase Reaction

This is very straightforward and has been used to prepare grams of fatty acid hydroperoxides. Initially it is useful to carry out an analytical scale incubation in an ultraviolet (UV) cuvette and check the reaction using a scanning spectrophotometer (Fig. 2, left).

The small-scale preparative procedure (Fig. 2, right) has the advantage that it is optimized for minimal formation of by-products and for purification using standard-sized high-pressure liquid chromatography (HPLC) columns (25 $\times$ 0.46 cm). To avoid secondary reactions that convert the initial hydroperoxide product to dihydroperoxides and oxodiene derivatives, low concentrations of enzyme (generally $\leq$2 μg/ml) should be employed. Also, the solution must remain aerobic or the enzyme will catalyze anaerobic oxodiene formation.[9] Normally, buffers in the laboratory contain $\approx$250 μM O$_2$ and, as O$_2$ is used on an equimolar basis in hydroperoxide synthesis, this sets an upper limit on the concentration of fatty acid substrate. Oxygenation of the buffer permits use of higher fatty acid concentrations (Fig. 2, right).

[9] G. J. Garssen, J. F. G. Vliegenthart, and J. Boldingh, *Biochem. J.* **122,** 327 (1971).

Extraction of the Product

Use of cold solutions, with the sample kept on ice, helps with the quantitative recovery of undegraded fatty acid hydroperoxide. Acidify the solution to pH 3–4 and extract once with 0.5 volume of methylene chloride (or chloroform, ethyl acetate, or diethyl ether). Collect the organic phase, wash once or twice with water (to remove traces of acid), evaporate to dryness, and dissolve the hydroperoxide product in EtOH (≈5–10 mg/ml). Keep on ice and store at −20° under nitrogen or argon.

Purification of Fatty Acid Hydroperoxide

A standard 5- or 10-μm silica HPLC column (25 × 0.46 cm) is suitable for the purification of amounts of ≈0.5–1 mg per injection. The HPLC system is essentially the same as shown under straight-phase (SP)-HPLC analysis of AOS-derived products (Fig. 4B). A semipreparative column (e.g., Alltech Econosil 10 μm silica, 25 × 1 cm) can be used with the same solvents and with sample load and flow rates scaled up fivefold.

Quantitation of Fatty Acid Hydroperoxide

The molar extinction coefficient of the hydroperoxy fatty acids is 23,000.[10] As the molecular weights are 308, 310, and 336 for hydroperoxy-C18.2, -C18.3, and -C20.4, respectively, 100 μM gives an absorbance at 235 nm of 2.3 absorbance units (AU) and corresponds to approximately 30 μg/ml of product.

Detection of Allene Oxide Synthase: UV Assay

The assay of activity in fractions from enzyme purification is easily carried out using the reduction in UV absorbance at 235 nm associated with the formation and almost instantaneous hydrolysis of the allene oxide mainly to non-UV-absorbing α-ketols (Fig. 3).[11]

In crude samples the UV assay is also extremely useful, although caution must be applied initially as there exists the possibility of interference from other pathways of degradation of the fatty acid hydroperoxide. In particular, disappearance of the conjugated diene chromophore of the hydroperoxide is also catalyzed by the hydroperoxide lyases that are abundant in some plant tissues.[12] To distinguish these activities requires HPLC analysis or

---

[10] M. J. Gibian and P. Vandenberg, *Anal. Biochem.* **163,** 343 (1987).
[11] W.-C. Song and A. R. Brash, *Science* **253,** 781 (1991).
[12] H. W. Gardner, *Biochim. Biophys. Acta* **1084,** 221 (1991).

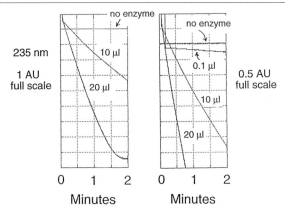

FIG. 3. UV assay of allene oxide synthase. Conditions: 13S-hydroperoxylinolenic acid substrate (20 μl of 1 mg/ml EtOH stock solution) is mixed with 0.96–0.97 ml of 50 mM phosphate buffer, pH 7, and absorbance at 235 nm is recorded for 2 min following the addition of enzyme (final reaction volume, 1 ml). (Left) Overlaid recordings (235 nm) from separate assays with no enzyme and after addition of 10 or 20 μl of an extract of 20 mg/ml acetone powder of flaxseed. Activity is quantified as the initial rate of decrease in 235 nm absorbance. (Right) Recordings on an expanded scale show that the assay has the sensitivity to detect reaction by 0.1 μl (added from a 10-fold dilution) of the same flaxseed extract. Assay precision is evidenced by almost indistinguishable duplicate assays of the 20 μl of extract.

another physicochemical technique such as the NADPH-coupled UV assay developed by Vick.[13]

Detection of AOS Products by HPLC

The synthesis of allene oxide is carried out essentially as in the UV assay (Fig. 3), although it is convenient to scale up considerably, e.g., react 1 mg fatty acid hydroperoxide, added in a small volume of EtOH (20–50 μl), to 1 ml of flaxseed acetone powder extract. After 5 min at room temperature, adjust pH to ≈3–4, extract into methylene chloride, wash the organic phase one to two times with water, and evaporate to dryness. Analyses of 2–5% aliquots of such an incubation are illustrated in Fig. 4A [reversed-phase (RP)-HPLC] and Fig. 4B (SP-HPLC). Formation of the different ketols from the allene oxide is pH dependent.[14]

[13] B. A. Vick, *Lipids* **26**, 315 (1991).
[14] A. N. Grechkin, R. A. Kuramshin, E. Y. Safonova, S. K. Latypov, and A. V. Ilyasov, *Biochim. Biophys. Acta* **1086**, 317 (1991).

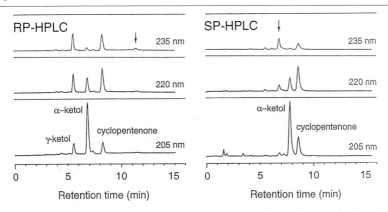

FIG. 4. HPLC analysis of allene oxide-derived products. (Left) Reversed-phase HPLC analysis. Column: Beckman 5-μm ODS Ultrasphere (25 × 0.46 cm) with Bio-Rad 5-μm ODS guard cartridge (5 × 0.4 cm) run with a solvent of methanol : water : glacial acetic acid (80 : 20 : 0.01, by volume) at a flow rate of 1 ml/min. (Right) Straight-phase HPLC analysis. Column: Alltech 5-μm silica column (25 × 0.46 cm) run with a solvent of hexane : isopropanol : glacial acetic acid (100 : 1.5 : 0.1, by volume) at a flow rate of 2 ml/min. The 205-, 220-, and 235-nm channels are set at the same sensitivity. The arrow in the 235-nm channels points to a small amount of unmetabolized 13S-hydroperoxylinolenic acid substrate. The γ-ketol is very strongly retained on SP-HPLC; it elutes at a retention time of 36 min (= 72 ml elution volume).

Quantitation of α-Ketol, γ-Ketol, and Cyclopentenone

The prominent α-ketol shows only end absorption in the UV and cannot be quantified reliably by this method; either weighing or, preferably, measurement of specific radioactivity starting from [14C]hydroperoxylinolenic acid is required. Nonetheless, a very useful approximation of the relative levels of the products on HPLC is given by the signals at 205 nm; this reflects fairly well the relative abundances of the hydroperoxide and all its derivatives.

A reasonable estimate of the molar extinction coefficients for the two conjugated ketones (cyclopentenone and γ-ketol) is 10,000–12,000, based on measurement[3] and the reported values for prostaglandin analogs (prostaglandin A$_2$ and 15-oxo prostaglandins).[15] In SP-HPLC solvent (98% hexane), the λ_{max} are observed at 217 nm (cyclopentenone) and 223 nm (γ-ketol). In more polar solvents, these chromophores are shifted to slightly longer wavelength, e.g., on changing to RP-HPLC solvent (MeOH : H$_2$O, 80 : 20, v/v), the bathochromic shifts are 6–7 nm (cyclopentenone) and 3–4 nm (γ-ketol).

[15] J. E. Pike, F. H. Lincoln, and W. P. Schneider, *J. Org. Chem.* **34,** 3552 (1969).

Isolation and Purification of AOS

The purification procedure we developed was optimized by monitoring enzymatic activity through several open column and fast protein liquid chromatography (FPLC) methods. The method may be applicable to purification of allene oxide synthase from other plant sources. In addition to the flaxseed AOS, we achieved purification of the AOS from maize and obtained N-terminal and CNBr peptide sequences using essentially the same chromatographic steps.

Preparation of the Initial Tissue Extract

An acetone powder preparation is convenient and the AOS activity is stable on storage at $-20°$ for years. Twenty grams of flaxseeds is homogenized using a Polytron in 200 ml of cold acetone (-10 to $-20°$). After allowing the solids to settle for about 10 sec, the tan-colored cloudy supernatant is decanted and the remaining solids are rehomogenized with fresh cold acetone. The pooled fine suspension is filtered on a Buchner funnel under vacuum and washed with 500 ml cold acetone. After removal of most of the acetone under a strong vacuum for 10–20 sec, the solids are scraped off the filter paper into a tube normally used for holding desiccants for gas drying (Aldrich, Cat. No. 37, 123-8). The flaxseed proteins are dried to a powder by passing through a stream of nitrogen for ≈ 15 min until there is no longer a smell of acetone in the effluent. About 2 g acetone powder is recovered and stored at $-20°$.

Ammonium Sulfate Precipitation

The acetone wash removes the microsomal lipids, and aqueous extraction of the acetone powder (50 mM phosphate buffer, pH 7, stirred with 50 mg/ml powder, for 30 min on ice) leaves the AOS activity in a 100,000-g supernatant. As shown originally by Zimmerman and Vick,[16] AOS activity with the most enriched specific activity is precipitated by 30% $(NH_4)_2SO_4$, although use of 42% saturation is required to recover most of the activity. We collected the 0–45% $(NH_4)_2SO_4$ pellet.[11] After this initial ammonium sulfate fractionation of this extract, the resuspended AOS exists in protein microaggregates and requires treatment with detergent prior to chromatography.

Solubilization of AOS Activity

We tested the solubilizing efficiency of various detergents by measuring the percentage recovery of AOS activity after passing a 1-ml aliquot through

---

[16] D. C. Zimmerman and B. A. Vick, *Plant Physiol.* **46**, 445 (1970).

a standard 0.2-μm HPLC sample filter (Acrodisc LC13 PVDF, Gelman Sciences).[11] Using this simple test in conjunction with a spectroscopic assay of enzyme activity, we quickly established the efficiency of various detergents to effect complete solubilization. This showed that Triton X-100 was unsuitable for solubilization of flaxseed AOS, whereas Tween 20 (1–2%), CHAPS (5–10 mM), and octylglucoside (20–40 mM) allowed quantitative recoveries. Use of 0.1 to 0.25% of the nonionic detergent Emulgen 911 not only allowed the free passage of the enzyme through the microfilters, it increased the specific activity by two- to threefold and this detergent was selected for routine applications.

In a typical preparation from 20 g flaxseed, the ammonium sulfate pellet is dissolved by stirring on ice for 30 min in 100 ml of 50 mM phosphate buffer, pH 7.0, containing 0.25% (w/v) Emulgen 911 (Kao-Atlas, Tokyo); this solution is adjusted to 0.5 M ammonium sulfate immediately prior to the next step.

Hydrophobic Interaction Chromatography

A load of ≈100 ml can be applied to an open bed column (10 × 2.5 cm, 50 ml bed volume) of octyl-Sepharose CL-4B (Pharmacia). Chromatography is carried out at room temperature; at all other times the enzyme is kept at 0°.

After loading the sample (flow rate ≈5 ml/min) and extensive washing with 50 mM phosphate, pH 7, containing 0.5 M (NH$_4$)$_2$SO$_4$ (and no detergent), the activity is eluted with the 50 mM phosphate buffer containing 0.25% Emulgen 911. Typically it may require ≈100 ml of solution to elute the majority of the activity. The most active fractions are visible as a brownish band on the column and generally are retained more strongly on gel used for the first time.

This open bed column greatly reduces the mass of sample, it effects a major enrichment in AOS activity, and for small-scale preparations (up to 20 g flaxseed, Table I) it leaves the sample sufficiently purified for FPLC. For larger preparations, an initial fractionation on Q-Sepharose Fast Flow (Pharmacia) followed by concentration and gel filtration will prepare the sample for FPLC.

Anion-Exchange FPLC (Mono-Q Column, or Equivalent)

The most active fractions are concentrated using Centriprep-30 concentrators (Amicon), and PD-10 (Pharmacia) columns were used for buffer exchange. The sample is then applied to a Mono-Q HR 5/5 column (Pharmacia) in 20 mM triethanolamine, pH 7.3, and 0.1% Emulgen 911, and the column is eluted at 1 ml/min with a linear gradient over 60 min of the same

TABLE I
PURIFICATION OF FLAXSEED ALLENE OXIDE SYNTHASE[a]

Purification step	Total protein (mg)	Specific activity (AU/mg/min)	Recovery (%)	Purification (fold)
Flax extract	3398	32	100	
Ammonium sulfate (0–45%)	1308	64	78	2
Octyl-Sepharose (CL-4B)	50	989	46	31
Anion exchange (Mono-Q)	4.2	9,990	39	315
Chromatofocusing (Mono-P)	0.9	15,480	13	491

[a] Values are from an experiment starting with 20 g acetone powder. Enzyme activity was measured using the spectrophotometric assay (Fig. 3). Protein was determined with the Bio-Rad Bradford method; compared to the results of amino acid analysis, this colorimetric assay significantly overestimated the protein levels in the final steps of purification.

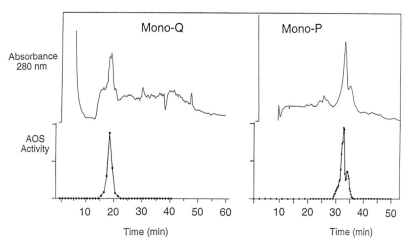

FIG. 5. Purification of flaxseed AOS on anion-exchange chromatography and chromatofocusing. (Left) Mono-Q column, HR 5/5, (Pharmacia): Conditions are described in the text. (Right) A Mono-P column, HR 5/20, (Pharmacia) was equilibrated with 25 mM bis–Tris, pH 6.7, 0.1% Emulgen 911, and then the sample was injected in 7 ml of the same buffer and the column was eluted with 40 ml of Polybuffer 74 (Pharmacia), pH 5.0, 0.1% Emulgen 911 at 1 ml/min. Enzyme activity eluted as two peaks at pI 5.5 and 5.4. (Upper traces) UV absorbance at 280 nm; the very large peak on injection in A (in B, not recorded) is mainly detergent concentrated from the previous chromatographic step. (Lower traces) AOS activity, measured by a decrease in 235 nm absorbance.

solution containing 0.5 M NaCl (buffer B). The AOS activity from flaxseed eluted at ≈20 min (Fig. 5A).

Chromatofocusing FPLC (Mono-P Column)

Finally, following buffer exchange using a gel filtration (PD-10 column, Pharmacia), the enzyme was purified on a Mono-P HR 5/20 chromatofocusing column (Pharmacia) (Fig. 5B and Table I).

The two isozymes differ in a single amino acid at the N terminus, and the major isozyme was cloned and sequenced.[1] The isozymes have indistinguishable UV-VIS spectra, typical of a P450 in the high-spin ferric state.[11] Each catalyzes allene oxide synthesis with a turnover number of ≥1000 sec^{-1}.[11]

Emulgen Detergent Removal

For microsequencing, removal of Emulgen 911 is desirable. It cannot be removed effectively by dialysis or with Centricon spin concentrators. Injection of the sample on a Mono-Q column, followed by extensive washing with detergent-free buffer, and then elution with a salt gradient including octylglucoside (20 mM) is effective. Octylglucoside can be dialyzed. The denatured enzyme can also be resolved from Emulgen 911 on RP-HPLC (Vydac C4 column and a water/acetonitrile gradient), although this is not effective on a preparative scale.

Acknowledgment

This work was supported by NIH Grant GM-49502.

[29] Cinnamic Acid Hydroxylase Activity in Plant Microsomes

By FRANCIS DURST, IRÈNE BENVENISTE, MICHEL SCHALK, and DANIÈLE WERCK-REICHHART

Introduction

Cytochrome P450 (P450) enzymes were characterized in plant tissues in the early 1970s. For almost two decades these enzymes attracted little attention and progress was slow. The two main reasons for this were (1) technical difficulties in dealing with poorly abundant membrane proteins

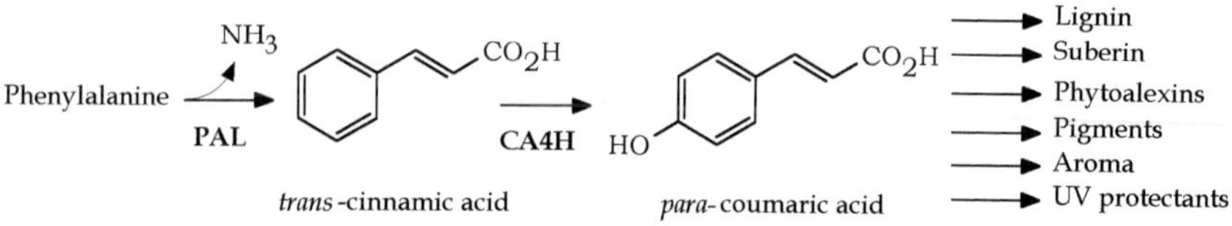

Fig. 1. The two first reactions in phenylpropanoid synthesis. PAL, phenylalanine ammonia lyase.

in tissues harboring numerous and active proteases and lipases, the presence of hydrophobic and extremely "sticky" pigments like chlorophyll, and a vacuole filled with inhibitory secondary metabolites that the plant cell produces but cannot excrete, and (2) the assumption that plant P450, being so similar to mammalian hepatic enzymes, was bound to catalyze the same reactions. In recent years it was realized that plant P450s catalyze unique reactions[1-3] on substrates that have no equivalent in animals. Since then, progress has been so rapid that at present the plant P450 families span from CYP71 to CYP96 and probably many more are to come.

The cinnamic acid 4-hydroxylase (CA4H) is the earliest and best studied P450 activity in plants. The discovery that CA4H activity was microsomal and required NADPH[4] was followed by the demonstration of the P450 nature of the enzyme.[5,6] The enzyme was first purified,[7] cloned,[8] and heterologously expressed[9] in our laboratory.

CA4H catalyzes the conversion of trans-cinnamic acid to trans-4-hydroxycinnamic acid (para-coumaric acid), committing the C6-C3 carbon skeleton from phenylalanine toward the phenylpropanoid pathway (Fig. 1). This is the second reaction, and the first oxidative step, in the general phenylpropanoid pathway, which is common to all plants. Phenylpropanoids, and their derivatives, constitute an extremely diversified family of molecules with important biological functions or activities: precursors for

---

[1] F. Durst, in "Frontiers in Biotransformation" (K. Ruckpaul and H. Rein, eds.), Vol. 4, p. 191. Akademie-Verlag, Berlin, 1991.

[2] F. Durst and I. Benveniste, Handb. Exp. Pharmacol. **105,** 293 (1993).

[3] P. Bolwell, K. Bozak, and A. Zimmerlin, Phytochemistry **37,** 1491 (1994).

[4] D. W. Russell, J. Biol. Chem. **246,** 3870 (1971).

[5] I. Benveniste and F. Durst, C. R. Hebd. Seances Acad. Sci., Ser. D **278,** 1487 (1974).

[6] M. Potts, R. Weklych, and E. E. Conn, J. Biol. Chem. **249,** 5019 (1974).

[7] B. Gabriac, D. Werck-Reichhart, H. G. Teutsch, and F. Durst, Arch. Biochem. Biophys. **288,** 302 (1991).

[8] H. G. Teutsch, M.-P. Hasenfratz, A. Lesot, C. Stoltz, J.-M. Garnier, J.-M. Jeltsch, F. Durst, and D. Werck-Reichhart, Proc. Natl. Acad. Sci. U.S.A. **90,** 4102 (1993).

[9] P. Urban, D. Werck-Reichhart, H. G. Teutsch, F. Durst, S. Regnier, M. Kazmaier, and D. Pompon, Eur. J. Biochem. **222,** 843 (1994).

lignin and suberin, pigments, aroma, defense molecules (phytoalexins), anti-oxidants, and UV protectants. Many of these compounds are specific to certain plant families or species, and a number of other P450 enzymes are involved in these reactions.[1-3]

The major difference between CA4H and most other plant P450s which appear to be involved in species specific reactions is that CA4H is probably present in all plants and in almost all tissues. Assaying CA4H is therefore a foolproof way to ascertain that microsomal preparations contain intact and active P450 electron transfer chains, and provides a sort of internal standard when extraction conditions are modified.

This chapter covers the following topics: (i) the nature of the enzyme source (plant species and type of tissue), (ii) induction of CA4H activity, (iii) preparation of microsomes, and (iv) two reliable and simple assays.

Nature of Plant and Tissue

Preparing microsomes from adult green tissues should be avoided when possible. At this stage, cells contain an enormous vacuole, and high mechanical forces are required to break the rigid walls. This leads to intensive mixing of membrane fractions with pigments and phenolics: activities are weak and difference spectra are marred by a high absorbance background and spurious redox shifts. It is better to use young (2–8 days) etiolated seedlings, root or tuber slices, and cell suspension cultures. Cell cultures are beyond the scope of this chapter. We will briefly describe a standard procedure to grow *Vicia* seedlings and prepare Jerusalem artichoke tuber slices.

Seedlings

Vicia sativa seeds are immersed for 10 min in a 0.3% sodium hypochlorite solution, washed with running tap water and distilled water, and soaked for 6 hr in distilled water at 27°. Seeds are then sown on six layers of water-soaked filter paper placed in covered 20 × 20-cm plastic boxes and incubated for 4 days at 26° in total darkness. Seedlings may be used as such or further induced. In any case, the seed teguments that are rich in tannins and other polyphenols are removed manually prior to microsome preparation.

Tuber Slices

Reproduction of *Helianthus tuberosus* is vegetative. We use a clone from variety 'blanc commun' isolated in 1948, which ensures great genetic homogeneity. Other varieties purchased from stores or markets, however, show similar activities. Tubers are harvested in the fall and stored in open polyethylene bags in darkness at 4°. In the case of storage tissues like tubers, CA4H is very low in intact tissue and is strongly enhanced by slicing, i.e., wounding, and by chemical treatments (see below). After harvest, a 3-week vernalization period is required before consistent and reproducible results are obtained.

Induction of CA4H

CA4H is a highly inducible enzyme,[1,3] and since the reaction product, *p*-coumarate, is a precursor to so many different compounds and involved in different processes, the enzyme is induced by a wide range of factors. This may be of great help in characterizing the activity in rare or difficult to sample tissues. On the other hand, it implies that strict standardization of growth and handling conditions is necessary to achieve reproducibility.

Wounding, Infection, and Light

Any kind of mechanical wounding or exposure to fungi, microbes, or viruses elicits the production of phytoalexins and tissue repair mechanisms. CA4H is involved in both. The elicitation of CA4H with fungal extracts will not be treated here, but it should be noted that an unnoticed infection may cause important activity variations.

Wound Induction of H. tuberosus Tuber. Tubers are peeled, washed with distilled water, and cut in 1-mm-thick slices. Slices are washed for 2 hr in running water, rinsed with distilled water, and aged in a 2-liter Erlenmeyer flask containing 1.5 liter distilled water for 50–120 g tissue. The slices are aerated by a 4-liter min^{-1} stream of hydrated and filtered air, and aged in the dark at 25–30°. It is important to consider that induction is a rapid and transient phenomenon, which reaches a peak after 18–24 hr and declines thereafter.[10] Long storage of the tubers at 4° before use (i.e., release of dormancy) delays and lowers maximum activity. Timing of tissue sampling for microsome preparation is thus essential for adequate activity and reproducibility.

Light Is a Known Inducer of CA4H Activity. This effect is mediated by the phytochrome photoreceptor,[11] which may be activated either by continuous irradiation with low intensity far-red (730 nm) light or pulses of red (660 nm) light. Under both conditions, CA4H is considerably enhanced, while chlorophyll synthesis remains minimal. As with other induction systems, it is essential to establish a time course of induction first, so sampling

[10] I. Benveniste, J.-P. Salaün, and F. Durst, *Phytochemistry* **16,** 69 (1977).
[11] I. Benveniste, J.-P. Salaün, and F. Durst, *Phytochemistry* **17,** 359 (1978).

of the seedlings is made at the plateau for activity measurements. Red light induces CA4H in the seedlings of many plant species, but does not increase the enzyme activity in tuber slices. Seedlings are first grown for 4–7 days in the dark before irradiation. For irradiation, a projector equipped with a 660-nm interference filter is automatically switched on for 2 min every 6 hr. Even a single 5-min red pulse followed by 18 hr darkness before preparation of microsomes may double CA4H activity.[11]

Chemicals

Early work from our laboratory had demonstrated that CA4H activity was increased when tuber slices or seedlings were exposed to certain chemicals like 2,4-dichlorophenoxyacetic acid,[12] several herbicides, phenobarbital, ethanol or to manganese ions,[13] and by aminopyrine.[14] The response of CA4H to these compounds varies considerably depending on plant species and tissue. For example, $MnCl_2$, which increases strongly the activity in tuber slices,[13] had deleterious effects on pea seedlings. We will describe the induction of CA4H in *H. tuberosus* slices.

The *H. tuberosus* tuber is sliced, washed, and aged in aerated water in the dark as described earlier for wound induction. Chemicals are dissolved directly into the aging water or, if needed, in a small volume of dimethyl sulfoxide added to the aging medium. The most effective inducers are 25 mM $MnCl_2$, 20 mM aminopyrine, and 4 mM phenobarbital. The pH of the $MnCl_2$ solution is adjusted to 7 with NaOH. As noted previously, time course studies are required to determine the peak of CA4H activity. For the previously mentioned inducers, maxima are observed at 48 to 72 hr after slicing.

Chemical induction of 3- to 5-day-old seedlings can be achieved in a similar way. After removal of the seed tegument, seedlings are aged in an aqueous solution as described for *H. tuberosus* slices.

Preparation of Microsomes

Standard Extraction Buffer for Tuber Slices

Just before extraction, add 100 mM sodium phosphate buffer, pH 7.4, containing 1 mM EDTA and 250 mM sucrose; 15 mM 2-mercaptoethanol, 40 mM sodium ascorbate. Check pH after adding sodium ascorbate.

[12] P. Adelé, D. Reichhart, J.-P. Salaün, I. Benveniste, and F. Durst, *Plant Sci. Lett.* **22**, 39 (1981).
[13] D. Reichhart, J.-P. Salaün, I. Benveniste, and F. Durst, *Plant Physiol.* **66**, 600 (1980).
[14] R. Fonné-Pfister, A. Simon, J. P. Salaün, and F. Durst, *Plant Sci.* **55**, 9 (1988).

Standard Extraction Buffer for V. sativa Seedlings

Just before extraction, add 100 mM sodium phosphate buffer, pH 7.4, containing 1 mM EDTA and 10% glycerol (v/v); 15 mM 2-mercaptoethanol, 40 mM sodium ascorbate, 10% (w/w of fresh tissue) insoluble PVP, and 1 mM phenylmethylsulfonyl fluoride (from a stock solution in methanol).

Additions

Modifications of the extraction buffer may improve enzyme recovery, depending on plant materials.

If the extraction medium turns brown after grinding with tissues and 2-mercaptoethanol no longer "smells," increase concentration or add dithiothreitol or glutathione.

Significant improvement can be achieved by adding:

1% (w/v) bovine serum albumin

0.6% (w/v) Amberlite XAD-4, 10 mM NaS$_2$O$_5$, and 1% (w/v) insoluble PVP as described for the preparation of enzymes involved in terpene synthesis[15]

0.5% Dowex 1 × 2 (w/fresh weight) or pulverulent active charcoal (1–3% w/fresh weight)

Dowex and charcoal are very effective in removing inhibitory polyphenols but they also adsorb protein! The effects of several of these and other additives have been recently compared.[16]

Grinding and Centrifugation

All operations are performed on ice. Tuber slices or *Vicia* seedlings are homogenized in 1.5 volume chilled extraction buffer with a Warring blender or with an Ultra-Turrax at 8000 rpm. Homogenates are filtered through a 50-μm nylon mesh and centrifuged for 15 min at 10,000 g. The supernatant is then centrifuged for 60 min at 100,000 g. Resuspending the microsome pellet in 100 mM sodium pyrophosphate buffer containing 10 mM 2-mercaptoethanol and running a second centrifugation may improve specific activity, but it decreases CA4H recovery. If an ultracentrifuge is not available or if large volumes are to be handled, add MgCl$_2$ (from a 1 M stock solution) to the 10,000-g supernatant to a final concentration of 30 mM, stir for 20 min, and centrifuge for 20 min at 40,000 g. The yield is lower, but specific activities are sometimes higher using this method.

[15] I. M. Rajaonariviny, J. Gershenzon, and R. Croteau, *Arch. Biochem. Biophys.* **296,** 49 (1992).
[16] A. E. Haack and N. E. Balke, *Pestic. Biochem. Physiol.* **50,** 92 (1994).

The 100,000-g pellet (or the MgCl$_2$ precipitation pellet) is resuspended in sodium phosphate, pH 7.4, containing 1.5 mM 2-mercaptoethanol and 30% (v/v) glycerol to a final concentration of 5–7 mg microsomal protein ml^{-1}. If stored at $-20°$, activity is stable for at least 2 months.

Two CA4H Assays

The two methods routinely used in our laboratory are described. HPLC determination has also been used successfully, but is more time-consuming.[9,17]

Using Radiolabeled Cinnamic Acid

The use of [3-^{14}C]*trans*-cinnamic acid (Isotopchim, B.P. 16, F-04310 Ganobie, France) provides an easy and sensitive assay for CA4H.

A 4-μCi ml^{-1} [3-^{14}C]cinnamate stock solution is prepared in 100 mM sodium phosphate, pH 7.4, and is stored as 2-ml aliquots at $-20°$. The exact cinnamate concentration of this stock solution is determined from its absorbance at 268 nm using the ε_M: 20,400. Depending on the CA4H activity in the microsomes to be studied, different amounts of cold substrate are added to this radiolabeled stock. A final cinnamate concentration ranging from 75 to 200 μM is often needed to ensure reaction linearity over 10–20 min with microsomes from induced plant tissues.

Time (2 to 30 min) and protein concentration (0.05 to 0.5 mg in 200 μl) are adjusted such that less than 50% of substrate is converted, ensuring saturation of the enzyme: the K_m for cinnamate is in the 1–8 μM range, depending on enzyme source. A control reaction, with HCl added before NADPH, is run to obtain the background and to make sure no contaminant running with an R_f comparable to that of coumarate is present in the substrate (it may happen occasionally).

A typical assay with induced *H. tuberosus* microsomes contains

175 μM *trans*-cinnamic acid
0.5 mM NADPH
0.1 mg microsomal protein
200 μl final sodium phosphate buffer 0.1 M, pH 7.4

When assays are incubated longer than 10 min, an NADPH regeneration system (10 mM glucose 6-phosphate and 0.5 units glucose-6-phosphate dehydrogenase) is added to the incubation mixture.

As the addition of NADPH to plant microsomes usually results in rapid P450 inactivation, the reaction is started by the addition of NADPH. After a 10-min incubation at 27°, the reaction is stopped with 20 μl 4 N HCl, and

[17] C. B. Stewart and M. A. Schuler, *Plant Physiol.* **90,** 534 (1989).

10 μl of a mixture of nonlabeled cinnamic and coumaric acids (100 μg each) is added. A 100-μl aliquot of reaction medium is spotted onto a precoated TLC silica gel plate 60 F_{254} (Merck) and is developed in the organic phase from a toluene/acetic acid/water (6/7/3) mixture. R_f is 0.54 for cinnamate and 0.3 for *para*-coumarate. Activity is directly determined with a thin layer analyzer (Berthold LB 2820-1). Alternatively, *para*-coumarate and cinnamate are visualized as purplish-blue fluorescent spots under 254-nm UV light, and silica is scraped and counted in a 5-ml Ready organic scintillation cocktail (Beckman).

The reaction is usually linear for at least 10 min, but the enzyme is inactivated at incubation times over 20 min at 27° or 10 min at 30°. The optimal pH for CA4H is around 7.4, with half-maximal activities at pH 6.2 or 8. A significant activity increase can be achieved by increasing the ionic strength of the incubation buffer (i.e., using 100 mM rather than 25 mM sodium phosphate or 100 mM Tris buffer). For the yeast-expressed enzyme, inhibition was observed when the CA4H concentration was increased above 6 pmol in the assay.[18]

Fluorescent Determination of 2-Naphthoate Hydroxylation

Radioactivity measurements are not always possible or desired. An alternative, stemming from our recent studies of the substrate specificity of yeast expressed CA4H,[18] is a very fast and sensitive fluorimetric method.[19]

We have demonstrated that 2-naphthoic acid perfectly mimics *trans*-cinnamate in the CA4H active site and is metabolized with an efficiency (K_m and V_{max}) comparable to that of cinnamate. The product of the reaction, 6-hydroxy-2-naphthoate, is strongly fluorescent. This property can be used for a rapid but reliable assay of the CA4H activity.

The optimal excitation wavelength of 6-hydroxy-2-naphthoate varies slightly with the composition of the buffer and with the concentration of NADPH and substrate in the assay. In 100 mM sodium phosphate, pH 7.4, and in the presence of 50 μM NADPH, its maximum excitation is 299 nm, with an emission peak at 443 nm. The substrate does not interfere with the readings, and a direct and continuous measurement of fluorescence in the incubation medium is possible. However, this assay cannot be used with the microsomes contaminated with pigments (e.g., chlorophyll).

A typical fluorescent CA4H assay contains

    100 μM 2-naphthoic acid (Sigma)

[18] M. Pierrel, Y. Batard, M. Kazmaier, C. Mignotte-Vieux, F. Durst, and D. Werck-Reichhart, *Eur. J. Biochem.* **224,** 835 (1994).
[19] M. Schalk, Y. Batard, A. Seyer, A. Zimmerlin, F. Durst, and D. Werck-Reichhart, to be published.

50 μM NADPH

1 mM glucose 6-phosphate

0.5 units glucose-6-phosphate dehydrogenase

0.1–0.5 mg microsomal protein

2 ml final sodium phosphate 100 mM, pH 7.4

The fluorescence of the reaction mixture without NADPH is recorded for approximately 3 min to allow temperature equilibration (30°) and to check for the activity in the absence of NADPH (this step can be shortened by using a preheated buffer). NADPH is then added to the reaction medium, and the increase in fluorescence (excitation: 299 nm; emission: 443 nm) is recorded for an additional 2 to 5 min. The reaction is usually linear for at least 10 min. Calibration of the assay is performed using known amounts of 6-hydroxy-2-naphthoate (Lancaster Synthesis, Morecambe, UK).

Some Typical CA4H Activities

Enzyme source	Treatment	Activity (pmol sec^{-1} mg^{-1})
H. tuberosus tuber	Dormant	0–1
	Slicing + 24 hr aging	40
	Slicing + 56 hr aging in 25 mM MnCl$_2$	200
Wheat seedlings	5 days old	2
Vicia seedlings	4 days old	1
	4 days old + 48 hr aging in amino-pyrine	6

When All Attempts Have Failed

As already discussed, with some plants it might be impossible or very difficult to obtain measurable and reproducible CA4H activity. It is our experience that CA4H protein or transcript levels measured by Western or Northern blotting do provide a reasonable estimate of CA4H activity even when working with green tissues from nonlaboratory plants, i.e., plants collected in the natural environment.[20] However, the use of antibodies requires the preliminary purification of a small amount of homogeneous enzyme, not an easy task, although several purification methods have been published.[7,21,22] It is further hampered, at least for use on Western blots, by their poor cross-reactivity when the boundaries of plant families are

[20] Y. Batard, Ph.D. Thesis, Université Louis Pasteur, Strasbourg, (1995).

[21] M. Mizutani, D. Ohta, and R. Sato, *Plant Cell. Physiol.* **34,** 481 (1993).

[22] D. Werck-Reichhart, I. Benveniste, H. Teutsch, F. Durst, and B. Gabriac, *Anal. Biochem.* **197,** 125 (1991).

crossed.[23] This makes cDNA probes particularly attractive. CA4H has now been cloned from a range of species: *H. tuberosus* (Z17369), *Phaseolus aureus* (L07634), *Medicago sativa* (L11046), *Catharanthus roseus* (X69788, Z32563), *Arabidopsis thaliana* (T04086), *Pisum sativum* (U29243), and *Zinnia elegans* (U19922). These may be obtained from the authors or recloned using PCR or RT-PCR methods. CA4H is well conserved (generally over 80% positional nucleotide identity between different species) and cross-hybridization on Northern blot seems more extended than with Western blotting.

[23] D. Werck-Reichhart, Y. Batard, G. Kochs, A. Lesot, and F. Durst, *Plant Physiol.* **102**, 1291 (1993).

[30] Isolation of Plant and Recombinant CYP79

By BARBARA ANN HALKIER, OLE SIBBESEN, and BIRGER LINDBERG MØLLER

Introduction

We have developed an isolation procedure for CYP79, also called P450tyr, which is a key enzyme in the biosynthesis of the cyanogenic glucoside dhurrin in *Sorghum bicolor* (L.) Moench.[1] Cyanogenic glucosides are secondary plant products which, on hydrolysis, release HCN. CYP79 is a multifunctional N-hydroxylase converting tyrosine into *p*-hydroxyphenyl-acetaldoxime by two consecutive N-hydroxylations, of which the first produces *N*-hydroxytyrosine and the second *N,N*-dihydroxytyrosine.[2] The latter dehydrates to 2-nitroso-3-(*p*-hydroxyphenyl)propionic acid which decarboxylates into the oxime. The multifunctional property of CYP79 has been confirmed by reconstitution experiments using recombinant CYP79 expressed at high levels in *Escherichia coli*.[3] This chapter focuses on the generally applicable procedures developed for the isolation of plant and recombinant CYP79. In addition, a strategy is suggested for successful expression of microsomal P450s in *E. coli*.

[1] O. Sibbesen, B. Koch, B. A. Halkier, and B. L. Møller, *Proc. Natl. Acad. Sci. U.S.A.* **91**, 9740 (1994).
[2] O. Sibbesen, B. Koch, B. A. Halkier, and B. L. Møller, *J. Biol. Chem.* **270**, 3506 (1995).
[3] B. A. Halkier, H. L. Nielsen, B. Koch, and B. L. Møller, *Arch. Biochem. Biophys.* **322**, 369 (1995).

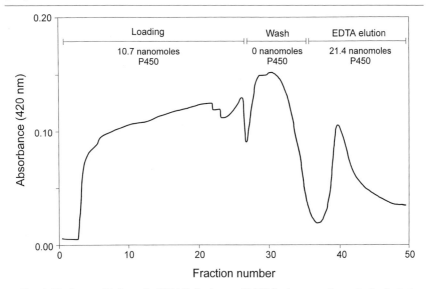

FIG. 1. Elution profile from the DEAE-Sepharose/S-100 Sepharose column during isolation of plant CYP79 as monitored at 420 nm. The strong absorption during column loading and washing was caused by yellow pigments present in the etiolated plant material. Total P450 contents were determined from CO spectra.

Isolation of Plant CYP79

Principles

The procedure for isolation of plant CYP79 from etiolated sorghum seedlings utilizes DEAE and dye affinity columns (Sigma). The general applicability of dye columns for purification and fractionation of P450s is presumably due to the structural similarity between the dye ligands and many P450 inhibitors. Using sorghum microsomal preparations, CYP79, cinnamic acid hydroxylase (C4H) (the ubiquitous plant P450), and P450ox (the second P450 in dhurrin biosynthesis) are sequentially eluted from the Cibacron Blue agarose column.[3] On the Yellow agarose column, CYP79 does not bind, whereas P450ox and C4H bind and are sequentially eluted.[4] It is important to emphasize that the selective binding of different P450 enzymes to the dye columns is achievable only after an initial fractionation step on a DEAE column. One effect of the DEAE step is removal of the yellow pigments present in etiolated plant material during the washing step (Fig. 1). This is of major advantage due to the gluey nature of these pig-

[4] R. Kahn, O. Sibbesen, and B. L. Møller, unpublished results.

ments. The P450s are only weakly bound to the DEAE column as evidenced by their gentle elution in one step by fortifying the buffer with 0.2% CHAPS and a 25-fold increase in the concentration of EDTA. A major problem encountered in the purification of low abundant plant P450s is an irreversible aggregation of proteins, which prevents their subsequent separation by column chromatographic techniques. This problem was overcome by constantly working with low protein concentrations, e.g., by replacing ultrafiltration steps with dilution steps and by diluting the DEAE-Sepharose with Sepharose S-100 gel filtration material in order to avoid high local concentrations of proteins on initial binding to the column.

Procedure

All processes are performed at 4°. Equilibration of column materials is carried out by diluting the required volume of material settled in water with an equivalent volume of 2× concentrated buffer. During fractionation, the total P450 content is continuously monitored by the absorption at 420 nm. Elution of CYP79 is monitored by substrate-binding spectra and reconstitution experiments.[2] To prolong the stability of dithiothreitol (DTT), all buffers are degassed *in vacuo* and flushed with argon three times prior to the addition of detergents and DTT. Five hundred grams of etiolated sorghum seedlings provides approximately 20 ml of microsomes (20 mg protein/ml). The microsomes are diluted to 100 ml with 8.6% glycerol, 2 mM DTT, 10 mM KPO$_4$, pH 7.9, and solubilized by the slow addition of 100 ml of the same buffer fortified with 2% (v/v) Renex 690 (J. Lorentzen A/S, Kvistgaard, Denmark) and 0.2% (v/v) reduced Triton X-100 (RTX-100). After stirring for 30 min, the mixture is centrifuged at 200,000 g for 30 min. The supernatant (190 ml) is applied to a DEAE-Sepharose fast flow/S-100 Sephacryl (1:4 wet volumes) column (5 × 5 cm) equilibrated in 8.6% glycerol, 0.2 mM EDTA, 2 mM DTT, 1% Renex 690, 0.1% RTX-100, 10 mM KPO$_4$, at pH 7.9 (buffer A). After the column is washed with 150 ml of buffer A, the total P450 proteins are eluted with 250 ml of 8.6% glycerol, 5 mM EDTA, 2 mM DTT, 1% Renex 690, 0.1% RTX-100, 0.2% 3-[(3-cholamidopropyl)dimethylammonio]-1-propanesulfonate (CHAPS), 20 mM KPO$_4$, pH 7.9 (buffer B). The P450-containing fractions are combined (ca. 130 ml) and adjusted to 1% CHAPS, stirred for 30 min, and then applied to a Cibacron blue–agarose column (2.8 × 8 cm) equilibrated in 8.6% glycerol, 5 mM EDTA, 2 mM DTT, 1% CHAPS, 0.05% RTX-100, 20 mM KPO$_4$, pH 7.9 (buffer C), fortified with 1% Renex 690. The column is depleted of Renex 690 by washing with buffer C and monitoring at 280 nm until a stable baseline is reached. Subsequently, the column is eluted with a linear gradient of 0–500 mM KCl (200 ml) in buffer C. CYP79-

containing fractions are combined and diluted fivefold with buffer C and applied to a Reactive Red 120–agarose column (0.9 × 5 cm) equilibrated in buffer C. The column is washed with buffer C, and CYP79 is eluted with a linear gradient of 0–1 M KCl (60 ml) in buffer C. The yield of CYP79 is typically 50–100 μg.

Isolation of Recombinant CYP79

Principles

We have optimized an earlier described procedure for extraction of recombinant P450 from $E.$ $coli$ spheroblasts.[5] The procedure is based on the observation that the presence of glycerol significantly lowers the cloud point (temperature of aggregation) for Triton X-114 (TX114), allowing phase partitioning of solubilized membrane proteins into detergent-rich and -poor phases at 25°.[6] The concentration of glycerol strongly affects the efficiency of the extraction method as well as the distribution of phases. In the presence of 30% glycerol at an initial detergent:protein ratio of 1, 60–70% of recombinant CYP79 was obtained in the detergent-rich upper phase[3] (Fig. 2). In the presence of 20% glycerol, the phase partitioning was very poor. At 10% glycerol only about 30% of recombinant CYP79 ended up in the detergent-rich phase, which at this concentration of glycerol formed the bottom phase. The sum of CYP79 in detergent-rich and -poor phases always accounted for the total amount of CYP79 present in the spheroblasts, demonstrating that the method is very gentle and efficient.

Procedure

Large-scale cultures of $E.$ $coli$ transformed with recombinant CYP79 from construct TYRΔ(1–25)$_{bov}$ (see later) are obtained by inoculating four 2-liter flasks containing 300 ml of Terrific broth containing 50 μg/ml ampicillin, 1 mM thiamine, 75 mg/liter δ-aminolevulinic acid, and 1 mM isopropyl β-D-thiogalactopyranoside with 0.6 ml of an overnight culture. After 60 hr of incubation at 28° at 125 rpm, the $E.$ $coli$ cells are pelleted (3000 g, 10 min). The pellet is resuspended in 100 ml 200 mM Tris–HCl, pH 7.6, 1 mM EDTA, 0.5 mM phenylmethylsulfonyl fluoride, and 0.5 M sucrose and is homogenized in a Potter–Elvehjem. Lysozyme (0.2 mg/ml) is added and diluted 1:1 with cold water. After a 30-min incubation at 4°, 10 mM

---

[5] H. J. Barnes, Ph.D. Thesis, The University of Texas Southwestern Medical Center, Dallas (1992).

[6] D. Werck-Reichart, I. Benveniste, H. Teutsch, F. Durst, and B. Gabriac, $Anal.$ $Biochem.$ **197,** 125 (1991).

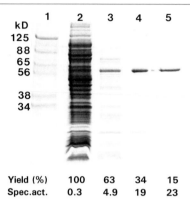

FIG. 2. Polypeptide profiles of the fractions obtained during isolation of recombinant CYP79 as monitored by SDS–PAGE and staining with Coomassie brilliant blue. (1) Molecular mass standards, (2) *E. coli* spheroblasts, (3) TX114-rich phase, (4) DEAE pool, and (5) concentrated Red pool. Identical amounts of CYP79 (10 pmol) were applied in each lane. Yields and specific activity (nanomoles P450/mg protein) of CYP79 obtained in a typical isolation experiment are indicated. The values of the specific activity are overestimated due to the presence of detergent in the samples.

$Mg(OAc)_2$ is added, and the spheroblasts are pelleted (3000 g, 10 min) and resuspended in 80 ml 10 mM Tris–HCl, pH 7.6, 14 mM $Mg(OAc)_2$, 60 mM KOAc, pH 7.4. DNase is added to a final concentration of 4 μg/ml, and the spheroblasts are stirred for 30 min at 4°, homogenized in a Potter–Elvehjem, made 30% with respect to glycerol, and stored at −80°.

CYP79 is extracted from the *E. coli* spheroblasts (137 ml, 6 mg protein/ ml) by slowly adding 6.25 ml 10% TX114 per 100 ml. After incubation for 30 min at 4° with stirring, the solubilisate is centrifuged (25 min, 23000 g), and the small reddish, upper phase is collected. The lower phase is incubated for 15 min at 25° and is centrifuged for 15 min at 30°. The small, reddish upper phase is collected, and the lower phase is reextracted by adding 6.25 ml 10% TX114 per 100 ml. The reddish phases are combined and diluted 100-fold in 10 mM KPO_4, pH 8.6, 0.05% reduced RTX-100, and 1 mM DTT and applied to a 70-ml DEAE-Sepharose (fast flow) column equilibrated in 10 mM KPO_4, pH 7.9, 0.2% TX114, 0.05% RTX-100, 10% glycerol, and 1 mM DTT (buffer D). The column is washed with buffer D. CYP79 is eluted with a 2 × 300-ml linear gradient composed of buffer D and buffer E containing 50 mM KPO_4, pH 7.9, 5 mM EDTA, 0.2% TX114, 0.05% RTX-100, 0.1% CHAPS, 10% glycerol, and 1 mM DTT. The CYP79-containing fractions are combined and directly applied to a 30-ml Reactive Red 120– agarose affinity column equilibrated with 50 mM KPO_4, pH 7.9, 5 mM EDTA, 1% CHAPS, 10% glycerol, and 1 mM DTT. The column is washed extensively to remove TX114 and is eluted with a 2×130-ml gradient from 0 to 1.5 M KCl in buffer D. The fractions containing homogenous CYP79 as judged by SDS–PAGE are combined and concentrated on an Amicon ultrafiltration cell fitted with a YM 30 filter to 2–5 nmol CYP79/ml in 50 mM KCl, 50 mM KPO$_4$, pH 7.9, and 1 mM DTT containing approximately 0.5% CHAPS. During purification, the elution of CYP79 is monitored by SDS–PAGE and CO difference spectroscopy. Approximately 2 mg of purified protein is obtained from 1.2 liters of cell culture.

Expression of CYP79 in E. coli

Principles

Native and N-terminally modified CYP79 have been expressed in *E. coli*. The N-terminal modifications of CYP79 were designed to accommodate the previously obtained results for high-level expression of P450s in *E. coli*, i.e., (1) enrich the first eight codons for A's and T's, (2) replace the second codon with GCT (alanine), and (3) "bovine-modify," i.e., replace the first eight codons of the P450 sequence with the first eight codons from bovine P45017α.[7,8] The N terminus of CYP79 is characterized by having a long hydrophobic core of 31 amino acid residues between the negatively charged residues of the N terminus and the positively charged stop-transfer signal and by having a long hydrophobic stretch of 18 amino acid residues between the positively charged region and the proline-rich region. A minimum length of 18 Å has been predicted for the hydrophobic core of signal sequences in bacteria[9] and corresponds to 12 amino acids in an α-helix, indicating a minimum length of 12 amino acids for the membrane-spanning region of the membrane anchor of P450s expressed in *E. coli*.

Constructs

Based on the principles described earlier, we designed four constructs for expression of CYP79 in *E. coli* using the expression vector pSP19g10L obtained from Dr. Henry Barnes (Immune Complex Incorporation, La Jolla, CA).[3] Two constructs contained full-length CYP79: TYR1, which had only silent mutations, and TYR$_{bov}$, which was bovine modified (see

[7] H. J. Barnes, M. P. Arlotto, and M. R. Waterman, *Proc. Natl. Acad. Sci. U.S.A.* **88,** 5597 (1991).
[8] C. W. Fisher, D. L. Caudle, C. Martin-Wixström, L. C. Quattrochi, R. H. Tukey, M. R. Waterman, and R. W. Estabrook, *FASEB J.* **6,** 759 (1992).
[9] H. Bedoulle and M. Hofnung, in "Intermolecular Forces" (B. Pullman, ed.), p. 361. Reidel, Dordrecht, The Netherlands, 1981.

earlier discussion). In the two other constructs, TYRΔ(1–14)$_{bov}$ and TYRΔ(1–25)$_{bov}$, the 31 amino acid hydrophobic core was reduced to 24 and 13 amino acids, respectively, and the first 8 amino acids were bovine modified. In TYRΔ(1–14)$_{bov}$, the lysine located in the middle of the hydrophobic core was mutated into a leucine. Native and various N-terminally modified, truncated forms of CYP79 were introduced into pSP19g10L using polymerase chain reaction mutagenesis.[3]

Comments

Spectral analysis of the cell cultures expressing different constructs demonstrated that only TYRΔ(1–25)$_{bov}$ and, to a lesser extent, TYRΔ(1–14)$_{bov}$ produced a CO difference spectrum. However, the level of expression of immunoreactive CYP79 from the different constructs was similar, and reconstitution assays showed that all constructs produced at least some functional CYP79.[3] The expression of CYP79 from TYRΔ(1–25)$_{bov}$ was typically between 200 and 500 nmol/liter culture. A characteristic peak at 420 nm was found in all cultures, including the vector-transformed culture. This peak may represent *E. coli*-derived heme proteins, e.g., the respiratory cytochromes *o* and *d*[10] or unattached heme groups produced due to the presence of a surplus of δ-aminolevulinic acid in the medium. All bovine-modified constructs have the same calculated potential free energy for RNA secondary structure formation from base region −26 to 2 with respect to the ATG start codon. This indicates that bovine modification per se does not guarantee high-level expression of a microsomal P450 in *E. coli*. However, our data would suggest that successful expression of microsomal P450s in *E. coli* can be obtained by bovine modification of the first 8 amino acids when combined with a proper length of minimum 12 amino acids for the hydrophobic core N-terminally to the positively charged stop-transfer signal.

[10] T. D. Porter and J. R. Larson, *Methods Enzymol.* **204**, 108 (1991).

[31] Cloning of Novel Cytochrome P450 Gene Sequences via Polymerase Chain Reaction Amplification

By Timothy A. Holton and Diane R. Lester

Introduction

Although cytochrome P450 enzymes are known to catalyze many important reactions within plants, the first plant P450 gene was isolated relatively recently.[1,2] This gene was (and still is) of unknown function. The first plant P450 gene of known function (cinnamate 4-hydroxylase) was isolated in 1993.[3,4] Since that time, the number of cytochrome P450 genes cloned from plants has increased rapidly, but is still much smaller than the number isolated from animals. Although many P450-catalyzed reactions have been identified in plants, there is no indication of the number of genes that exist.

Most of the plant P450 genes of known function have been isolated via protein purification strategies except for flavonoid 3',5'-hydroxylase genes which have been isolated via a polymerase chain reaction (PCR) amplification strategy.[5] This chapter describes PCR strategies that have successfully been used to isolate novel cytochrome P450 gene sequences from petunia. The same methods and primers have also been used successfully to isolate P450 genes from other plant species.

Amplification of P450 Sequences Using a Single P450-Specific Primer

Rationale

The work described in this chapter was initiated before any published plant P450 sequences were available. However, there were hundreds of sequences available from other (mostly mammalian) sources,[6] and four

---

[1] D. P. O'Keefe and K. J. Leto, *Plant Physiol.* **89,** 1141 (1989).

[2] K. R. Bozak, H. Yu, R. Sirevag, and R. E. Christofferesen, *Proc. Natl. Acad. Sci. U.S.A.* **87,** 3904 (1990).

[3] M. Mizutani, E. Ward, J. DiMaio, D. Ohta, J. Ryals, and R. Sato, *Biochem. Biophys. Res. Commun.* **190,** 875 (1993).

[4] H. G. Teutsch, M. P. Hasenfratz, A. Lesot, C. Stoltz, J. M. Garnier, J. M. Jeltsch, F. Durst, and D. Werck-Reichhart, *Proc. Natl. Acad. Sci. U.S.A.* **90,** 4102 (1993).

[5] T. A. Holton, F. Brugliera, D. R. Lester, Y. Tanaka, C. D. Hyland, J. G. T. Menting, C.-Y. Lu, E. Farcy, T. W. Stevenson, and E. C. Cornish, *Nature (London)* **366,** 276 (1993).

[6] D. W. Nebert, D. R. Nelson, M. J. Coon, R. W. Estabrook, R. Y. Feyereisen, Y. Fujii-Kuriyama, F. J. Gonzalez, F. P. Guengerich, I. C. Gunsalus, E. F. Johnson, J. C. Loper, R. Sato, M. R. Waterman, and D. J. Waxman, *DNA Cell Biol.* **10,** 1 (1991).

P450 conserved domains

PCR amplification using one P450 primer

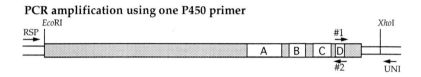

PCR amplification using two P450 primers

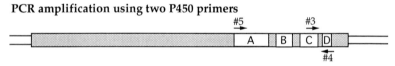

FIG. 1. Strategies for amplification of cytochrome P450 gene sequences. A, B, C, and D represent the approximate positions of four conserved cytochrome P450 domains. D corresponds to the heme-binding domain. PCR amplification of P450 sequences from a directional cDNA library was achieved using the primers #1/UNI or #2/RSP. PCR amplification of P450 sequences from cDNA or genomic DNA was obtained using the primers #3/#4 or #5/#4.

regions of sequence conservation between different P450 families have been identified.[7] The approximate positions of these regions (A–D) are shown in Fig. 1. The most highly conserved region (D) is around the cysteine residue involved in heme binding. The amino acid sequence F(G,S)XGXRXCXG is present in the heme-binding domain of nearly all microsomal cytochromes P450 sequenced to date. This consensus sequence was compared with the NBRF protein database, using the FASTA[8] program to determine the frequency of occurrence of amino acids around this area for all of the microsomal cytochrome P450 sequences in the database. This comparison showed that the most common amino acid sequence for each position around the heme-binding domain was F<u>MPFGAGXRXCLG</u>. An oligonucleotide was designed to hybridize to a gene encoding the underlined sequence and similar sequences. This oligonucleotide, designated oligo #1, is shown in Table I. The underlined portion is an additional sequence which includes a *Hin*dIII recognition site to facilitate the directional cloning of PCR products. The inclusion of deoxyinosine (I) covered the different possibilities for codon usage where more than two codons could encode

[7] V. F. Kalb and J. C. Loper, *Proc. Natl. Acad. Sci. U.S.A.* **85,** 7221 (1988).
[8] W. R. Pearson and D. J. Lipman, *Proc. Natl. Acad. Sci. U.S.A.* **85,** 2444 (1988).

TABLE I
PCR PRIMERS USED TO AMPLIFY AND CLONE CYTOCHROME P450 GENE FRAGMENTS[a]

Primer	Sequence
UNI	5′-GTAAAACGACGGCCAGT-3′
RSP	5′-AACAGCTATGACCATG-3′
Oligo #1	5′-GGAAGCTTATICCITT(T,C)GGIGCIGG-3′
Oligo #2	5′-CCIGG(A,G)CAIATIC(G,T)(C,T)(C,T)TICCIGCICC(A,G)AAIGG-3′
Oligo #3	5′-AGGAATT(C,T)(A,C)GICCIGA(A,G)(A,C)GITT-3′
Oligo #4	5′-CCIGG(A,G)CAIATIA(G,T)(C,T)(C,T)TICCIGCICC(A,G)AAIGG-3′
Oligo #5	5′-TT(C,T)(G,T)(G,T)IG(C,G)IGGI(A,T)(C,T)IGAIAC-3′

[a] Restriction enzyme recognition sites useful for cloning PCR fragments are underlined.

the same amino acid. Deoxyinosine base pairs with similar efficiency to A, T, G, and C.[9,10]

Construction of a Directional cDNA Library

RNA was extracted from flower buds of petunia,[11] and a directional cDNA library was made in the λZAP vector (Stratagene) using conditions recommended by the manufacturer. This method results in the production of cDNAs flanked by *Eco*RI and *Xho*I restriction sites at the 5′ and 3′ ends, respectively. Plasmid clones can be readily excised from the λZAP clones with the aid of helper phage. To provide a template for PCR amplification, plasmid clones are excised from the total cDNA library.

3′-End Amplification of P450 cDNAs

Plasmid DNA obtained from the directional cDNA library is used as a template for the amplification of a 360-bp cytochrome P450-related sequence using oligos #1 and UNI (universal sequencing primer). The PCR fragment is cloned into pBluescript, and the resulting plasmid is designated pCGP450. The 5′ region of pCGP450 encodes a polypeptide sequence with significant homology to previously sequenced cytochrome P450 molecules. DNA from pCGP450 is radiolabeled and used as a probe to isolate the corresponding cDNA clone (Pet450-1) from the cDNA library using high stringency hybridization conditions. The cDNA library is probed with PetP450-1 under low stringency hybridization conditions to isolate the cytochrome P450 clones PetP450-2, PetP450-3, and PetP450-4.

[9] F. M. Martin, N. M. Castro, F. Aboula-ela, and I. Tinoco, *Nucleic Acids Res.* **13,** 8927 (1985).
[10] E. Ohtsuka, S. Matsuki, M. Ikehara, Y. Takahashi, and K. Matsubara, *J. Biol. Chem.* **260,** 2605 (1985).
[11] T. H. Turpen and O. M. Griffith, *BioTechniques* **4,** 11 (1986).

PCRs contain 5 to 100 ng of excised DNA, 10 mM Tris–HCl (pH 8.3), 50 mM KCl, 1.5 mM MgCl$_2$, 0.01% (w/v) gelatin, 0.2 mM each dNTP, 0.4 μM of each primer, and 1.25 units of *Taq* polymerase (Cetus). Reaction mixes (50 μl) are cycled 30 times among 94, 48, and 72° for 1 min at each temperature. The amplified products are gel purified using Geneclean (Bio 101 Inc.) and reamplified to obtain sufficient material for cloning. DNA amplified using oligos #1 and UNI is digested with *Hin*dIII and *Xho*I prior to cloning into pBluescript.

5'-End Amplification of P450 cDNAs

Sequence information from around the putative heme-binding domain of the petunia clones PetP450-1, PetP450-2, and an avocado cytochrome P450 sequence[1,2] is used to design a second "P450-specific" oligonucleotide (oligo #2). This oligonucleotide is used to amplify related sequences by PCR using the directional cDNA library as the template and RSP (reverse sequencing primer) as the second primer. Reaction products are cloned into the ddT-tailed pBluescript vector.[12] The cloned PCR fragment is sequenced and encodes a fifth cytochrome P450 homolog (PetP450-5).

Screening of a cDNA Library with P450 Clones

A mixed probe of ^{32}P-labeled DNA fragments that included the coding regions of the five cytochrome P450 homologs was used to screen 50,000 clones from the petal cDNA library for related sequences. A total of 152 hybridizing clones were detected under low stringency hybridization and washing conditions. Fifteen different cytochrome P450 genes were identified by sequence analysis of these cDNA clones (Table II).

Duplicate plaque lifts are hybridized and washed as follows: high stringency conditions [hybridization in 50% (v/v) formamide, 6× SSC, 1% (w/v) sodium dodecyl sulfate (SDS) at 42° for 16 hr and washing in 2× SSC, 1% (w/v) SDS at 65° for 2× 15 min followed by 0.2× SSC, 1% (w/v) SDS at 65° for 2× 15 min] are used to detect sibling clones whereas low stringency conditions [hybridization in 20% (v/v) formamide, 6× SSC, 1% (w/v) SDS at 42° for 16 hr and washing in 6× SSC, 1% (w/v) SDS at 65° for 1 hr] are used to detect related sequences.

Two P450 Primer Amplification of cDNAs and Genomic DNAs

We had shown that petunia petals expressed at least 15 different P450 genes. However, we were interested in determining how many different

[12] T. A. Holton and M. W. Graham, *Nucleic Acids Res.* **19,** 1156 (1991).

TABLE II

AMINO ACID SEQUENCES AROUND THE HEME-BINDING DOMAIN OF PETUNIA CYTOCHROME
P450 CLONES ISOLATED FROM A PETUNIA PETAL cDNA LIBRARY

P450 clone	Translated sequence
PetP450-1	FENSPVEFIGNHFELVPFGAGKRICPG
PetP450-2	NVWAIGRDPKYWDDAESFKPERFEH---NSLNFAGNNFEYLPFGSGRRICPG
PetP450-3	NEWAIAYDPKIWGDPLSFKPERFID---SKIDHKGQNFEYFPFGSGRRICAG
PetP450-4	NAWAIGRDPKYWEKPLEFMPERFLK---CSLDYKGREFEYIPFGAGRRICPG
PetP450-5	NAWTMGRDPLTWENPEEYQPERFLN----RDTVKGVNFEFIPFGAGR
PetP450-6	PDRFWG---SKMEVRGQDFELIPFGXGRRICPG
PetP450-7	RTWENPEEYQPERFLN---SDIDVKGLNFELIPFGAGRRVCPG
PetP450-8	NAWWLANNPAHWKNPEEFRPERFFE-EEKHVEANGNDFRYLPFGVGRR
PetP450-9	NAWAIGRDPKLWTEAEMFNPERFLD---STVDYMGNDFHFIPFGAGRRI
PetP450-10	NVWAIGRDPLIWEEPQKFRPQRFLS---SNMDYKGNDFEFLPFGAGRRICPG
PetP450-11	NVWAIGRDPLIWEEPQKFRPQRFLK---LQYGLQGMTFEFLPFGAGRRICPG
PetP450-12	NIWAIGRDPQVWENPLEFNPERFLSGRNSKIDPRGNDFELIPFGAGRRICAG
PetP450-13	NIWAIGRDPEVWENPLEFYPERFLSGRNSKIDPRGNDFELIPFGAGRRICAG
PetP450-14	NVWAIQRDPSIWEKSYMFHPERFLE---NRLDYSGNDFNYFPFGTGRRICAG
PetP450-15	NVWAIGRNSDLWENPLVFKPERFWE---SEIDIRGRDFELIPFGAGRRICPG
Consensus[a]	N Waigrdp We p f PeRFl s d GndFe iPFGaGrRiCpG

[a] Oligos #3 and #4 were designed to encode the underlined consensus sequences.

cytochrome P450 gene sequences could be isolated using a PCR amplification approach. Making use of the sequence conservation of the petunia P450 genes, new oligomers were designed to regions C (oligo #3) and D (oligo #4) (Fig. 1 and Table I). These two primers were used to amplify petunia P450 sequences via PCR using (1) cDNA made from petal RNA, (2) cDNA made from leaf RNA, and (3) genomic DNA as templates. The expected PCR product size was approximately 96 bp.

Preparation of Template DNA

cDNA Synthesis. Fifty micrograms of total RNA is ethanol precipitated, pelleted, and resuspended in 11.5 μl of DEPC-treated distilled water. One-half microliter of oligo(dT) (1 μg/μl) is added, and the mixture is incubated at 70° for 10 min, then chilled on ice. Each tube receives:

4 μl 5× reaction buffer (Superscript reverse transcriptase)
2 μl 0.1 M dithiothreitol
1 μl 10 mM dNTPs

Tubes are then warmed to 37°, and 1 μl of Superscript reverse transcriptase (Bethesda Research Laboratories) is added. The reaction is incubated at 37° for 1 hr and then diluted to 300 μl with TE buffer.

Genomic DNA. DNA is isolated from leaves of petunia seedlings using the method of Dellaporta *et al.*[13]

PCR Amplification of P450 Sequences

Each amplification reaction contains:

5 μl 10× reaction buffer [100 mM Tris–HCl (pH 8.3), 500 mM KCl, 15 mM MgCl$_2$, 0.1% (w/v) gelatin]
8 μl dNTPs (1.25 mM each of dATP, dCTP, dGTP, and dTTP)
3 μl 25 mM MgCl$_2$
5 μl cDNA template (or 1 μg genomic DNA)
2.5 μl FRPERF (~200 ng/μl)
2.5 μl PETHAEM (~300 ng/μl)
24 μl water
1.25 units *Taq* polymerase (Cetus)

The PCR conditions are 40 cycles: 95° for 50 sec, 45° for 50 sec, and 72° for 45 sec. PCR products are electrophoresed in a 1.5% agarose gel, and DNA fragments in the expected size range are isolated and reamplified to obtain enough product for cloning.

PCR products are ligated with a *Eco*RV-cut, ddT-tailed vector.[12] The ligation mixture is used to transform competent *E. coli* DH5α cells[14] and is plated on LB plates containing 100 μg/ml ampicillin. Clones are grown up, sequenced, and compared with previously characterized P450 sequences using FASTA or TFASTA.[8] The cloning and sequencing of PCR products result in a large number of new P450-type sequences (Table III), in addition to many of those listed in Table II (Nos. 2, 4, 8, 10, 13, 15).

Without isolating more extensive regions of the corresponding genes, it is not possible to confirm that each sequence belongs to a P450 gene. However, the sequence conservation between the PetP450-16 to PetP450-70 clones (and to previously characterized P450 genes) and the fact that the reading frame continues from oligo #3 through to oligo #4 provide strong evidence that these sequences belong to P450 genes. Clones 71–73 do not show much sequence similarity to the other petunia P450 clones or to P450 sequences from other organisms. These clones may be non-P450 or may belong to previously uncharacterized P450 families. The number of P450 genes isolated from petunia is certainly not exhaustive: many sequences were only obtained once and many of the clones obtained by low stringency screening of the cDNA library were not obtained by PCR using oligo #3 and oligo #4.

---

[13] S. J. Dellaporta, J. Wood, and J. B. Hick, *Plant Mol. Biol. Rep.* **1,** 19 (1983).
[14] H. Inoue, H. Nojima, and H. Okayama, *Gene* **96,** 23 (1990).

TABLE III
CYTOCHROME P450 CLONES OBTAINED FROM
PCR AMPLIFICATION OF PETUNIA cDNA AND
GENOMIC DNA

P450 clone	Translated sequence[a]
PetP450-16	FLTLVILVKGLNFELIP
PetP450-17	FIGNNVDLRGHDFHLLP
PetP450-18	FLEEDVDTKGHDYRLLP
PetP450-19	FIKEDVDMKGHDYRLLP
PetP450-20	FWESEIDVRGQNFELIP
PetP450-21	FIGSSNILRGRDFQLLP
PetP450-22	FLNSDIDVKGLNFGLIP
PetP450-23	FWGSDIDVCGQDFELIP
PetP450-24	FLEEDVDMKGHDYRLLP
PetP450-25	FWGSELDVRGWDFVFIP
PetP450-26	FEHNSMDFVGNNFEYLP
PetP450-27	FLNSNIDYKGQDFELLP
PetP450-28	FEHNSVDFAGNNFEYLP
PetP450-29	FLNSKIDFKGQCFEFIP
PetP450-30	FLENNIDIKGQNFTLLP
PetP450-31	FIGKNVDLRGHDFHLLP
PetP450-32	FLNSSLDYKGRDFEYIP
PetP450-33	FLSSKTDFKGQNFELLP
PetP450-34	FLSSKTDFKEQNFELIP
PetP450-35	FLESEIDFRGRDFELNP
PetP450-36	FENTKVDYKGHDFEYIP
PetP450-37	FLENDIDMKGQNFTLLP
PetP450-38	FSESEVDFRGRDFELIP
PetP450-39	FMEKEIDISGQNFNLLP
PetP450-40	FFEEDVDMKGHDYRLLP
PetP450-41	FLGEDVDMKGHDYRLLP
PetP450-42	FKSSWVDFNGYHYQFIP
PetP450-43	FENSSIDFRGNHFEFIP
PetP450-44	FIGSKIDFKGLNYELIP

(*continued*)

Cloning of Differentially Expressed P450 Genes

The results shown in this chapter show that PCR is a very powerful method for isolating large numbers of novel P450 gene sequences. However, it is often desirable to isolate specific P450 genes rather than all P450 genes expressed in a particular organ or tissue. The PCR approach can be modified in a number of ways to clone P450 genes that are expressed in one tissue (or genotype) but not in another. The simplest method is to isolate RNA separately from the positive and negative tissues, make cDNA, and amplify

TABLE III (*continued*)

P450 clone	Translated sequence[a]
PetP450-45	FSWLKIDIKGQHYELMP
PetP450-46	FLNSDIDVRGLNFGLIP
PetP450-47	FLESEVDFRGRDFELIP
PetP450-48	LRQLGTDFKGTDFRYIP
PetP450-49	FLGSKIDIFGQHYELMP
PetP450-50	FLSSKTDLKGQNFELIP
PetP450-51	FENNSVDFVGNNFEYLP
PetP450-52	FWESEIHVRGQNFELIP
PetP450-53	FWESEIDVRGQNFQLIP
PetP450-54	FLSSNVDYKGNDFEFLP
PetP450-55	FLKCRKDYKGRDFEYIP
PetP450-56	FLNSNMGYKGNDFPFLP
PetP450-57	FWGSELDVRGQDFVFIP
PetP450-58	FNGKNVDLRGHDFHLLP
PetP450-59	FLSSTDLKGQNFELIP
PetP450-60	FFKLVDIDVRGLNFVLIP
PetP450-61	FLEEEKMLRLNGNDFRYLP
PetP450-62	FLSQSTNLNYAGNSFKYMP
PetP450-63	FLNLSSKTTDFKGQNFEFIP
PetP450-64	FMVSEGSANIDVRGGDLRLAP
PetP450-65	FTTDKDVDVRGFELIP
PetP450-66	FLMERDGQPFDITGSREIKMMP
PetP450-67	FLSVSGGDIEAFDITGSREIKMIP
PetP450-68	FLSVSGGDIEAFDITGSGEIKMMP
PetP450-69	FLSVSGGDFEAFDITGSREIKMIP
PetP450-70	FLNNNEEFALTGSREIKMMP
PetP450-71	FMNIENNVDANKDITSAFTP
PetP450-72	FMDIENNVDANKDITCAFTP
PetP450-73	FMNIENNLHHNKDITSAFTP

[a] The first and last amino acid residues of each sequence are derived from the primer sequences (oligo #3 and #4).

the P450 genes in the two different samples. The amplified P450 fragments can then be radiolabeled and used to screen a cDNA library to identify clones of P450 genes that are expressed in the positive tissue and not the negative tissue. A similar differential PCR screening approach has been previously described for the isolation of a specific dioxygenase gene from petunia.[15] Such an approach has also been used to identify differentially expressed P450 genes in petunia. It is important that high stringency hybridization conditions are used so that related P450 sequences do not hybridize

[15] T. A. Holton, F. Brugliera, and Y. Tanaka, *Plant J.* **4,** 1003 (1993).

to each other. When using probes derived from small PCR products, stringent washing conditions can be used to distinguish between closely related sequences.[16]

Limitations

It is obvious from our work that the number of P450 genes in plants is very large: petunias contain at least 70 different genes. It is unlikely that one pair of PCR primers will allow amplification of every P450 gene, but the use of a few different primer combinations will probably amplify most of them. The most difficult step lies in the identification of gene function of a newly isolated gene. However, the availability of a genetic map and/or genetically defined mutants of known P450 genes can greatly facilitate the identification of gene function.[5]

[16] C. Hassett, R. Ramsden, and C. J. Omiecinski, *Methods Enzymol.* **206,** 291 (1991).

Section V

Insect P450s

[32] Preparation of Microsomes from Insects and Purification of CYP6D1 from House Flies

By JEFFREY G. SCOTT

Introduction

The P450 monooxygenases of insects have several functional roles, including growth, development, resistance to pesticides, and tolerance to plant toxins.[1,2] Monooxygenases are intimately involved in fatty acid metabolism and in the biosynthesis of insect hormones and pheromones including 20-hydroxyecdysone and juvenile hormone.[1,2] Historically, the study of insect P450s has been hampered by the small size of the organisms and the relative instability of microsomal preparations. This chapter details methods for the isolation of microsomes that have been successfully used for several species and the procedure for the purification of CYP6D1 from house fly microsomes.

Buffers

Homogenization buffer: 10% glycerol, 1 mM phenylmethylsulfonyl fluoride (PMSF), 1 mM ethylenediaminetetraacetic acid (EDTA), 1 mM phenylthiourea (PTU), 0.1 mM dithiothreitol (DTT) in 0.1 M phosphate buffer, pH 7.4[3]

Resuspension buffer: prepared the same as homogenization buffer except glycerol is increased to 20% and PTU is omitted[3]

HIC buffer A: 2 M KCl and 20% glycerol in 80 mM phosphate buffer, pH 7.5

HIC buffer B: 0.4% emulgen 911 (Kao Atlas, Tokyo), 0.6% sodium cholate, and 20% glycerol in 80 mM phosphate buffer, pH 7.5

Exchange buffer: 40 mM Tris (tris[hydroxymethyl]aminomethane) pH 7.2, 20% glycerol and 1 mM PMSF, 0.2% sodium azide, 3 μg/ml leupeptin, 1 μM pepstatin A, 1 mM EDTA, 0.1 mM DTT, and 0.2% emulgen 911

[1] M. Agosin, in "Comprehensive Insect Physiology, Biochemistry and Pharmacology" (G. A. Kerkut and L. I. Gilbert, eds.), Vol. 12, p. 647. Pergamon, New York, 1985.
[2] E. Hodgson, Insect Biochem. 13, 273 (1983).
[3] S. S. T. Lee and J. G. Scott, J. Econ. Entomol. 82, 1559 (1989).

IEX buffer A: 0.4% emulgen 911 and 20% glycerol in 20 mM Tris acetate, pH 7.5

IEX buffer B: 0.5 M sodium acetate in IEX buffer A

Preparation of Microsomes

Standard Method

Unless specified otherwise, all procedures are conducted at 4°. Flies are anesthetized by chilling at −20° for 3 min, transferred to precooled plastic freezer containers, each containing three 7-ml screw-capped plastic tubes (to facilitate separation of abdomens from thoraces), and frozen at −80° for as little as 2 hr or as long as 5 days. Immediately upon removal from the freezer, the containers are shaken vigorously for 30 sec to separate all the body parts. Body parts are then sieved across a No. 12 (1.7-mm apertures) brass sieve (Fisher Scientific, Rochester, NY). Detached heads and legs passed through the sieve, while most detached abdomens and thoraces are retained. Wings are removed using a gentle stream of air from underneath the sieve. These abdomens and thoraces are transferred onto a flat ice surface. Typically, 100–500 abdomens are sorted and immediately placed into a Wheaton Potter glass–Teflon homogenizer containing 1.0 ml of homogenization buffer per 50 abdomens. Care is taken to avoid contaminating the abdomens with heads and thoraces because of the inhibitory effect of eye pigment (xanthommatin)[4] and the spectral interference from flight muscle cytochromes in thoraces.[5] The abdomens are thoroughly checked for the presence of fly heads or thoraces (which usually stay on top of the solution) in the homogenizing tube before homogenization. The contents are homogenized for 30 sec (10–15 strokes) at 120 rpm using a motor-driven homogenizer and filtered through four layers of cheesecloth. Homogenization buffer (10 ml) is added to the residues, and they are homogenized for another 15 sec, filtered, and combined into the centrifuge tube. Homogenization buffer (5 ml) is used to rinse the homogenizer and is then added to the homogenate. The concentration of the homogenate is 20 abdomens per ml. Homogenates are centrifuged at 10,000 g for 20 min. The 10,000-g supernatant is filtered through two layers of cheesecloth and centrifuged at 100,000 g for 1 hr. The microsomal pellets are resuspended by homogenizing in 1.0 ml of resuspension buffer per 100 abdomens, diluted to a final protein concentration of 2 mg/ml, and stored at −80°. Microsomes prepared in this manner are stable for months and are suitable for all types of common

[4] R. D. Schonbrod and L. C. Terriere, *Pestic. Biochem. Physiol.* **1,** 409 (1972).
[5] M. J. J. Ronis, E. Hodgson, and W. C. Dauterman, *Pestic. Biochem. Physiol.* **32,** 74 (1988).

assays. These general methods have been used to successfully isolate micro-somes from several insect species (Table I).[6]

Large-Scale Method

One limitation to this method is that individual abdomens have to be picked by hand. This presents a problem when large numbers of abdomens are needed for enzyme purification. To overcome this problem, a method for rapid isolation of large numbers of abdomens was developed based on a suggestion from Dr. E. Hodgson.[7] The abdomens and thoraces retained after sieving (see earlier) are placed in a liter beaker and are gently swirled with 400 ml ice-cold 0.1 *M* phosphate buffer, pH 7.4. After a few minutes, thoraces float to the surface and are poured off. The operation is repeated with additional washes of 250 and 200 ml, whereupon the remaining abdo-mens are drained, weighed, and placed in the homogenization buffer (5 ml buffer per gram abdomens). Any remaining thoraces or heads are removed before homogenization.

Small-Scale Method

To quickly prepare microsomes from small insects, insect cell or tissue cultures, specific tissues, or individual insects, a micro method was devel-oped.[8] To prepare microsomes using the micro method, 10 dissected fat bodies, or as few as 5 abdomens from house flies, are homogenized by hand in 200 μl homogenization buffer in a 1 ml Teflon–glass tissue grinder. The homogenate is pooled with a 25-μl rinse of the homogenizer and centrifuged in a 500-μl Eppendorf tube using a microcentrifuge for 17 min at 13,000 rpm. The supernatant from the first centrifugation is collected and centrifuged at 100,000 rpm for 10 min in an A-100 rotor using an Airfuge ultracentrifuge (Beckman). After centrifugation, the supernatant is poured off, and the microsomal pellet is resuspended with 25 μl of resuspension buffer by repeated gentle aspiration into a 200-μl plastic disposable pipette tip and stored at $-80°$.

Purification of CYP6D1

The strategy for the purification of house fly cytochrome P450 was to first separate the cytochrome P450s from unwanted proteins without fractionating the cytochrome P450s, and then to resolve the different cyto-

[6] G. D. Wheelock, Y. Konno, and J. G. Scott, *J. Biochem. Toxicol.* **6,** 239 (1991).
[7] G. D. Wheelock and J. G. Scott, *Insect Biochem.* **19,** 481 (1989).
[8] G. D. Wheelock and J. G. Scott, *Entomol. Exp. Appl.* **61,** 295 (1991).

TABLE I
P450 CONTENT IN MICROSOMES PREPARED FROM VARIOUS INSECTS[a]

Species	Strain[b]	Life stage[c]	Tissue[d]	Cytochrome P450 (nmol/mg protein)
Diptera				
House fly	LPR	A	AB	0.61 ± 0.05
House fly	S+	A	AB	0.18 ± 0.03
Face fly	LB	A	AB	0.018 ± 0.002
Face fly	LB (+PB)	A	AB	0.12 ± 0.02
Flesh fly	LB	A	AB	0.071 ± 0.009
Stable fly	LB	A	AB	0.043 ± 0.003
Stable fly	LB (+PB)	A	AB	0.14 ± 0.04
Fruit fly	Hikone-R	A	AB + TH	0.18 ± 0.02
Fruit fly	Canton-S	A	AB + TH	0.13 ± 0.03
Fruit fly	Canton-S	L	W	0.058 ± 0.003
Fruit fly	Canton-S (+PB)	L	W	0.069 ± 0.002
Fruit fly	Hikone-R	L	W	0.060 ± 0.001
Fruit fly	Hikone-R (+PB)	L	W	0.090 ± 0.002
Hymenoptera				
Honey bee	WT	A	MG	ND[e]
Honey bee	WT (+PB)	A	MG	0.092 ± 0.040
Carpenter ant	LB	A (queen)	W	0.11 ± 0.03
Carpenter ant	LB	A (male)	W	0.29 ± 0.04
Lepidoptera				
Cabbage looper	LB	L	MG	0.11 ± 0.02
Tobacco hornworm	LB	L	MG	0.015 ± 0.003
Tobacco hornworm	LB (+PB)	L	MG	0.020 ± 0.002
Orthoptera				
German cockroach	Dursban-R	A (male)	AB	0.22 ± 0.02
German cockroach	CSMA	A (male)	AB	0.22 ± 0.02
German cockroach	CSMA (+PB)	A (male)	AB	0.36 ± 0.01
Coleoptera				
Colorado potato beetle	LB	L	W	0.025 ± 0.006

[a] From Wheelock *et al.*[6] and J. G. Scott, unpublished data.
[b] LB, reared under laboratory conditions; WT, derived from wild populations; PB, diet included phenobarbital.
[c] A, adult; L, larvae.
[d] AB, abdomen; TH, thorax; W, whole body; MG, midgut.
[e] None detected.

chrome P450s at the next stage. Thus, it was reasoned that hydrophobic chromatography would be the optimal first step if cytochrome P450s have similar hydrophobic natures that are different from other proteins. Further advantages of a polarity separation as the first step are the high capacity of this type of column, tolerance of high ionic strength buffers, and large

loading volumes. In this way preliminary sample manipulation and conse-
quent additional losses are minimized.

Microsomes (5–6 ml, 3 mg protein/ml) are thawed and 10% 3-[3-chol-
amidopropyl)dimethylammonio]-1-propanosulfonate (CHAPS) in water is
added with brief shaking to a final concentration of 0.5%, left on ice for
up to 4 hr, then 40% (w/v) polyethylene glycol 8000 (PEG) is added to a
final concentration of 8% and is incubated on ice for 30 min. Precipitates
are removed by centrifuging at 10,000 g for 10 min and the supernatant is
saved. PEG precipitates a large amount of protein, while most of the
cytochrome P450 remains in solution. KCl is dissolved in the supernatant
to a final concentration of 2 M and insoluble material is pelleted at 10,000 g
for 10 min. The supernatant (2 ml) is diluted 1 : 1 with HIC buffer A and
HPLC purified with an analytical phenyl-5PW column (Beckman, Fullerton,
CA) equilibrated with HIC buffer A. The elution is monitored at 280 and
417 nm. Proteins are eluted with a linear 30-min gradient to HIC buffer B
at 0.5 ml/min and two main fractions are seen. One fraction, unretained
proteins, as judged by absorbance at 280 nm, elutes soon after injection.
This peak contains a minority of the total 417-nm absorbing material. The
second 417 nm peak elutes at 13 min. This 417-nm peak is the major 417-
nm peak observed and it contains cytochrome P450, cytochrome b_5, and
cytochrome P450 reductase activity. Three such runs are performed, and
the major 417-nm peak from each is pooled and frozen at $-80°$ overnight.

Coeluting with the 417-nm peak is an off-scale 280-nm peak. This 280-
nm peak is likely due to emulgen 911 from HIC buffer B, which absorbs
strongly at this wavelength (>2 OD at 0.4%). Confirmation of this is revealed by making an identical injection without the sample, which shows
a similar pattern.[7] In addition to obscuring any protein peaks in this region,
this pattern of emulgen elution shows that it was initially binding to the
column. The coincident elution of peak A and the emulgen breakthrough
suggest that this phenomenon may be important to the elution pattern seen.

Excess salt is removed by dilution with an equal volume of exchange
buffer and is concentrated fivefold in an ultrafiltrator (Amicon, Danvers,
MA) using a 30-kDa (YM 30) cutoff membrane at 4°. The retained sample
is then diluted fourfold with exchange buffer and is concentrated fivefold.
This procedure is repeated one more time. The resulting concentrate (2
ml) is diluted with an equal volume of IEX buffer A and HPLC purified
with an analytical DEAE-5PW (or DEAE-5SW) column (Beckman) equilibrated with IEX buffer A. A gradient (0.5 ml/min) to 100% IEX buffer B
in 30 min is used.

Ion exchange[9] resolves the sample into three major hemoprotein

[9] S. Imaoka and Y. Funae, *J. Chromatogr.* **375,** 83 (1986).

peaks[7,10]: A (elution volume ~3–6 ml), B (elution volume ~12 ml), and C (elution volume ~15 ml). Peak B contains CYP6D1. Peak C contains substantial levels of cytochrome b_5 and a trace of cytochrome P450. This ion-exchange step is a key step in resolving the three monooxygenase components from each other.

Yield and Specific Content of the Isolated Enzymes

The described isolation procedure gives a yield of 17% for CYP6D1 having a specific content of 8–14 nmol/mg protein.[7,10] The phenyl-5PW column gives high yields (91%) of cytochrome P450, while the ion-exchange step results in the largest increase in cytochrome P450-specific content.

Acknowledgment

Supported by grants from the National Institutes of Health and the United States Department of Agriculture.

[10] J. G. Scott and S. S. T. Lee, *Arch. Insect Biochem. Physiol.* **24,** 1 (1993).

[33] Quantification of Ecdysteroid Biosynthesis during Short-Term Organ Culture

By TIMOTHY J. SLITER, KOMSUN SUDHIVORASETH, and JOHN L. MCCARTHY

Introduction

During the larval development of insects, molting and pupariation are triggered by ecdysteroids synthesized by the prothoracic gland in response to stimulation by the prothoracicotropic hormone (PTTH). Methods for the *in vitro* culture and activation of prothoracic glands were initially developed in the lepidopteran insect *Manduca sexta* as a sensitive bioassay for use in the purification of neuropeptides with PTTH activity.[1,2] Subsequently, *in vitro* methodologies have been used to investigate second messenger pathways mediating PTTH action on the prothoracic gland and the enzymology of ecdysteroid biosynthesis, particularly the role of putative cytochromes P450 in the multistep conversion of dietary sterols to ecdyste-

[1] W. E. Bollenbacher, N. Agui, N. Granger, and L. I. Gilbert, *Proc. Natl. Acad. Sci. U.S.A.* **76,** 5148 (1979).
[2] D. P. Muehleisen, E. J. Katahira, R. S. Gray, and W. E. Bollenbacher, *Experientia* **50,** 159 (1994).

roids.[3,4] Cytochromes P450 have been implicated in several discrete steps in the biosynthesis of ecdysteroids by the prothoracic gland (Fig. 1).[4–7]

More recently the development of highly sensitive ecdysteroid radioimmunoassays has made it possible to reliably measure *in vitro* ecdysteroid synthesis by single larval ring glands of small dipteran insects such as *Drosophila melanogaster*.[8] This has led to the identification of lethal mutations of *Drosophila* that cause ecdysteroid deficiency by directly impairing the ability of larval ring glands to synthesize ecdysteroids.[9,10] These mutations are the insect equivalent of congenital abnormalities in adrenal and gonadal steroid synthesis in vertebrates. The primary biochemical defect is not known for any such insect mutation. Possible sites include ecdysteroid biosynthetic cytochromes P450 or components of the second messenger pathway mediating PTTH action.

This chapter describes methods for the quantification of ecdysteroid biosynthesis by ring glands of dipteran larvae during short-term organ culture. We first describe a sensitive radioimmunoassay (RIA) for measuring femtomole quantities of ecdysteroids. Then we present detailed protocols for the *in vitro* culture of larval ring glands from two dipteran species, *D. melanogaster* and the blowfly *Sarcophaga bullata*.

Ecdysteroid Radioimmunoassay

Overview

Total ecdysteroids secreted by ring glands *in vitro* are measured using a competitive radioimmunoassay that employs [^{3}H]ecdysone as the radiolabeled ecdysteroid and ecdysone as the unlabeled competitor in the standard binding curve.[11,12] The antiserum used is directed against the A ring of ecdysone. Antibody-bound label is separated from unbound label by adsorption to excess formalin-fixed *Staphylococcus aureus* (protein A) and is quantified by liquid scintillation counting.

---

[3] W. A. Smith, W. L. Combest, and L. I. Gilbert, *Mol. Cell. Endocrinol.* **47**, 25 (1986).

[4] M. L. Grieneisen, J. T. Warren, and L. I. Gilbert, *Insect Biochem. Mol. Biol.* **23**, 13 (1993).

[5] S. L. Smith, *in* "Comprehensive Insect Physiology, Biochemistry and Pharmacology" (G. A. Kerkut and L. I. Gilbert, eds.), Vol. 7, p. 295. Pergamon, New York, 1985.

[6] J. T. Warren, S. Sakurai, D. B. Rountree, L. I. Gilbert, S.-S. Lee, and K. Nakanishi, *Proc. Natl. Acad. Sci. U.S.A.* **85**, 958 (1988).

[7] S. Sakurai, J. T. Warren, and L. I. Gilbert, *Arch. Insect Biochem. Physiol.* **10**, 179 (1989).

[8] C. P. F. Redfern, *J. Insect Physiol.* **29**, 65 (1983).

[9] V. C. Henrich, R. L. Tucker, G. Maroni, and L. I. Gilbert, *Dev. Biol.* **120**, 50 (1987).

[10] T. J. Sliter and L. I. Gilbert, *Genetics* **130**, 555 (1992).

[11] J. T. Warren and L. I. Gilbert, *Insect Biochem.* **16**, 65 (1986).

[12] J. T. Warren, W. A. Smith, and L. I. Gilbert, *Experientia* **40**, 393 (1984).

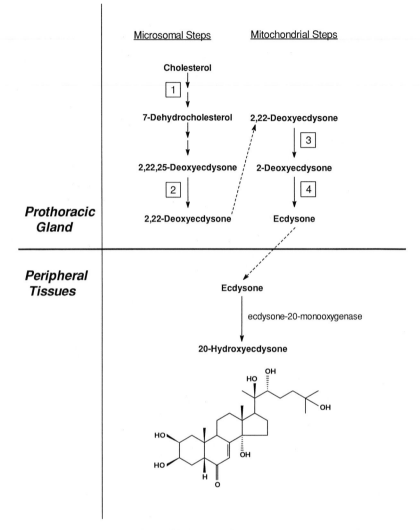

FIG. 1. Core scheme for ecdysteroid biosynthesis from cholesterol. Conversions occurring in the prothoracic gland and catalyzed by putative cytochromes P450 are numbered (1–4). Those conversions are (1) 7,8-dehydrogenation via P450-catalyzed hydroxylation and subsequent dehydration; (2) 25-hydroxylation; (3) 22-hydroxylation; and (4) 2-hydroxylation. All of the identified reactions are suppressed by specific inhibitors of cytochromes P450, and all except the 2-hydroxylation step are sensitive to carbon monoxide.[4] The conversion of ecdysone to the biologically active 20-hydroxyecdysone by the confirmed cytochrome P450 enzyme ecdysone-20-monooxygenase occurs in peripheral tissues, most notably the fat body.[5] Numerous variations on the core scheme are known to occur. In some species, 24-alkylated sterols are utilized to form biologically active 24-alkylated ecdysteroids. Lepidopteran insects produce a 3-oxo intermediate leading to the formation and secretion of 3-dehydroecdysone which is subsequently converted to ecdysone by a hemolymph 3β-reductase.[6,7]

Protocols for the quantitation of ecdysone-20-monooxygenase activity in isolated tissues and mitochondrial preparations have been described.[5] While 20-monooxygenase could be measured by RIA using an antiserum specific for the side chain of 20-hydroxyecdysone, standard protocols are based on the conversion of radiolabeled ecdysone to the 20-hydroxylated form, separation by thin-layer or high-performance liquid chromatography (HPLC), and quantitation by liquid scintillation counting.

Protocol

Preparation of Ecdysone Standards. Ecdysone standards for construction of the standard binding curve are prepared by twofold serial dilution of a 1-ng/10 μl (2.16 pmol/10 ml) ecdysone solution that is prepared by dilution of a 5-mg/ml ecdysone stock solution in 70% ethanol. The concentration of the stock solution is determined by UV absorbance at 243 nm ($\varepsilon = 11600$). Ecdysone standards are diluted in Grace's insect culture medium.[13] Both the ethanol stock solution and the standards can be stored indefinitely at $-20°$ and are stable to repeated thawing and refreezing.

Ecdysteroid Antiserum. In the results presented here, a rabbit antiserum (H-22) raised against ecdysone-22-syccinylthyroglobulin amide[14] was used. This antiserum binds strongly to ecdysteroids that have the same A ring configuration as ecdysone, including 20-hydroxyecdysone, and Makisterone A (24-methylecdysone) and 20-deoxy Makisterone A.[11] A 1:250 dilution of the antiserum in borate buffer was used in the assay. This dilution was sufficient to precipitate approximately 33% of the total radioactivity in the B_{max}-binding reaction.[11]

Preparation of Protein A Solution. Antibody is quantitatively precipitated in the assay by binding to excess formalin-fixed *S. aureus* (Cowan I strain).[12] *S. aureus* is grown to saturation at $37°$ in tryptic soy broth. The cells are pelleted by centrifugation for 15 min at 2000 g, and the bacterial pellet is washed twice in phosphate-buffered saline (PBS). Following resuspension in PBS, 37% formalin is added to a final concentration of 4%, and the cells are fixed for 90 min with constant stirring to prevent clumping. After fixation, the cells are washed twice, then resuspended in PBS at a concentration of 10% (w:v).

To determine the concentration of protein A that is optimum for the assay, the protein A solution is titrated in the B_{max}-binding assay (see below). With freshly prepared protein A, maximum precipitation of label is observed with a 2–3% (w:v) solution. There is a tendency during long-

[13] T. D. C. Grace, *Nature* (*London*) **195**, 788 (1962).
[14] D. H. S. Horn, J. S. Wilkie, B. A. Sage, and J. A. O'Connor, *J. Insect Physiol.* **22**, 901 (1976).

term storage for some protein A to become soluble, thereby reducing the amount of label precipitated in the B_{max} reaction. This problem can be overcome by a brief centrifugation (10 min at 2000 g) to precipitate the cells, followed by resuspension in PBS. Diluted protein A solution is stored at $-20°$.

Standard Binding Curve. The standard binding curve consists of nonspecific binding tubes (NSB, four replicates), maximum binding tubes (B_{max}, four replicates), and ecdysone standard binding tubes (B_{std}, three replicates each). Nine ecdysone standards are used to construct the standard binding curve: 1, 0.5, 0.25, 0.125, 0.0625, 0.0313, 0.0156, 0.0078, and 0.0039 ng. The binding reactions are set up as follows. NSB tubes contain 10 μl Grace's medium, 50 μl borate buffer, 50 μl labeled ecdysone, and 20 μl protein A solution. B_{max} tubes contain 10 μl Grace's medium, 50 μl antiserum, 50 μl labeled ecdysone, and 20 μl protein A solution. B_{std} tubes contain 10 μl ecdysone standard, 50 μl antiserum, 50 μl labeled ecdysone, and 20 μl protein A solution.

Unknowns. Assay tubes for unknown samples (UNK) contain the unknown in 10 μl Grace's medium, 50 μl diluted antiserum, 50 μl labeled ecdysone, and 20 μl protein A.

Detailed Protocol. NSB, B_{max}, B_{std}, and UNK-binding reactions are set up in 0.25-ml polypropylene microcentrifuge tubes and are vortexed well. Following overnight incubation at 4°, tubes are vortexed for several seconds to resuspend settled protein A. They are then centrifuged for 15 min (10,000 g at 4°) in a refrigerated microcentrifuge. Immediately following centrifugation, the supernatant is removed by gentle vacuum aspiration. Care is taken not to disturb the protein A pellet. Aspiration must be done without delay because the protein A pellet becomes fragile following centrifugation. This tendency increases if the tubes are allowed to warm above 4°.

Following centrifugation, the protein A pellet is resuspended completely in 20 μl distilled water by vortexing. Tubes may be accumulated at this stage prior to the addition of scintillation cocktail. Liquid scintillation cocktail (180 μl) is added to each RIA tube, and the tubes are vortexed. Radioactivity is counted on the tritium channel of a liquid scintillation counter.

Analysis of Data and Typical Results. Many computer-controlled liquid scintillation counters are equipped with resident software programs for RIA data analysis. Alternatively, count data can be exported to a personal computer and analyzed using a standard statistical analysis software package. We typically observe B_{max} values of 1100–1200 cpm and NSB values of 100–150 cpm. When plotted as a logit transformation against the log of the ecdysone concentration, the standard curve is linear ($r^2 > 0.95$) across a range from 8 fmol to 2 pmol (Fig. 2).

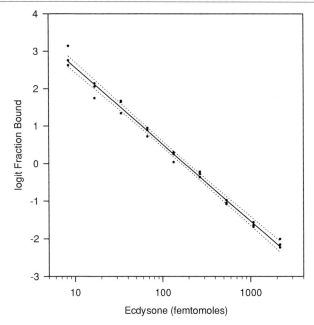

FIG. 2. An ecdysteroid radioimmunoassay standard curve. Binding reactions were carried out overnight at 4° in the presence of standard amounts of unlabeled ecdysone, as described in the text. Radioactivity (cpm) recovered in the protein A pellet is plotted as the logit transformation of the fraction bound [fraction bound = $(B_{std} - NSB)/(B_{max} - NSB)$]. The solid line shows the best-fit linear regression line for the data ($r^2 > 0.99$). The dotted lines show the 99% regression confidence limits. Data were analyzed using SigmaPlot for Windows, version 2.0.

RIA Solutions

Grace's medium: Grace's insect culture medium (GIBCO), supplemented with 0.02% sodium azide

Borate buffer: 100 mM sodium borate, 25 mM sodium tetraborate, 75 mM sodium chloride, 0.05% bovine serum albumin (fraction V), 0.02% sodium azide (pH 8.0)

Phosphate-buffered saline: 100 mM sodium chloride, 44 mM sodium phosphate (monobasic), 8 mM sodium phosphate (dibasic), 0.02% sodium azide (pH 7.5)

Labeled ecdysone: 0.12 μCi/ml α-[23,24-^{3}H]ecdysone (50–90 Ci/mmol, New England Nuclear) in borate buffer

Antiserum: 1 : 250 dilution of the H-22 serum in borate buffer

Protein A: 2–5% solution of formalin-fixed *S. aureus* (Cowan I strain) in PBS

Short-Term Culture of Larval Ring Glands

Ecdysteroid synthesis during the larval development of diptera is episodic. High *in vivo* rates of ecdysteroid biosynthesis and elevated whole body ecdysteroid levels are observed only during relatively brief intervals of time (ca. 4–8 hr) preceding larval molts and immediately prior to the transformation of the larva into the puparium at the initiation of metamorphosis.[15] High rates of ring gland activity are typically found during the wandering stage, an interval of time at the end of larval life when larvae migrate away from the food substrate in preparation for metamorphosis.

The transient character of ecdysteroid synthesis *in vivo* can sometimes make it difficult to obtain preparations that are reliably active *in vitro*. However, given sufficient familiarity with the timing of the morphological and behavior changes that occur in larvae coincident with changes in endocrine activity, it is usually possible to identify individuals that have a high likelihood of showing elevated levels of ecdysteroid synthesis.

The following section describes methods of *in vitro* culture of larval ring glands from *D. melanogaster* and *S. bullata*. These are followed by two supplemental protocols. The first is a simple and rapid method of evaluating tissue function *in vitro* based on incorporation of radiolabeled uridine into acid-precipitable RNA. The second is a protocol for the preparation of a crude neural extract with prothoracicotropic activity *in vitro*.

D. melanogaster

D. melanogaster larvae are raised on an artificial media supplemented with live baker's yeast. There are a number of variations of such media, recipes for which can be found in collections of *Drosophila* laboratory methods.[16] *D. melanogaster* ring glands secrete three ecdysteroids *in vitro:* ecdysone, 20-deoxy Makisterone A (24-methylecdysone), and an unidentified ecdysteroid that elutes after 20-deoxy Makisterone A in reverse-phase HPLC.[17,18] Changes in the sterol content of the diet can alter the relative amounts of these ecdysteroids.[17]

The best *in vitro* results are obtained when larvae are reared under uncrowded conditions (20–30 larvae/cm^2) at 25°. When raised in this fashion, larvae attain a maximum fresh weight of approximately 1.5 mg. Wan-

[15] B. Roberts, L. I. Gilbert, and W. E. Bollenbacher, *Gen. Comp. Endocrinol.* **54,** 469 (1984).
[16] M. Ashburner, "Drosophila: A Laboratory Handbook." Cold Spring Harbor Lab. Press, Cold Spring Harbor, NY, 1989.
[17] C. P. F. Redfern, *Proc. Natl. Acad. Sci. U.S.A.* **81,** 5643 (1984).
[18] V. C. Henrich, M. D. Pak, and L. I. Gilbert, *J. Comp. Physiol., B* **157,** 543 (1987).

dering and pupariation occur 5 days following egg laying. High rates of ecdysteroid synthesis are observed throughout the wandering period, which occupies the final 6–8 hr of larval life.

To set up *in vitro* cultures, larvae are dissected under Grace's insect culture medium (GIBCO) in a glass well slide. The use of plastic dishes for dissection is not recommended as the tissue tends to adhere to the plastic, making transfer difficult. All dissection is done using extremely fine forceps (e.g., Dumont No. 5). Animals can be immobilized by brief (1–3 min) immersion in ice water.

The brain–ring gland complex is exposed by gently ripping open the dorsal surface of the larva at the boundary of the third and fourth segments. This action exposes the brain–ventral ganglion complex, which is easily recognized by the large number of nerves that radiate from the ventral ganglion. Most of these nerves are broken as the animal is opened. The ring gland lies between and slightly anterior to the brain lobes, and is connected to the brain lobes by paired nerves and respiratory trachea that also extend anterior to the ring gland.[19] The ring gland is attached to the dorsal aorta and to the eye–antennal imaginal discs by lateral connective ligaments that are fairly inconspicuous.

The brain–ring gland complex consists of the brain and ventral ganglion, the ring gland with intact nervous connections to the brain lobes, and a short section of the dorsal aorta that is attached to the ring gland. Additionally we usually leave the eye–antennal discs attached to the brain and the ring gland, as this minimizes the possibility of trauma to the gland. Long tracheal branches are left attached to the ring gland for the same reason.

Following dissection, the brain–ring gland complex can be transferred to a drop of medium for culture with forceps, using the radiating nerves from the ventral ganglion as a handle. Alternatively, the ring gland can be isolated by carefully removing the attached aorta and severing the connections to the brain and the eye–antennal discs. The ring gland can be moved about in the dissection dish with forceps, using the attached tracheal branches as handles. However, the small size of the ring gland precludes the use of forceps to transfer the gland between the dissecting dish and a culture vessel. This transfer can be readily performed with negligible trauma to the tissue by using a miniature transfer loop fashioned from fine platinum wire.

Single glands or complexes are cultured individually in 10- or 20-μl drops of Grace's culture medium. Incubations can be performed as hanging drop cultures in 35 × 10-mm petri dishes containing 1 ml of medium to maintain high humidity inside the vessel and minimize evaporation of the

[19] S. K. Aggarwal and R. C. King, *J. Morphol.* **129,** 171 (1967).

culture drop. Equivalent results are obtained using culture vessels consisting of the tops of 0.5-ml microcentrifuge tubes placed on filter paper moistened with medium and enclosed in a petri dish. Multiple ring glands or complexes may be cultured together in volumes up to 100 μl. To terminate cultures the tissue is removed with a fine wire hook fashioned from tungsten or platinum wire. The medium is then transferred to a 0.25-ml microcentrifuge tube, quick frozen on dry ice, and stored at $-20°$. No further processing of the culture medium is required before assaying by RIA.

S. bullata

S. bullata larvae are reared on raw calf or beef liver (100 larvae/80 g meat) at 25°. Mature larvae achieve a weight of 150–200 mg. Larvae enter the wandering stage 4–5 days after egg laying. Pupariation takes place 32–36 hr later. Approximately 6–8 hr prior to formation of the definitive puparium, larvae begin to tan the posterior respiratory spiracles as a consequence of a transient rise in ecdysteroid levels (Fig. 3). This is referred to as the red spiracle stage.

Larvae are reared in shallow dishes loosely covered with aluminum foil and are placed on a bed of sawdust approximately 5 cm deep. Cohorts of newly wandering larvae are recovered from the sawdust by sifting through a wire-mesh screen. They are then transferred to a plastic box containing sawdust and are kept at 25° until needed.

The anatomical layout of the S. bullata brain–ring gland complex is similar to that described earlier for D. melanogaster. The dissection and the transfer of the brain–ring gland complex are done in a manner essentially identical to that described earlier for the D. melanogaster complex. To expose the brain–ring gland complex, the larva is cut in half longitudinally with dissecting scissors. The anterior half is then cut along the dorsal midline and is pinned out (using either fine dissecting pins or 23-gauge syringe needles) under Grace's medium in a small dissecting dish. Fine scissors are used to cut all connections. The complex is transferred to the culture vessel with forceps, using the numerous nerves attached to the ventral ganglion as a handle.

Isolated ring glands are dissected without first isolating the brain–ring gland complex. Transfer to culture medium is performed with a pair of forceps, using an attached trachea as a handle.

In vitro culture and culture termination are performed as described earlier for D. melanogaster. The synthetic activity of S. bullata ring glands is high compared to D. melanogaster. To bring the amount of ecdysteroid within the range of the assay, the culture medium is diluted two- or fourfold with Grace's medium, and 10-μl aliquots are subjected to RIA.

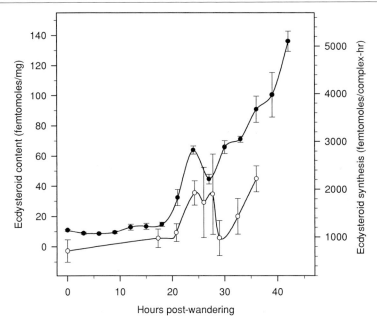

FIG. 3. Ecdysteroid synthesis during the wandering period of *S. bullata*. Groups of larvae were collected within 3 hr of the initiation of wandering behavior and held at 25° for varying periods of time. Closed circles show the developmental profile of RIA-detectable ecdysteroids recovered in a methanol extract of whole animals. A peak is observed at 24 hr postwandering which immediately precedes the appearance at 25–30 hr of tanned posterior respiratory spiracles (red spiracle stage). Whole body ecdysteroid levels increase significantly following formation of the white puparium at 32–36 hr postwandering. To determine the profile of ecdysteroid synthesis by the ring gland (open circles), brain–ring gland complexes were incubated for 1 hr at 25° in 10 ml, and ecdysteroids released into the medium were quantified by RIA. A transient peak of ring gland activity is seen at 24–28 hr, coincident with the rise in whole body ecdysteroid levels at this stage. Error bars indicate SEM.

RNA Synthesis during *in Vitro* Culture

The incorporation of radiolabeled uridine into acid-precipitable RNA can be used as a rapid and sensitive indicator of overall tissue function *in vitro*. Brain–ring gland complexes are incubated in 10-μl drops of Grace's medium containing 5 μCi/ml [5-^{3}H]uridine. Following incubation, each complex is rinsed in ice-cold Grace's, and individual complexes or pooled groups of complexes are transferred to a microcentrifuge tube containing 100 μl denaturing solution [4 M guanidinium thiocyanate, 25 mM sodium citrate (pH 7), 0.5% sarcosyl, 0.1 M 2-mercaptoethanol].[20] The tissue is

[20] P. Chomczynski and N. Sacchi, *Anal. Biochem.* **162,** 156 (1987).

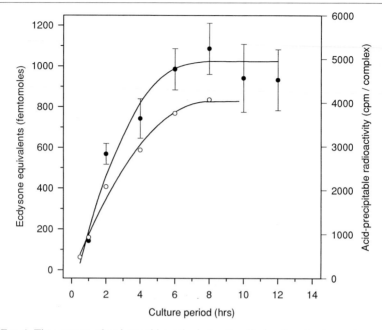

FIG. 4. Time course of ecdysteroid synthesis *in vitro*. Brain–ring gland complexes were dissected from wandering stage *D. melanogaster* larvae and were incubated in 10-μl cultures as described in the text. Closed circles show the amount of RIA-detectable ecdysteroids present in the culture medium after varying periods of culture. Ecdysteroid secretion ceases after 6–8 hr *in vitro*. New RNA synthesis also ceases at this time, as evidence by a reduced rate of incorporation of [^{3}H]uridine into PCA-precipitable material (open circles). Error bars indicate SEM.

dissociated by vigorous vortexing. RNA is precipitated by the addition of 1 ml ice-cold 0.5 M perchloric acid (PCA) and centrifugation at 5000 g (10 min). The pellet is washed three times with 1 ml 0.3 M PCA to remove unincorporated label. The pellet is then resuspended in 1 ml 0.3 M potassium hydroxide and is incubated at 37° for 1 hr with occasional vortexing to hydrolyze the RNA.[21] The amount of radioactivity is measured by liquid scintillation counting.

Activation of Ring Glands by Brain–Ventral Ganglion Extracts

Ecdysteroid synthesis is activated *in vivo* by the brain neuropeptide PTTH. Isolated ring glands can be activated *in vitro* by PTTH-containing

[21] L. M. Riddiford, A. C. Chen, B. J. Graves, and A. T. Curtis, *Insect Biochem.* **11,** 121 (1981).

aqueous extracts of larval brain–ventral ganglion complexes.[1] Neural extracts are made by homogenizing brain–ventral ganglion complexes in Grace's medium using a glass-on-glass tissue grinder (200-μl capacity, Kontes). The homogenate is transferred to a microcentrifuge tube and boiled for 10 min. It is then chilled on ice for 10 min and centrifuged for 10 min at 5000 g. The supernatant is transferred to a fresh tube and diluted as required with Grace's medium. A typical neural extract dose–response curve is shown in Fig. 5 for *S. bullata*. Ring glands are typically not activated by extracts at concentrations less than 0.1 brain equivalents/10 μl and are maximally activated by extracts at approximately 1 brain equivalent/10 μl.

Characteristics of *in Vitro* Ring Gland Activity

Ring glands and brain–ring gland complexes that have been activated by PTTH continue to synthesize and release ecdysteroids in a close to

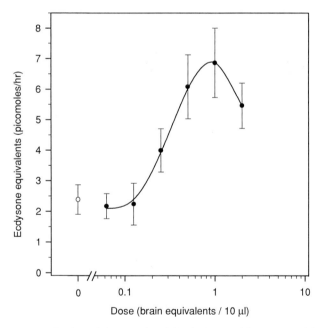

FIG. 5. *In vitro* activation of ring gland activity by heat-stable extracts of larval brains. Ring glands were dissected from *S. bullata* larvae 0–12 hr postwandering and were incubated for 1 hr in 10-μl cultures containing varying concentrations of boiled brain extract prepared as described in the text. No activation was observed at concentrations below 0.2 brain equivalent/10 μl. Maximal activation was obtained with extracts at a concentration of 1 brain equivalent/10 μl.

linear fashion for several hours following explantation to culture (Fig. 4). In *D. melanogaster,* glands from wandering stage larvae initially secrete ecdysteroids at a rate of approximately 250 fmol/hr. Secretion ceases after 6–8 hr of culture. This decrease in biosynthetic activity mirrors a general decrease in cellular function during culture as indicated by a reduced rate of incorporation of uridine into acid-precipitable RNA by brain–ring gland complexes (Fig. 4).

Significantly higher rates of ecdysteroid synthesis are seen in *S. bullata.* Brain–ring gland complexes release ecdysteroids at an initial rate of approximately 2 pmol/hr during the peak of activity that immediately precedes the red spiracle stage (Fig. 3). This represents a two- to threefold increase in ecdysteroid synthesis compared to ring glands from early wandering larvae. This degree of activation is comparable to the threefold increase observed when early wandering ring glands are maximally activated by PTTH-containing neural extracts *in vitro* (Fig. 5).

Acknowledgments

H-22 antiserum was generously provided by L. I. Gilbert. The research described in this chapter was supported by a grant from the U.S. Department of Agriculture (CSRS 93-37302-9080).

[34] Sampling P450 Diversity by Cloning Polymerase Chain Reaction Products Obtained with Degenerate Primers

By MARK J. SNYDER, JULIE A. SCOTT, JOHN F. ANDERSEN, and RENÉ FEYEREISEN

Introduction

Cytochrome P450 proteins are particularly difficult to purify from insects or other invertebrates because limitations on total quantities isolated are compounded by limitations on the stability of the solubilized proteins. Purification from single tissues often represents a daunting task when the whole organism itself is very small. Pooling large numbers of whole organisms or body parts (e.g., abdomens of flies) has been used in most published attempts at P450 purification from insects. This leads to the often irremediable introduction of hydrolytic enzymes (e.g., digestive enzymes) or inhibitors (e.g., xanthommatin from pigments). Molecular cloning of P450 genes and their heterologous expression and reconstitution has so far provided the best method for the biochemical characterization of insect P450s. Molecular probes may also be needed to investigate a myriad of biological problems, such as pesticide resistance, induction by foreign chemicals, and hormonal or developmental control.

The first molecular probes for insect P450s were obtained serendipitously or by directed but arduous procedures.[1] The combined availability of a few P450 sequences from insects and of the polymerase chain reaction (PCR) has now allowed the development of rapid and efficient methods to clone P450 fragments from any source, including very small insects (such as the whitefly *Bemisia tabaci*, a major agricultural pest) or very small endocrine glands (such as the corpora allata, the source of insect juvenile hormone). Techniques used in our laboratory to sample P450 diversity by cloning/sequencing are described in this chapter.

PCR and Cloning New P450s

Use of the PCR with specific primers to amplify P450 fragments of known identity is very useful in many situations, i.e., to document the expression of P450s in tissues (e.g., brain[2]), developmental stages (e.g., fetus[3]), or closely related species[4] that had not been studied previously. Specific primers are also used to study P450 induction by reverse transcriptase (RT)-PCR.[5] When the specific primers are located on each side of an intron of known length, RT-PCR can provide initial evidence for expression of the gene.[6] PCR has also been used to verify positive signals when screening libraries for P450 clones.[7]

To obtain *new* P450 gene sequences, the assumption is made that consensus P450 sequences from one or more species and representing one or more P450 subfamilies will be represented in the unknown P450s being targeted for cloning. This assumption is generally valid for any members of large multigene families, as demonstrated by the ease with which odorant recep-

[1] R. Feyereisen, *in* "Cytochrome P450" (J. B. Schenkman and H. Greim, eds.), p. 311. Springer-Verlag, Heidelberg, 1993.

[2] M. Warner, R. Ahlgren, P. G. Zaphiropoulos, S. I. Hayashi, and J. A. Gustaffsson, *Methods Enzymol.* **206,** 631 (1991); A. V. Hodgson, T. B. White, J. W. White, and H. W. Strobel, *Mol. Cell. Biochem.* **120,** 171 (1993).

[3] J. Hakkola, M. Pasanen, R. Purkunen, S. Saarikoski, O. Pelkonen, J. Maenpaa, A. Rane, and H. Raunio, *Biochem. Pharmacol.* **48,** 59 (1994).

[4] D. R. Bell and T. L. Ribam, *Biochem. J.* **294,** 173 (1993).

[5] M. J. Fasco, C. Treanor, and L. Kaminsky, *Methods Enzymol.* **272,** Chap. 43, 1996 (this volume).

[6] M. B. Cohen and R. Feyereisen, *DNA Cell Biol.* **14,** 73 (1995).

[7] C. Hassett, R. Ramsden, and C. Omiecinski, *Methods Enzymol.* **206,** 291 (1991).

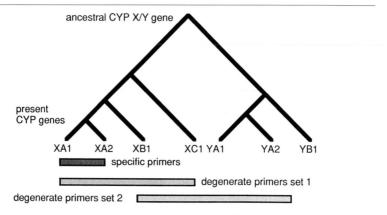

FIG. 1. Specific primers amplify genes orthologous to CYPXA1 in different species or very closely related P450s. Degenerate primers 1 amplify most P450 in the CYPX family. The sequence of full-length clones from the P450s most distant from CYPXA1 helps design degenerate primer set 2 which would extend the range of P450s that can be amplified.

tors were cloned.[8] However, in the case of P450 proteins, domains that are highly conserved are difficult to find beyond the family level, and even the heme-binding region consensus (F--G---C-G) is no longer a universal P450 "filter." The Phe is replaced by Trp in CYP8 and CYP10, and by Pro in CYP74. The first Gly is replaced by Ala or Glu in allene oxide synthases (CYP74), and the second Gly is replaced by Ala in a number of P450s from various sources. Thus, a PCR method has to be iterative, with redesign of amplification primers to "walk" in the sequence space occupied by all extant P450s (Fig. 1).

Strategies

One strategy is to use the conserved region around the conserved cysteine liganding the heme as a forward primer. Vector primers (for library screening) or oligo(dT) primers (for cDNA) are used as reverse primers. This 3′-RACE strategy was successfully used by several authors.[9–11] For instance, combined PCR amplification and screening identified 18 different genes from a petunia cDNA library.[9] This strategy is described in detail by Holton and Lester.[12]

[8] L. Buck and R. Axel, *Cell (Cambridge, Mass.)* **65,** 175 (1991).
[9] T. A. Holton, F. Brugliera, D. R. Lester, Y. Tanaka, C. D. Hyland, J. G. T. Menting, C. Y. Lu, E. Farcy, T. W. Stevenson, and E. C. Cornish, *Nature (London)* **366,** 276 (1993).
[10] A. H. Meijer, E. Souer, R. Verpoorte, and J. H. Hoge, *Plant Mol. Biol.* **22,** 379 (1993).
[11] P. B. Danielson, R. J. MacIntyre, and J. C. Fogleman, *J. Cell Biochem.* **21A,** 196 (1995).
[12] T. A. Holton and D. R. Lester, *Methods Enzymol.* **272,** Chap. 31, 1996 (this volume).

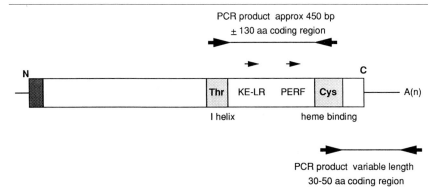

Fig. 2. Two strategies used to amplify P450 gene fragments. The target P450 sequence is boxed. Arrowheads indicate primers for PCR. (Top) The conserved Cys region is used in the design of a degenerate reverse primer, and the conserved Thr region (or other family or subfamily-specific regions) provides a forward primer sequence. (Bottom) The conserved Cys region is used in the design of a degenerate forward primer in a 3′-RACE strategy.

Another strategy described in this chapter is to use the region surrounding the conserved cysteine to design a reverse primer and to use as forward primer a sequence derived from another conserved region in the P450 protein. A comparison of the two strategies is given in Fig. 2.

CYP4 Family Strategy

Our strategy was initially based on the observation that sequences of CYP4C1 from the cockroach, *Blaberus discoidalis*, and of CYP4D1 from *Drosophila melanogaster* share *two* highly conserved stretches of amino acids with members of the CYP4 family from vertebrates, while being otherwise quite distinct from these vertebrate CYP4 proteins (30–36% identity).[13,14] In addition to the region that surrounds the universally conserved cysteine,

CYP4C1, CYP4D1   PFSAGPRNCIGQKFA
CYP4A1   PFSGGARNCIGKQFA
CYP4B1   PFSAGPRNCIGQQFA
                 *** * * ***** **

another highly conserved region precedes a hallmark residue of P450 proteins: the threonine that corresponds to Thr 252 of P450cam (CYP101).

[13] J. Y. Bradfield, Y. H. Lee, and L. L. Keeley, *Proc. Natl. Acad. Sci. U.S.A.* **88,** 4558 (1991).
[14] R. Gandhi, E. Varak, and M. L. Goldberg, *DNA Cell Biol.* **11,** 397 (1992).

CYP4C1   EVDTFMFEGHDTT
CYP4D1   EVDTFMFKGHDTT
CYP4A1, CYP4B1 EVDTFMFEGHDTT
                   * * * * * *   * * * * *

This sequence, particularly favorable for primer design, corresponds to a portion of helix I in the tertiary structure of P450cam, which is distal to the heme and is thought to be important in the activation of molecular oxygen and catalysis. This observation opened the possibility that a combination of two degenerate primers would allow amplification of all P450s that also possessed these two regions of very high sequence conservation. Because of the great evolutionary divergence of cockroaches and flies from mammals and from each other, it was expected that these sequences would be shared by P450s in a very wide diversity of animals.

Design of Oligonucleotide Primers

Two sets of partially degenerate primers (A and B) were designed based on these two conserved regions (Fig. 3). Both forward primers chosen encode for Glu instead of Lys, as in CYP4D1; however, this difference is actually only a single nucleotide mismatch (G to A) and should not prevent the primers from annealing to their target sites. In fact we did clone CYP4D1 from *Drosophila*, as well as eight other CYP4D subfamily-related sequences from both the house fly and the mosquito using these primer sets. Primers of set A are shorter and more degenerate (29- and 23-mers, 192- and 384-

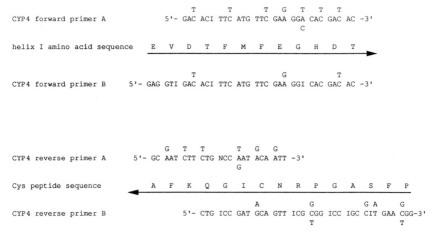

Fig. 3. Forward and reverse primers used to amplify P450 fragments of the CYP4 family (N = A, C, G, T; I = inosine).

fold degeneracy for forward and reverse primers, respectively), whereas primers of set B are longer and less degenerate (35- and 36-mers, 8- and 72-fold degeneracy for forward and reverse primers, respectively), but use 3,5-inosines and make several choices based on estimates of codon usage in *Anopheles gambiae* genes. If a codon usage table can be constructed for the organism being studied, then it may be possible to reduce the degeneracy of the primers without decreasing their potential to anneal to the target sequences. The presence of inosines did not appear to reduce the effectiveness of our primers.

Template Isolation and Amplification Conditions

The template may be either genomic DNA or cDNA. Poly(A)$^+$ RNA is isolated either by using the Micro-Fast Track Kit (Invitrogen) or by oligo(dT) column purification of total RNA. Poly(A)$^+$ RNA is reverse transcribed to single-stranded cDNA using the Superscript RNase H- Reverse Transcriptase Kit (GIBCO-BRL) and an 18-mer oligo(dT) primer or the degenerate reverse primer. Initial template amplification is accomplished with the following program: 75° for 2 min, 94° for 2 min, then 35 cycles of 94° for 1 min, 54° (primer set A) or 45° (primer set B) for 1 min, and 72° for 1 min, followed by 75° extension for 5 min. The annealing temperatures chosen are intentionally well below the respective T_m for each primer set to allow for mismatches to occur and to increase the probability of finding more distantly related genes. Each 30-μl reaction contains 50 mM KCl, 10 mM Tris (pH 8.3), 2.5 mM MgCl$_2$, 0.2 mM of each dNTP, 0.01% gelatin, 3 μM (set A) or 1 μM (set B) of each oligonucleotide primer, 1.5 U AmpliTaq polymerase (Perkin Elmer Cetus), and approximately 1 μg genomic DNA or cDNA. The combination of reducing the MgCl$_2$ concentration to 1.5 mM and the primer concentration to 0.15 μM does not appear to compromise results. The gelatin may be omitted if the dNTPs are added to the reaction mix just prior to amplification without altering results. To reduce the formation of "primer dimers," the primers can be spotted separately on the sides of the reaction tube and mixed with the other components by quick centrifugation just before the 75° start of the reaction. It cannot be emphasized enough that the usual precautions must be taken to prevent PCR contamination, particularly in laboratories that work with templates from a number of sources.

Amplification products are visualized on an ethidium bromide-stained 1% agarose gel, and distinct bands range in size from approximately 425 to 450 bp. Longer bands can be obtained from genomic DNA if an intron(s) is present. With primer set B we find that shorter bands are usually false positives. The bands are cut from the gel and DNA is isolated using the

Sephaglas BandPrep Kit (Pharmacia Biotech). Isolated products may be reamplified using the same reaction conditions and program and isolated as previously described if more product is needed for cloning.

Cloning and Sequencing

Because each band of DNA will undoubtedly contain several different P450 gene fragments, direct sequencing of the PCR product should not even be contemplated. Instead, purified amplification products are quantitated and cloned into the pCR II Vector plasmid (InVitrogen). It should be noted that thermostable polymerases which possess 3′ to 5′ exonuclease activity will remove the 3′-A overhang and make TA cloning inefficient, in which case vectors accepting blunt-ended DNA should be used. DNA from putative clones is isolated, and the presence of an insert is verified either by PCR amplification of the plasmid with primers which anneal to the vector regions flanking the insert or by reaction with an appropriate restriction endonuclease. Clones containing inserts are then sequenced.

DNA Sequence Analysis

The number of clones obtained is obviously dependent on many variables, but isolation of a large number of clones (or of *all* clones, when numbers permit) is highly advisable: 60% of the clones were P450s in our study on *Manduca sexta*, 73% in *Anopheles albimanus*, and 79% in *Drosophila melanogaster*.[15–17] In our experience, numerous sequence variants, possibly alleles of the same P450 gene, will be obtained. However, because these fragments are generated via PCR, the possibility of PCR errors cannot be disregarded and results must be interpreted with this possibility in mind. Nucleotide sequences are compared to protein sequence databases using the BLASTX program on the NCBI BLAST Server. Only those clones that have statistically significant similarity scores to numerous P450s and to no other genes are considered to be P450 genes.

To evaluate the percentage identity of the new P450s to known P450 proteins, we use a function[16] that relates the percentage identity of whole length P450 protein sequences with the percentage identity of the portion of these proteins corresponding in length and location to the PCR product.

[15] M. J. Snyder, J. L. Stevens, J. F. Andersen, and R. Feyereisen, *Arch. Biochem. Biophys.* **321,** 13 (1995).

[16] J. A. Scott, F. H. Collins, and R. Feyereisen, *Biochem. Biophys. Res. Commun.* **205,** 1452 (1994).

[17] B. C. Dunkov, R. Rodriguez-Arnaiz, B. Pittendrigh, R. ffrench-Constant, and R. Feyereisen, *Mol. Gen. Genet.* (in press).

Validation of the Method

The sequence of the PCR amplification products provides only about one-third of the total P450 amino acid sequence. Almost all the sequences are characterized by a $\phi\phi$KE-LRϕ-P motif (where ϕ is a hydrophobic residue) and a PERF motif, or closely similar sequences. These characteristics are found in many P450 proteins. Full-length clones obtained in subsequent studies, e.g., library screening, can verify the sequence of the PCR products (e.g., CYP4M2[15]). Alternatively, P450 genes that are already known from other work are isolated again by the PCR method and the sequence of these fragments is compared to the known sequence (e.g., CYP4D1,[17] CYP4D2,[17] CYP4E2[17,18]). Such known genes can clearly serve as positive controls for the PCR method.

Isolating Other P450 Families

The strategy described in this chapter has also been used to isolate P450 fragments from other families. In some cases this was done with forward primers that are derived from the sequence surrounding the I helix conserved Thr residue. Two PCR products were obtained in the yeast *Candida apicola* with CYP52-derived primers for the I helix and the Cys peptide. One P450 of the CYP52 family was cloned and shown to contain the PCR fragment which was used as a probe.[19]

Other conserved regions have also been used in conjunction with the Cys peptide. A 110-bp PCR product obtained with inosine-rich (10/32-mer and 15/35-mer) oligonucleotide primers for the "PERF" and Cys peptide regions of CYP71A1 was used to clone CYP71B1 from a crucifer, the field pennycress.[20] In addition, primers designed for the sequences LPLLRAA (putative steroid-binding site) and RQCLGRR (Cys peptide) of the vertebrate CYP11B were reported to amplify a fungal cDNA of about 270 bp, but no details on the sequence of the PCR product were given.[21]

These examples show that the strategy of using two P450 regions in the design of oligonucleotide primers can often result in the selective amplification of P450 fragments that share a somewhat predictable level of sequence similarity.

---

[18] M. Amichot, M., A. Brun, A. Cuany, C. Helvig, J. P. Salaun, F. Durst, and J. B. Berge, *in* "Cytochrome P450" (M. C. Lechner, ed.), p. 689. John Libbey Eurotext, Paris, 1994.

[19] K. Lottermoser, O. Asperger, and W. H. Schunck, *in* "Cytochrome P450" (M. C. Lechner, ed.), p. 643. John Libbey Eurotext, Paris, 1994.

[20] M. K. Udvardi, J. D. Metzger, V. Krishnapillai, W. J. Peacock, and E. S. Dennis, *Plant Physiol.* **105**, 755 (1994).

[21] R. Komel, D. Rozman, M. Vitas, K. Drobic, and S. L. Kelly, *in* "Cytochrome P450" (M. C. Lechner, ed.), p. 805. John Libbey Eurotext, Paris, 1994.

Whereas the conserved Cys peptide is always easy to locate in P450 sequences, other regions of similarity can be more difficult to discern. P450 alignments can be obtained with several computer programs for sequence alignment. We have found the alignment prepared by Dr. David R. Nelson particularly useful in the design of degenerate primers. This alignment is accessible on the World Wide Web at the following location:

http://drnelson.utmem.edu/nelsonhomepage.html

Dr. Nelson's homepage provides access to all P450 sequences currently in the public domain, as well as lists of accession numbers and literature citations for all P450 sequences.

Conclusion

Versatile methods are now available to clone P450 fragments from virtually any organism. The 3'-RACE method based on the conserved cysteine region and the method described here both have advantages and drawbacks. The first probably yields a greater number of P450 fragments, but their identification may be problematic because of the limited amount of sequence information obtained. Also, it may be difficult to implement the first method to isolate P450 fragments from genomic DNA. The second is probably more selective in the types of P450 fragments obtained, but is perhaps more "user-friendly" for subsequent identification, or probe design for Northern hybridization or ribonuclease protection assays. It has success-fully generated P450 fragments from genomic DNA.

Acknowledgments

This work was supported by NIH Grants GM 39014 and ES06694 and by USDA Grants 89-37263-4960 and 92-37302-7789.

Section VI

Analysis of P450 Structure

[35] Structural Alignments of P450s and Extrapolations to the Unknown

By SANDRA E. GRAHAM-LORENCE and JULIAN A. PETERSON

Introduction

With the crystallization and structural resolution of four soluble, bacterial P450s—P450cam (Cyp101),[1] P450terp (Cyp108),[2] P450BM-P (Cyp102),[3] and P450eryF (Cyp107A)[4]—we are able to compare and contrast P450s to learn which regions of P450s are structurally conserved and which regions are variable. Additionally, from these structures, we are able to make "structural alignments" which, unlike "sequence alignments," are not related to sequence identity between the proteins, but rather to their structural similarity. At the time of this writing, the three X-ray crystal structures which are generally available, along with their structural alignment, have given us a basis for predicting the structure of other P450s, soluble and insoluble, and prokaryote as well as eukaryote.

By superimposing the three available P450 structures, we have determined that there are some highly conserved structural elements (i.e., helices, β-sheets, and loops or coils) which form what we have termed the "core structure" of P450s,[5] and there are regions that are more variable in length and position which are believed to be associated with substrate recognition and binding, membrane association, and redox partner binding. While there is some sequence similarity and identity in this core structure, it cannot be identified solely on the basis of sequence homology, i.e., primary structure, rather it requires some prediction of secondary and tertiary structures. Thus, one of the most important tasks in understanding the structural relationships among P450s or in building a molecular model is the construction of a proper structural alignment based on secondary structure as well as conserved sequences. While there are many available generic programs to align proteins based on sequence homology, they are generally not

---

[1] T. L. Poulos, B. C. Finzel, and A. J. Howard, *J. Mol. Biol.* **195,** 687 (1987).
[2] C. A. Hasemann, K. G. Ravichandran, J. A. Peterson, and J. Deisenhofer, *J. Mol. Biol.* **236,** 1169 (1994).
[3] K. G. Ravichandran, S. S. Boddupalli, C. A. Hasemann, J. A. Peterson, and J. Deisenhofer, *Science* **261,** 731 (1993).
[4] J. R. Cupp-Vickery and T. L. Poulos, *Struct. Biol.* **2,** 144 (1995).
[5] C. A. Hasemann, R. G. Kurumbail, S. S. Boddupalli, J. A. Peterson, and J. Deisenhofer, *Structure* **3,** 41 (1995).

refined enough to align P450s adequately enough to predict structural elements or their beginnings and ends. Several investigators have been able to construct good alignments of the P450 superfamily,[6-8] but one should further refine the alignment manually when attempting structure/function studies.

Alignments and the Core Structure

The structural alignments for P450cam, P450terp, and P450BM-P were obtained by overlaying the atoms of the heme of the three proteins.[9] Alternatively, these proteins may be aligned by superimposing the Cα backbone as was done by Hasemann et al.[5] and Ravichandran et al.[3] Overall, the differences are small between these two methods of superimposition. The alignments of Hasemann et al.[5] and Ravichandran et al.[3] were constructed to optimally overlay the overall protein structures to attain the best RMS deviation, whereas the ultimate goal here in aligning P450s is to optimally align the individual structural elements of the proteins, which we have found is best done by overlaying the heme groups, most probably because of the proximity of the heme to the conserved structural elements, e.g., the I helix, the heme-binding region, etc.

As can be seen for P450BM-P in Fig. 1, all three of these proteins have two domains: a predominately α-helical domain accounting for approximately 70% of the protein and a predominately β-sheet domain accounting for 22% of the protein. When the heme groups of the three crystal structures are overlaid, one can readily identify the conserved "core structure" which spans both domains. As is shown in Fig. 2, the most highly conserved structures are a four-helix bundle consisting of helices D, I, and L, and the anti-parallel helix E, helices J and K, β-sheets 1 and 2, the heme-binding region, and the "meander" just N-terminal of the heme-binding region. We believe that this "core structure" should be the basis for sequence alignments and model building of structures from the structurally based alignments. In addition, these conserved structures may play a role in folding and certainly in heme binding.

[6] A. I. Archakov and K. N. Degtyarenko, *Biochem. Mol. Biol. Int.* **31,** 1071 (1993).

[7] D. R. Nelson, T. Kamataki, D. J. Waxman, F. P. Guengerich, R. W. Estabrook, R. Feyereisen, F. J. Gonzalez, M. J. Coon, I. C. Gunsalus, O. Gotoh, K. Okuda, and D. W. Nebert, *DNA Cell Biol.* **12,** 1 (1993).

[8] O. Gotoh, *J. Biol. Chem.* **267,** 83 (1992).

[9] J. A. Peterson and S. E. Graham-Lorence, *in* "Cytochrome P450: Structure, Mechanism, and Biochemistry" (P. R. Ortiz de Montellano, ed.), 2nd Ed., p. 151. Plenum, New York, 1995.

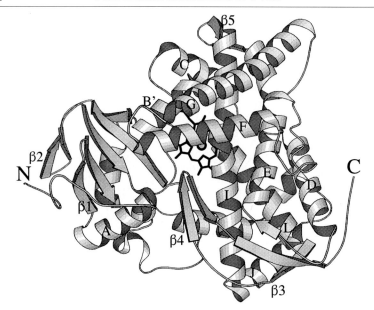

FIG. 1. A comparison of a ribbon representation of the three-dimensionally determined structure of P450BM-P with visible helices and sheets labeled appropriately, and with the N and C terminus labeled with a large N and C, respectively. The right half of the structure is the α-helical domain, and the left half is the β-sheet domain.

Generating Alignments for P450s

To obtain a structural alignment of ones favorite P450, one should initially computer generate a sequence alignment with P450cam, P450terp, or P450BM-P; however, in most cases, we believe that P450BM-P is the best model for eukaryotic P450s since on alignments of over 300 different P450s, P450BM-P clusters with the eukaryotic proteins rather than with the prokaryotic proteins. We routinely use the University of Wisconsin GCG "Pileup" program for sequence alignments[10] and include representative P450s from most of the P450 families, e.g., Cyp1, Cyp2, Cyp3, and Cyp4, in generating our initial alignments. This program clusters proteins using a pairwise alignment of the proteins with the highest sequence similarity, and then generates the final alignment using the Needleman–Wunsch algorithm.[11] Following the computer-generated sequence alignment, the P450 can be aligned with the structural alignment of P450cam, P450BM-P, and P450terp, and thereafter it can be refined manually using the absolutely

[10] "GCG Package." Genetics Computer Group, Madison, WI, 1991.
[11] S. B. Needleman and C. D. Wunsch, *J. Mol. Biol.* **48**, 443 (1970).

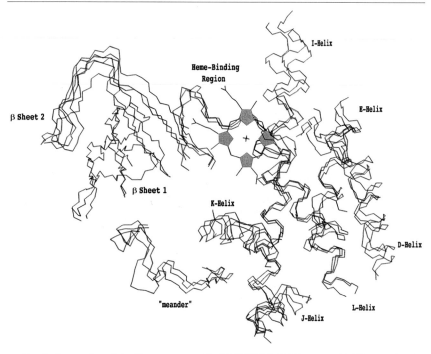

FIG. 2. An overlay of the "core structure" of P450BM-P, P450cam, and P450terp with the structural elements labeled.

conserved or highly conserved sequences, the amphipathicity of helices, or potential turns as guideposts; when all else fails, sequence similarity can be used to optimize the alignments.

An example of just such a sequence alignment can be found for five different eukaryotic P450s in Fig. 3 with highly conserved residues, as well as hydrophobic residues in amphipathic helices and certain potential turns in β-strands highlighted in bold letters. As can be seen in Fig. 3, there are three absolutely conserved residues identified in P450s: the conserved Cys in the heme-binding region which is the fifth ligand to the heme iron, and a Glu and an Arg in the K helix which are one helical turn apart and form a salt bridge. The highly conserved sequences include the consensus sequence in the center of the I helix, in the "meander," the heme-binding region, and a basic residue forming the heme propionate ligand and a Trp in the C helix. Potential turns are composed of residues that are frequently referred to as helix breakers, N-capping residues (residues of helices to balance the helix dipole at the N terminus), or folding blockers, such as Pro, Gly, Thr, Asp, Asn, and, less frequently, Glu, Gln, and Ser. We also believe that clusters of two or more positively charged amino acids (i.e., of Arg, Lys, or His) indicate potential turns or ends of helices (i.e., "C-capping" to balance the helix dipole). Also, above the prokaryotic P450s and, by comparison, above the eukaryotic P450s are letters representing a consensus for the sequences, which are explained in Table I.

The C-Terminal Alignment

Since the C-terminal half of the molecule starting at the I helix contains most of the structurally conserved core sequences (i.e., helices I, J, K, and L, part of β-sheet 1 and β-sheet 2, the meander, and the heme-binding region), it is easily aligned using the GCG pileup program with only slight adjustments. The hydrophobic I helix, which contains the consensus sequence (A/G)Gx(E/D)T in the center of the helix, lies directly over pyrrole ring B of the heme. The I helix frequently starts with Asp or Thr and ends with Pro where the J helix begins. The J helix is amphipathic and usually ends with a helix breaker. In P450BM-P, an additional helix called J' is present as compared to P450cam and P450terp. It appears to be present in most of the eukaryotic P450s and begins after a small random coil consisting of three or four residues and a potential turn. Further along is the K helix with the absolutely conserved sequence ExxR, where the E and R form a salt bridge and face into the "meander." The beginning of the K helix is not obvious, i.e., there is no N-capping. In our alignments, however, there was adequate sequence homology of P450arom (Cyp19) and P450c17 (Cyp17) with P450BM-P to align the K helices; on the other hand, there is N-capping of the K helix by residues Gly-Asp for P450scc (Cyp11a), and a Pro for P4502E1 (cyp2e1) and thromboxane synthase (Cyp5a1). The K helix ends with a helix breaker, usually Pro, that subsequently leads into the next structural elements—a cluster of β-strands beginning with β1-4.

At the beginning of the cluster of β-strands, there is usually three to five residues followed by a potential turn (e.g., D, G, P, S, or T). β1-4 contains five or six residues, and then there is a potential turn within two or three residues, usually an acidic residue, followed by β2-1. Following β2-1 are three or four residues which encode a hairpin turn, usually containing a Gly or Asp. There are three to five residues of β2-2 followed frequently by (P/E/K)(K/R)(G/N). Last is β1-4, followed by a potential turn leading into a five- or six-residue helix: the K' helix. This helix was not identified by Poulos in his structure of P450cam,[1] but has been defined subsequently as a helix when using recent structure-defining programs.[12]

---

[12] "Ribbons 2.0 Manual" (computer program), Carson M. University of Alabama at Birmingham, Center for Macromolecular Crystallography, 1991.

```
           .....leader.........membrane spanning region.........Omura hinge.................
scc   MLAKGPPRSVLVKGCQTFLSAPREGLGRLRVPTGEGAGISTR.......SPRPFNEIPSPGDNGWLNLYHPWRET....... 68
5A1   MM.EALGFLKLEVN......GPMVTVALSVALLALLKWY....STSAFSRLKLGLRHPKPSFIGNLTFFRQ.......... 60
2E1   MAV................LGITVALLGWMVILLFISVWKQIH.....SSWNLPPGPFPLPIIGNLLQLDLK.......... 51
17α   M.WE...............LVALLLLTLAYLFWPKRRCPGAKYPRSLPSLPLV.GSLPFLPRHGHMHNN........ 52
arom  MVLEMLNPIHYNITSIVPEAMPAATMPVLLLTGLFLLVWNYE......GTSSIPGPGYCMGIGP.LISHGRFLWM.... 68

           .........................................................................
BMP   ...............................................TIKEMPQPKTFGELKNLPLLTDKP..... 24
cam   ...........................................TTETIQSNANLAPLPPHVPEHLVFDFDMYNPSNLSA 36
terp  ...........................................DARATIPEHIARTVILPQGYA... 21

                                                   A' HELIX


           (.............hpp.aG).t.....h...tGttc.phhh.p..(...phhp.h...).....pattR............
scc   (.GTHKVHLHHVQNFQK.YG).PIYREKL....GNVESVYVI.DP(..EDVALLFK..)..SEGPNPERFLIP.......... 123
5A1   (.....GFWESQMELRKLYG).PLCGYYL....GRRMFIVIS.EP(...DMIKQVLV.).ENFSNFTNRMAS.......... 112
2E1   (.....DIPKSFGRLAERFG).PVFTVYL....GSRRVVVLH...(.GYKAVREMLL.).NHKNEFSGRGEI.......... 103
17α   (..FFKLQKKYGPIYSVRMG).TKTTVIV....GH.HQLA.K...(..EVLIKK.....).GKDFSGRPQMATL....... 102
arom  (.....GIGSACNYYNRVYG).EFMRVWI...SGEETLIISK.SS(...SMFHIMKH..)....NHYSSRFGSKL....... 120

           (.t...th..hhphhpc. t).t.........tttcphhh.p...(..phhp.h...).......tcap...........
BMP   (......VQALMKIADEL G).EIFKFEA...PGRVTRYLS...(.SQRLIKEAC..).......DESRFDKNL....... 71
cam   (.....GVQEAWAVLQES)NVP.DLVW....TRCNGGH.WIAT...(.RGQLIREAYED)......YRHFSS.ECPFI... 88
terp  (DDEVIPYP.AFKWLRDE Q).PLAMA..HIEGYDPMWIAT ...(KHADVMQIGKQ.)........PGLFSNAEGSEILY... 80

           A Helix        ß1-1         ß1-2          B Helix            β1-5


           (.............).....ttthhh.tt..(tp.WpppRphh.....).........(......c.hppphp.h.pp.hcc...)
scc   (WVAYHQYYQR...)...PIGVLLKK...(SAAWKKDRVALNQEV.).MAPEATK.(NFLPLLDAVSRDFVSVLHRRIKKAG.) 188
5A1   (.GLEFKSVA....)......DSVLFLR...(DKRWEEVRGALMSAF.)....SPEK..(.LNEMVPLISQACDLLLAHLKRYAE) 170
2E1   (..PAFREFK....)......DKGIIFNNG..(.PTWKDTRRFSLTTLR)..DYGMGK..(.QGNEDRIQKEAHFLLEE.LRK...) 160
17α   (.DIASNNR.....)....KGIAFADS..(GAHWQLHRRLAMA...).TFALFKD.(GDQKLEKIICQEISTLCDMLATH...) 160
arom  (GLQCIGMH.....).......EKGIIFNNN..(PELWKTTRPFF.....).MKALSG.(.PGLVRMVTVCAESLKTH.LDR...) 174

           (.............).......hhphhtt.(ttp.p.hcph.....).........(.....hc..h.chh...hpphc....)
BMP   (SQALKPVRDFAG.)......DGLFTSWTH.(EKNWKKAHNILLPSF.)..SQQAMK.(....GYHAMMVDIAVQLVQKWER...) 132
cam   (.PREAGEAY....)......DFIPTSMDH.(.PEQRQFRALANQVVG)..MPVVDK.(....LENRIQELACSLIESLR....) 145
terp  (DQNNEAFMRSIS.)GGCPHVIDSLTSMDP.(.PTHTAYRGLTLNWF.).QPASIRK.(.....LEENIRRIAQASVQRLL....) 144

           B' Helix                  C Helix         C` Helix        D Helix


           .............(....ph...h.phhp.hh......)..............(.......h..h....h..h....)...
scc   SGNYSGDI.......( .SDDLFRFAFESITNVIFGERQ.) GMLEEVVN....(.....PEAQRFIDAIYQMFH .).... 240
5A1   ..SGDA........( .FDIQRCYCNYTTDVVASV...)..PFGTPVDSWQA.(...PEDPFVKHCKRFFEFCI.)..... 220
2E1   .TQGQPFD.......(.PTFVIGCTPFNVIAKILF.....).NDRFDYKDKQ...(...ALRLMSLFNENFYL.LST.).... 212
17α   NGQSIDISF......(...PVFVAVTNVIS.LICF...).NTSYKNGD....(....PELNVIQNYNEGIIDNLSK.).... 211
arom  ......LEEV....( ...TNESGYVDVLT.LLRRVML ) .DTSNTLFLRI...(.PLDESAIVVKIQGYFDAWQAL).... 227

           .............(....pcc...hthc..hh..hh..)..............(.....h..h...h..h...)...
BMP   .LNADEHIEV....( ...PEDMTR.LTLDT.IGLCGFN.) .YRFNSFYRDQP..(.HPFITSMVRALDEAMNKLQR.).... 190
cam   PQGQCN.........(..FTEDYAEPFPIRI.FMLLAGL.).....PEED.....(.....IPHLKYLTDQMTR....).... 188
terp  DFDGECD........(..FMTDCALYYPLHV.VMTALG..).YAESVPEDD....(.....EPLMLKLTQDFF .).... 191

           ß3-1                     E Helix                          F Helix


           .............(.............hc..p-h.p..hh..+p+).............................
scc   ..TSVPMLNLPPDLFRLFR..(TKTWKDHVAAWDVIFSKADIYTQNFYWELRQK).........GSVHHD.......... 295
5A1   ..PRPILVLLLSFPSIMV..(.....PLARILPNKNRDELNG.FFNKLIRHK).NVIALRDQQAAEERRD......... 277
2E1   ..PWLQVYNNFSNYLQYM..(.....PGSHRKV IKNVSEIKE.YTLARVKEH).........SLDPSCPRD.......... 262
17α   .....DSLVDLVPWLKIF...(.PNKTLEKLKSH.VKIRNDLLNK.ILENYKEK).........FRSDSIT.......... 259
arom  .....LIKPDIFFKISW....(......LYKKYEKSVKDLKDAIE.VLIAEKRCR).....ISTEEKLEECMD.......... 277

           .............(.............ch....hp-h....hh..+pp).............................
BMP   ..ANPDDP..........(.AYDENKRQFQEDIKVMNDLVDK.IIADRKAS)........GEQSDD............. 232
cam   .....PDGSM..........(...................TFAEAKEALYDYLIP.IIEQRRQK)........PGT............ 219
terp  .GVHEPDEQAVAAPRQSADE.(......AARRHETIATFYDYFNG.FTVDRRS).........CPKDD........... 239

           F-G Loop                 G Helix
```

FIG. 3. The structural alignments of P450BM-P, P450cam, and P450terp along with the alignments of P450scc (Cyp11a), thromboxane synthatase (Cyp5A1), P4502E1 (Cyp2E1), steroid 17α-hydroxylase/lyase (Cyp17A), and P450arom (Cyp19). Helices and β-strands in the structurally determined P450s are underlined as are the residues believed to be in the "core structure" of all the proteins in this alignment. The helices and strands are also labeled

```
          (.hh..hh.......).....................(ttcph..phh.hhhAt.-Tpp..h.a.hhhhh+ )........
scc       (D.YRGILYRLL...).........GDSKM.....(SFEDIKANVTEMLAGGVDTTSMTLQWHLYEMAR )........ 343
5A1       (DFLQMVLDARHSAS).......PMGVQDFDL....(TVDEIVGQAFIFLIAGYEIITNTLSFATYLLAT )........ 333
2E1       (DFIDSLLIEM....).........EKDKHSTEPLY.(TLENIAVTVADMFFAGTETTSTTLRYGLLILLK )........ 316
17a       (NMLDTLMQAKM...).NSDNGNAGPDQDSELLS..( DNHILTTIGDIFGAGVETTTSVVKWTLAFLLH )........ 319
arom      (DFATELILA....).........EKRGDL......( TRENVQCILEMLIAAPDTMSVSLFFMLFLIAK )........ 324
          (.hh........).....................(ttcphp.p.h.hhhAt.-Tpp....a.h.hh.+ )........
BMP       (.LLTHML.......).....NGKDPETGEPL....(DDENIRYQIITFLIAGHETTSGLLSFALYFLVK )........ 281
cam       (DAISIVAN......).........GQVNGRPI...(TSDEAKRMCGLLLVGGLDTVVNFLSFSMEFLAKS)........ 269
terp      (.VMSLLA.......).....NSKLDGNYID.....(.DKYINAYYVAIATAGHDTTSSSSGGAIIGLSR )........ 287


          H Helix                          I Helix



          (ttphpc.h.cEh....).ttttt.(..hp...t.)(.h.hhc..hpE.hRh.).tt....ht+.h..D..h.........
scc       (NLKVQDMLRAEVLAAR).HQAQG.(DMATMLQLV)(..PLLKASIKETLRLH)..PISVTLQRYLVNDLVLR........ 404
5A1       (NPDCQEKLLREVDVFK).EKHMA.(PEFCSLEE.)(GLPYLDMVIAETLRMY)..PPAFRFTREAAQDCEVL........ 395
2E1       (HPEIEEKLHEEIDRVI).GPSRM.(PSVRDRVQ.)(.MPYMDAVVHEIQRFI).DLVFSNLPHEATRDTTFQ........ 378
17α       (NPQVKKKLYEEIDQNV).GFSRT (PTISDRNR.)(.LLLLEATIREVLR..)(.PVAPMLIPHKANVDSSIGE....... 380
arom      (HPNVEEAIIKEIQTVI).GERD..(IKIDDIQK.)(.LKVMENFIYESMRYQ).PVVDLVM.RKALEDDVIDGY....... 386
          (ttc...+hhc......)..ttttt.(..hp...t.)(....ht.h.pEhhR.h).tth.......h..D..h.........
BMP       (NPHVLQKAAEEAARVL).VDPVP (.SYKQVKQ.)(.LKYVGMVLNEALRLW)..PTAPAFSLYAKEDTVL........ 341
cam       (.PEHRQELIE.....)..RPE..(.........)(...RIPAACEELLRRF).SLV.ADGRILTSDYEFH........ 308
terp      (NPEQLA.LAKS.....)..DPAL..(.........)(....IPRLVDEAVRWT)..APVKSFMRTALADTEVR........ 330


          J Helix          J' Helix          K Helix          ß1-4   ß2-1



          .....htttt.h.h..(......).........P..F.PE+a..............tFG.G.R.Ch.(G..hs.hch.h.h..h....)
scc       ..DYMIPAKTLVQVA..(IYAL..)GREPTFFFDPENFDPTRWLSKD...KNITYFRNLGFGWGVRQCL (GRRIAELEMTIFLINMLE..).480
5A1       ..GQRIPAGAVLEMAV.(GALHHD)...PEHWPSPETFNPERFTAEARQQHR..PPFTYLPFGAGPRSCL (GVRLGLLEVKLTLLHVLH..).472
2E1       ..GYVIPKGTVVIP.(TLDSLLYD)...KQEFPDPEKFKPEHFLNEEGKFKYSDYFK..PFSAGKRVCV (GEGLARMELFLLLSAILQ..).455
17α       ....FAVDKGTEVIIN..(LWALH.).HNEKEWHQPDQFMPERFLNPAGTQLISPSVSYLPFGSGPRSCI (GEITLAR.ELFLIMAWLLQ..).457
arom      ....PVKKGTNIILN..(IGRMH..).RLEFFPKPNEFTLENPAKNV......PYRYFQPFGSGPRACG (GKYIAMVMMKAILVTLLR..).456
          ....htttt.hhh.........P..F..R..............tFG.G.+hCh..GpphA..Eh.h.h..h....
BMP       GGEYPLEKGDELMVL..(IPQLH.)RDKTIWGDDVEEFRPERFEN .PSAIPQHAFKPFGNGQRACI (GQQFALHEATLVLGMMLK..).419
cam       ..GVQLKKGDQILLP..(QMLSGL)..DERENACPMHVDFSRQKVSH..........TTFGHGSHLCL (GQHLARREIIVTLKEWLTRI).378
terp      ..GQNIKRGDRIMLS..(YPSAN.).RDEEVFSNPDEFDITRFPNRH..........LGFGWGAHMCL (GQHLAKLEMKIFFEEL....).396


          ß2-2   ß1-3   K' Helix         meander          | Heme-Binding|   L Helix



          cF........ht.........................................
scc       NFRVEIQHLSDVGTTFNLILMPEKPISFTFWPFNQEATQQ.............. 520
5A1       KFRFQACPETQVPLQLESKSALGPKNGVYIKIVSR.................. 507
2E1       HFLKPLVDPEDIDL.RNITVGFGRVPPPRYKLCVI PRS.............. 491
17α       RFDLEVPDDGQLPSLEGIPKVVFLIDSFKVKIKVRQAWREAQAEST........ 503
arom      RPHVKTLQGQCVESIQKIHDLSLHPDETKNMLEMIFTPRNSDRCLEH........ 503
          .........................................................
BMP       ..HFDFEDHTNYELDIKETLTLKPEGFVVKAKSKKIPLGGIPSPSTEQSAKKVR. 471
cam       ...PDFSIAPGAQIQHKSGIVSGVQALPLVWD...PATTKAV............. 414
terp      LPKLKSVELSGP..PRLVATNFVGGPKNVPIRFTKA.................. 429


          ß3-3       ß4-1     ß4-2   ß3-2
```

as defined by Hasemann et al.[5] Residues shown in parentheses are believed to be in helices and those in bold are defined in the text. A consensus sequence is shown above the structurally determined P450s and above the eukaryotic sequences with the lowercase letters defined in Table I and the uppercase letters representing highly conserved or absolutely conserved residues. Residues with double underlining are those residues that are in the active site of the crystallized proteins.

The "meander," which is just C-terminal of the K' helix, was so named because at first glance this region appears to make a random walk from the K' helix to the heme-binding region, but, in fact, it does not. In the overlay of the three crystallized structures, the conformation and the position in three-dimensional space are sufficiently similar to consider this

TABLE I
LETTER CODE FOR CONSENSUS SEQUENCE[a]

Letter designation	Characterization	Residues included
h	Hydrophobic residues	I, L, M, V, F, W, Y, A, G
a	Aromatic residues	F, Y, W
t	Potential turns	P, G, T, D, N
	Less common turns	E, Q, S, and 2 positive Charges, e.g., KK, HK, RK ...
c	Charged residues	R, H, K, D, E
p	Polar residues	R, H, K, D, E, N, Q, T, S, C
+	Basic residues	R, H, K
−	Acidic residues	D, E
s	Small residues	A, G

[a] Please note that Gly and Ala are not only typed as small, but also are considered here as null residues in that they may be present in any set without a penalty against a consensus.

region conserved and a part of the core structure.[5] When comparing P450s in this region, the sequence identity seems to vary somewhat depending on which class of P450 it is, i.e., which redox partner it requires, if any, and whether it is eukaryotic or prokaryotic. In microsomal P450s, there is a consensus of xPcxFxPE+a, where "x" represents any amino acid, "a" for aromatic, "c" for charged residues, and "+" for positively charged residues.

The heme-binding region (or the cysteinyl loop) contains the absolutely conserved cysteine which is the fifth coordinating ligand of the heme iron and is responsible for the characteristic absorption spectrum of P450s, namely an absorption band with a 450-nm maximum upon binding of CO. Along with the cysteine (C) in the heme-binding region is the highly conserved consensus sequence F(G/S)xGx(H/R)xCxGxx(I/L/F)A. The last few residues of the heme-binding region consensus sequence form the beginning of the L helix starting with Gxx(I/L/F)A and ending with a helix breaker (e.g., Q, N, RR). The L helix is amphipathic and easily aligned using the defined letter and with conserved amino acids in capital letters, e.g.,

```
SCC        GRRIAELEMTIFLINMLE
17α        GEILAR.ELFLIMAWLLQ
arom       GKYIAMVMMKAILVTLLR
BM-P       GQQFALHEATLVLGMMLK
           012345678901234567
consensus  Gp.hA.hEh.hhh..h..
```

Generally, in the L helix of *class I* mitochondrial P450s, there are two positive charges at residues 2 and 3 in the example just given; whereas in *class II* microsomal P450s, the second residue is negative. Since the N-terminal portion of the L helix faces into the redox partner-binding site, these changes may well be related to recognition of the redox partner. In Cyp5A1, which is a *class III* P450 and has no redox-binding partner, it is not surprising that the first residue is hydrophobic.

Finally, at the C terminus, there is another cluster of β-strands which comprise β-sheets 3 and 4; however, this region should be thought of more as a loop when comparing the three known structures and aligning them with other P450s. That is, even though in all three proteins the overall β-cluster traverses essentially the same route from the end of the L helix, over the end of the I helix, up to the active site, and back again, the β-strand elements are not conserved in either the secondary or the tertiary structure, and may require a large amount of gapping in structural alignments. Thus, from our perspective, the most important feature to define here is what residues are inserted into the active site rather than which residues form a β-strand. It appears that the region 16 to 19 residues after the helix breaker at the end of the L helix is that which is inserted into the active site.

The N-Terminal Alignment

Unfortunately, the N-terminal portion of P450s has fewer conserved residues and therefore is more difficult to align. It is known from the crystal structures of P450cam, P450terp, and P450BM-P that the N-terminal region encodes a large portion of the sequences involved in substrate recognition and binding, and redox partner binding. Therefore, the alignment of ones favorite P450 with these sequences should be done thoughtfully, paying attention to conserved residues, if any, to potential turns and helix capping, and here, as compared to the C-terminal half, one also has to utilize the amphipathicity of the helices.

To aid in the alignment (e.g., in Fig. 3), the highly conserved residues in helix C and G are in bold letters, as are the hydrophobic residues in the amphipathic helices A through H (except helix G), and the potential turns in β-strands. The only other helix in the N terminus is B′ which is nondescript. Above the aligned structurally determined P450s and by comparison above the aligned eukaryotic P450s is a code to indicate the similarity of the residues if any, i.e., whether hydrophobic, charged, polar, or aromatic residues, or encoding potential turns as discussed previously.

To align ones chosen P450, it is best to start with the B′-C loop through the D helix—more specifically the C helix—then align helices E through H, and finally the N terminus through the B′ helix. The C helix contains a conserved Trp, and four residues downstream there is a conserved Arg which is charge-paired with the heme propionate group. The B′-C loop, which begins three or four residues N-terminal of the conserved Trp, has a characteristic signature with residues encoding turns, hydrophobic residues, then turns again. Going toward the C terminus from the C helix, the C-D loop (or in some cases the C′ helix) is short, ending with potential turns and leading into the amphipathic D helix, which usually is "C-capped" with basic residues.

In the region from helix E through H, it is best to align helix G first. (Note that the D-E and H-I loop are determined by default after aligning the helices.) When looking N-terminal from the start of the I helix and for the moment skipping the H helix, the amphipathic helix G is generally "C capped" with two or more basic residues shown in bold letters (e.g., RR, RK RHK) or with one intervening polar or charged residue (e.g., RQK, KEH, etc.). Ten or 11 residues upstream from these residues, and also shown in bold letters, is an acidic residue—either Asp or Glu. Looking N-terminal of that, the G helix is variable in length from P450 to P450, beginning with a strong series of potential turns. The F-G loop begins and ends with potential-turn residues and contains at least one set of turns within the loop (e.g., a Pro or AsnAsn). The F helix is amphipathic (with hydrophobic residues highlighted in bold) and generally begins with a Pro or Asp. Of this group the E helix is the hardest to define. It generally appears to be more hydrophobic at the C-terminal end and more polar at the N-terminal end. It is amphipathic and generally begins with a residue involved in turns. Finally, the H helix is a short amphipathic helix which begins with N-capping or one of the potential turn residues.

The last group to align is the span from the N terminus through the B′ helix. Of course at the N terminus of membrane-bound eukaryotic P450s is the leader region and the membrane-spanning region which is beyond the scope of this chapter and is arbitrarily aligned in Fig. 3. We also have attempted to align the hinge region described by Omura,[13] and it is at this point that we start our alignment strategy. The hinge region, which permits flexibility and close association of the protein with the membrane, is composed generally of noncharged residues which may induce or be involved in potential turns (e.g., the "t" group), especially Pro, Gly, Ser, and Asn. Between the Omura hinge region and the B′ helix, the easiest region to

[13] S. Yamazaki, K. Sato, K. Suhara, M. Sakaguchi, K. Mihara, and T. Omura, *J. Biochem.* (*Tokyo*) **114,** 652 (1993).

initially align is β-strands 1-1 and 1-2. The turn between these two strands generally contains a Gly followed by charged or polar residues, e.g., GEE, GSR, etc. This turn leads into the β1-2 strand which is polar at either end and hydrophobic in the middle. This region spans the width of the access channel mouth and is hydrophobic to accommodate the hydrophobic substrate. Strand β1-1 begins with potential-turn residues (e.g., GP or GT) and is more polar than β1-2. Helix A just N-terminal of β1-1 is amphipathic and appears to be variable in length from one P450 to another beginning with potential-turn residues and ending with a Gly. The B helix is a short amphipathic helix that begins and ends with potential turns. Following that is sheet β1-5 which has been defined here by first determining which regions could encode a potential turn. These regions are highlighted in bold letters, including the turn residue which may be present at the N terminus of the B' helix. The residues in this span which potentially form the β-strand are underlined and are bracketed by the turn residues. If there is an additional span of more than two residues which does not contain a potential turn in the B'-C loop such as is found in the P450-17α sequence, this region and the residues contained therein may protrude into the active site as they do in P450cam and P450terp. Finally, the B' helix begins at a potential turn and ends at the beginning of the B'-C loop which was defined earlier.

Summary

To obtain a sequence alignment that allows model building and identification of potential active site residues or redox partner-binding residues, one should first generate a computer alignment of ones favorite P450 with one of the structurally defined P450s, usually with P450BM-P as it is the most eukaryotic-like P450 of the three available structures. The C-terminal half of the generated alignment will align fairly well due to the large number of sequence similarities and conserved regions in this half of the molecule, i.e., the I helix consensus, the J and K helix, the meander, the heme-binding region, and the L helix. One should then justify the C-terminal half of the P450 with the potential turns found in β-sheets and between other structural elements of P450BM-P. After optimizing the alignment at the C-terminal half of the molecule, attention should be turned toward the more variable N terminus, starting the alignment with conserved residues in the C helix, in the G helix, and in β-strands 1-1 and 1-2. Using these sequences to initiate the N-terminal alignments, one should further refine the remaining half by identifying potential turns at the beginnings and ends of structural elements, and then by identifying hydrophobic residues in the amphipathic helices to determine the orientation of the helix and its length. However, no matter how many times the same sequence is aligned, one should never feel that ones work is done until a crystal structure has been determined for the protein!

Acknowledgment

This work was supported in part by Research Grants GM43479 and GM50858 from the NIH.

[36] Predicting the Rates and Regioselectivity of Reactions Mediated by the P450 Superfamily

By JEFFREY P. JONES and KENNETH R. KORZEKWA

Introduction

If the rate of a given CYP-mediated oxidation of a substrate can be predicted, a number of properties could be anticipated, including the half-life of drugs, the toxicity of a xenobiotic, and the amounts of different metabolites that would come from a given substrate. Studies in our laboratories are aimed at developing computational tools to predict the rates of CYP-mediated oxidations. To this end we have exploited two different computational methodologies, molecular dynamics (MD) and quantum mechanics (QM), to predict steric or electrostatic effects, respectively. In our studies to date, we have been able to predict the rates and regioselectivity when either steric factors and electronic factors are the predominant influence on the outcome of the reaction by using the appropriate method. It is our hope that these two methodologies can be merged to predict the rates and regioselectivity of CYP-mediated oxidations in general. While others have used similar methodology, we will focus on the protocols established in our laboratories.

While both QM and MD computational methods have been used in the study of other enzymic systems, the cytochrome P450 enzyme family is relatively unique in a number of ways. (1) At least for xenobiotic metabolism, P450 substrates have been shown to bind in a number of different orientations and with relatively low affinity. Thus, a given substrate can give a large number of different products. While most enzymes are stereospecific, xenobiotic metabolism by the cytochrome P450 superfamily is usually stereoselective, resulting in varying degrees of enantiomeric excess. (2) The substrate oxidation step is not catalyzed in the classic sense by the enzyme.

Instead, the enzyme catalyzes the activation of molecular oxygen,[1,2] and this active oxygen species acts like a chemical in a nonpolar solution. The transition state for the oxidation of substrate does not appear to be stabilized by the enzyme to significant extent.[3,4] (3) The substrate oxidations occur at very slow rates (<100 min^{-1}) for all but a few enzymes in the superfamily.

These three characteristics of the P450 superfamily have an effect on the design and interpretation of computational methods for these enzymes. The first characteristic, the lack of strong binding forces, makes the use of QM methods to predict rates possible since the product formed is in some cases dependent entirely on the reactivity at a given position and is not influenced by the enzyme. However, this same characteristic can make MD methods difficult to use since no specific binding orientation can be assumed and very long simulations may be required to approximate the free motion of substrate in the active site. The second characteristic, the chemical-like nature of the oxidation step, again favors QM methods, but will also make MD methods easier to apply in the exploration of enzyme interactions with transition states and intermediates. The third characteristic, the slow rates of product formation, hinders the use of both QM and MD methods. The slow reaction rate is due to rate-limiting steps prior to substrate oxidation. These steps act to mask the electronic effects on the rates of reaction. This characteristic also means that MD will have to be extremely long to simulate the complete space available for substrate motion prior to oxidation. Thus, the choice of substrate, based on the compounds reactivity and freedom motion in the active site, can have a profound effect on the success of a QM or MD simulation. Our success to date, in part, reflects our judicious choice of substrates.

Structural Models of Mammalian P450s Based on P450cam and P450bm3 Crystal Structures

In an attempt to predict the effects of structure on the rates of CYP-mediated reactions, we have used two of the four bacterial CYP enzymes for which crystal structures have been published.[5,6] From our studies of benzo[a]pyrene and [R]- and [S]-nicotine, it would appear that certain

[1] K. R. Korzekwa and J. P. Jones, *Pharmacogenetics* **3**, 1 (1993).
[2] J. Aikens and S. G. Sligar, *J. Am. Chem. Soc.* **116**, 1143 (1994).
[3] J. P. Jones, A. E. Rettie, and W. F. Trager, *J. Med. Chem.* **33**, 1242 (1990).
[4] S. B. Karki, J. P. Dinnocenzo, J. P. Jones, and K. R. Korzekwa, *J. Am. Chem. Soc.* **117**, 3657 (1995).
[5] T. L. Poulos, B. C. Finzel, and A. J. Howard, *J. Mol. Biol.* **195**, 687 (1987).
[6] K. G. Ravichandran, S. S. Boddupalli, C. A. Hasemann, and J. A. Peterson, *Science* **261**, 731 (1993).

structural features are conserved between the mammalian and the bacterial systems. If this is true, it is likely to be the regions of the protein responsible for heme binding and orientation, as suggested by Poulos.[7] However, the low primary sequence homology and the fact that a single amino acid change can alter substrate selectivity[8] means that predicting structural effects for mammalian CYP enzymes will require crystal structures for the mammalian enzymes. In the absence of crystal structures for mammalian CYP enzymes, we are forced to use the bacterial enzymes as surrogates for the mammalian enzymes and, through a process we call functional homology, make deductions about what factors may influence binding. These studies serve at least two purposes: (1) they produce hypotheses about the active site that can be tested through experiment and (2) they help us sharpen the tools that we will use when crystal structures become available for mammalian enzymes.

The following strategy is used to prepare for a MD run. (1) The compound of interest is tested to determine if it is a substrate for CYP101 or CYP102. (2) Point charges are determined for, and a force field assigned to, the substrate molecule. (3) The enzyme is prepared for dynamics. (4) The molecule is docked in the active site. (5) Dynamics runs are made to equilibrate the molecule in the active site. (6) Production dynamics runs are made. (7) Statistics are collected and analyzed. (8) The results are compared with experiment.

To date, we have studied three substrates, benzo[a]pyrene,[9] [S]-nicotine, and [R]-nicotine,[10] in detail. We have also performed preliminary studies with norcamphor, a series of dimethylanilines,[4] and a series of toluenes. In all cases a large amount of CYP101 or CYP102 was required to see substantial turnover. Normally we must use between 1 and 10 nmol of enzyme to see enough product for accurate quantitation. To ensure that the turnover occurs in the active site, we perform inhibition experiments with camphor or palmitic acid for CYP101 or CYP102, respectively. Since molecular dynamic calculations are usually run with a single substrate molecule, the experiments should be run to determine V/K values or to determine some intramolecular ratio since these values reflect turnover at low substrate concentrations. For example, for the nicotine enantiomers we made pseudoracemic mixtures [the (R) enantiomer was labeled with 1 dueterium in the $2'$ position] so that the levels of metabolism could be compared using an intermolecular competitive type of experiment. For benzo[a]pyrene we looked at the intramolecular competition between oxidation on two sepa-

[7] T. L. Poulos, *Methods Enzymol.* **206,** 11 (1991).
[8] R. L. P. Lindberg and M. Negishi, *Nature (London)* **339,** 632 (1989).
[9] J. P. Jones, M. Shou, and K. R. Korzekwa, *Biochemistry* **34,** 6956 (1995).
[10] J. P. Jones, W. F. Trager, and T. J. Carlson, *J. Am. Chem. Soc.* **115,** 381 (1993).

rate faces of the molecule. These intermolecular and intramolecular competition types of experiments are usually required since the low level of turnover with the bacterial enzymes makes the use of high substrate concentrations desirable, but the theoretical construct predicts behavior at low substrate concentrations.

The second step, assignment of point charges to the substrate, remains problematic in that no consensus has been established on how to make these assignments. The standard method for determining point charges is to do a quantum chemical calculation and then perform a Mulliken population analysis on the resulting wave function. However, Mulliken population analysis has a number of problems that have been described in the literature.[11] In our laboratory we normally use the semiempirical AM1 Hamiltonian[12] to optimize our structure since this method gives very good geometries for a number of different functional groups. We then do a single point calculation (a calculation with no geometry optimization) using the semiempirical MNDO Hamiltonian.[13] The MNDO Hamiltonian has been shown to give a good wave function for charge determination.[14] The point charges are assigned to this wave function using the ESP program of Besler and Merz.[14] This program probes the wave function with a charge and assigns point charges based on a least-squares fit to the resulting potential. All of these methods are available in either the Gaussian 94 or Mopac 93 quantum chemical packages.

Preparation of the protein involves downloading the protein structure from the Protein Data Bank at Brookhaven (a PDB file) and loading charges and force constants for the various amino acids. For the CYP superfamily of enzymes, an added complication is that the parameters for the iron–protophophyrin IX–sulfur complex are not available in the standard databases and needs to be constructed. Our method of construction is outlined in Jones *et al.*[10] We tested our parameters using a thermodynamic cycle for [R]- and [S]-nicotine binding and found that we could quantitatively reproduce the experiment.[10] While our results are encouraging, parameters for the heme-thiol group can likely be improved. After the input parameters are constructed, the protein must be minimized. The starting crystal structure will be at a very high energy and should be relaxed before docking a molecule in the active site. For CYP101 we used the 3CPP structure from the PDB at Brookhaven. To relax the protein the following

[11] A. E. Reed, J. L. Curtiss, and F. Weinhold, *Chem. Rev.* **88**, 899 (1988).

[12] M. J. S. Dewar, E. G. Zoebisch, E. F. Healy, and J. T. Stewart, *J. Am. Chem. Soc.* **107**, 3902 (1985).

[13] M. J. S. Dewar and W. Thiel, *J. Am. Chem. Soc.* **99**, 4899 (1977).

[14] B. H. Besler, K. M. Merz, and P. A. Kollman, *J. Comput. Chem.* **11**, 431 (1990).

protocol was used. (1) The waters of crystalization were minimized using the steepest descent method in the MINMD module of AMBER[15] without any electrostatic energy terms and holding the entire protein fixed. (2) The electrostatic terms were included for the water and minimized. (3) All hydrogen atoms were minimized by steepest descent. (4) The side chains were minimized holding the C_α backbone fixed. (5) All atoms were minimized by steepest descent. (6) All atoms were minimized by the conjugate gradient method. After complete minimization the RMS deviation from the crystal structure was 0.57 Å.

Docking the substrate in the enzyme active site of the bacterial enzymes is an art. At least two problems are encountered when docking a substrate: (1) the small active site of P450cam makes docking difficult and (2) the starting orientation can have a profound effect on the results. In docking our substrates, every effort is made to keep the protein as close to the crystal structure as possible. Thus, we fix the protein when we dock the substrate, minimize the substrate, and perform dynamics with the protein fixed. Only after the substrate has low energy are the side chains of the protein allowed to move. Finally, the whole protein is minimized. After docking, the entire enzyme–substrate complex should maintain a significant negative energy. The effect of starting orientation on the subsequent dynamics runs was apparent in both our nicotine and B[a]P studies. The nicotine enantiomers were docked in a number of orientations in CYP101. (Note: Nicotine was not metabolized by CYP102 so no studies were done with this enzyme.) In each case, the pyridine ring nitrogen established a hydrogen bond with TYR96 during equilibration. However, the pyrolidine ring would adapt one of two different conformations depending on the starting structure. These two conformations exchanged very slowly at 310 K. Thus, the starting conformation influenced the sampling statistics. While the slow interchange is consistent with our experimental results,[16] it confounds analysis of the data. An even larger dependence on the starting orientation was observed for B[a]P. In this case, no interchange was observed among four possible starting orientations. Again, these results are consistent with experimental results, but it means that predictions about the amount of each oxidation product are difficult to make.

After docking, the enzyme–substrate complex must be equilibrated for the dynamics runs. We use the SANDER module in AMBER for our dynamics runs. Initially, velocities are randomly assigned and the system

[15] D. A. Pearlman, D. A. Case, J. C. Caldwell, G. L. Seibel, U. C. Singh, P. Weiner, and P. A. Kollman, "AMBER 4.0." University of California, San Francisco, 1991.
[16] T. J. Carlson, J. P. Jones, L. Peterson, N. Castagnoli, K. R. Iyer, and W. F. Trager, *Drug. Metab. Dispos.* **23**, 749 (1995).

is slowly heated to 310 K over 10 psec. The temperature is maintained by loosely coupling the system to a 310 K constant temperature bath. Dynamics runs are made at 310 K until the system maintains a relatively constant temperature, potential energy, and kinetic energy. For the majority of our calculations we have used the belly routine in AMBER to limit the number of residues that are allowed to move. This has two effects: (1) the overall structure of the molecule will remain close to the crystal structure and (2) the computational effort is greatly reduced. The first effect may be deleterious if the crystal structure is not a substrate-bound structure. Preliminary reports from Ornstein indicate that this may be the case for P450BM3.[17] However, if the structure is close to the substrate-bound one, this method provides the opportunity to explore more of the substrate–enzyme interaction surface since fewer amino acids are in motion and the protein does not need to be solvated. We establish which amino acids should be included in the calculation by trial and error. Initially we use a sphere around the substrate of ca. 12 Å. After the system is reasonably equilibrated, the dynamics runs are observed and different amino acids are either added or deleted based on proximity to the substrate. If allowing an amino acid to move alters the simulation by causing a significant move from equilibrium, it is added in the simulation. We usually do not allow the amino acids on the proximal side of heme to move and the heme itself is held rigid.

Production dynamics are run after equilibration. Thus far, the length of our dynamics runs has been dictated by CPU and disk space constraints. Simulation of around 200 psec can be performed in about 1 week of CPU time on today's workstations.

The analysis of production runs involves compiling statistics about the motion of the substrate in the active site. The assumption is that the oxidizable positions will be in close proximity to the active oxygen species for a reaction to occur. We have used two methods to predict reactivity: (1) the average distance between the position of oxidation and the active oxygen species is determined and (2) the number of times a position is within a given distance, during the simulation, is counted. The first is qualitative and presents a simple picture of the predicted regioselectivity. The second can be used to quantify the simulation's predicted regioselectivity. Neither takes into account the differences in the reactivity of different positions in the molecule. Post facto inclusion of reactivity is extremely complicated since the rise in energy as a hydrogen atom approaches the active oxygen species is 40–50 kcal/Å in quantum chemical calculations. Thus, to a first approximation it would appear that distance would overwhelm reactivity

[17] M. D. Paulsen and R. L. Ornstein, *Proteins* **21**, 237 (1995).

differences. However, White and co-workers[18] have shown that enzymatic reactivity is similar to that expected for Boltzman distributions of orientations, at least for some substrates.

Electronic Models of Mammalian P450s Based on Semiempirical Methods

QM models for cytochrome P450 oxidations are models that use the electronic properties of the P450 enzymes and their substrates to predict the rates of oxidations. The rationale for this approach has intuitive and physical chemical origins, and has been used qualitatively for many years. For example, the tendency for oxidation of a certain functional group generally follows the relative stability of the radicals that are formed, e.g., N-dealkylation > O-dealkylation > 2° carbon oxidation > 1° carbon oxidation. This is a consequence of the generally broad regioselectivity of the P450 enzymes. If an enzyme has access to several positions of a substrate, the most easily oxidized position will be metabolized. The more formal relationship between stability and reaction rates can be described as variations of the Bronsted relationship.[19] Within certain classes of reactions, the activation energy of a reaction is proportional to the heat of reaction. In other words, the more stable the products, the faster the rate of reaction.

Our first quantitative QM model (the PNR model)[20] was based on hydrogen abstraction reactions using the p-nitrosophenoxy radical (PNR) and the semiempirical AM1 Hamiltonian. The model uses the calculated (AM1) enthalpies of reaction and the ionization potentials of the resultant radicals to predict the AM1 activation enthalpies (H_{act}). This model oxygen radical was chosen because it gave a thermodynamically symmetrical reaction ($H_{reac} = 0$) for the abstraction of a hydrogen atom from primary methyl groups in alkanes, within the AM1 formalism. Isotope effect experiments for cytochrome P450-mediated oxidations suggested that the P450-catalyzed reaction is also symmetrical.[21] The p-nitrosophenoxy radical was used to abstract hydrogen atoms from a series of 20 substrates, providing transition state geometries and thermodynamic properties. Although a moderate correlation was obtained for the pure Bronsted relation ($r^2 = 0.86$), the correlations could be greatly improved by including resonance parameters. When compared to inductive effects, a smaller fraction of resonance stabilization

[18] R. E. White, M. McCarthy, K. D. Egeberg, and S. G. Sligar, *Arch. Biochem. Biophys.* **228,** 493 (1984).

[19] R. P. Bell, *in* "Correlation Analysis in Chemistry" (N. D. Chapman and J. Shorter, eds.), p. 55. Plenum, New York, 1972.

[20] K. R. Korzekwa, J. P. Jones, and J. R. Gillette, *J. Am. Chem. Soc.* **112,** 7042 (1990).

[21] J. P. Jones and W. F. Trager, *J. Am. Chem. Soc.* **109,** 2171 (1987).

of the product radical is present at the transition state.[22] Since resonance parameters are not available for most cytochrome P450 substrates, the ionization potential of the radical (IP) was used as a measure of the resonance contribution for stabilizing the radical. Inclusion of the ionization potential gave excellent correlations between calculated heats of reaction and calculated activation energies.

Equation (1) can be used to predict the AM1 activation energy for hydrogen atom abstraction by the p-nitrosophenoxy radical:

$$H_{act.} = 2.60 + 0.22 \, H_{reac.} + 2.38 \, IP. \tag{1}$$

To predict the AM1 activation energy for hydrogen abstraction, it is only necessary to model the substrate and resultant radicals. Any standard implementation of the AM1 semiempirical program can be used, e.g., AMPAC, MOPAC, etc. Standard RHF electronic calculations should be used for closed shell calculations, and the UHF formalism should be used for open shell calculations. No post-Hartree–Fock methods such as configuration interaction should be used when generating electronic structures. Substrates and radical geometries should be minimized to default tolerances. Care should be taken that the conformation of the radical corresponds to a geometry accessible by the abstraction of a hydrogen atom from the substrate. This can usually be accomplished using the substrate geometry, less a hydrogen atom, as the starting geometry for the radical. Relaxation of this geometry will usually result in the conformation of the radical that is accessible from the transition state for hydrogen atom abstraction. Standard output for most programs includes the heat of formation and ionization potentials, which are used to estimate the AM1 activation energy for the hydrogen abstraction reaction by Eq. (1). This process is far less time-consuming than generating the reaction coordinates necessary to model the actual transition states.

The values of these calculated energies do not correspond to absolute rates since semiempirical calculations overestimate transition state energies. Instead, these values should be used on a relative basis. When calculating $H_{act.}$ values, the heats of formation for the nitrosophenoxy radical and nitrosophenol are 19.04 and -14.65 kcal/mol, respectively. The value for the radical is for one of two stable geometries for two different electronic states of the radical. The other geometry is lower in energy, but was not used to generate the reaction coordinates for the model. Since only relative activation energies are obtained, the actual values used in the calculation

[22] C. F. Bernasconi, in "Nucleophilicity" (J. M. Harris and S. P. McManus, eds.), p. 115. American Chemical Society, Washington, DC, 1987.

are not important, as long as they are consistent within the reactions being compared.

The PNR model has been used to describe the reaction rates of several cytochrome P450-mediated reactions. In our first attempt to use an electronic model for cytochrome P450 oxidations, the PNR model and calculated hydrophobicities were used to predict the experimental toxicities of a series of nitriles.[23] The acute toxicities of most nitriles are due to cyanide release, mediated by the hydroxylation of nitriles adjacent to the cyano functionality. The resultant cyanohydrin chemically decomposes, releasing cyanide. We used the calculated thermodynamic properties of 26 nitriles and their radicals to predict the relative rates for oxidation at all possible positions. Since oxidations by the cytochrome P450 enzymes can result in either detoxification or activation, it was necessary to include all potential oxidations in the correlations. Significant correlations were obtained between the experimental LD_{50} values and $k\alpha$ corr, the rate constants corrected for the fraction of the metabolism to the toxic metabolite:

$$k\alpha_{corr} = k\alpha[k\alpha/(k\alpha + k\beta + k\gamma \dots)]. \qquad (2)$$

Better correlations were obtained with k corr than either k or the fraction $(k\alpha/k\alpha + k\beta + k\gamma \dots)$ alone, suggesting that both rates and regioselectivity of metabolism are important.

Hydrophobicity parameters are often included in QSAR studies, presumably to account for substrate transport and binding[24] In the nitrile study, the CLOGP program[24] was used to predict the hydrophobicities of the various substrates. However, recent studies with an extended set of nitrile LD_{50} values suggest that using the Syracuse Program[25] to calculate the hydrophobicities provide improved correlations (unpublished results).

Using the PNR model, we have obtained excellent correlations with several other sets of experimental data. Using data provided by White and McCarthy[26] for the metabolism of substituted toluenes by CYP2B4, a fit of the experimental k_{cat} values to the predicted enthalpy of activation and the log P of each compound gave a regression coefficient of 0.94. The AM1 model was used to calculate the predicted activation energies for P450-mediated hydrogen abstraction from a series of pentahaloethanes.[27] *In vivo*

[23] J. Grogan, S. C. DeVito, R. S. Pearlman, and K. R. Korzekwa, *Chem. Res. Toxicol.* **5,** 548 (1992).

[24] "CLOGP Version 3.42." Pomona College Medicinal Chemistry Project, Clarmont, CA.

[25] W. Meylan and P. Howard, "KOWWIN Program." Syracuse Research Corporation Environmental Sciences Center, Merrill Lane, Syracuse, NY.

[26] R. E. White and M. B. McCarthy, *Arch. Biochem. Biophys.* **246,** 19 (1986).

[27] J. W. Harris, J. P. Jones, J. L. Martin, A. C. LaRosa, M. J. Olson, L. R. Pohl, and M. W. Anders, *Chem. Res. Toxicol.* **5,** 720 (1992).

administration to rats of the pentahaloethanes in this study resulted in the trifluoroacylation of liver proteins, presumably by trifluoroacylhalides formed by P450 oxidations. Both the degree of trifluoroacylation, as determined by immunoblotting, and the amount of trifluoroacetic excreted in the urine paralleled the activation energy for hydrogen abstraction as predicted by the AM1 model. The agreement between theory and experiment indicates that the electronic characteristics of the halogenated hydrocarbon are the important determinants in the rate of trifluoroacetic acid produced. Finally, the predicted activation energies for a series of inhalation anesthetics was calculated with our model. An excellent correlation is obtained between the predicted rates of metabolism and the percentage *in vivo* metabolism in humans.[28]

Unfortunately, the PNR model cannot be extended to other (nonhydrogen atom abstraction) P450-mediated reactions. The nitrosophenoxy radical, within the AM1 formalism, does not remain on the same potential energy surface for some calculations, including addition to aromatic and olefinic compounds. Although other small oxygen radicals remain on the same potential energy surface, excessive amounts of spin contamination suggest that the calculations will not be size consistent. Using RHF AM1 calculations with partial configuration interaction prevents the spin contamination problem, but these calculations are also not size consistent. Therefore, we are in the process of using *ab initio* calculations to develop expanded electronic models for P450 oxidations.

Conclusions

Thus, while we have had a number of successes with both QM and MD calculations, our goal of obtaining a general predictive methodology has not been obtained. Our PNR model appears to give a good quantitative prediction of rates for hydrogen atom abstractions, but is inappropriate when applied to reactions that do not occur by hydrogen atom abstraction or to substrates, such as nicotine and benzo[*a*]pyrene, which have limited motion in the active site. In turn, MD models can provide information on binding factors and motion of the substrate in the active site, but mammalian structural models do not exist and regioselectivity can only be predicted in a semiquantitative way, especially when the positions of interest have different chemical reactivity.

---

[28] H. Yin, M. W. Anders, K. R. Korzekwa, L. Higgins, K. E. Thummel, E. D. Kharasch, and J. P. Jones, *Proc. Natl. Acad. Sci. U.S.A.* (in press).

[37] Substrate Docking Algorithms and the Prediction of Substrate Specificity

By JAMES J. DE VOSS and PAUL R. ORTIZ DE MONTELLANO

Introduction

The cytochrome P450 enzymes are notable not only for the wide range of oxidative transformations that they catalyze but also for the broad substrate specificity that many of them display.[1] Little is known, however, about the factors that actually determine the substrate specificity of these enzymes. An understanding of these factors, coupled with a knowledge of P450 active site structure, should eventually allow one to predict whether a given molecule will be a substrate for a given P450. This ability is clearly crucial to the construction of novel P450 enzymes as biooxidative catalysts with engineered activities. Such an ability will also be useful in understanding, predicting, and modulating the physiological metabolism of pharmaceuticals and other xenobiotics.

We have recently explored the utility of computer-assisted molecular docking for the prediction of cytochrome P450 substrates.[2] This approach has been used extensively in the rational design of enzyme inhibitors, but had not previously been employed for substrate prediction.[3-7] Cytochrome P450 enzymes appear to be particularly amenable to this approach to substrate prediction because (i) their broad substrate range implies that there are few specific interactions between the substrates and the protein and (ii) the highly reactive nature of the enzymic oxidizing species means that there is little requirement for chemical reactivity in the substrate.

[1] P. R. Ortiz de Montellano, ed., "Cytochrome P450: Structure, Mechanism, and Biochemistry," 2nd Ed. Plenum, New York, 1995.
[2] J. J. De Voss and P. R. Ortiz de Montellano, *J. Am. Chem. Soc.* **117,** 4185 (1995).
[3] R. L. DesJarlais, R. P. Sheridan, G. L. Seibel, J. S. Dixon, I. D. Kuntz, and R. Venkataraghavan, *J. Med. Chem.* **31,** 722 (1988).
[4] R. L. DesJarlais, G. L. Seibel, I. D. Kuntz, P. S. Furth, J. C. Alvarez, P. R. Ortiz de Montellano, D. L. DeCamp, L. M. Babé, and C. S. Craik, *Proc. Natl. Acad. Sci. U.S.A.* **87,** 6644 (1990).
[5] B. K. Shoichet, R. M. Stroud, D. V. Santi, I. D. Kuntz, and K. M. Perry, *Science* **259,** 1445 (1993).
[6] C. S. Ring, E. Sun, J. H. McKerrow, G. K. Lee, P. J. Rosenthal, I. D. Kuntz, and F. E. Cohen, *Proc. Natl. Acad. Sci. U.S.A.* **90,** 3583 (1993).
[7] D. L. Bodian, R. B. Yamasaki, R. L. Buswell, J. F. Stearns, J. M. White, and I. D. Kuntz, *Biochemistry* **32,** 2967 (1993).

We have used P450cam as our model P450 enzyme because several high resolution crystal structures are available for it,[8,9] although the approach should be applicable to any other isoform of P450 for which an active site structure is available.

Background

The P450cam reaction cycle starts with the ferric enzyme which, upon binding of a substrate, undergoes a change in oxidation potential from -300 to -170 mV.[10] This poises the oxidation potential of the hemoprotein in a range that is much more suitable for reduction by putidaredoxin, the iron–sulfur protein that provides electrons to P450cam. The redox potential of putidaredoxin is -196 mV when it is bound to the hemoprotein. Substrate binding thus facilitates electron transfer to the heme and binding of molecular oxygen to the reduced iron atom. Addition of a second electron from putidaredoxin results in the formation of a molecule of water and a heme-bound activated species equivalent to a ferryl ($Fe^V{=}O$) complex. This ferryl-like species is the powerful oxidant at the heart of cytochrome P450 chemistry. In the case of P450cam, its normal function is to hydroxylate camphor at an unactivated methylene position to produce 5-*exo*-hydroxy-camphor. With unnatural substrates, hydroxylation competes to varying extents with uncoupling reactions that result in the reduction of molecular oxygen to hydrogen peroxide or water. The uncoupling reactions are themselves a measure of substrate binding in the sense that they also depend on the change in the redox state triggered by productive substrate binding.

The interaction of a potential substrate with cytochrome P450 can be separated into three stages: (a) determination of whether a compound binds within the active site and triggers the required change in redox state of the heme iron atom ("productive binding"), (b) determination of the degree of coupling of oxygen and NAD(P)H consumption to substrate oxidation rather than to uncoupling reactions, and (c) determination of the detailed regio- and stereochemistry of substrate oxidation. The docking algorithms dealt with in this chapter primarily address the first step of this process: substrate binding and activation of catalytic turnover. Computational approaches to analysis of the two subsequent steps, the degree of uncoupling and the site specificity of the oxidation process, are therefore only briefly reviewed.

---

[8] T. L. Poulos, B. C. Finzel, and A. J. Howard, *J. Mol. Biol.* **195,** 687 (1987).
[9] T. L. Poulos and R. Raag, *FASEB J.* **6,** 674 (1992).
[10] S. G. Sligar and R. I. Murray, *in* "Cytochrome P450: Structure, Mechanism, and Biochemistry" (P. R. Ortiz de Montellano, ed.), p. 429. Plenum, New York, 1986.

Procedure

There are a number of programs available that can be used to predict whether a given small molecule will bind to a macromolecular receptor.[11] We chose to use DOCK,[3] a receptor constrained, three-dimensional screening program that has been developed by, and is available from,[12] I. D. Kuntz and his group at UCSF. DOCK has been used extensively for the prediction of small molecules that bind to and inhibit a variety of enzymes,[3–7] but this is the first instance of its use to predict the suitability of compounds as enzyme substrates.[2]

The first step in the DOCKing process is the creation of a molecular surface of the active site (or any other area of interest) of the protein.[13] This is done computationally by rolling a theoretical probe the size of a water molecule across the van der Waal's surface of the relevant residues of the protein.[14] The surface created resembles a van der Waal's surface but the small gaps between atoms are smoothed over and any invaginations too small to accommodate the probe are eliminated. Of course, for DOCKing molecules into the active site, one is initially concerned with both the size and the shape of the volume encompassed by the surface rather than with the surface itself. One of the suites of programs that accompany DOCK (sphgen) relates the molecular surface to the active site volume by generating a set of overlapping spheres of varying size that describes the shape and volume of the site (Fig. 1). The centers of these spheres serve as potential atom positions for the DOCKing program. In this program, atoms of a potential ligand are matched with sphere centers in the active site until at least four pairs of atom/sphere centers that are matched for internal distances are found. This uniquely defines a potential orientation of the ligand within the active site. As the process is repeated until all the possible orientations have been identified, the program generates a family of potential binding orientations. Each of these orientations is checked to ensure that the molecule is within the active site and does not overlap or intersect the protein structure. If the compound is within the active site, its goodness of fit is scored in one of a number of different ways.[15] In the present studies, we have used the simplest method, contact scoring, which rewards ligands for atoms that are within interaction distance of the receptor residues. This scoring algorithm, developed for inhibitor studies, assigns

[11] I. D. Kuntz, E. C. Meng, and B. K. Shoichet, *Acc. Chem. Res.* **27,** 117 (1994).

[12] I. D. Kuntz, Department of Pharmaceutical Chemistry, University of California, San Francisco.

[13] F. M. Richards, *Annu. Rev. Biophys. Bioeng.* **6,** 151 (1977).

[14] M. L. Connolly, *Science* **221,** 709 (1983).

[15] I. D. Kuntz, E. C. Meng, and B. K. Shoichet, *Acc. Chem. Res.* **27,** 117 (1994).

the highest scores to compounds that most completely fill the target volume. The list of potential ligands is therefore ranked by DOCK on their (calculated) steric complementarity to the active site surface.

DOCK itself can be used in one of two ways: either to screen a large (three dimensional) database of molecules (in SEARCH mode) for potential ligands or to search more intensively the binding orientations of a single ligand (in SINGLE mode). In our work, we have screened a 20,000 compound subset of the Available Chemicals Directory (ACD),[16] formerly the Fine Chemicals Directory. Compounds that were identified as potential ligands were then minimized with the default molecular mechanics parameters in the SYBYL[17] modeling package to give a "better" three-dimensional structure for the molecule. These can then be reDOCKed into the active site to check the SEARCH results and examine some of the possible orientations available for ligand/protein interaction. To give an idea of the time required for a database search, a typical experiment in this work screened approximately 15,000 to 20,000 orientations for each of 5000 compounds in 7 hr of cpu time on a Silicon Graphics Iris Indigo XS24-4000.

We initially used DOCK in SEARCH mode to select the top 500 compounds based on the contact score of a 20,000 compound subset of the ACD. These were then visually screened,[18] and 10 compounds were chosen as potential substrates that (i) would not bear any formal charge at neutral pH (the active site of P450cam is hydrophobic), (ii) represented a variety of structural types unrelated to camphor, and (iii) would hopefully give oxidized products amenable to assay by GC/GCMS or HPLC. A set of control compounds was also selected that were predicted not to be substrates for P450cam. This was done in such a way that we could also examine the precision with which these predictions were made. The structural model of the P450cam active site was modified by simply replacing the isobutyl side chain of leucine 244 with a methyl group (i.e., converting it to an alanine) to create a model for the L244A mutant. This change results in an increase in the active site volume by approximately the size of an isopropyl group (from 244 to 305 Å^3). The 20,000 compound subset was again screened using DOCK after computing a surface and sphere cluster for the enlarged active site of the L244A mutant. Six compounds were then chosen that would fit into the L244A but not the wild-type active site.

---

[16] MDL Information Systems, Inc., San Leandro, CA.

[17] SYBYL 6.0 Tripos Inc., 1699 South Hanley Road, St. Louis, MO.

[18] Molecular graphics images were produced using the MidasPlus program from the Computer Graphics Laboratory, University of California, San Francisco (supported by NIH RR-01081): T. E. Ferrin, C. C. Huang, L. E. Jarvis, and R. Langridge, *J. Mol. Graphics* **6,** 13 (1988).

A

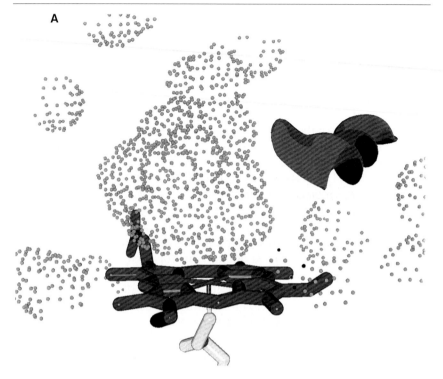

FIG. 1. (A) The molecular surface of the active site of P450cam. The thiolate-bound heme and the I helix are also shown. (B) The active site of P450cam as represented by the overlapping spheres generated by DOCK. The solid spheres are the sphere centers to which atoms are matched in the DOCKing process. Dots represent the surface defined by atoms located at the sphere centers.

The 16 structures chosen were then minimized with SYBYL and re-DOCKed in SINGLE mode into the enzymes. As a result, two compounds (**10** and **11**; see Table I) migrated from the "nonsubstrate" to the "substrate" list and one (**12**, see Table II) in the other direction. This underscores two important limitations inherent in the DOCKing process at the present time.

First, the conformation of the DOCKed compound is clearly important in determining whether it fits within the active site; even the small changes made with SYBYL minimization of our 16 compounds resulted in a change in the substrate/nonsubstrate prediction for 3 of them. The three-dimensional databases we have employed were constructed using CONCORD,[19]

[19] A. Rusinko, R. P. Sheridan, R. Nilakatan, K. S. Haraki, N. Bauman, R. Venkataraghavan, *J. Chem. Inf. Comput. Sci.* **29**, 251 (1989).

B

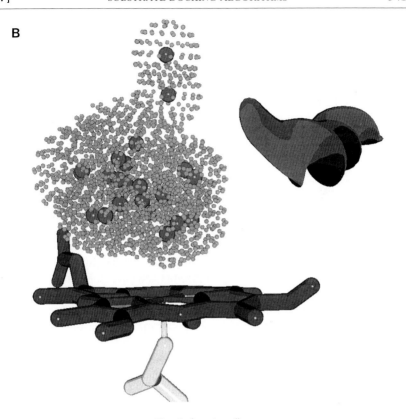

FIG. 1. (*continued*)

a rule-based program for converting two-dimensional to three-dimensional structures. SYBYL uses a force field to calculate the energy of different structures during minimization. These two approaches can and do lead to different "answers" for the structure of a given compound and neither of them address the issue of multiple low energy conformers or differences between binding and solution conformers, both of which can be particularly important for flexible compounds. These are computationally intensive problems and are currently being addressed by many groups, but at the present time some care should be exercised when excluding flexible compounds as ligands on the basis of DOCK results.

The second issue concerns the scoring protocol. Contact scoring, as employed here, depends heavily on the distance(s) chosen as resulting in unfavorable van der Waal's interaction between the ligand atoms and the protein surface. In DOCK, for computational purposes, this is actually taken as the distance between nonhydrogen atoms on the ligand and nonhydrogen atoms on the receptor. This distance, which is one of the parameters that can be adjusted in the algorithm, is often set at less than that at which unfavorable interactions would actually occur. This strategy is taken to make some allowance for self-adjusting motions of both the receptor and the ligand. More sophisticated protocols than contact scoring are available within DOCK, but the accurate prediction of binding energies of disparate ligands is still a challenge for computational chemistry.

Results

What can one expect from the DOCK database search? Compounds that are predicted to fit within the active site ("bind") should be ligands for the receptor, but the relative (and absolute) affinities predicted by the DOCK scoring algorithm should be treated with caution. These compounds should be substrates in the sense that they initiate enzyme turnover, whether that leads to substrate oxidation or simply to an uncoupled reduction of molecular oxygen. Conversely, compounds that are calculated to be too large to fit in the active site should not bind to the enzyme in a manner that initiates catalytic turnover. Compounds that are predicted to be too large may, however, interact with an open conformation of the active site or the channel that leads into it without allowing the channel to close. Such compounds could therefore inhibit substrate binding without themselves being substrates. It should be kept in mind, furthermore, that only one conformer of a molecule will be examined in the DOCK search (unless steps are taken to include explicitly multiple conformations) and that this conformational exclusion may dramatically influence the results, especially with conformationally flexible molecules.

The experimental results for the compounds predicted by DOCK to be substrates of P450cam (**1–11**) are shown in Table I, and those of the compounds predicted not to be substrates (**12–16**) are shown in Table II. Two questions can be addressed immediately with these results: (i) How well does DOCK predict small molecules that bind to the P450? (ii) How successful is the hypothesis that binding to the P450 active site is enough to make a compound a substrate for a P450?

Binding of ligands to P450cam can be monitored spectroscopically either directly by measuring the change in the Soret absorption as the iron changes from low (417 nm) to high spin (392 nm), or indirectly by measuring the ability of the compound to inhibit the spin state transition caused by the binding of camphor. The first yields a spectroscopic dissociation constant (K_S) and the latter an inhibitory constant (K_I). The existence of a linear-free energy relationship between the iron spin state and the oxidation potential of P450cam suggests that the ability of a compound to alter the spin state is a valid measure of productive substrate binding. The results in Table I show that DOCK does reasonably well in predicting compounds that bind to P450cam. Seven (**1–6, 11**) of the 11 compounds predicted to be ligands do bind to the enzyme as judged by spin state changes of the iron; the other 4 compounds inhibit the binding of camphor to some degree (vide infra). The observed K_S values vary greatly, ranging from approximately 4 μM to 4 mM. As expected, there is little relationship between the measured K_S values for the compounds and the binding energies or rank orders assigned by DOCK. Impressively, none of the predicted non-substrates (Table II) brings about the typical low to high spin transition, although some alter the Soret absorption profile and most inhibit camphor binding. This is not surprising, as there would be a clear energetic preference for association of small hydrophobic molecules with hydrophobic protein sites which sequester them from the aqueous environment; DOCK simply predicts that the whole molecule cannot fit within the active site, a fact that seems to be related to the absence of the low to high spin transition.

Binding is not simply a function of the volume of the substrate. Although the nonligands tend to be larger than the ligands, this is not always true (compare, for example, compounds **6** and **16**). It is therefore clear that DOCK, as desired, considers both the size and the shape of the molecules in assigning scores. It appears, in sum, that DOCK is not only relatively successful at predicting the ability of small molecules to bind to P450cam but does so with a fair degree of precision, as the molecules predicted not to bind were calculated to fit an active site that was only larger by the volume of an isopropyl group.

How does the hypothesis that binding equates to turnover, and therefore that binding predictions by DOCK translate into predictions of the activity of compounds as P450 substrates, hold up? Of the 11 compounds predicted to be substrates for P450cam, 8 (**1–7, 11**) stimulated catalytic turnover, as judged by NADH consumption and/or organic product formation. Some of these were quite good substrates, turning over at 10–20% the rate of camphor, while the turnover of others was only a little above background. Furthermore, none of the compounds predicted not to bind to P450cam increased NADH consumption over background. As expected, substrate-mediated increases in NADH consumption provide a measure of the ability of the compound to activate enzyme turnover, and therefore of their activity as "substrates." As expected, however, NADH consumption does not correlate with substrate hydroxylation due to the partition between hydroxylation and uncoupling processes.

The results indicate that the analysis of substrate fit by DOCK is a promising method for the identification of substrates for P450 enzymes for which (crystal) structures are available. These substrate predictions are

TABLE I

PARAMETERS FOR THE INTERACTION OF DOCK-PREDICTED SUBSTRATES WITH P450cam

Entry	Compound	DOCK[a] maximum score (number) rank	Volume (Å^3)	K_S or K_I (μM)	% high spin	NADH (nmol/min/ nmol P450)[b]	% H_2O_2 from O_2
Camphor		126 (5000)	160	2.0 ± 0.1	100	530 ± 10	0–8
1		141 (84) 267	194	83 ± 5	8	8 ± 1	100
2		129 (56) 143	205	4.3 ± 0.3	91	110 ± 5	5–10
3		127 (1) 233	210	280 ± 70	6	12 ± 1	100
4		123 (28) 414	189	4400 ± 600	13	41 ± 3	31
5		138 (430) 60	141	530 ± 100	22	27 ± 1	50
6		129 (2) 120	249	27 ± 4	30	6 ± 1[f]	100
7		132 (368) 52	199	1200 ± 200[c]	ND[d]	80 ± 5	11

TABLE I (*continued*)

Entry	Compound	DOCK[a] maximum score (number) rank	Volume (Å³)	K_S or $K_I(\mu M)$	% high spin	NADH (nmol/min/ nmol P450)[b]	% H_2O_2 from O_2
8		125 (2) 39	248	74 ± 25^c	<2	4 ± 1	100
9	OAc	146 (1) 164	232	$\gg 300^c$	<1	5 ± 1	100
10	O, OEt, OEt, O	159 (1)^e	258	500 ± 260^c	<1	6 ± 1	100
11	OH	125 (3)^e	221	228 ± 32	<2	8 ± 1	100

[a] DOCK results given as (i) the maximum binding energy score in "DOCK kcals" from DOCKing of each compound in SINGLE mode; (ii) the number of orientations generated in the SINGLE DOCK run; and (iii) the DOCK rank of each compound after the SEARCH run of the ACD subset.
[b] Background NADH consumption is 4–6 nmol/min/nmol P450.
[c] K_I determined by inhibition of camphor binding.
[d] Not determined because the UV/visible absorption of the compound interfered with spectrophotometric assay.
[e] These compounds were not ranked as they were not identified as substrates in the original DOCK search.
[f] An organic product is formed even though there is little stimulation of NADH consumption.

subject to a number of limitations, however. Thus, although binding in the active site is a prerequisite for a substrate, neither the affinity of this association nor the ability of the compound to induce the iron spin state transition ("productive" binding) is a measure of how well the compound is metabolized by the enzyme (cf. **6** and **7**). The finding that compounds that have very few docking orientations within the active site (see Table I) are relatively poor substrates suggests, furthermore, that good substrates

TABLE II
PARAMETERS FOR THE INTERACTION OF DOCK PREDICTED NONSUBSTRATES WITH P450cam

Entry	Compound	Volume (Å^3)	K_S or $K_I(\mu M)$	% high spin	NADH (nmol/min/ nmol P450)[a]	% H_2O_2 from O_2
12		289	Not detected	>8	4 ± 1	100
13		238	75 ± 23	>6	5 ± 1	100
14		244	4300 ± 420	>1	5 ± 1	100
15		231	Not detected	>1	5 ± 1	100
16		226	72 ± 26	>1[b]	6 ± 1	100

[a] Background NADH consumption is 4–6 nmol/min/nmol P450.
[b] A decrease is seen in the Soret at 417 nm with a corresponding increase at 430 nm.

must not only be sterically accommodated but must retain a degree of mobility within the active site. Physically, this may reflect the need to accommodate the binding of oxygen or a requirement for some motion of protein residues that assist in the cleavage of the dioxygen bond. For example, compound **6** may be a poor substrate even though it binds tightly and well (30% maximum spin state change) because it completely fills the active site whereas **7** may be a good substrate because it binds well enough to initiate catalytic turnover but does not fill the site to the extent that it interferes with catalysis.

Future developments in this area should include utilization of the continuing improvements in ligand docking programs, i.e., conformationally flexible searching, minimization of ligand/receptor complexes during searches, and more sophisticated scoring algorithms. The group of compounds reported here, for which the protein interactions have been experimentally determined, provide a data set that can be used to optimize the parameters of the DOCK program to improve its ability to predict P450 substrates. For example, preliminary results indicate that the distance chosen for prohibitive van der Waal's interactions in the scoring algorithm affects the degree of mobility (i.e., number of docking orientations available) attributed to potential ligands by DOCK. Further refinements of the present protocol, in conjunction with computational approaches to the prediction of uncoupling and substrate regiospecificity,[20–23] should eventually allow one to approach the prediction of substrate specificty with some degree of confidence.

Acknowledgments

We thank I. D. Kuntz and his research group for providing access to DOCK and for help in its utilization. We particularly thank D. Gschwend for providing a program for determination of the molecular and active site volumes. The work reported here was supported by National Institutes of Health Grant GM25515.

[20] J. A. Fruetel, J. R. Collins, D. L. Camper, G. H. Loew, and P. R. Ortiz de Montellano, *J. Am. Chem. Soc.* **114,** 6987 (1992).
[21] D. Harris and G. Loew, *J. Am. Chem. Soc.* **117,** 2738 (1995).
[22] J. P. Jones, W. F. Trager, and T. J. Carlson *J. Am. Chem. Soc.* **115,** 381 (1993).
[23] P. J. Loida, S. G. Sligar, M. D. Paulsen, G. E. Arnold, and R. L. Ornstein, *J. Biol. Chem.* **270,** 5326 (1995).

[38] Using Molecular Modeling and Molecular Dynamics Simulation to Predict P450 Oxidation Products

By MARK D. PAULSEN, JOHN I. MANCHESTER, and RICK L. ORNSTEIN

Introduction and Overview

Both the ubiquity of P450s in nature and the wide variety of molecules that are oxidized in their active sites have focused interest in them as targets for drug design and as biotechnological tools in chemical manufacture and waste treatment. A large amount of work has been directed at understanding the mechanistic details of the P450 reaction cycle and controlling substrate and product specificity. A variety of theoretical techniques have been used to address questions at each stage of the P450 reaction cycle. These computational methods have met with varying degrees of success, depending on the theoretical rigor and the amount and nature of available experimental data. Several contributions to this volume examine a range of state of the art applications of computational methods to understanding the mechanistic details of P450. This chapter, briefly overviews the application of computational methods to P450s and discusses in more detail the use of molecular dynamics simulations for the accurate assessment of oxidation specificity by P450 systems.

Considerable effort has been directed toward understanding factors that control the metabolism of drugs and carcinogens by mammalian P450s. Because of the absence of high resolution structural data for these membrane-bound proteins, most of these studies are confined to detailed analyses of the isolated substrates. Insights into substrate reactivity in the presence of P450s have come from analyses of the substrate electrostatic potentials,[1,2] or studies of their conformational flexibility by molecular modeling alone[3,4] or in conjunction with X-ray crystallography,[5] or NMR spectroscopy.[6,7] One approach to combine these properties into predictive models is the formulation of QSARs. For example, Lewis and co-workers developed a relationship[8] that incorporates structural parameters obtained via molecular modeling and frontier molecular orbital (MO) energies via semiempirical calculations. The authors observed that substrates for the cytochromes P450I and P450II subfamilies cluster into two groups: substrates for the former subfamily are more planar in shape and exhibit smaller energy differences between the highest occupied and lowest unoccupied molecular orbitals (HOMO and LUMO, respectively) than did substrates

[1] F. Sanz, F. Manaut, J. Rodriguez, E. Lozoya, and E. Lopez-de-Brinas, *J. Comput.-Aided Mol. Des.* **7**, 337 (1993).
[2] U. Fuhr, G. Strobl, F. Manaut, E. M. Anders, F. Sorgel, E. Lopez de Brinas, D. T. Chu, A. G. Pernet, G. Mahr, and F. Sanz, *Mol. Pharmacol.* **43**, 191 (1993).
[3] C. A. Laughton, R. McKenna, S. Neidle, M. Jarman, R. McCague, and M. G. Rowlands, *J. Med. Chem.* **33**, 2673 (1990).
[4] R. A. Lubet, J. L. Syi, J. O. Nelson, and R. W. Nims, *Chem.-Biol. Interact.* **75**, 325 (1990).
[5] C. A. Laughton and S. Neidle, *J. Med. Chem.* **33**, 3055 (1990).
[6] J. Gharbi-Benarous, P. Ladam, M. Delaforge, and J.-P. Girault, *J. Chem. Soc., Perkin Trans.* **2**, 2303 (1993).
[7] S. P. Lam, D. J. Barlow, and J. W. Gorrod, *J. Pharm. Pharmacol.* **41**, 373 (1989).
[8] D. F. V. Lewis, *Front. Biotrans.* **7**, 90 (1992).

for the latter subfamily.[9] They also observed that the overall reactivity of the substrate correlated with the energy difference between the HOMO and LUMO. More sophisticated MO calculations have been employed to probe not just overall substrate reactivity, but the reactivity of specific functional groups on the molecule. Several studies have employed semiempirical MO calculations to characterize the regiospecificity of hydrogen atom abstraction from camphor, camphor analogs, and other substrates by P450cam.[10] Although these methods provide a reasonable degree of accuracy in return for modest cpu requirements, they are generally not rigorously suited to reproduce the properties of unfilled molecular orbitals or excited states that may be of interest. Recently, Harris and Loew[11] used unrestricted Hartree–Fock *ab initio* calculations to evaluate the radical stability for camphor and compared this with semiempirical results. They found that the semiempirical calculations tended to overestimate the difference in radical stability between the primary and the secondary radicals of camphor.

The second approach to draw inferences about factors controlling substrate reactivity or binding is the construction of pharmacophore models. These models are obtained by consideration of features common to numerous substrates or inhibitors, and include factors such as hydrogen bond donors or acceptors, the site of attack by the ferryl intermediate, and key distance criteria among these functional groups.[12] The resulting molecular template can be used to evaluate the reactivity or inhibitory potency of other chemicals of interest. Sternberg and coworkers[13] developed such a model for P450 2D6, which has been implicated in the activation of procarcinogens. Based on this model, the authors found that a potent carcinogen in tobacco smoke did not exhibit the requirements of a substrate for this P450, consistent with experimental data. This observation supports the usefulness of pharmacophore models when little or no structural data are available for the enzyme. However, the applicability of these models to potential substrates that differ substantially in size, or in the nature and number of functional groups present, is limited by their ability to accurately model the explicit interactions with residues in the P450-binding pocket.

High-resolution crystallographic coordinates are now available for four

[9] D. F. V. Lewis, H. Moereels, B. G. Lake, C. Ioannides, and D. V. Parke, *Drug. Metab. Rev.* **26,** 261 (1994).
[10] K. R. Korzekwa and J. P. Jones, *Pharmacogenetics* **3,** 1 (1993).
[11] D. Harris and G. Loew, *J. Am. Chem. Soc.* **117,** 2738 (1995).
[12] T. Wolff, G. Strobl, and H. Greim, *Handb. Exp. Pharmacol.* **105,** 195 (1993).
[13] S. A. Islam, R. C. Wolf, M. S. Lennard, and M. J. E. Sternberg, *Carcinogenesis* (*London*) **12,** 2211 (1991).

cytosolic, but no membrane-bound P450s.[14-17] This relative paucity of structural data contrasts strongly with the numerous additional members of the superfamily whose sequences have been reported during the last several years. The imbalance between structural and sequence data has led many researchers to use homology modeling to try and gain a better understanding of the relationship between structure and function for several members of the P450 superfamily. Several investigators have modeled the active sites of P450s based on sequence-derived structural homology with known crystal structures, most notably P450cam.[18] In several instances, structure–function insights gained by computer-graphics docking of substrate in the model have allowed rationalization of experimental results.[19-22] However, these modeling efforts are very sensitive to the sequence alignment on which they are based, producing misleading results from a nonoptimal alignment. This problem becomes more acute as the isozyme under scrutiny diverges from the template, as is the case for most mammalian P450s.[23] In general, extreme care should be exercised in attempts to dock substrates to enzyme models built by homology. Even in cases where high-resolution structural data are available, simple docking studies are susceptible to the fact that they do not allow the enzyme binding pocket to relax under the influence of the substrate.

Structural and Dynamics Insights from Molecular
  Dynamics Trajectories

If a detailed three-dimensional model of the enzymes is available from X-ray crystallography or potentially from homology modeling, then it is possible to calculate molecular dynamics trajectories of several hundred picoseconds to a few nanoseconds in length. Such simulations can complement the available experimental data. For both P450cam and P450BM-3, simulations starting from the reported crystal structures have been reported.

[14] C. A. Hasemann, K. G. Ravichandran, J. A. Peterson, and J. Deisenhofer, *J. Mol. Biol.* **236,** 1169 (1994).
[15] K. G. Ravichandran, S. S. Boddupalli, C. A. Hasemann, J. A. Peterson, and J. Deisenhofer, *Science* **261,** 731 (1993).
[16] T. L. Poulos, B. C. Finzel, and A. J. Howard, *J. Mol. Biol.* **195,** 687 (1987).
[17] J. R. Cupp-Vickery and T. L. Poulos, *Nat. Struct. Biol.* **2,** 144 (1995).
[18] S. D. Black, *Handb. Exp. Pharmacol.* **105,** 155 (1993).
[19] G. D. Szklarz, R. L. Ornstein, and J. R. Halpert, *J. Biomol. Struct. Dyn.* **12,** 61 (1994).
[20] G. M. Morris and W. G. Richards, *in* "Cytochrome P450" (A. I. Archakov and G. I. Backmanova, eds.), p. 692. INCO-TNC, Moscow, 1992.
[21] D. F. V. Lewis and H. Moereels, *J. Comput.-Aided Mol. Des.* **6,** 235 (1992).
[22] J. A. Braatz, M. B. Bass, and R. L. Ornstein, *J. Comput.-Aided Mol. Des.* **8,** 607 (1994).
[23] T. L. Poulos, *Methods Enzymol.* **206,** 11 (1991).

In the case of P450cam, several simulations have reported that Phe 87 can adopt an alternate conformation which results in a significant increase in the volume of the substrate-binding pocket.[24,25] Thus significantly larger substrates might be accommodated in the active site of P450cam than would be anticipated from an examination of the crystal structure. Simulations have also shown high mobility in the I helix between residues Gly 248 and Thr 252, a region in which a kink is reported in the crystal structure of P450cam.[24,25] Flexibility in this region may be important in controlling solvent access to the active site pocket. In addition, a recent simulation of the ferrous dioxygen-bound species by Harris and Loew[26] indicated that the side chain of Thr 252 can adopt an alternate conformation which may promote bond cleavage of dioxygen through a charge relay mechanism in agreement with the kinetic solvent isotope studies of Sligar and co-workers.[27,28] Recent simulations of P450BM-3 indicate significant mobility of the residues lining the active site pocket, again indicating that attempts to dock novel substrates into the crystal structure might lead to significantly different results from those obtained using the simulated time-averaged structure.[29,30]

Using Molecular Dynamics to Obtain Product Profiles

Since 1991, a number of studies have appeared in which the product distribution of novel P450 substrates has been predicted.[31–38] These studies have examined both regio- and stereospecificity. Attempts have also been made to predict the effect of active site mutations on the observed product

[24] M. D. Paulsen and R. L. Ornstein, *Proteins: Struct. Funct. Genet.* **11,** 184 (1991).
[25] M. B. Bass, M. D. Paulsen, and R. L. Ornstein, *Proteins: Struct. Funct. Genet.* **13,** 27 (1992).
[26] D. L. Harris and G. H. Loew, *J. Am. Chem. Soc.* **116,** 11671 (1994).
[27] N. C. Gerber and S. G. Sligar, *J. Am. Chem. Soc.* **114,** 8742 (1992).
[28] J. Aikens and S. G. Sligar, *J. Am. Chem. Soc.* **116,** 1143 (1994).
[29] M. D. Paulsen and R. L. Ornstein, *Proteins: Struct. Funct. Genet.* **21,** 237 (1995).
[30] H. Li and T. L. Poulos, *Acta Crystallogr., Sect D* **D51,** 21 (1995).
[31] J. P. Jones, W. F. Trager, and T. J. Carlson, *J. Am. Chem. Soc.* **115,** 381 (1993).
[32] J. P. Jones, M. Shou, and K. R. Korzekwa, *Biochemistry* **34,** 6956 (1995).
[33] D. Filipovic, M. D. Paulsen, P. Loida, R. L. Ornstein, and S. G. Sligar, *Biochem. Biophys. Res. Commun.* **189,** 488 (1992).
[34] J. Fruetel, Y.-T. Chang, J. Collins, G. Loew, and P. R. Ortiz de Montellano, *J. Am. Chem. Soc.* **116,** 11643 (1994).
[35] J. R. Collins, D. L. Camper, and G. H. Loew, *J. Am. Chem. Soc.* **113,** 2736 (1991).
[36] P. J. Loida, S. G. Sligar, M. D. Paulsen, G. E. Arnold, and R. L. Ornstein, *J. Biol. Chem.* **270,** 5326 (1995).
[37] P. R. Ortiz de Montellano, J. A. Fruetel, J. R. Collins, D. L. Camper, and G. H. Loew, *J. Am. Chem. Soc.* **113,** 3195 (1991).
[38] M. D. Paulsen and R. L. Ornstein, *J. Comput.-Aided Mol. Des.* **6,** 449 (1992).

profile.[39] These product profile predictions all start by assuming that the experimentally observed hydroxylation or epoxidation pattern for a given potential substrate will be determined by a combination of the intrinsic reactivity of a particular site on the substrate and the accessibility of that site to the presumed ferryl intermediate. To address the accessibility issue, molecular dynamics simulations can be used to generate an ensemble of substrate orientations in the binding pocket of a P450. Due to the requirement of a complete set of starting coordinates in order to proceed with the calculation of an MD trajectory, all of the product profiles studies prior to 1995 have examined only substrates of P450cam. The reports since 1993 of three additional crystal structures of members of the cytochrome P450 superfamily now permit the use of this approach for these additional P450s, and in principle the approach could also be used with an appropriate homology-derived model. Several issues need to be addressed in doing a calculation of this type, including the starting conformation of the substrate in the enzyme–substrate complex, the force field used to describe the protein and substrate, the type and length of simulation to perform, and the criteria for evaluating the ensemble of configurations generated.

Substrate and Force Field

The studies reported to date have examined both substrates for which crystal structures of the P450–substrate complex have been reported and substrates for which no detailed structural information is available for the complex. In the absence of detailed structural information it is necessary to dock the novel substrate into the P450-binding pocket. The most common approach for generating a starting conformation in these product profile studies is to first optimize the geometry of the substrate in the absence of the enzyme and then manually dock the substrate into the binding pocket of P450cam. Energy minimization can then be used to optimize the starting geometry. For the case of highly flexible substrates, the preferred conformations when bound to the enzyme may not be the same as those deduced for the isolated molecule in the gas phase. In most cases, there will be more than one energetically favorable way of docking the substrate into the active site; in this case, it is important to run trajectories starting from each of the different orientations found by whatever docking procedure is used. Even in the case where structural information is available, it must be remembered that the reported structure is for the binary complex with a ferric heme, not for the putative ferryl intermediate which is the catalytically relevant structure for which the MD trajectory will be calculated. In the

[39] M. D. Paulsen, D. Filipovic, S. G. Sligar, and R. L. Ornstein, *Protein Sci.* **2**, 357 (1993).

case of the substrate analog thiocamphor, it is clear that the reported crystal conformations in the binary complex are sterically incompatible with the formation of the ferryl intermediate and that the substrate conformation(s) in the reactive intermediate will be significantly different from the crystallographically observed conformations.[40,41]

After constructing the initial model, choices concerning the force field to be used in the calculation will need to be made. A variety of force fields have been developed to model proteins.[42–47] Of these, both AMBER and CVFF have been used successfully in published product profile studies of cytochrome P450cam. Regardless of which protein force field is used, additional force field parameters will be required to describe the heme cofactor. Two sets of heme parameters have been published for use with AMBER,[11,32] both of which have been used to succesfully predict product profiles. The product profile simulations calculated with the CVFF force field have all used a single set of heme parameters.[24,33] If the substrate being examined has a novel functional group, it will be necessary to develop additional force field parameters to describe the bonded interactions in the substrate as, for example, in the recent study of thioansole hydroxylation by Loew and co-workers.[34] Even if no novel functional groups are present in the substrate so that existing force field parameters are available for all the bonded interactions, it is still necessary to derive a set of partial atomic charges to describe the electrostatic interaction of the substrate with the enzyme and with solvent. Researchers have used a variety of methods for assigning substrate partial charges, including Mulliken populations,[34] charges derived by analogy with other identical functional groups for which partial charges are already available,[33,38] charges obtained by fitting a quantum mechanical electrostatic potential,[11] or by using a natural bond order analysis of a quantum mechanical wave function.[31] In their study of thioanisole and *p*-methylthioanisole sulfoxidation, Loew and co-workers[34] tested

[40] R. Raag and T. L. Poulos, *Biochemistry* **30**, 2674 (1991).
[41] M. D. Paulsen and R. L. Ornstein, *Protein Eng.* **2**, 357 (1993).
[42] B. R. Brooks, R. E. Bruccoleri, B. D. Olafson, D. J. Slater, S. Swaminathan, and M. Karplus, *J. Comput. Chem.* **4**, 187 (1983).
[43] S. J. Weiner, P. A. Kollman, D. T. Nguyen, and D. A. Case, *J. Comput. Chem.* **7**, 230 (1986).
[44] W. D. Cornell, P. Cieplak, C. L. Bayly, I. R. Gould, K. M. Merz, Jr., D. M. Ferguson, D. C. Spellmeyer, T. Fox, J. W. Caldwell, and P. A. Kollman, *J. Am. Chem. Soc.* **117**, 5179 (1995).
[45] P. Dauber-Osguthorpe, V. A. Roberts, D. J. Osguthorpe, J. Wolff, M. Genest, and A. T. Hagler, *Proteins: Struct., Funct., Genet.* **4**, 31 (1988).
[46] A. T. Hagler, *Peptides (N.Y.)* **7**, 213 (1985).
[47] W. F. van Gunsteren and H. J. C. Berendsen, "Groningen Molecular Simulations (GROMOS) Library Manual." Biomos, Nijenborgh, 16, 9747. AG Groningen, The Netherlands, 1987.

two sets of partial charges for the substrates and found that the differences in the calculated product profiles were negligible.

Model System Issues: Size and Constraints

At this point, several choices must be made about the type of MD simulation to run. Two related questions are the amount of explicit solvent to include in the simulation and whether to simulate the motions of the entire enzyme or, instead, to freeze the atomic coordinates of most of the enzyme and simulate the motions of only those residues close to the substrate. The inclusion of significant amounts of explicit solvent will greatly increase the amount of cpu time required to calculate a given length of trajectory. Because of the large size of the enzyme itself, the product profile studies reported to date have used only crystallographically determined solvent molecules or a small layer of added explicit solvent in the simulations. In general, a simulation with only crystallographic waters (~200 in the case of P450) will result in a slightly smaller, more compact time-average structure for the enzyme than an analogous simulation with a significant amount of added solvent. When examining the binding pocket of P450cam, which is deeply buried, this phenomenon is likely to have only a minor effect. The good agreement in most of the studies published to date between experimental and calculated product profiles indicates that the use of a minimal amount of explicit solvent is an acceptable approximation, particularly since it enables a researcher to do either much longer simulations or repeated simulations from different starting conformations in order to improve the sampling of reactive conformations. Two different predictions of the product profile for d-norcamphor have been reported using the CVFF force field: one in which only crystallographic waters were included and a distance-dependent dielectric was used, and one in which a 3-Å layer of additional explicit solvent was included and a dielectric of 1.0 was used in calculating electrostatic interactions.[36,38] The two studies gave similar predictions for the ratio of 5-alcohol to 6-alcohol formation. Clearly, however, if the substrate-binding pocket was not well sequestered from bulk solvent as is the case for P450BM-3 or P450eryF, it will be necessary to include significantly more explicit solvent to ensure that artifactual rearrangements of the substrate-binding pocket do not occur during the trajectories.

For the reaction profile studies reported to date, about an equal number have used a mobile binding pocket approximation versus simulating the dynamics of the entire enzyme. In the binding pocket only simulations, the mobile region has generally been defined as a sphere of residues surrounding the substrate of 12 Å or more in radius. Such an approximation reduces the number of atoms included in the simulation by about 75% and thus results in a significant time savings. This approximation rests on the assumption that the binding pocket undergoes no significant rearrangement in response to the binding of different substrates. The available crystal data indicate that this is a reasonable assumption for P450cam, but for other isozymes like P450BM-3, this assumption may not be at all warranted. Clearly those residues that are frozen will have artificially low mobilities, as will any residue bonded to a frozen residue. The residues in nonbonded contact with frozen residues will also have damped motions. Thus a sufficiently large buffer region is required between the substrate-binding pocket and the fixed portions of the enzyme if reliable predictions are to be obtained. The good agreement between experimental and calculated product profiles seen in studies of styrene and β-methylstyrene epoxidation and thioanisole sulfoxidation, which used a 12-Å sphere of moving residues, suggests that substrate dynamics are not significantly damped or skewed due to freezing most of the enzyme in the simulations.[35,38] However, in the same study with the thioanisole results, much poorer agreement was seen with p-methylthioanisole which the authors argue may have been due to the frozen protein approximation. In a recent study by Harris and Loew[11] on a series of camphor analogs, a direct comparison of product profiles calculated with full protein dynamics and with a 16-Å sphere of moving residues was made with camphane as a substrate. The percentage of conformations with C5 of camphane as the reactive carbon was significantly lower in the full protein simulations than in the mobile-binding pocket simulations. Studies in our laboratory on the substrate analog d-norcamphor have also compared the results of simulations with differing fractions of the enzyme included in a mobile-binding pocket. A comparison of five different trajectories is summarized in Table I. Simulations with a 6-, 9-, or 12-Å

TABLE I
COMPARISON OF SUBSTRATE MOBILITY AND PREDICTED PRODUCT PROFILE FROM
SIMULATIONS OF d-NORCAMPHOR-BOUND P450cam

Simulation description	Number of residues	Substrate B-factor ($Å^2$)	Product profile C5:C6:C3	Relative timing
6 Å sphere of residues	23	3.63	100:0:0	1.0
9 Å sphere of residues	61	3.68	100:0:0	1.9
12 Å sphere of residues	113	5.07	100:0:0	3.5
15 Å sphere of residues	177	21.19	10:90:0	4.8
Full protein	405	53.03	54:41:4	10.9
Experimental		33.5[a]	67:31:2[b]	

[a] Raag and Poulos.[40]
[b] Loida et al.[36]

sphere of moving residues all show very small substrate mobility. The predicted product profiles for these simulations also indicate that the substrate is only sampling the crystal conformation. There is clearly a qualitative change in the dynamics of the substrate when the sphere of moving residues is increased to 15 Å. The calculated B-factor for the substrate increases by a factor of four, although it is still only half its value in a free protein trajectory. In addition, both conformations favoring C5 hydroxylation and C6 hydroxylation are observed, although perhaps due to the relatively short length of the trajectories, the predicted ratio is not in good agreement with the experimentally observed product profile.

Substrate–Enzyme Interactions: Sampling and Product Formation

Most of the product profile studies have used multiple shorter trajectories instead of a single longer trajectory. For any substrate that has multiple energetically favorable binding orientations in the active site of a P450, it is much better to use multiple shorter trajectories, starting from different binding orientations, than a single longer trajectory. Because of the barriers to rotation in the active site of P450cam, even a simulation of 500 or more picoseconds is unlikely to result in a substrate sampling several minima. A recent study by Ornstein and co-workers[36] on the hydroxylation of *l*- and *d*-norcamphor used four separate trajectories of each enantiomer and saw significant variations in the predicted product profile from trajectory to trajectory. The same phenomenon was also observed by Harris and Loew[11] in their study of the *l*- and *d*-norcamphor system in which they did two simulations of each enantiomer. In one trajectory, no C3 product was predicted for *l*-norcamphor while in the other trajectory one-third of the product was predicted to be the C3 alcohol. The fewer specific contacts between substrate and binding pocket, the more likely it is that there will be significant variation in the predicted product profile among trajectories.

The final question that must be answered is how to transform the raw trajectories into a product profile prediction. The published studies have used geometric criteria (distances to the ferryl oxygen and/or angles involving the substrate and the ferryl oxygen) to classify each configuration examined as either reactive or unreactive and, for the reactive configurations, to determine which site on the molecule will be involved in the reaction. In the study by Paulsen and Ornstein[38] on hydroxylation of *d*-norcamphor, the use of only a distance criterion or both an angle and distance criteria resulted in a minor change in the predicted product profile. In the same study, little effect was also seen in the predicted product profile for thiocamphor when just a distance criterion was used compared to the results obtained when both a distance and angle criterion were used. In the recent study by Harris and Loew[11] on the hydroxylation of camphor analogs, little difference was seen in the distribution of reactive conformations for either camphane or thiocamphor when the cutoff distance for reactive conformations was increased from 2.7 to 3.4 Å. For many substrates, the vast majority of configurations examined will be classified as unreactive. For instance, in recent simulations of norcamphor-bound P450cam, more than 75% of the configurations examined were classified as unreactive.[36] Thus it would be necessary to simulate for four times as long in order to get the same number of reactive conformations as in the case of camphor where over 90% of the conformations generated were classified as reactive. Likewise, in a recent study of thioanisole sulfoxidation, some of the trajectories had as few as 2% of the configurations examined in a reactive conformation.[34] For the entire set of 12 simulations, about 40% of the configurations were classified as reactive. For some substrate analogs, such as thiocamphor and camphane, a large fraction of the configurations generated will be reactive, but for most novel substrates, behavior similar to that of norcamphor and thioanisole will be typical.

Advances in cpu processor technology over the past several years have made simulations of systems the size of P450cam much more tractable and have moved such calculations from the realm of the supercomputer to that of a desktop workstation. The studies published to date by a number of different researchers indicate that if due attention is given to ensuring proper sampling of the important substrate conformations, accurate product profiles can be predicted for a number of reactions of P450, including hydroxylations, epoxidations, and sulfoxidations. These studies suggest that under the appropriate conditions, many time-saving approximations such as the inclusion of little or no added explicit solvent, or simulating the dynamics of only the substrate binding pocket can still lead to reliable results.

Acknowledgments

Pacific Northwest National Laboratory is a multiprogram laboratory, operated for the U.S. Department of Energy by Battelle Memorial Institute under Contract DE-AC06-76RLO 1830. JIM has a graduate student fellowship (in the Department of Biophysics, Roswell Park Division, State University of New York at Buffalo) administered by Associated Western Universities. This work was supported by a grant (KP0402) from the Health Effects and Life Sciences Research Division within the Office of Health and Environmental Research of the Office of Energy Research of the U.S. Department of Energy (RLO). Computer resources were provided in part by the National Energy Research Supercomputer Center (Livermore, CA) of the U.S. Department of Energy.

[39] Approaches to Crystallizing P450s

By Thomas L. Poulos

Introduction

The main bottleneck in any crystal structure determination is very often the initial crystallization. The usual trial and error method has been considerably streamlined with the recent development of commercially available crystallization kits. The author's laboratory, as well as many other protein crystallography laboratories, has had considerable success with these kits. The primary advantage is that the kits are so simple to use that anyone with a modest amount of wet laboratory experience can try their hand at crystallizing their favorite protein, hopefully a P450. The more experienced individual is needed to take the initial crystals, which most often are not good enough for diffraction work, to that next level of quality for high resolution studies. This requires some dexterity skills in manipulating small crystals but is something most people with a steady hand can learn rather quickly. Protein crystallization also is becoming more of a science than an art, although a "green thumb" still helps. International meetings are held on a regular basis where the latest information is shared and published.[1] There also is a crystallization data base[2] which consolidates in one place all the known information on protein crystallization, the "bible" of protein crystallization[3] and a more recent treatment of membrane protein crystallization.[4]

In this chapter, a summary of methods which have been successfully used to crystallize P450s is described with more details provided on those P450s being worked on in the author's laboratory. Since most of the P450 community is interested in P450s that are membrane bound, some strategies for crystallizing these P450s will be considered.

---

[1] The proceedings from International Conferences on Crystallization of Biological Macromolecules usually are published. The proceedings from the 3rd meeting were published in the *Journal of Crystal Growth*, Vol. 110 and proceedings from the 5th meeting were published in *Acta Crystallogr, Sect. D*, Vol. D50, Part 4.

[2] G. L. Gilliland, M. Tung, D. M. Blakeslee, and J. E. Ladner, *Acta Crystallogr., Sect. D* **D50,** 408 (1994).

[3] A. McPherson, "The Preparation and Analysis of Protein Crystals." Wiley, New York, 1982.

[4] H. Michel, "Crystallization of Membrane Proteins." CRC Press, Boca Raton, FL, 1991.

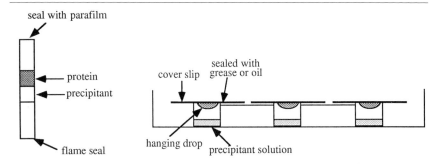

seal with parafilm

protein
precipitant

flame seal

cover slip

sealed with
grease or oil

hanging drop   precipitant solution

Free interface diffusion   *Vapor diffusion*

FIG. 1. Schematic representation of the free interface diffusion and hanging drop methods of protein crystallization.

Crystallization of P450s

P450cam. P450cam was first crystallized by Yu *et al.*[5] in 1974 but this form, which was later termed orthorhombic I,[6] was not suitable for the initial structure determination. A new crystal form was obtained, orthorhombic II, by switching to high protein concentrations and using the free interface diffusion technique (Fig. 1).[7] In this method, the protein is layered over an equal volume of precipitant (ammonium sulfate, PEG, etc.). In theory, nucleation sites form at the interface between the protein and precipitant where the concentrations of both are very high. Over a period of time, the two phases mix and, hopefully, crystals form.

During the early days of crystallization trials with P450cam, it was known that P450cam had a tendency to dimerize through an S–S bond. Given the high concentration of P450cam used, it was deemed necessary to use thiols to protect the SH groups. However, dithiothreitol (DTT) forms a bis–thiolate complex with P450cam and it was reasoned that the thiol protectant might compete with the substrate, camphor, for the active site, giving a heterogeneous mix. Therefore, it was decided to work with the substrate-free form of P450cam. To remove camphor, the enzyme is first chromatographed over a Sephadex G25 column equilibrated with 0.05 M Tris–HCl, pH 7.4, followed by a second and immediate desalting step using a 0.05 M potassium phosphate buffer, pH 7.0, containing 0.25 M KCl. Both columns can be run at room temperature. The reason for the two desalting steps is

[5] C.-A. Yu, I. C. Gunsalus, M. Katagari, K. Suhara, and S. Takemori, *J. Biol. Chem.* **249,** 94 (1974).
[6] T. L. Poulos, M. Perez, and G. C. Wagner, *J. Biol. Chem.* **257,** 10427 (1982).
[7] F. R. Salemme, *Arch. Biochem. Biophys.* **151,** 533 (1972).

the well-known effect of potassium promoting the binding of camphor.[8] Hence, the first column removes potassium which lowers the camphor affinity, thereby ensuring that the enzyme is substrate free. The second column adds potassium back for enzyme stability. Solid DTT is added to 50 mM and the protein is concentrated using Centricon devices to 1 mM (about 45 mg/ml). The spectral properties indicated, and the crystal structure later proved, that the DDT diffuses into the substrate pocket and forms a bis–thiolate complex, as expected. This likely will happen with other P450s so care should be taken on how thiols are utilized.

A series of ammonium sulfate solutions ranging from 60 to 70% saturation in 2% increments in 0.05 M potassium phosphate, 0.25 M KCl, pH 7.0, are adjusted to pH 7.0 with 2 N KOH. Initially, small test tubes were used to layer 50 μl of 1 mM P450cam onto 50 μl of the ammonium sulfate solution. The method has been adapted to capillary tubes in order to save protein. In this method, 100- to 200-μl capillary micropipettes are first treated with an organosilane surface agent like Prosil 28. The micropipettes are broken to about 6–7 cm in length and flame sealed at one end. Using a Hamilton syringe, 10 μl of ammonium sulfate is delivered about half way down the micropipette, taking care not to form drops on the sides. If drops do form, the micropipette is discarded. Ten microliters of P450cam is layered over the ammonium sulfate, the open end is sealed with Parafilm, and the micropipettes are placed in an incubator at 10–13°. Crystals usually appear within 1–2 days and reach a maximum size within 4–5 days. A delicate balance exists between protein and ammonium sulfate concentrations and temperature. All three are difficult to strictly control and the micro-free interface diffusion method is difficult to repeat exactly the same each time. Therefore, it is necessary to use a range of ammonium sulfate concentrations for each crystallization setup. Typically a range of 62–68% in 2% increments is sufficient. The yields are not good and generally about five very good crystals are obtained from four to six tubes.

To recover crystals, the ends of the micropipettes are opened and a stream of 40% ammonium sulfate, 0.05 M potassium phosphate, and 0.25 M KCl, pH 7.0, is squirted from a syringe into one end of the micropipette which forces the crystals out the other end. Crystals that adhere to the sides of the capillary can be nudged free by a thin piece of Parafilm or a syringe needle. The softer Parafilm is preferred. Normally, the 40% ammonium sulfate mother liquor will contain whatever ligand, substrate, or inhibitor is being investigated. Therefore, the crystals are never in the substrate-free form. To obtain the substrate-free form requires extensive washing to ensure that the DTT is removed. Insufficient washing allows DTT to remain

[8] J. A. Peterson, *Arch. Biochem. Biophys.* **144,** 678 (1971).

as experience demonstrates since the author's laboratory unintentionally solved the structure of the bis–thiolate DTT complex.

P450scc. A brief report appeared in 1988[9] describing crystallization of P450scc, the cholesterol sidechain cleavage P450 from bovine adrenocortical mitochondria. Prior to crystallization trials, the P450scc was modified with pyridoxal phosphate (PLP) and NaBH$_4$. The PLP appeared to covalently modify the protein by attaching to lysine groups. The chemistry likely involves initial formation of the imine, protein–N$=$CH–PLP, between the lysine side chain NH$_2$ group and the pyridoxal phosphate followed by reduction with NaBH$_4$ to the secondary amine, protein–NH–CH$_2$–PLP. Dialysis against water gave thin plate-like crystals that apparently were not of sufficient quality for X-ray diffraction studies.

P450BM-3 Heme Domain. The initial crystallization of the heme domain proved rather straightforward.[10] The hanging drop vapor diffusion method (Fig. 1) was used to screen a large number of conditions. The following protocol was the first to work. The heme domain in 0.1 M potassium phosphate, 1 mM DTT, 0.1 mM EDTA, pH 7.0, is concentrated to 15–20 mg/ml. The reservoir solution for the vapor diffusion contained 54–58% saturated ammonium sulfate and 3% methyl pentane diol in 0.05 M HEPES, pH 7.0. Protein (5 μl) is mixed with 5 μl of the reservoir solution on the coverslip. Crystals grew at both 20 and 4°. As shown in Table I, the crystals do not diffract well but are probably suitable for the initial structure determination to 3.0 Å. The main problem with this crystal form was the large unit cell edge (Table I) and the uncertain number of molecules in the asymmetric unit. This made conventional data collection and interpretation of heavy atom difference Pattersons difficult.

Boddupalli *et al.*[11] then described the heme domain crystals that ultimately led to the solution of the structure.[12,13] The important variables appear to be very high concentrations of protein to initialize nucleation, 5 mM (about 275 mg/ml), and the inclusion of 40 mM MgSO$_4$. The hanging drop vapor diffusion method yielded the P2$_1$ crystal form that diffracts to at least 2.0 Å (Table I). Our laboratory obtained the same crystal form by a slightly different procedure. The two main differences were a shortened version of the heme domain consisting of residues 1–455 rather than 1–471 and the exclusion of DTT. Initially, the high concentrations of enzyme used

[9] Y. Iwamoto, M. Tsubaki, A. Hiwatashi, and Y. Ichikawa, *FEBS Lett.* **233,** 31 (1988).

[10] H. Li, K. Darwish, and T. L. Poulos, *J. Biol. Chem.* **266,** 11909 (1991).

[11] S. S. Boddupalli, C. A. Hasemann, K. G. Ravichandran, J.-Y. Lu, E. J. Goldsmith, J. Deisenhofer, and J. A. Peterson, *Proc. Natl. Acad. Sci. U.S.A.* **89,** 5567 (1992).

[12] K. G. Ravichandran, S. S. Boddupalli, C. A. Hasermann, J. A. Peterson, and J. Deisenhofer, *Science* **261,** 731 (1993).

[13] H. Li and T. L. Poulos, *Acta Crystallogr., Sect. D* **D51,** 21 (1995).

gave dense precipitates plus a few useful crystals. These crystals were removed and washed with 20% PEG 8000, 50 mM $MgSO_4$ in 0.1 M PIPES, pH 6.8. The crystals are ground into microcrystals and used as seeds. For preparing diffraction quality crystals, the reservoir solution in the hanging drop setup contained 18% PEG 8000, 50 mM $MgSO_4$ or 16% PEG 8000 and 100 mM $MgSO_4$, both in 0.1 M PIPES, pH 6.8. The protein concentration was also lowered closer to 1 mM (40–50 mg/ml) so the final concentration of protein after mixing with an equal volume of reservoir solution was 20–25 mg/ml. After equilibration for at least 1 day at 4–7°, the drop was touch seeded. This method involves using a thin wire or a cat wisker to touch a solution containing the microcrystal seeds to the surface of the equilibrated drop. Two crystal habits appear, one plate shaped and one rod shaped. These crystals are isomorphous with respect to space group and cell dimensions, but the rods are more ideally shaped for data collection. The main advantage of the seeding method is that much lower concentrations of protein are required so less protein is consumed per crystallization trial. In addition, the number of crystals obtained in each drop can be controlled by adjusting the concentration of seeds in the microcrystal seeding solution.

P450terp. Boddupalli *et al.*[11] obtained crystals of P450terp suitable for determining the structure[14] by the hanging drop method. Again a high concentration of protein was required, about 50 mg/ml. The crystals do not diffract as well as either the heme domain of P450BM-3 or P450cam (Table I) but nevertheless were good enough to obtain the 2.3-Å crystal structure.

P450nor. This is a relatively new P450 with an unusual activity. The enzyme catalyzes the NADH-dependent reduction of nitric oxide without the need for another redox partner[15]:

$$2NO + NADH + H^+ \rightarrow N_2O + NAD^+ + H_2O.$$

It appears that NADH can deliver electrons directly to the P450, a most unusual and expected feature. A recent brief report describes the crystallization of the enzyme in a form suitable for structure determination[15] (Table I).

P450eryF. This P450 is from the actinomycete *Saccharopolyspora eythaea* where it participates in the biosynthesis of erythromycin.[16] The substrate for P450eryF is 6-deoxyerythronolide (6-DEB), a complex 14 member

[14] C. A. Hasemann, K. G. Ravichandran, J. A. Peterson, and J. Deisenhofer, *J. Mol. Biol.* **236,** 1169 (1994).
[15] K. Nakahara, H. Shoun, S.-I. Adachi, and Y. Shiro, *J. Mol. Biol.* **239,** 158 (1994).
[16] J. F. Andersen and R. Hutchinson, *J. Bacteriol.* **174,** 725 (1992).

TABLE I
A SUMMARY OF P450s THAT HAVE BEEN CRYSTALLIZED

P450	Space group	Unit cell parameters						Diffraction limit (Å)
		a(Å)	b (Å)	c (Å)	α(°)	β(°)	γ(°)	
cam	P2$_1$2$_1$2$_1$	108.5	104.4	36.4	90	90	90	1.7
BM-3	?	306	108	65	90	90	90	3.0
terp	P6$_1$22	68.9	68.9	458.7	90	90	120	2.3
BM-3	P2$_1$	59.4	154	62.2	90	94.7	90	2.0
scc	—	—	—	—	—	—	—	—
nor	P2$_1$	74.7	86.7	62	90	97	90	≈2.5
eryF	P2$_1$2$_1$2$_1$	54.2	79.7	99.5	90	90	90	2.0

compound with 21 carbon and 6 oxygen atoms. Initial crystallization screens were carried out with the hanging drop method using one of the commerically available crystallization kits (Hampton). The final protocol is as follows.[17] The protein is incubated for 5 days at 10 mg/ml in 24% PEG 4000, 0.2 M sodium acetate, and 0.1 M Tris, pH 8.5. A considerable amount of protein precipitate forms which is removed by centrifugation. The 6-DEB substrate is added to the clarified protein solution to a concentration of 0.10 mM and vapor diffused against the 24% PEG 4000 reservoir. Small crystals are collected and washed in the same buffer except that the PEG 4000 concentration is lowered to 21%. The crystal is added to a fresh hanging drop containing 0.1 mM substrate and vapor diffused against the 24% PEG 4000 solution. The small seed crystals double in size over a period of 4 days to give crystals suitable for determining the crystal structure.[18]

What should be evident from what has been described so far is that the current methods used for crystallizing P450s are much simpler than we used for P450cam. This is primarily due to the increased popularity of the hanging drop method, the availability of crystallization kits, and the increasing use of recombinant proteins which enables a large amount of protein to be committed to crystallization trials. It probably is the case that a new crystal form of P450cam would be found by using one of the kits, thus avoiding the technical problems with the capillary-free interface diffusion methods. The author's laboratory has used the kits to successfully obtain a new crystal form of substrate-free P450cam, although these are not yet good enough for diffraction work (H. Li, unpublished).

[17] J. R. Cupp-Vickery, H. Li, and T. L. Poulos, *Proteins* **20**, 191 (1994).
[18] J. R. Cupp-Vickery and T. L. Poulos, *Nat. Struct. Biol.* **2**, 144 (1995).

Strategies for Membrane-Bound P450s

The P450BM-3 Lesson. P450s can be broadly classed as those that are very specific and those that are not. P450cam and P450eryF belong to the specific class while P450BM-3 is a prototype for the less specific microsomal P450s. Experiences gained with the P450BM-3 crystals then may be of some use in working with microsomal P450s. The heme domain crystals have two molecules in the asymmetric unit.[11] The access channel that connects the surface of the enzyme to the active site through which substrates must pass is wide open in one of the molecules and partially closed in the other. Energy minimization[10] and subsequent molecular dynamics simulations[19] show that the access channel can undergo a rather large open/close motion. This shows that there is essentially no energy barrier in going between the two extreme open and closed conformations. The intermolecular contacts in the crystal lattice hold the access channel in the open conformation[10] which is why soaking crystals in fatty acid substrates leads to no binding or crystal cracking. The access channel is locked in a conformation incompatible with substrate binding.

Recently our laboratory has succeeded in cocrystallizing the heme domain with a fatty acid substrate, but the crystals diffract to only about 3.2 Å and have several molecules in the asymmetric unit (H. Li, unpublished). Such poor crystal quality compared to the substrate-free form could reflect heterogeneity in the substrate-bound complex. That is, the access channel may be able to adopt one of several conformations between the extremes of completely open to completely closed, all of which are compatible with substrate binding. Such heterogeneity could reflect the fact that P450BM-3 is able to utilize fatty acids of variable length and saturation,[20] which also makes sense based on some molecular modeling work carried out in our laboratory. The methyl and methylene groups of the substrate near the heme should be rigidly fixed, but the remainder of the fatty acid chain could be viewed as a rope that can adopt a variety of conformations as long as the chain remains dehydrated in the hydrophobic crevice of the access channel. Such flexibility might well be compatible with a variety of protein conformers. While this is only a working hypothesis, it is consistent with what has been seen in the crystal structure and from our experiences in crystallizing the substrate free and bound forms.

The main value of this view is its relevance to eukaryotic P450s that also are relatively nonspecific and might be expected to also experience a large range of motion in the access channel. It is the opinion of the author

[19] M. D. Paulsen, and R. L. Ornstein, *Proteins* **21**, 237 (1995).
[20] Y. Mirua and A. J. Fulco, *Biochim. Biophys. Acta* **388**, 395 (1975).

that investigators might have more success in crystallizing eukaryotic P450s by focusing on specific P450s such as the steroid hydroxylases. One might anticipate that these P450s form more homogeneous complexes and would, therefore, have a better chance of crystallizing.

The Membrane. A major factor that must be considered with eukaryotic P450s is that these enzymes are membrane bound while the examples discussed earlier, with the exception of P450scc, are soluble enzymes. However, because P450s do not appear to be integral membrane proteins, the N-terminal hydrophobic sequences anchor the protein to the membrane.[21] The strategy that has been employed to crystallize integral membrane proteins is not, therefore, likely to be applicable to P450s. It should not be taken as fact that all P450s are not integral membrane proteins so one should not dismiss any potential crystallization protocol unless there is hard evidence on the membrane-binding mode of whatever P450 one is attempting to crystallize. A preferred approach is to "create" soluble P450s. Removing the leader peptide does increase the soluble fraction in recombinant systems,[21,22] but it also is clear that a significant amount of the protein remains attached to the membrane. Hence, there are likely other "sticky" regions apart from the N-terminal sequences. In this case, some of the exotic cocktails that have worked with integral membrane proteins[4] might be useful for P450s.

Another factor that must be considered is the state of the protein once separated from the membrane. Figure 2 outlines the prevailing view on how the P450 is oriented relatively to the membrane.[23] The substrate access channel is oriented toward the lipid bilayer such that lipophilic substrates in the microsomal membrane can diffuse into the P450 access channel. This leaves free the Cys–ligand surface available for interaction with the reductase. Such a picture leads to the possibility that the lipid bilayer itself might form part of the access channel. Certainly it is true that many P450s are quite unstable when separated from the membrane and, with many proteins, including P450s, this instability can be overcome using glycerol.

The Potential of Glycerol and Polyols in P450 Crystallization. Glycerol is widely used as a protein-stabilizing agent but only recently has glycerol and other polyols received attention as agents for promoting crystal growth. These are now a number of examples where glycerol has been employed to promote crystal growth.[24] The physiochemical processes involved are

[21] S. J. Pernecky, J. R. Larson, R. M. Philpot, and M. J. Coon, *Proc. Natl. Acad. Sci. U.S.A.* **90,** 2551 (1993).

[22] S. J. Pernecky, N. M. Olken, L. L. Bestervelt, and M. J. Coon, *Arch. Biochem. Biophys.* **318,** 446 (1995).

[23] D. R. Nelson and H. W. Strobel, *J. Biol. Chem.* **263,** 6038 (1988).

[24] R. Sousa, *Acta Crystallogr., Sect. D* **D51,** 271 (1994).

electron transfer

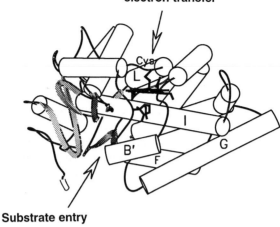

Substrate entry

Membrane

FIG. 2. A model of P450cam shown relative to the lipid bilayer according to how P450s are thought to be oriented on the membrane.[23] The substrate access channel is oriented toward the membrane while the "proximal" surface containing the Cys heme ligand is oriented toward the solvent. This region then is available to interact with redox partners.

not fully understood, but the mechanism of glycerol stabilization appears to be in altering protein hydration. High concentrations of glycerol ($\approx$20–30% v/v) lower the amount of water available for hydrating the protein surface whereas glycerol is repelled from hydrophobic surfaces but attracted to polar regions. The net effect of these combined forces is to stabilize regions of the protein that would otherwise be disordered, leading to a conformationally homogeneous protein. In the case of P450s, one could think of the conformationally flexible access channel as becoming more homogeneous in one of the available isoenergetic conformational states due to the change in solvent activity which promotes stabilizing intramolecular interactions. Therefore, if membrane-bound P450s belonging to the less specific group are conformationally "floppy" molecules, glycerol could be a useful agent to include in crystallization trials.

Engineering and Chemical Modification. Several eukaryotic P450s have been cloned and expressed in recombinant systems. The importance of the N-terminal hydrophobic leader region noted earlier clearly is important for partitioning the P450 in the cytosol or membrane.[21,22] Nevertheless, it appears that other regions of the protein also have a propensity for membrane binding and that N-terminal engineering might not be sufficient to obtain monodisperse soluble samples. Here it is useful to consider a recent success story where a hydrophobic protein with a tendency to form aggregates was engineered to crystallize. The central core of the HIV-1 integrase has a very low solubility which hindered crystallization efforts. To circumvent this problem, a systematic replacement of hydrophobic amino acids was carried out and an attempt was made to crystallize each variant.[25] Replacing a single Phe with Lys resulted in a protein that was soluble up to 25 mg/ml and crystallized. This dramatically illustrates the power of mutagenesis to promote crystallization as well as the influence that one surface hydrophobic group can have on the solubility of a protein. A similar effort might work with P450s. The problem may not be as formidable as one might first suppose because, unlike the HIV-1 integrase, we already know a good deal about what the structure of membrane-bound P450 should look like. This provides a reasonably good estimate of where loops and helical connections will be located, and with many of the helices we will have a good estimate of which parts of the sequence are oriented toward solvent and which parts of the sequence form the hydrophobic core. Those regions thought to be oriented toward the solvent would be the regions that one would target for mutagenesis. Such an effort would be an experimental tour de force but might well be worth the effort.

Chemical modification of surface residues also has been used to improve crystallization with the most recent example being myosin.[26] In this case, methylation of surface lysines resulted in a myosin that crystallized. As noted earlier, the modification of lysines in P450scc also promoted crystallization. These examples illustrate that both mutagenesis and chemical modification are avenues worth exploring for those P450s that have proven difficult to crystallize due to solubility, aggregation, or stability problems.

Conclusions

Although the number of examples is limited, it appears that a promising approach for crystallizing eukaryotic P450s is to use site-directed mutagenesis for the purpose of producing soluble, monodisperse P450s. The inclusion of glycerol or other polyols also should be included in crystallization trials. The commercial crystallization kits are highly recommended. Another new and important tool which shows some promise is light scattering.[27,28] New commercial devices are available that enable a rapid screen of protein

[25] F. Dyda, A. B. Hickman, T. M. Jenkins, A. Engelman, R. Craigie, and D. Davies, *Science* **266,** 1981 (1994).
[26] I. Rayment, W. R. Rypniewski, K. Schmidt-Base, R. Smith, D. R. Tomchick, N. M. Benning, D. A. Winkelman, G. Wesenberg, and H. M. Holden, *Science* **261,** 50 (1993).
[27] B. Bishop, J. C. Martin, and W. M. Rosenblum, *J. Cryst. Growth* **110,** 164 (1991).
[28] S. Veesler, S. Marco, S. Lafont, J. P. Astier, and R. Boistelle, *Acta Crystallogr., Sect. D* **D50,** 355 (1994).

solutions to check the state of aggregation and whether or not the sample is monodisperse. Since membrane-bound P450s may have a tendency to aggregate, a check on the state of aggregation prior to setting up a large number of crystallization trials could save a considerable amount of time and frustration.

[40] Crystallization Studies of NADPH–Cytochrome P450 Reductase

By Jung-Ja Park Kim, David L. Roberts, Snezana Djordjevic, Miug Wang, Thomas M. Shea, and Bettie Sue Siler Masters

Introduction

NADPH–cytochrome P450 reductase (CPR) is an essential component of the microsomal cytochrome P450 monooxygenase system. It is an integral membrane protein that catalyzes the transfer of electrons from NADPH to cytochrome P450 in the oxidative metabolism of both endogenous (steroids, fatty acids, and prostaglandins) and exogenous (therapeutic drugs, environmental toxicants, and carcinogens) substrates.

Despite the intensive studies carried out since its discovery by Horecker in the 1950s,[1] the exact molecular mechanism by which CPR transports electrons from NADPH through its two prosthetic flavins (FAD and FMN) to the heme proteins is still not known. The three-dimensional structure of CPR by X-ray analysis will reveal, at the molecular level, the interactions among the pyridine nucleotide NADPH, the flavins, and the polypeptide chain of CPR, as well as the spatial relationship among the different domains of CPR that allow for efficient electron transfer. In addition, the nature of the binding site for cytochrome P450 can be identified, providing the structural basis for the ability of CPR to interact with scores of different cytochromes P450. Obtaining crystals of CPR establishes the basis for the full structural analysis of CPR by X-ray diffraction methods. Furthermore, the crystallization conditions obtained from the CPR studies can be used as a starting point for crystallization of other FMN- and FAD-containing proteins, such as nitric oxide synthase.

[1] B. L. Horecker, *J. Biol. Chem.* **183**, 593 (1950).

Cloning, Expression, and Purification of CPR

The rat liver CPR was overexpressed in *Escherichia coli* utilizing the plasmid pOR263, containing the entire coding sequence for rat CPR cloned into the pIN-IIIA3 expression vector.[2] This cloning procedure adds the *ompA* signal peptide to the N-terminal end of the CPR protein, which directs transport of the expressed protein out of the cytoplasm into the periplasmic space of *E. coli*, where proteolytic activity is reduced.[3] Although the cloned CPR contains an additional eight amino acid residues (GIPGDPTN) prior to the N-terminal methionine resulting from this cloning procedure, detailed analyses have shown that the cloned CPR behaves identically to the native CPR, indicating that these additional residues have no effect on CPR function.[2]

E. coli cells containing the plasmid pOR263 are grown in 12 liters of LB media and induced after 2.5 hr (OD_{550} = 1.2) with 0.5 mM isopropyl β-D-thiogalactoside (IPTG) and allowed to grow an additional 15 hr at 37° prior to harvesting. The cell pellet is resuspended in 3600 ml of 75 mM Tris, pH 8.0, containing 0.5 M sucrose, 0.5 mM EDTA, and 12 mg lysozyme and incubated at 4° for 20 min. The spheroplasts (containing CPR) are then isolated by centrifugation at 2500 g for 2 hr. The pellet is then resuspended in 50 mM Tris, pH 8.0, containing 0.5 mM EDTA and 10 μg/ml aprotinin, after which it is sonicated. The mixture is centrifuged at 2500 g for 40 min to remove unsonicated material and is then ultracentrifuged at 100,000 g for 1 hr. At this stage, the CPR is still associated with the membrane fraction. The pellets are then homogenized in 50 mM Tris, pH 7.7, containing 0.1 mM EDTA, 0.05 mM dithiothreitol and 10% glycerol. The homogenized pellets are sonicated briefly and then treated with 0.1% Triton X-100 at 4° for 1 hr to solubilize the CPR protein. The CPR is finally removed from the cell debris by ultracentrifugation at 100,000 g for 1 hr. The supernatant is then directly applied to a 2′,5′-ADP Sepharose 4B affinity column and elution is achieved with Tris buffer, pH 7.7, containing 5 mM 2′-AMP and 0.1% Triton X-100.[4]

As isolated from the affinity column, there is a mixture of fully oxidized and semiquinone forms of the reductase. Therefore, the reductase isolated from the affinity column is oxidized by the addition of small aliquots of 2 mM ferricyanide until the semiquinone is no longer detected by visible spectroscopy (peak between 550 and 650 nm). Typically, for 30 mg of CPR, 0.5 to 0.75 ml of 2 mM ferricyanide is required.

Following ferricyanide oxidation, the CPR is exhaustively dialyzed

---

[2] A. L. Shen, T. D. Porter, T. E. Wilson, and C. B. Kasper, *J. Biol. Chem.* **264,** 7584 (1989).
[3] G. Duffaud, P. March, and M. Inouye, *Methods Enzymol.* **153,** 492 (1987).
[4] Y. Yasukochi and B. S. S. Masters, *J. Biol. Chem.* **251,** 5337 (1976).

against 50 mM phosphate buffer, pH 7.7. Immobilized trypsin bound to beaded agarose (Sigma Chemical) is added to the CPR solution at a ratio of 5 units of immobilized trypsin per 0.25 mg of the protein per ml of buffer solution (protein concentration critical), and the entire mix is rotated at 4° for 1 hr. The immobilized trypsin is then removed by filtration. The soluble domain of CPR is then purified using a 2′,5′-ADP Sepharose affinity column in the absence of any detergent. The enzyme is eluted with Tris buffer, pH 7.7, containing 5 mM 2′-AMP.

Final purification is achieved by applying CPR to a Bio-Rad Ceramic HPHT FPLC column and eluting with a linear potassium phosphate gradient from 10 mM, pH 7.0, to 200 mM, pH 7.7. The cleaved CPR is assayed for its ability to reduce the artificial electron acceptor, cytochrome c. To assay, 0.35 μg of purified CPR is added to a cuvette containing 0.1 mM NADPH, 40 μM cytochrome c, and 0.1 mM EDTA in 50 mM phosphate buffer, pH 7.7. Immediately after the addition of CPR to the cuvette, the change in absorbance at 550 nm is followed against a cuvette containing the assay mix in the absence of CPR as previously described.[5] A typical turnover number for the purified CPR is approximately 1400 per min per flavin with cytochrome c as the acceptor.[4]

Crystallization of CPR

The purified, clipped CPR is buffer-exchanged against 50 mM HEPES, pH 7.5, by repeated dilution/concentration using an Amicon Centracone. After the first round of dilution/concentration, the concentration of CPR is determined from absorbance at 455 nm (ε = 10.8 mM^{-1} flavin^{-1} cm^{-1}), and a 5× molar excess of purified FMN and a 10× molar excess of NADP$^+$ are added to the CPR solution and allowed to incubate at 4° for 1 hr. The CPR is then diluted with 50 mM HEPES, pH 7.5, and the dilution/concentration procedure is continued for an additional four rounds. Prior to the final concentration, a visible spectrum is recorded. The CPR is then concentrated to approximately 25 mg/ml as determined by A_{455}.

Crystallization experiments are performed by the vapor diffusion method using a combination of hanging-drop, sitting-drop, and macroseeding techniques.[6,7] Initial crystallization of CPR is carried out with the hanging-drop vapor diffusion technique in 24-well Linbro culture plates at 19°. An incomplete factorial experiment using a variety of precipitants, salts, and pH is performed.[8] Poor crystals were obtained from a solution con-

[5] B. S. S. Masters, C. H. Williams, and H. Kamin, *Methods Enzymol.* **10,** 565 (1967).
[6] A. McPherson, "Preparation and Analysis of Protein Crystals." Wiley, New York, 1982.
[7] E. A. Stura and I. A. Wilson, *Methods, Companion Methods Enzymol.* **1,** 38 (1990).
[8] C. W. Carter, Jr. and C. W. Carter, *J. Biol. Chem.* **254,** 12219 (1979).

taining 25% polyethyleneglycol 4500 (PEG 4500, Fluka Chemical) and 0.85 M NaCl in 50 mM HEPES, pH 7.0.

The crystallization conditions were further refined using macroseeding techniques with minor modifications of the crystallization solution. The final crystallization involved mixing equal volumes of CPR (15 mg/ml) and precipitant containing 21% PEG 4500, 150 mM HEPES, pH 6.5, 5 mM MgCl$_2$, and 0.8 M NaCl with equilibration at 19° overnight prior to seeding. Seed crystals (0.05 × 0.05 × 0.05 mm) that had been washed three times in artificial mother liquor containing 21% PEG 4500 are then introduced into the crystallization sitting drop. Crystals routinely grow to approximately 0.3 × 0.5 × 0.6 mm within 2–3 weeks.

X-Ray Analysis

The space group was determined by the precession method, and subsequent data collection was performed on an R-AXIS II image plate system using CuK$_\alpha$ radiation with a graphite monochrometer from a Rigaku rotating anode X-ray generator (RU200) operated at 50 kV and 100 mÅ. Prior to data collection, the crystals are first equilibrated at 4° for about 15 hr. After equilibration, the crystals are then mounted in thin-walled glass capillary tubes. Diffraction data sets are collected at 4° using a crystal to detector distance of 140 mm with $2\theta = 0°$. Data collection and reduction are carried out using the R-AXIS software.[9] For low temperature data collection, the crystals are first mounted in a loop made of hair and then are immersed in a mineral oil solution that serves as a cryoprotectant. The crystals are then frozen in liquid N$_2$, and the data collection is carried out at $-120°$ using a liquid N$_2$ cold stream.[10]

Discussion

Refinement of Crystallization Conditions

Rat liver CPR has a molecular mass of 78,225 Da and shares approximately 90% sequence identity among various mammalian reductases.[11] Although trypsin-solubilized, 72-kDa CPR cannot interact with cytochromes P450, it is capable of electron transfer to cytochrome c and other artificial electron acceptors, and therefore retains most of the functionality of the intact protein. From the initial screen using a modified incomplete factorial

[9] T. Higashi, *J. Appl. Crystallogr.* **23**, 253 (1990).
[10] H. Hope, *Acta Crystallogr., Sect. B: Struct. Sci.* **B44**, 22 (1988).
[11] A. L. Shen and C. B. Kasper, *Handb. Exp. Pharmacol.* **105**, 35 (1993).

experiment by Jancarik and Kim,[12] one condition (25% PEG 4500, 0.85 M NaCl, 50 mM HEPES, pH 7.0) yielded "poor" crystals of CPR (Fig. 1B) and established the starting point for all crystallization results to follow.

As is true in all crystallization experiments, purity of the sample is of the utmost importance for reproducibility and quality of crystals grown. In the case of CPR, it is even more so (First Purify, then Crystallize. "Don't Waste Clean Thinking on Dirty Enzymes."—E. Racker). CPR contains two noncovalently bound prosthetic groups, FAD and FMN, as well as a binding site for NADP$^+$, one of the natural substrates for CPR. With these many sites available for binding nucleotides, one must pay close attention to cofactor integrity and homogeneity during crystallization experiments. From preparation to preparation, it has been observed that the purified CPR has a different ratio of A_{280}/A_{454}, which can be used as a measure of the cofactor content of the purified protein. Since the last step of the purification procedure involved affinity chromatography with 2',5'-AMP, and elution was achieved with 2'-ADP (containing the 2'-phosphate which distinguishes NADP$^+$ from NAD$^+$), an initial hypothesis was that the enzyme must contain, at least with a partial occupancy, a nucleotide analog in the NADP$^+$-binding site. Ratios for the purified protein range anywhere from 5.6 to 7.2, indicating differences in the amounts of FAD/FMN/NADP$^+$ (analogs) in the purified protein. In addition, the redox states of these cofactors can be modulated, and therefore care must be taken to ensure that only one species with one redox state exists for each cofactor. One final problem that must be addressed is trypsinolysis of the holo-CPR protein to form "clipped" CPR. Since only 64 residues (including the *ompA* octapeptide) make up the difference between the two enzymes, careful separation must be achieved to ensure that all the protein is clipped and that all the clipped protein begins at the same residue. All of these have been found at one point to be important factors that influence the CPR crystal quality, and will be addressed in the following section.

The FMN cofactor has been shown to be easily removed during purification and can be added back to regain full activity.[13] FAD, on the other hand, is tightly bound to the enzyme and consequently is very difficult to remove. Therefore, initial concerns regarding cofactor integrity focused on FMN rather than FAD, and attempts were made to simply add FMN at different ratios to the crystallization cocktail. Commercially available FMN contains a significant level of riboflavin contaminant and must first be purified by DE52 chromatography. Systematic studies have shown that simply adding FMN and/or NADP$^+$ (or NADP$^+$ analogs) to the crystalliza-

[12] J. Jancarik and S.-H. Kim, *J. Appl. Crystallogr.* **24**, 409 (1991).
[13] J. L. Vermilion and M. J. Coon, *J. Biol. Chem.* **253**, 8812 (1978).

tion cocktail is not sufficient to support crystal growth (Fig. 1A), but rather the enzyme must be pretreated with each nucleotide prior to the crystallization experiment as follows: The enzyme is initially concentrated against 50 mM HEPES, pH 7.0, to ensure reproducibility. Then, a 5× molar amount of FMN (50–100 μM) and a 10× molar amount of NADP$^+$ (100–200 μM) are added to the enzyme and allowed to incubate for 1 hr on ice. Any excess ligands are removed by four additional concentration/dilution steps. Prior to setting up the crystallization experiment, the A_{280}/A_{454} ratio is typically 6.5–6.7. Only by this method have we been able to successfully grow crystals of CPR that are suitable for X-ray diffraction studies.

Ionic strength and divalent cations also play important roles in the crystallization of CPR. Optimization with NaCl indicated that high ionic strengths (0.8 M) are required for crystal growth. At low concentrations of NaCl or in the absence of 5 mM MgCl$_2$, thin, layered star-like clusters are the only crystals that form (Fig. 1B).

Crystals obtained at 21% PEG were in the form of layered plates. Although they often looked perfect, even under polarized light, diffraction analysis often indicated that twinned or multiple crystals were present. In an attempt to slow the crystallization process, simply lowering the temperature under the same conditions did not improve the crystal quality since only small needle-shaped crystals formed at either 4 or 10°. The only way to alleviate the problem of twinning was by using macroseeds that had been grown overnight in crystallization solution containing 25% PEG 4500 (Fig. 1C). Single crystals free of twinning and other crystallites were washed by serially transferring to the three washing solutions; the first one containing 25% PEG 4500, then 23% PEG 4500, and finally 21% PEG 4500. Washed seed crystals were then transferred to the crystallization dips, one seed in each dip, that had been equilibrated with the 21% PEG 4500 solution. The macroseeds took approximately 3 weeks to reach a size suitable for diffraction analysis (0.3 × 0.5 × 0.6 mm; Fig. 1D).

SDS-PAGE analysis of the crystalline protein showed a single band with a molecular mass identical to that of the trypsin-cleaved CPR, indicating that no proteolysis occurred during crystal growth. N-terminal sequence analysis confirmed these results, showing a single polypeptide that begins at residue Ile-57, the identical residue found at the beginning of the trypsin-solubilized CPR. Furthermore, the crystals of CPR exhibited catalytic activity. When a crystal of CPR was immersed in a solution of 10 mM NADPH, the yellow color of the crystal gradually diminished, indicating that the crystalline CPR is capable of being reduced by NADPH. The crystals were also dissolved and assayed for their ability to reduce cytochrome c. Immediately upon the addition of a dissolved crystal into a cuvette containing the assay mixture, reduction of cytochrome c was detected by an

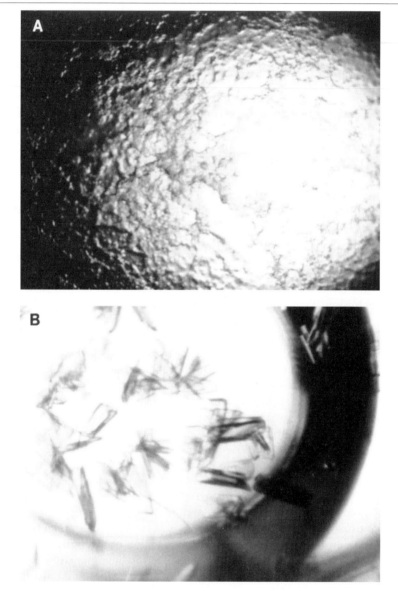

Fig. 1. Photomicrographs of CPR crystals. Deviations from the standard crystallization conditions [21% polyethyleneglycol 4500 (PEG 4500), 150 mM HEPES, pH 6.5, 0.85 M NaCl, and 5 mM MgCl$_2$] are indicated. The crystallization conditions are as follows: (A) Normal conditions except no pretreatment of the CPR protein with FMN and NADP$^+$; (B) initial

crystals grown in 25% PEG 4500, 0.10 M NaCl, and 50 mM HEPES, pH 7.0; (C) macroseeds grown in 25% PEG 4500, 50 mM HEPES, pH 6.5, 0.8 M NaCl, and 5 mM MgCl$_2$; and (D) diffraction quality crystals obtained by macroseeding.

TABLE I
DATA COLLECTION STATISTICS FOR CPR CRYSTALS

Parameter	Data set	
Collection temperature	4°	−120°
Resolution	2.60	2.32
Unique reflections	37,409	47,711
Observations ($I/\sigma > 1$)	106,500	118,607
$R_{sym}{}^{*a}$	7.02%	7.89%
Completeness	82.7%	81.9%
Cell dimension		
a =	103.3 Å	99.3 Å
b =	116.1 Å	115.5 Å
c =	120.4 Å	115.6 Å

^{a} $R_{sym}* = \Sigma|I - \langle I \rangle|/\Sigma\langle I \rangle$.

increase in the absorbance at 550 nm, thus further supporting the contention that crystalline CPR is unaltered and remains in the catalytically active state.

Preliminary X-Ray Analysis of CPR Crystals

Since the crystals were grown at 19°, our initial data collection was performed at the same temperature. Symmetry and systematic absences in the diffraction pattern of precession photographs are consistent with the orthorhombic space group $P2_12_12_1$, with unit cell parameters a = 103.32 Å, b = 116.12 Å, and c = 120.37 Å. Based on the assumption of two molecules per asymmetric unit (M_r = 72,000 per molecule), the calculated value of V_m is 2.54 Å^3/Da, which is well within the range expected for protein crystals.[14] Still photographs indicate that the crystals diffracted beyond 2.6 Å resolution. However, there was considerable radiation damage when data were collected at 19°, and therefore data collection was performed at 4°, at which temperature the crystals exhibited only a 20% decay after irradiation for 15 hr. Low temperature data collection at −120° improved the resolution to at least 2.3 Å, with no apparent decay after irradiation for 30 hr.

For 4° data collection, the CPR crystals must be preequilibrated overnight at 4° prior to data collection. Without equilibration, the crystals undergo phase transformations during data collection, causing the unit cell to change and making data analysis difficult. During these transformations, the a axis ranges anywhere from 101.1 to 107.4 Å, while the b and c axes remain relatively stable. After equilibration of the crystals at 4° for 24 hr,

---

[14] B. W. Matthews, *J. Mol. Biol.* **33**, 491 (1968).

diffraction analysis shows that the unit cell dimensions remain constant and data can be routinely collected. Lower temperature data collection was performed to minimize crystal damage due to X-ray exposure and to enhance the resolution. At $-120°$, the unit cell dimensions were found to be smaller (Table I), but there were no difficulties with phase transformations during data collection.

A complete data set was collected at $4°$ from a single CPR crystal (Fig. 1D), and the statistics are given in Table I. This data set includes 82.7% of the theoretically possible reflections to a resolution of 2.6 Å, with intensities greater than 1σ (Table I). A second data set was collected at $-120°$, with similar statistics and enhanced resolution (Table I). Initial studies have been published from our laboratories.[15]

Future Plans

Although CPR has FNR-like and flavodoxin-like domains, these two proteins constitute only $\sim$45% of the total mass of CPR, and therefore are not suitable models for molecular replacement studies for structure determination. Therefore, in order to solve the phasing problem, we are currently screening for heavy atom derivatives for use in multiple isomorphous replacement methods. Once the structure of CPR has been solved to 2.6 Å using $4°$ data, a higher resolution structure will be determined utilizing data collected at cryo-temperatures, at which the crystals diffract to 2.1 Å. Ultimately, the structure of CPR, together with the structures compiled for various cytochrome P450 enzymes, will enhance our knowledge on the interactions between CPR and its physiological redox partners, cytochromes P450, leading to a detailed understanding of the molecular mechanism of the microsomal monooxygenase system. Furthermore, the CPR structure will give insights into other members of the FNR family of flavoproteins, including nitric oxide synthase.

Acknowledgments

This research was supported by grants from the National Institutes of Health, GM29076 (JJPK) and HL30050 (BSSM), and Grant AQ-1192 from The Robert A. Welch Foundation (BSSM).

[15] S. Djordjevic, D. L. Roberts, M. Wang, T. Shea, M. G. W. Camitta, B. S. S. Masters, and J.-J. P. Kim, *Proc. Natl. Acad. Sci. U.S.A.* **92,** 3214 (1995).

Section VII

Regulation

[41] Sphingolipid-Dependent Signaling in Regulation of Cytochrome P450 Expression

By Edward T. Morgan, Mariana Nikolova-Karakashian, Jin-Qiang Chen, and Alfred H. Merrill, Jr.

Introduction

Stimulation of the immune system results in profound changes in the pattern of P450 gene expression in the liver.[1,2] Cytokines and interferons released from monocytes, macrophages (including Kupffer cells), and various other cell types can act directly on the hepatocyte to modulate P450 gene expression.[3,4] The majority of P450 gene products thus far studied are down-regulated by immunomodulators, but members of the CYP4A family are induced under some circumstances.[5]

Recent work in several laboratories has shown that hydrolysis of sphingomyelin (SM) to ceramide plus phosphorylcholine (Fig. 1), catalyzed by sphingomyelinase (SMase), is an early event in cellular signaling pathways for the cytokines interleukin-1 and tumor necrosis factor-α in various cell types.[6,7] This chapter describes procedures and approaches that we have used to investigate the role of ceramide and other sphingolipids in the suppression of CYP2C11 expression by interleukin-1 in primary rat hepatocytes.

Cell Culture

To obtain constitutive expression of CYP2C11, hepatocytes from male Sprague–Dawley rats (200–300 g) are cultured on 60-mm plates coated with 0.5 ml of Matrigel (6.3 mg/ml) for 5 days in the absence of growth hormone according to the procedure published by Liddle *et al.*,[8] except

---

[1] E. T. Morgan, *Mol. Pharmacol.* **36,** 699 (1989).

[2] A. E. Cribb, E. Delaporte, S. G. Kim, R. F. Novak, and K. W. Renton, *J. Pharmacol. Exp. Ther.* **268,** 487 (1994).

[3] Z. Abdel-Razzak, P. Loyer, A. Fautrel, J.-C. Gautier, L. Corcos, B. Turlin, P. Beaune, and A. Guillouzo, *Mol. Pharmacol.* **44,** 707 (1993).

[4] J. Chen, A. Ström, J.-Å. Gustafsson, and E. T. Morgan, *Mol. Pharmacol.* **47,** 940 (1995).

[5] M. B. Sewer, D. R. Koop, and E. T. Morgan, *Drug Metab. Dispos.* **24,** 401 (1996).

[6] Y. A. Hannun, *J. Biol. Chem.* **269,** 3125 (1994).

[7] R. Kolesnick and D. W. Golde, (*Cambridge, Mass.*) *Cell* **77,** 325 (1994).

[8] C. Liddle, A. Mode, C. Legraverend, and J.-Å. Gustafsson, *Arch. Biochem. Biophys.* **298,** 159 (1992).

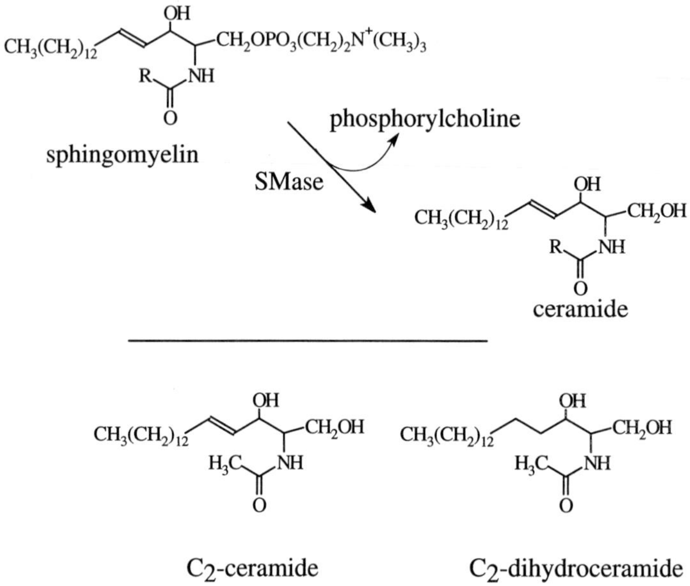

FIG. 1. Reaction catalyzed by sphingomyelinase, and structures of C_2-ceramide and C_2-dihydroceramide. R = -(CH_2)_{14}-CH_3.

that standard Waymouth's 752 medium (Life Technologies Inc., Bethesda, MD) is used and medium is changed at 4, 48, 96, and 120 hr. Additions of cytokines and sphingolipids are made at the time of the last medium change. This general design may be applicable for the study of other P450s, but this should be checked in each case. Matrigel can be prepared from Engelbreth–Holm–Swarm tumors as described[9] or purchased from Collaborative Biomedical Products (Bedford, MA).

Effects of Elevating Intracellular Ceramide

If ceramide formed from SM hydrolysis is involved in the modulation of P450 gene expression by a hormone or cytokine, then P450 expression should respond to other manipulations that elevate cellular ceramide levels.

Water-Soluble Ceramides

Natural ceramides (Fig. 1) are highly hydrophobic and are difficult to deliver to the cells. Ceramides with a shorter chain on the N-acyl group

[9] E. G. Schuetz, D. Li, C. J. Omiecinski, U. Müller-Eberhard, H. K. Kleinman, B. Elswick, and P. S. Guzelian, J. Cell. Physiol. **134,** 309 (1988).

are more hydrophilic, with critical micellar concentrations in aqueous media of about 10 μM.[10] Short-chain ceramides containing a 4,5-*trans* double bond appear to mimic many of the biological activities of cellularly derived ceramide.[6] Moreover, these compounds are relatively resistant to hydrolysis by ceramidase. For studies on CYP2C11 expression, we have utilized *N*-acetyl-D-*erythro*-ceramide (C_2-ceramide), which we synthesized from sphingosine.[11] However, pure preparations of C_2-, C_6-, and C_8-ceramide and C_2-dihydroceramide are available from Matreya, Inc. (Pleasant Gap, PA). C_2- and C_6-ceramide can also be purchased from Sigma Chemical Co. (St. Louis, MO). Effects of these agents should be observed in the 1–10 μM range[12]; however, they should be checked at several concentrations since the potency is affected by many factors, including cell number.

Delivery of Ceramide to the Cells

An ethanolic stock solution of 60 mM C_2-ceramide is prepared and added rapidly to sterile culture medium by syringe injection to give an aqueous stock solution of 250 μM. Immediately after addition of the ethanolic solution, the medium should undergo several cycles of repeated vigorous vortex mixing and bath sonication. The aqueous solution is then diluted in medium to give the desired concentration of ceramide. The final concentration of ethanol does not exceed 0.05%. The aqueous stock solution should be prepared fresh for each experiment.

SMase

Bacterial SMase can be used to elevate cellular levels of endogenous ceramide. SMase is presumed to hydrolyze SM on the outer leaflet of the membrane. Because of the specificity of the enzyme for SM, such treatment does not disrupt the lipid bilayer, and the enzyme does not enter the cell. Since the ceramide formed is highly hydrophobic, it can traverse the membrane and thus access intracellular target molecules. We have used *Staphylococcus aureus* SMase (100–300 U/mg, Sigma). When dissolved in medium at a concentration of 0.1 U/ml, this affects the hydrolysis of about 30% of hepatocyte SM, with generation of a near stoichiometric amount of ceramide.

[10] Y. A. Hannun, L. M. Obeid, and R. A. Wolf, *Adv. Lipid Res.* **25,** 43 (1993).
[11] S. Nimkar, D. Menaldino, A. H. Merrill, and D. C. Liotta, *Tetrahedron Lett.* **29,** 3037 (1988).
[12] R. A. Wolf, R. T. Dobrowsky, A. Bielawska, L. M. Obeid, and Y. A. Hannun, *J. Biol. Chem.* **269,** 19605 (1994).

Specificity and Interpretation

One target of ceramide that appears to mediate some of its cellular actions is a ceramide-activated protein phosphatase (CAPP).[6,12] CAPP has a high degree of structural and enantio-selectivity in that the 4,5-*trans* double bond of ceramide is required for activity, and the D-*erythro*- and L-*threo*- stereoisomers of C_2-ceramide are more potent activators than the L-*erythro*- and D-*threo*- compounds.[12] Thus, a differential effect of ceramide and dihydroceramide (lacking the double bond, Fig. 1) may indicate participation of CAPP in a given signaling pathway; lack of a differential effect may mean that a ceramide-activated protein kinase is involved.[7,13]

Analysis of P450 Gene Expression

Preparation and Analysis of Hepatocyte RNA

Total hepatocyte RNA is prepared by the acid-phenol extraction method.[14] We have recently developed a modification that allows the simultaneous harvesting of large numbers of plates: Medium is removed, and the cells are washed twice with ice-cold phosphate-buffered saline (PBS). A solution of 0.75 ml of 4 M guanidinium thiocyanate containing 25 mM sodium citrate, pH 7, 0.5% Sarkosyl, and 0.1 M 2-mercaptoethanol[14] is added directly to each plate at room temperature; and the plate is rocked gently to lyse all of the cells. After 5 min at room temperature, the solution is mixed by pipetting and transferred to a 1.5-ml microcentrifuge tube.

RNA is then purified by acid-phenol extraction exactly as described[14]: the solution is acidified with 50 μl of 2.0 M sodium acetate, pH 4.0, and 0.5 ml of water-saturated phenol is added. The solution is mixed gently by inversion after each addition, 0.1 ml of chloroform:isoamyl alcohol (49:1, v/v) is added, and the tube is shaken vigorously for 10 sec. After 15 min at 4°, the phases are separated by centrifuging at 14,000 g on a microcentrifuge for 15 min. The upper (aqueous) phase is carefully transferred to an RNase-free microcentrifuge tube. An equal volume of cold isopropanol is added; the solution is mixed by vortexing and placed at −20° for at least 1 hr.[15] RNA is recovered by pelleting at 14,000 g for 20 min at 4°. The pellet is washed three times with 1.0 ml of cold 75% ethanol, with sequential mixing and centrifuging at 14,000 g. The washed pellet is allowed to dry for 5 min at room temperature and is dissolved in 0.1 ml of RNase-free

[13] T. Goldkorn, K. A. Dressler, J. Muindi, N. S. Radin, J. Mendelsohn, D. Menaldino, D. Liotta, and R. D. Kolesnick, *J. Biol. Chem.* **266**, 16902 (1991).
[14] P. Chomczynski and N. Sacchi, *Anal. Biochem.* **162**, 156 (1987).
[15] The solution can be left at −20° for several days with no apparent RNA degradation.

water. One microliter of 10% sodium dodecyl sulfate is added to inhibit RNases, and the RNA is stored at $-80°$. Expression of specific mRNAs can then be measured by any of several standard methods, such as Northern blotting or RNase protection.

Preparation and Analysis of Hepatocyte Microsomes

In our experience, Western blots of hepatocyte lysates frequently give unacceptable levels of nonspecific staining. Therefore, hepatocyte microsomes are prepared. Cells from two plates are harvested by scraping into 10 ml of cold PBS containing 1 mM EDTA. The tubes are incubated on ice for 20 min to dissolve the Matrigel, pelleted, resuspended in 1 ml of PBS, and transferred to a 1.5-ml microcentrifuge tube. Cells are pelleted again and resuspended in 0.5 ml of 0.1 M phosphate buffer (pH 7.0). The cells are lysed by sonicating twice for 10 sec with a probe-type sonicator. The lysate is centrifuged at 1300 g for 5 min on the microcentrifuge at 4° and the speed is increased to 14,000 g for 13 min. The supernatant is centrifuged at 100,000 rpm (417,000 g) for 15 min at 4° on a Beckman TLK tabletop ultracentrifuge with a TLA 100.4 rotor. The microsomal pellet is resuspended in 10 mM Tris acetate, pH 7.4, containing 0.1 mM EDTA and 23% (v/v) glycerol and is stored at $-20°$. Expression of specific P450s in the microsomes is then examined by Western blotting.

Measuring Hydrolysis of Sphingomyelin and Formation of Ceramide

To conclude that a cytokine or other hormone regulates P450 expression through the SM/ceramide pathway, it is also necessary to show that the hormone can elicit the hydrolysis of cellular SM, with a concomitant generation of ceramide.

Extraction and Separation of Lipids

Cells from each 60-mm dish (3.5×10^6 cells) are harvested in 1 ml of PBS. A 20-μl aliquot is taken for protein determination, and 6 ml of chloroform:methanol (1:2, v/v) is added immediately. The solution is incubated for 1 hr at 37°, and the lipids are recovered by adding 1 ml of chloroform and 5 ml of slightly basic water (pH 8–8.5, adjusted with ammonium hydroxide). After mixing, the phases are separated and the upper layer is removed by aspiration. The lower (chloroform) phase is washed twice with water and dried through a small column (e.g., prepared from a Pasteur pipette) containing anhydrous, granular sodium sulfate. The solvent is evaporated under reduced pressure. The lipids are dissolved in 50 μl of chloroform:methanol (1:2) and applied on glass TLC plates

(20 × 20 cm, Silica gel H). The plates are developed in chloroform : methanol : triethylamine : 2-propanol : 0.25% KCl (30 : 9 : 18 : 25 : 5, v/v) until the solvent is 4–5 cm from the top of the plate. The lipids are identified after visualization with I_2 vapor by comparison with adjacent lanes of the plate that have been spotted with standards for SM, phosphatidylcholine, and ceramide (Sigma). The regions of interest are scraped from the plate and quantitated by liquid scintillation counting (in the case of [14C]choline-labeled cells) or by phosphate or HPLC assay (for mass measurements) as detailed later.

Measurement of SM by Labeling with [14C]Choline chloride

To label cellular SM, hepatocytes are cultured in the presence of medium containing 0.5 μCi/ml of [14C]choline chloride (55 mCi/mmol, Amersham, Arlington Heights, IL) for 48 hr, before treatment with cytokines. To avoid dilution of the radiolabel, modified Waymouths medium lacking choline chloride is used. This method allows rapid and sensitive determination of changes in levels of SM.[16] Since the SM is labeled in the phosphorylcholine moiety, it is not possible to directly measure the formation of ceramide by this method.

Measurement of SM by Microphosphate Assay

This is the most direct and reliable way to determine the mass of cellular SM.[17] Regions of the TLC plate corresponding to standard SM and phosphatidylcholine are scraped into glass tubes (new or cleaned with 1.0 N HCl), 0.3 ml of 1.0 N H_2SO_4 is added, and the uncapped tubes are incubated in an oven at 150° overnight. After cooling to room temperature, 2–3 drops of 30% H_2O_2 are added, and the incubation is continued for another hour or until the dark color has cleared. If the samples are still brown, more H_2O_2 is added and the solution is heated for another hour. After cooling to room temperature, 0.3 ml of 30% H_2O_2, 0.1 ml of 10% ascorbic acid (made fresh in the dark), and 0.6 ml of 0.42% ammonium molybdate (made fresh in the dark) are added. The tubes are incubated for 20 min at 50° in the dark, and the absorbance is read at 820 nm. A reagent blank is prepared from an equivalent amount of silica from a clean region of the TLC plate. The assay is calibrated with a standard curve using 0.05 ml of 5–500 μM K_2HPO_4.

---

[16] The labeled pool represents only SM synthesized from labeled phosphatidylcholine, and SM formed by other routes is not measured.

[17] B. N. Ames, *Methods Enzymol.* **8,** 115 (1966).

Measurement of Ceramides by HPLC

In the TLC system just described, ceramide and dihydroceramide run close to the front of the solvent, together with free fatty acids. This area is scraped, and the lipids are eluted by vigorous vortexing for 1 min with 3×1.5-ml aliquots of chloroform : methanol (1 : 1, v/v) followed by sedimentation. The pooled eluates are evaporated under reduced pressure, and 1 ml of 0.5 M HCl in methanol and 5 nmol of N-acetyl-C_{20}-sphinganine (as an internal standard) are added. The mixture is sonicated briefly and incubated for 16 hr at 63°. The samples are cooled to 20°, neutralized with 125 μl of 4 N KOH, and 1 ml of chloroform and 1 ml of water are added. After extraction, the lower (chloroform) layer contains the long-chain bases, sphingosine and sphinganine, that can be quantitated by HPLC as described later.

Further Metabolites of Ceramide

In our experience, the cytokine-induced turnover of SM in hepatocytes is not accompanied by stoichiometric formation of ceramide under all experimental conditions. This might be due to the further metabolism of ceramide to sphingosine and sphingosine-1-phosphate (also potent biologically active molecules),[18] to its direct phosphorylation to ceramide-1-phosphate, or to reutilization of ceramide in synthesis of more complex sphingolipids. At least one possibility is easily addressed: hydrolysis of ceramide to sphingosine. Determination of sphingosine is as follows: Cells are harvested, and 600 pmol of C_{20}-sphinganine is added as an internal standard. The lipids are extracted as described earlier, and the dried total lipid extracts are subjected to mild base hydrolysis in 1 ml of 0.5 M methanolic KOH for 1 hr at 37°. Under these conditions, glycerolipids are degraded while sphingolipids and some ether phospholipids are stable. The samples are neutralized, and 2 ml of chloroform and 5 ml of water are added. After phase separation, the organic phase is dried on a sodium sulfate column, and the solvent is evaporated under reduced pressure.

Free long-chain bases are determined as o-pthalaldehyde (OPA) derivatives by HPLC on a C_{18} reverse-phase column.[19] The samples are dissolved in 500 μl of HPLC buffer (methanol : 5 mM potassium phosphate buffer, pH 7.0, solution, 91 : 9, v/v). To this is added 250 μl of OPA reagent [99 ml of 3% (w/v) boric acid, pH 10.5, plus 1 ml of ethanol containing 50 mg of OPA and 50 μl of 2-mercaptoethanol], and the samples are incubated for 5 min at room temperature (thereafter, they should be kept chilled to

[18] S. Spiegel, A Olivera, and R. O. Carlson, *Adv. Lipid Res.* **25,** 105 (1993).
[19] A. H. Merrill, Jr. and E. Wang, *Methods Enzymol.* **209,** 427 (1992).

stabilize the OPA derivatives). The solution is clarified by brief centrifugation, and aliquots of the samples are separated by isocratic HPLC on a Waters (Milford, MA) Radial Pak C_{18} column (4 μm), using the HPLC buffer at a flow rate of 2 ml/min. The OPA derivatives are detected with a fluorimetric detector with an excitation wavelength of 340 nm and an emission wavelength of 455 nm.

The internal standard is used to correct for losses during the extraction procedure.

Acknowledgment

This work was supported by United States Public Health Service Grants GM46897 and GM46368.

[42] Use of Human Hepatocytes to Study P450 Gene Induction

By Stephen C. Strom, Liubomir A. Pisarov, Kenneth Dorko, Melissa T. Thompson, John D. Schuetz, and Erin G. Schuetz

Introduction

A central goal of most medical research is to understand the aspects of human biology, biochemistry, or physiological processes. While understanding the basis of human disease or the manner in which a human subject would metabolize a specific xenobiotic may be the goal of a research project, the human, as a species, is the least studied laboratory subject. The vast majority of medical research relies on data collected with more easily controlled animal model systems which serve as surrogates for humans. The utility of the model system studied is determined by the manner in which the chosen animal model recapitulates the precise manner of the human response. Although all species examined have cytochrome (CYP) P450 function and genetic loci which control these activities,[1,2] the examples of differences between species in the metabolism of xenobiotics are so numerous as to make extrapolation of data to different species unwise. The CYP3A family is the major steroid-inducible cytochrome P450s in rats, rabbits, and humans. At the genetic, protein, and metabolic level, the rat

[1] S. A. Wrighton and J. C. Stevens, *Crit. Rev. Toxicol.* **22,** 1 (1992).
[2] F. J. Gonzalez, *Pharmacol. Rev.* **40,** 243 (1988).

(CYP3A1), rabbit (CYP3A6), and human (CYP3A3/4) cytochromes are not identical. The gene products do not have identical substrate specificities or metabolic activities, nor are they regulated in the same manner in the different species.[1] In our experience from the investigation of the induction of CYP3A genes, by chemical exposure in primary culture, there are differences among rats, rabbits, and humans in their responses. With certain inducing agents, rat and human hepatocytes behave in a similar manner while rabbit hepatocyes respond differently. With other compounds, human and rabbit but not rat hepatocytes behave in a similar manner. With still other compounds, including pregnenolone 16α-carbonitrile (PCN) and phenobarbital, the induction response in human hepatocytes is different from either that observed with rat or rabbit cells in primary culture.[3-11] Thus, we have concluded that no single animal model faithfully predicts the responses observed in human hepatocyte preparations. Because of the relative importance of cytochrome P450s in the biotransformation of a variety of exogenous and endogenous compounds, it is crucial to gain a complete understanding of the genetic structure, regulation, and functional activity of human cytochrome P450s.

Methods

Human Liver Acquisition

Data Are Only as Good as the Model. To get an accurate representation of the manner in which the human liver responds to a chemical exposure, either with respect to metabolism of xenobiotics or the induction of P450s, one needs large, stable, and divergent sources of liver. The most useful

[3] S. C. Strom, R. L. Jirtle, R. S. Jones, D. L., Novicki, M. R. Rosenberg, A. Novotny, G. Irons, J. R. McLain, and G. Michalopoulos, *JNCI, J. Natl. Cancer Inst.* **68,** 771 (1982).

[4] S. C. Strom, D. L. Monteith, K. Manoharan, and A. L. Novotny, *in* "The Isolated Hepatocyte: Use in Toxicology and Xenobiotic Transformation" (E. J. Rauchman and G. Padilla, eds.), p. 265 (1987).

[5] E. G. Schuetz, J. D. Schuetz, S. C. Strom, M. T. Thompson, R. A. Fisher, D. T. Molowa, D. Li, and P. S. Guzelian, *Hepatology* **18,** 1254 (1993).

[6] T. A. Kocarek, E. G. Schuetz, S. C. Strom, R. A. Fisher, and P. S. Guzelian, *Drug Metab. Disp.* **23,** 415 (1995).

[7] E. G. Schuetz, J. D. Schuetz, M. T. Thompson, R. A. Fisher, J. R. Madariaga, and S. C. Strom, *Mol. Carcinog.* **12,** 61 (1995).

[8] V. E. Kostrubsky, S. C. Strom, S. G. Wood, S. A. Wrighton, P. R. Sinclair, and J. F. Sinclair, *Arch. Biochem. Biophys.* **322,** 516 (1995).

[9] J. D. Schuetz, S. C. Strom, and E. G. Schuetz, *J. Cell Physiol.* **165,** 261 (1995).

[10] J. D. Schuetz, E. G. Schuetz, J. V. Thotassery, P. S. Guzelian, S. C. Strom, and D. Sun, *Mol. Pharmacol.* **49,** 63 (1996).

[11] E. G. Schuetz, K. N. Furuya, and J. D. Schuetz, *J. Pharmacol. Exp. Ther.* **275,** 1011 (1995).

estimates of drug metabolism, P450 gene expression, or induction in humans must come from a large number of observations made on tissues derived from a representative sampling of the population. Humans, by nature out-bred, with varied dietary and drug exposures, express vastly different P450 levels. An example of the variability observed in the expression of CYP3A, the multidrug resistance gene, P-glycoprotein (Pgp), and CYP1A1/2 was recently published by Schuetz et al.,[11] where a 40-fold difference in the expression of these genes among the 23 different human cases was observed. The results indicate that there is more variability between individuals in the expression of CYP1A1/2 or Pgp than in the expression of CYP3A; however, there are individuals that seem to express little to none of each of the genes examined. Because of the inherent variability in the human population, in order to get an accurate estimation of the range of responses one could get from human liver a large sample size is needed from which to draw the observations. Most liver tissue which becomes available for cell isolation comes from either of two sources. One is tissue that is removed during surgical resection of the liver; however, the major source of liver results from organ donation. Excellent cells can be isolated from either source. Liver resections are usually performed for metastatic neoplasms, usually originating in the colon. The typical patient undergoing liver resec-tion would be of advanced age, but generally in good health. Aside from the concern for the effects that the tumor burden may have on liver function, if resected liver was the only source of liver tissue, one would worry about the applicability of the data generated with those cells to the age groups not represented in the tissue samples. Organ donors, on the other hand, come from all age groups and are disease free. Donor livers become avail-able from normal individuals for a variety of reasons, including anatomic defects, no suitable recipients, surgical errors in tissue harvest, cold ischemic time exceeding 20 hr, and patients dying awaiting transplantation. Donor livers are harvested for transplantation by in situ perfusion with University of Wisconsin solution, the normal transport media for whole organ trans-plantation. Thus, the tissue is cold-preserved using the best available method prior to becoming available for experimental use. A review of the demographics of the last 200 liver cases acquired by our laboratory over a 22-month period from 1/26/94 through 11/5/95 is presented in Fig. 1. The tissue donors were approximately evenly divided between males and fe-males and fell into the age groups shown in Fig. 1. The histogram of the ages of the tissue donors roughly parallels the percentage of the population in those age groups. Significant among these donors are the livers from patients under age 15, a group not usually well represented in other tissue sources. If one was trying to model a drug exposure for a compound normally administered to pediatric or geriatric patients, the most useful

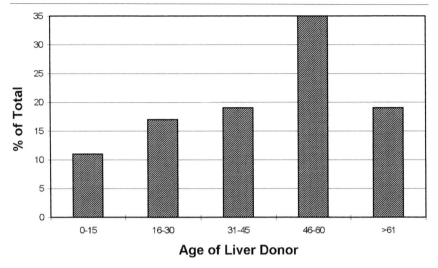

FIG. 1. Histogram showing the age distribution of human liver donors accumulated over a 22-month period.

data concerning that exposure would come from the individuals in that age group. Having large numbers of donors in all age groups ensures that the observations made with cells from those individuals are representative of the human population and directly relevant to the problem being studied.

Hepatocyte Isolation

The highest viability hepatocytes are isolated from liver by perfusion methods rather than simple slicing and digestion with proteases. The perfusion technique used depends on the type of tissue being perfused. For specimens of resected liver, perfusion through multiple catheters works best.[3,4] If the whole liver is available, as from organ donors, one has the option of perfusion of the whole liver or only a single lobe. We perfuse the left lobe of the human liver when the whole liver is available because the lobe is smaller, usually less than half the size of the right lobe. The smaller size makes it easier to perfuse the piece of tissue completely, and the whole right lobe is available for subsequent cell isolation or for cryopreservation of liver tissue for future use as microsomes or for molecular analysis. After the anatomic dissection of the right from left lobe, liver perfusion is conducted by tying off all of the major vessels on the small cut surface, which was the right/left lobe interface, with surgical sutures. Purse string sutures are used to encircle the large hepatic vein. A single catheter is inserted into the hepatic vein and the sutures are drawn tight around the catheter. With the catheter tied in place, the liver is placed in a sterile plastic bag and warmed in a water bath maintained at 37–39°, and the perfusate is introduced. Liver perfusion buffers are those described by Dorko et al.[12] and consist of a calcium-free buffer followed by a second buffer containing calcium, collagenase, and 0.5% bovine serum albumin. The calcium-free step helps separate intercellular junctions (which require calcium) while the cells are completely separated by the exposure to collagenase. Important considerations are the temperature and the digestion conditions. The liver must be maintained at 37° to fully activate the collagenase. Thus, the initial calcium-free perfusion step must be continued until the liver has warmed from ice temperatures during transport to 37°. Normally 30 min is sufficient for each step. The digestion conditions must be worked out for each lot of collagenase and is different if albumin is deleted from the second perfusion buffer.[3,4,12] Although more collagenase is required, a more gentle digest is obtained in the presence of albumin, thereby optimizing the yield of viable cells. Following isolation, hepatocytes are maintained in rather simple culture conditions.[3–10] Cultures are initiated by plating hepatocytes in William's medium E or a more enriched chemically defined media (CDM)[13] supplemented with dexamethasone and insulin at 0.1 μM concentrations. Cells are plated in media containing 10% calf serum but are changed to serum-free media as soon as possible (4–12 hr) and are maintained in serum-free media. Culture dishes are usually precoated with type 1 collagen[14] or gelatin (50 μg/10-cm culture dish) 30 min prior to the addition of cells. As indicated earlier, hepatocytes can be isolated from all age groups; we have isolated cells from donors as young as 1 day old up to age 77. Representative photographs of the morphology of hepatocytes isolated from a 1-year-old and a 72-year-old donor are provided in Fig. 2. At our current rates of tissue acquisition, we isolate and culture over 1 billion viable human hepatocytes/week.

With large numbers of human hepatocytes available from successful perfusions there is a great need for the cryopreservation of cells for later use. We have determined that human hepatocytes can be successfully cryopreserved following exposure to 2% dimethyl sulfoxide (DMSO) in the culture media for 24 hr. The DMSO exposure allows the reproducible cryopreservation of hepatocytes from virtually every human case examined in a freezing media containing 10% DMSO, 10% fetal calf serum, and 80% culture media. Cells are frozen rapidly, approximately 5°/min, by placing

[12] K. Dorko, P. D. Freeswick, F. Bartoli, L. Cicalese, B. A. Bardsley, A. Tzakis, and A. K. Nussler, Cell Transplant. 3, 387 (1994).
[13] H. C. Isom and I. Georgoff, Proc. Natl. Acad. Sci. U.S.A. 81, 6378 (1984).
[14] S. C. Strom and G. Michalopoulos, Methods Enzymol. 82, 544 (1982).

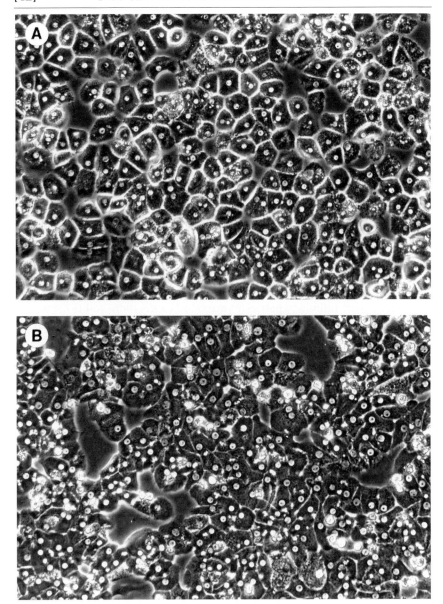

FIG. 2. Morphology of human hepatocytes after approximately 24 hr in culture. (A) Hepatocytes isolated from a 1-year-old donor. (B) Cells isolated from a 72-year-old donor.

the vials directly in a freezer maintained at $-70°$. Cryopreserved cells are stored in the liquid phase of a liquid nitrogen storage tank for months or years until needed. Following cryopreservation, hepatocytes can be rapidly thawed by placing the vials directly in a water bath maintained at $37°$. The viability of the hepatocytes drops approximately 10–20% through the freeze/thaw process. Hepatocytes are pipetted immediately after the thaw into 4 volumes of prewarmed media. Cells are recovered by centrifugation at 100 g for 4 min and are resuspended in fresh plating media. Hepatocytes that are cryopreserved following isolation and subsequently thawed and reestablished into culture display a normal basal level of CYP1A1 and respond to 3-methylcholanthrene (MC) or 2,3,7,8-tetrachlorodibenzo-p-dioxin (TCDD) exposure with increased gene expression (Case 107).[7] The availability of cryopreserved cells has made it possible to support liver function in human patients by liver cell transplantation[15] and should make collaborative efforts between laboratories with access to human tissue and those researchers with interests in human hepatocytes for basic science purposes much easier.

Although it is harder to procure human liver tissue and to isolate cells, once established, human hepatocytes are significantly more stable in culture than those from rodents. While P450 gene expression is rapidly lost in cultured rodent hepatocytes, human cells maintain significant levels of P450 genes in culture for at least 5 days. Western blots showing the relative levels of CYP3A, 1A2, 2B, 2C8, and 2E1 are provided in Fig. 3. Although CYP1A2 was rapidly lost, the other isoforms were well maintained and all forms can be readily induced in culture by supplementation of the media with known inducing agents. The relative long-term maintenance of the CYP2 (2B, 2C, 2E) and CYP3A family members affords the opportunity to examine the regulation of the expression of these isoforms.

Regulation of the Expression of Human Cytochromes P450

Because of the difficulties with access to human liver tissue and human hepatocyte isolation, one strategy has been to investigate the expression of cytochromes in liver-derived cell lines such as the human hepatocellular carcinoma, HepG2, or Tong HCC.[5] Primary human hepatocytes express CYP3A3/4, at the mRNA and protein levels, and gene expression is dramatically increased by the exposure of cells to typical inducing agents such as lovastatin (lov), dexamethasone (dex), rifampicin (rif), phenobarbital (PB), or clotrimazol (CTZ). The replicating liver lines only express the fetal form

[15] S. C. Strom, R. A. Fisher, M. T. Thompson, A. J. Sanyal, P. E. Cole, J. M. Ham, and M. A. Posner, *Transplantation* (in press).

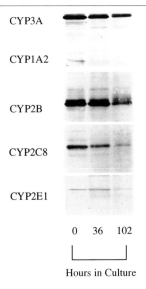

FIG. 3. Maintenance of human liver cytochromes CYP3A, CYP1A2, CYP2B, CYP2C8, and CYP2E1 over the first 102 hr of culture.

of CYP3A (3A7, HepG2) or CYP3A5 (Tong HCC). Hepatocytes from two of six donors expressed CYP3A5 but attempts to induce gene expression with dex, rif, or lov failed to enhance mRNA levels. Expression of CYP3A5 in Tong HCC was also not inducible. These results suggest that all mature forms of CYP450 are not represented in the replicating liver cell lines and that they cannot reproduce the full spectrum of responses observed in primary human hepatocytes.

A comparison of the induction of CYP3A genes was conducted by Kocarek et al.[6] in primary cultures of rat, rabbit, and human hepatocytes. Differences were observed among the species in their responses to prototypical inducing agents. In agreement with their known effects in vivo, PCN, or rif induced CYP3A expression exclusively in rat or rabbit hepatocytes, respectively. All human hepatocyte cases examined responded with increased CYP3A expression following exposure to rif or dex, but the response to PCN was variable with only two of four individuals showing a significant induction. Lovastatin was an effective inducer of CYP3A in all three species, whereas spironolactone, cyproterone acetate, and clortrimazol were effective only in human and rat hepatocytes. These results demonstrate the intrinsic variability of responses between hepatocytes from the different species and preclude the use of a single laboratory animal as a surrogate for humans.

Schuetz *et al.*[7] investigated the induction of P-glycoprotein (*MDR1*) and CYP1A1 mRNA by aromatic hydrocarbons (AH) in 15 different cases of primary human hepatocytes. While MC or TCDD induced CYP1A1 mRNA in all individuals tested, MC or TCDD only induced *MDR1* expression in slightly over 50% of the individuals. The results indicate that *MDR1* and CYP1A1 induction by aromatic hydrocarbons proceeds through different pathways and that *MDR1* expression is not controlled by the classical AH receptor. The divergence between *MDR1* and CYP1A1 induction may help identify certain individuals as potentially at high risk of AH-associated toxicity. Since AH are transported out of the cells by *MDR1*, individuals with low *MDR1* expression may be at higher risk from AH. AH exposure induces CYP1A1 expression and AH metabolism to toxic metabolites. In those individuals in which *MDR1* was not coordinately induced by AH exposure, higher levels of the toxic intermediates may accumulate.

In these experiments, and those described below,[8] hepatocytes were cultured in Waymouth media and on matrigel basement membrane or in simpler conditions in CDM or Williams media on plates coated with rat tail collagen. The results indicate that qualitatively the same results are obtained in both culture conditions. Induction of CYP1A1 by AH was slightly greater in cells maintained on matrigel, whereas *MDR1* expression was slightly greater in cells maintained on type-1 collagen. These results,[7] those presented below,[8] and those from other laboratories[14–16] indicate that the specialized culture media and substrate conditions needed to preserve an *in vivo*-like induction pattern of P450 gene expression with rat hepatocytes[17–20] are not necessary for human hepatocytes.

Ethanol was found to be an efficient inducer of CYP3A mRNA and protein expression in human hepatocytes.[8] In addition to CYP3A, ethanol exposure induced CYP2E1 expression as well. Both ethanol and isopentanol were more potent and effective inducers of CYP3A in cultured human hepatocytes than in cultured rat hepatocytes. Concentrations of ethanol in the range of 5–25 mM were more effective inducers of CYP3A expression in human cells than 200 mM ethanol in rat hepatocytes. The strong induction of CYP3A and CYP2E1 by ethanol may help explain the differences in drug

[16] L. Pichard, I. Fabre, G. Fabre, J. Domergue, B. S. Aubert, G. Mourad, and P. Maurel, *Drug Metab. Dispos.* **18**, 595 (1990).

[17] M. Daujat, L. Pichard, I. Fabre, T. Pineau, G. Fabre, C. Bonfils, and P. Maurel, *Methods Enzymol.* **206**, 345 (1991).

[18] M. Maurice, L. Pichard, M. Daujat, I. Fabre, H. Joyeux, J. Domergue, and P. Maurel, *FASEB J.* **6**, 752 (1992).

[19] E. G. Schuetz, D. Li, C. Omiecinski, U. Müller-Eberhard, H. K. Kleinman, B. Elswick, and P. S. Guzelian, *J. Cell. Physiol.* **134**, 309 (1988).

[20] E. G. Schuetz, J. D. Schuetz, B. May, and P. S. Guzelian, *J. Biol. Chem.* **265**, 1188 (1990).

metabolism and toxicities noted in humans consuming alcoholic beverages. Current investigations will determine if acetaminophen toxicity, known to be mediated through CYP3A and CYP2E1 pathways, is potentiated by ethanol exposure.

Novel inducers of CYP3A or CYP2B have been identified using the human hepatocyte culture system. Although well recognized as substrates for P450s, the compounds reserpine, nifedipine, and verapamil are potent inducers of CYP3A expression in cultured human hepatocytes (Fig. 4). Also, the well recognized CYP3A-inducing agents, dexamethasone and rifampicin, were found to induce expression of CYP2B6 (Fig. 5), indicating that xenobiotic exposures in humans result in the increased expression of multiple P450 isoforms.

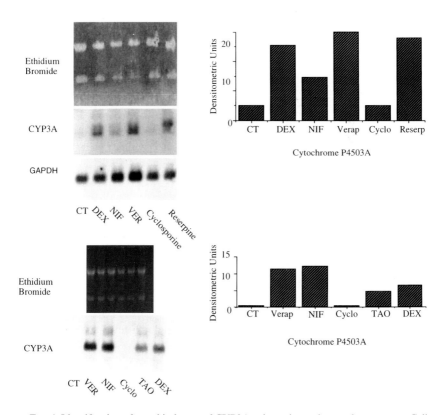

FIG. 4. Identification of novel inducers of CYP3A using primary human hepatocytes. Cells from two individuals were exposed to the indicated compounds for 48 hr in CDM media. The results indicate that previously unrecognized compounds such as reserpine, nifedipine, and verapamil are potent inducers of CYP3A expression.

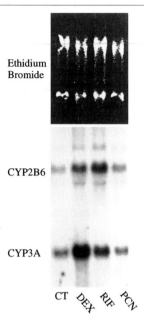

Ethidium
Bromide

CYP2B6

CYP3A

CT   DEX   RIF   PCN

Fig. 5. Coinduction of CYP3A and CYP2B6 by dexamethasone and rifampicin in primary human hepatocytes. PCN was very weak or without activity in this case. These results indicate that agents other than phenobarbital can induce the expression of CYP2B6 in human hepatocytes.

The human hepatocyte culture systems have been useful for the investigation of mechanisms which regulate gene expression. Hepatocyte RNA was isolated from three individuals selected for differences in their response to CYP3A expression in response to the prototypical inducer dex. Northern blots (Fig. 6) demonstrate that hepatocytes from Case 3 showed no basal expression of CYP3A and were not inducible by exposure to dex. Hepatocytes from Case 9 showed both a high basal level of CYP3A and a robust response to dex, while Case 12 showed a lower basal level CYP3A, but a strong induction in response to dex exposure. To investigate a potential basis for the variability of basal and inducible levels of CYP3A, we examined the levels of the glucocorticoid receptor (GCR) mRNA in the same cases. The results indicate that the hepatocytes which express high basal levels and respond to dex exposure with CYP3A induction are also the cases which express significantly higher levels of the GCR (Fig. 6). Future experiments with GCR-deficient mice or antisense experiments with human

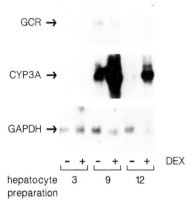

GCR →

CYP3A →

GAPDH →

- + - + - + DEX

hepatocyte   3   9   12
preparation

FIG. 6. Examination of the relationship between the expression of the glucocorticoid receptor and CYP3A expression in human hepatocytes. Cells in culture were exposed to 100 μM dexamethasone for 48 hr where indicated. These results demonstrate that the expression of hepatic GCR may be a determinant of the basal and inducible levels of CYP3A.

hepatocytes may help to discern the actual role of GCR in controlling CYP3A expression.

Human hepatocyte cultures have proved useful in the direct investigations of the molecular events controlling CYP3A gene expression. Regions 5' to the CYP3A5 were used in reporter systems to determine that the CYP3A5 promoter region does not contain the traditional 15-bp glucocorticoid response element (GRE), but instead contains two GRE half sites separated by 160 bp.[10] Mobility shift assays indicated that the CYP3A5 5'-GRE specifically binds GCR. These investigations were the first to reveal that human CYP3A genes contain sequences which bind the glucocorticoid receptor and result in enhanced gene expression.

Investigations using the reporter gene constructs revealed a problem that may be peculiar to the human liver. Mammalian liver is known to contain a chloramphenicol deacetylase which can be heat inactivated. With primary human hepatocytes, however, an endogenous chloromphenicol acetyltransferase (CAT)-like activity was not inactivated after 10 min at 65° which rendered authentic CAT activity produced by the reporter vector undetectable. However, when luciferase was substituted for CAT as the reporter system, transcriptional activity was easily detectable from either CYP3A5 or simian virus 40 (SV40) promoters (Table I). Because human hepatocytes contain a heat-stable CAT activity, the use of CAT reporter

TABLE I
COMPARISON OF SV40 AND CYP3A PROMOTERS IN
HUMAN HEPATOCYTES TRANSIENTLY
TRANSFECTED WITH A LUCIFERASE
REPORTER CONSTRUCT

Experiment	Luciferase values[a]
1	
pGL2-promoter[b]	827 ± 175[c]
CYP3A5-LUC[d]	154 ± 50
2	
pGL2-promoter	151.3 ± 1.2
CYP3A5-LUC	10.3 ± 3.8

[a] The mock-transfected hepatocytes had no measurable luciferase activity.
[b] The pGL2-promoter (Promega, Madison, WI) contains the SV40 promoter driving the luciferase reporter gene.
[c] Values represent SEM.
[d] The CYP3A5-LUC contains the CYP3A5 promoter (bases -306 to $+14$) subcloned into the pGL2Basic vector (Promega).

vectors is contraindicated and the use of the luciferase system appears to provide a sensitive method to detect CYP450 gene transcription.

Conclusions

Human hepatocytes have been available for research purposes since 1983. Despite ample evidence of interspecies variations in CYP450 gene structure, regulation, and enzymatic activity, the human hepatocyte culture system remains the least studied model for P450 investigations. While most investigators would agree that detailed information of the human liver CYP450 system would be valuable, problems with human liver tissue acquisition and cell isolation have prevented many interested investigators from utilizing human hepatocytes in their research. Recent efforts in our laboratory have demonstrated that hundreds of millions to billions of human hepatocytes can be isolated and cultured on a weekly basis. Following isolation, human hepatocytes can be shipped to collaborating institutions either in suspension on wet ice or at ambient temperatures in sealed flasks, and upon arrival the cells retain most if not all of their original liver function. We have accumulated nearly 100 instances where human hepatocytes shipped from our laboratory have been used by researchers at academic institutions or in pharmaceutical industries in this country 18–36 hr later. Successful international shipments of cultured and/or cryopreserved cells have also been accomplished. Nonprofit organizations such as the Anatomic Gift Foundation have set up systems to provide interested investigators with isolated human hepatocytes and liver tissue. Further research has determined that the hepatocytes can be cryopreserved and reestablished into culture, and that the frozen/thawed cells still respond to P450-inducing compounds with enhanced gene transcription and drug-metabolizing capacity.[7,15] Information provided here and in the accompanying references indicate that virtually all of the experiments that can be conducted with rat or rabbit hepatocytes can be completed with human cells as well. The human hepatocyte model system has been found to be extremely useful for investigations of interspecies and interindividual differences in P450-related activities, including induction of gene expression in response to xenobiotic exposure and the molecular events controlling gene expression through the transfection of reporter gene constructs. All of the recent improvements in the isolation culture and distribution of human hepatocytes should make it easier for investigators interested in this culture model system to acquire cells or tissue or to collaborate with laboratories who routinely do hepatocyte isolation. The results will provide much useful information concerning the human liver CYP450 system.

Acknowledgments

This work was supported by National Institutes of Health Grants ESO5851 and ESO4628 and American Cancer Society Grant CN43C and grants from the Schering-Plough Corporation and The Anatomic Gift Foundation.

[43] Cytochrome P450 mRNA Induction: Quantitation by RNA–Polymerase Chain Reaction Using Capillary Electrophoresis

By MICHAEL J. FASCO, CHRISTOPHER TREANOR, and LAURENCE S. KAMINSKY

Introduction

A highly sensitive, quantitative method for the determination of P450 mRNAs would provide a solid basis for many studies of P450 regulation. Such a method would be invaluable in investigating P450 transcriptional regulation, particularly in human systems in which tissue availability is limiting, and would permit assessments of which of the P450 mRNAs, and in what concentrations, are present in tissue of limited availability. These latter studies, while only providing indirect proof of P450 protein expression in these tissues, could provide the only accessible information on the P450 composition and inducibility. Quantitative reverse transcription and polymerase chain reaction (RNA–PCR) provides a potentially valuable technique for achieving these goals.

Wang et al.[1] were the first to realize the potential of RNA–PCR as a quantitative measure of specific cellular mRNAs present in low copy numbers. A predetermined quantity of generic cRNA internal standard, which contained the same primer sites as the mRNA target sequence, and tissue RNA were reverse transcribed and amplified in the presence of ^{32}P-labeled 5' primer. The two products of the reaction, which differed in size, were separated by agarose gel electrophoresis and visualized by staining with ethidium bromide, and their quantity was determined by scintillation counting of the excised bands. The concentration of the target mRNA was calculated, within the range of exponential amplification, by proportion to the internal standard.

Subsequently, other quantitative RNA–PCR methods were reported, each with some advantageous feature. Murphy et al.[2] used β_2-microglobulin mRNA as a "housekeeping" external standard to compensate for variations in the quality of different RNA preparations. Gilliland et al.[3] utilized internal standards containing the same primer sequences as the target mRNA as competitors of target cDNA amplification. A major advantage of competitive PCR is that the products do not have to be measured exclusively during the exponential portion of the amplication reaction. More recently, Apostolakos et al.[4] combined competitive RNA–PCR with the use of housekeeping mRNA in multiplex competitive RNA–PCR. In this method, the mRNA of glyceraldehyde-3-phosphate dehydrogenase (GAPDH) and a target mRNA were simultaneously reverse transcribed and dilutions of the cDNA were amplified in the presence of their appropriate competitive sequences.

All quantitative RNA–PCR methods are labor-intensive and require careful preparation and analysis of many samples per data point. The

[1] A. M. Wang, M. V. Doyle, and D. F. Mark, *Proc. Natl. Acad. Sci. U.S.A.* **86,** 9717 (1989).
[2] L. D. Murphy, C. E. Herzog, J. B. Rudick, A. T. Fojo, and S. E. Bates, *Biochemistry* **29,** 10351 (1990).
[3] G. Gilliland, S. Perrin, K. Blanchard, and H. F. Bunn, *Proc. Natl. Acad. Sci. U.S.A.* **87,** 2725 (1990).
[4] M. J. Apostolakos, W. H. T. Schuermann, M. W. Frampton, M. J. Utell, and J. C. Willye, *Anal. Biochem.* **213,** 277 (1993).

difficulty of sample analysis is compounded by the need for a method that can rapidly and accurately quantitate the low amounts of RNA–PCR products formed. Although several methods for determining competitor and target ratios have been described, each involves an agarose gel separation step, coupled to either scintillation counting, radioimaging, or densitometric scanning of autoradiographs or photographs of ethidium bromide-stained gels. Each of these methods has distinct disadvantages which limit its utility.

Capillary electrophoresis employing liquid or solid polymers as molecular sieves is highly efficient in separating DNA fragments in the size range frequently used in quantitative RNA–PCR (100 to 800 bp),[5–7] however, the UV detection methods used are not adequately sensitive for RNA–PCR products derived from low copy numbers. Recently, Srinivasan et al.[8] reported femtogram detection by laser-induced fluorescence (LIF) of PCR products, separated by capillary electrophoresis, using the highly fluorescent, dimeric intercalating agents of oxazole yellow (YOYO-1) and thiazole orange (TOTO).

The protocol described by Srinivasan et al.[8] incorporated premixing of the PCR–DNA with YOYO-1 or TOTO at a molar ratio of 5 bp to one dye prior to separation by capillary electrophoresis. Since this method requires prior knowledge of the DNA concentration in the sample, it is not well suited to the large numbers of samples generated during quantitative RNA–PCR studies. We report here a capillary electrophoresis method, using LIF detection of DNA–YOYO-1 complexes, that can analyze low levels of multiple DNA species generated during competitive or multiplex-competitive RNA–PCR without prior pretreatment or concentration of the sample. An application of this approach to P450 1A1 transcription is provided.

Methods and Results

RNA Isolation

The isolation of RNA from cell cultures or tissue samples is accomplished with TRI reagent as recommended by the manufacturer (Molecular

[5] H. E. Schwartz, K. Ulfelder, F. J. Sunzeri, M. P. Busch, and R. G. Brownlee, *J. Chromatogr.* **559**, 267 (1991).

[6] A. Guttman, B. Wanders, and N. Cooke, *Anal. Chem.* **64**, 2348 (1992).

[7] J. P. Landers, R. P. Oda, J. A. Spelsberg, J. A. Nolan, and K. J. Ulfelder, *BioTechniques* **14**, 98 (1993).

[8] K. Srinivasan, S. C. Morris, J. E. Girard, M. C. Kline, and D. J. Reeder, *Appl. Theor. Electrophor.* **3**, 235 (1993).

Research Center, Inc. Cincinnati, OH). RNA purity and concentration are determined, respectively, from the 260/280-nm absorbance (greater than 1.6) and an extinction coefficient of 40 mg/ml per absorbance unit at 260 nm. RNA is dissolved in and stored at $-85°$ in diethylpyrocarbonate-treated water.

RNA–PCR

Apparatus. RNA–PCR reactions are performed without oil in a Perkin Elmer 9600 thermal cycler (Applied Biosystems, Foster City, CA). Purification of DNA internal standards is done by high-performance liquid chromatography (HPLC) using a Gen-Pak Fax column (Waters Associates, Millipore Corporation, Milford, MA). The HPLC unit is from Waters Associates and consists of a 600E Pump/System controller, WISP, 996 detector, fraction collector, and Millenium software for visualization of eluted peaks. Oligonucleotides are synthesized at our Molecular Genetics Core Facility (Wadsworth Center, Albany, NY) using a Milligen 8750 DNA synthesizer (Millipore Corp.).

Method. Reverse transcription and PCR amplification reactions are performed with a RNA–PCR Core Kit essentially as described in the supplier's instructions (Applied Biosystems). The exceptions are (i) the RNA is pretreated with 1 unit of DNase I (Boehringer-Mannheim Corp., Indianapolis, IN) to remove any traces of contaminating DNA and (ii) a final concentration of 1% formamide is included in the PCR amplification reaction to enhance primer specificity.[9]

1. DNA as a contaminant of RNA is removed by incubating 1 unit of DNase I with 1 μg RNA in 20 μl of reverse transcription mixture (minus the reverse transcriptase) at $37°$ for 30 min followed by thermal inactivation of the DNase I at $75°$ for 5 min. Competitive PCR and related studies by this laboratory have demonstrated that contaminant DNA is completely removed by this treatment, without affecting the mRNA.

2. Following cooling to $4°$, the reverse transcriptase is added and reverse transcription is normally accomplished with oligo(dT)$_{16}$ priming for 30 min at $42°$. Random hexamers or sequence-specific primers may be substituted for the oligo(dT)$_{16}$. The reverse transcriptase is then inactivated by heating at $95°$ for 5 min. Samples of the cDNA are either used immediately after cooling or stored at $-20°$.

[9] G. Sarkar, S. Kapelner, and S. S. Sommer, *Nucleic Acids Res.* **18,** 7465 (1990).

3. For 0.1-ml competitive multiplex or competitive PCR reactions, the reverse-transcribed RNA (cDNA) is diluted with modified reverse transcription mix (water is substituted for the volumes of reverse transcriptase, RNase inhibitor, and oligo(dT)$_{16}$ primer) such that 10 μl of the cDNA included in the reactions is derived from 0.5, 0.25, 0.1, 0.05, 0.025, 0.01, or 0.005 μg of original RNA. Alternatively, the cDNA concentration is single and the competitor concentration(s) is varied and/or used at a single concentration. The primer concentrations are 50 pmol each/0.1-ml of reaction. Most PCR reactions are scaled down to 25 μl. Cycling conditions for the amplification reaction are a single, 1-min heating step at 94°, followed by 30 cycles of denaturation at 94° for 10 sec, annealing at 65° for 15 sec, and amplification at 72° for 15 sec. The reaction is terminated by cooling to 4°. Aliquots (10 to 20 μl) are mixed with 5 μl of actin internal standard (25 ng), and the mixture is transferred to a microvial for direct analysis by capillary electrophoresis.

Primers, Competitors, and Internal Standards. Forward and reverse primers for rat P450 1A1 and rat/human GAPDH were designed from their cDNAs with the aid of the computer program DNAsis and, for GAPDH, the literature.[10] Primer sets span an intron when possible.

Competitors are made from cDNA as described by Förster.[11] cDNA from the livers of rats treated with β-naphthoflavone (BNF) is used as the template for the preparation of P450 1A1 and 1A2 competitors and from HepG2 cells for the GAPDH competitor.

Forward primer for 1A1 (948–968): 5'-TGACCTCTTTGGAGCT-GGGTT

Reverse primer for 1A1 (1352–1370): 5'-CCAATGCACTTTCG-CTTGC

Linker primer for 1A1: 5'-TTTCGCTTGCTGTCCAGAGTGCCA-CTGGA

Target sequence is 422 bp; competitor sequence is 405 bp

Forward primer for GAPDH (15–38): 5'-GGTCGGAGTCAACG-GATTTGGTCG

Reverse primer for GAPDH (781–802): 5'-CCTCCGACGCCTGCT-TCACCAC

Linker primer for GAPDH: 5'-GCTTCACCACCAGGGATGATGT-TCTGGAGA

Target sequence is 787 bp; competitor sequence is 630 bp

---

[10] G. S. Dveksler, A. A. Basile, and C. W. Dieffenbach, *PCR Methods Appl.* **1**, 283 (1992).
[11] E. Förster, *BioTechniques* **16**, 18 (1994).

The human actin internal standard is prepared from DNA isolated from HepG2 cells using the protocol described by the manufacturer of the TRI reagent.

> Forward primer for actin (2104–2125): 5′-GCGGGAAATCGTGCG-
> TGACATT
> Reverse primer for actin (2409–2432): 5′-GATGGAGTTGAAGG-
> TAGTTTCGTG
> Size of internal standard is 328 bp

The competitor sequences and the actin internal standard are prepared in 10 repetitive 0.1-ml amplification reactions for 35 cycles. All reaction mixtures are combined, partially desalted in a Centricon 30 spin filter (Amicon, Beverely, MA), and purified by HPLC using the conditions for nucleic acid separation provided with the column. Our normal elution gradient is 30% solvent B to 80% solvent B over 30 min at a flow rate of 0.75 ml/min.

Concentrations of stock actin internal standard solutions are determined by mixing an aliquot with a known quantity of PhiX 174 RF DNA *Hae*III (PhiX) and, following separation by capillary electrophoresis, comparison of its integrated area with those of the PhiX components (see later). Concentrations of stock GAPDH and P450 1A1 competitor solutions are determined from integrated area comparison with the calibrated actin standard.

Results. Prerequisites for competitive RNA–PCR are that (i) the competitor and target sequences be amplified with equal efficiency,[3,12] and (ii) the RNA preparations be DNA free. In a log–log plot of an area competitor/area target versus competitor or RNA concentration, the theoretical slope of the line is 1, i.e., equal amplification efficiencies of the competitor and target sequences. Once the amplification efficiency is established for each competitor/target pair, relative concentrations can be calculated from the ratios of the competitor to target sequences obtained at a single concentration of competitor. Instead of employing a series of reaction mixtures that differ in RNA or competitor concentration, replicate samples at a single concentration of each can be run and significant differences accurately determined. DNA as a contaminant of purified RNA (we find this to be a common occurrence, particularly with human tissues) is a souce of potentially significant error because it can act as a second competitor in the PCR reaction. We have used quantitative, competitive RNA–PCR to develop conditions where any contaminating DNA is totally degraded by RNase-free DNase I and then thermally denatured without loss of mRNA that can be reverse transcribed. Using conditions cited in one published report[13]

---

[12] L. Raeymaekers, *Anal. Biochem.* **214,** 582 (1993).
[13] D. D. Dilworth and J. R. McCarrey, *PCR Methods Appl.* **1,** 279 (1992).

for DNase I treatment of RNA followed by heat inactivation, we experienced a loss of approximately 80% of the available mRNA.

Apostolakos et al.[4] described a variation of competitive RNA–PCR that the authors termed multiplex competitive RNA–PCR. The advantage of multiplex competitive RNA–PCR over conventional competitive RNA–PCR is that two target cDNAs and their respective competitor sequences are amplified in a single tube, and then separated and quantitated. Errors introduced by differences in the efficiency of the reverse transcription reaction and tube to tube variation in the PCR amplification are sharply reduced. One of the mRNAs is normally a "housekeeping" gene, against which other target mRNAs are compared. For many of our studies, GAPDH is used as the reference mRNA because it is not inducible by agents known to affect P450 concentrations.[10] One disadvantage of multiplex PCR is that the quantity of certain, but not all, PCR products produced are markedly reduced in the presence of a second primer set (unpublished results). GAPDH amplification from BNF-induced rat liver RNA, at least with the primers listed earlier, is readily detectable under competitive PCR conditions, but is barely detectable when the primers for P450 1A1 are also added. Fortunately, however, the competitor to target relationship remains constant even at the diminished rate of formation. The quantity of P450 1A1 amplified is not noticeably affected by the presence of the GAPDH primers.

Data from a typical multiplex PCR reaction are plotted in Fig. 1. In this experiment, total RNAs isolated from the livers and enterocytes of rats (pool of three) induced with BNF (40 mg/kg in corn oil administered orally 24 hr before sacrifice) were used to quantitate the amount of P450 1A1 mRNA present in each tissue. The quantities of both reverse-transcribed RNA and GAPDH competitor were fixed whereas the quantity of the P450 1A1 competitor was varied from 0.1 to 50 amol/μg RNA. For this experiment, inclusion of the GAPDH competitor served as a marker of potential tube to tube variation. The dotted lines in Fig. 1 are the average of the GAPDH competitor/target radios obtained at each P450 1A1 competitor concentration for each tissue. The open symbols in the figure illustrate the close proximity of the values to the average, enhancing the reliability of the P450 1A1 values. In addition to their usefulness in detecting tube to tube variation, "housekeeping" mRNAs can also be used to determine relative efficiencies of the reverse transcription reaction (see Application). The slopes of the regression lines for P450 1A1 in Fig. 1 are 1.12 ($r^2 = 0.999$) for liver and 1.06 ($r^2 = 0.999$) for enterocytes, well within estimated experimental limits of the theoretical value of 1. In competitive RNA–PCR, the quantity of mRNA is normally determined from the point where areas of the competitor (known concentration) and target (unknown concentration) bands, separated by gel electrophoresis, are determined to

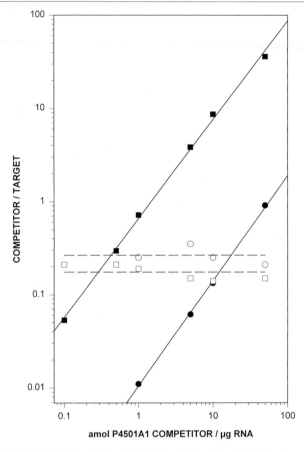

FIG. 1. MULTIPLEX RNA–PCR of P450 1A1 mRNA with GAPDH mRNA as a "house-keeper." Total RNA was isolated from pooled livers (circles) and enterocytes (squares) of three rats that received BNF. The quantity of reverse-transcribed [oligo(dT)$_{16}$priming] RNA was fixed at 0.1 μg/0.1 ml of the PCR reaction mixture. The quantity of the GAPDH competitor was 10 amol/μg RNA, and the quantity of the P450 1A1 competitor was varied from 0.1 to 50 amol/μg RNA. Dotted lines represent the average of all competitor/target ratios obtained for GAPDH (open symbols) for each tissue. Solid lines are linearly regressed for all P450 1A1 competitor/target ratios (solid symbols) for each tissue.

be equal. The accuracy of the method described here permits us to use any single ratio, combination of ratios, or value on the x axis, corresponding to a point where a vertical line from it crosses the competitor/target value at 1, to determine concentration. This is a definite advantage, as a single competitor concentration can be used to quantitate a wide range of mRNA concentrations. The overall benefit is that far fewer data points have to be analyzed to obtain a valid concentration.

Capillary Electrophoresis

Apparatus. The capillary electrophoresis unit used is a P/ACE 2200 with an argon laser (Beckman Instruments, Fullerton, CA), which was selected on the basis of its sensitivity. The capillary is a coated dsDNA 1000, 47 cm × 100 μm (40 cm to the detector window).

Materials and Reagents

A. 89 mM Tris, 89 mM boric acid, 2 mM EDTA, pH 8.5 (TBE)

B. Hydroxypropylmethylcellulose (HPMC) (Sigma Chemical Co., St. Louis, MO). Dissolve 0.5 g HPMC in 100 ml TBE by stirring overnight at room temperature. Filter through paper and then through a 0.45-μm sterilization filter.

C. Oxazole yellow (YOYO-1) (Molecular Probes, Eugene, OR). Stock dimethyl sulfoxide solution (1 mM). Dilute 2.5 μl with 250 μl of reagent A.

D. Separation/wash buffer: Add 150 μl of reagent C to 15 ml of reagent B and mix gently for 10 to 15 min. Solutions are stable for use over a period of 2 days.

E. YOYO-1 plug solution: dilute 0.1 ml of reagent C with 0.1 ml of reagent A

F. PhiX 174 RF DNA *Hae*III (Beckman Instruments)

G. 5-Carboxyfluorescein (Sigma Chemical Co.)

Procedure

1. High pressure wash the capillary in the reverse direction with fresh separation wash buffer for 4 min before each injection.

2. Forward load YOYO-1 plug solution under high pressure for 12 sec.

3. Forward load sample (see earlier discussion) under low pressure for 12 to 18 sec.

4. Electrophorese at 200 V/cm using reverse polarity and the separation/wash buffer.

Results. Capillary electrophoresis analyses for quantitation of RNA–PCR products were developed using PhiX digests and fluorescence of YOYO-1 intercalation of DNA fragments. Reproducible results were obtained by loading a concentrated plug of YOYO-1 onto the capillary immediately prior to sample loading. An electropherogram of the components of this commercially available mixture of DNA fragments is shown in

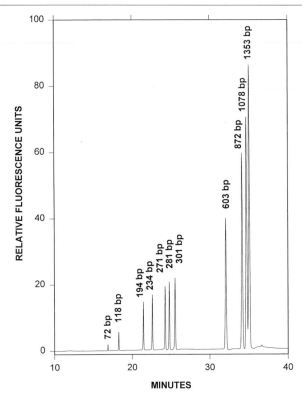

FIG. 2. Capillary electropherogram of a PhiX 174 *Hae*III digest. Detection is by laser-induced fluorescence of YOYO-1 complexes. Conditions are described under *Capillary Electrophoresis*.

Fig. 2. The integrated areas under the component peaks bore a linear relationship to the number of base pairs (over the range of 100 to 1353 bp) and the concentrations (ng/μl) of the DNA fragments. This permits calculation of an amount per area value for any DNA fragment in this size range. To correct for day-to-day variations in YOYO-1 intercalation and fluorescence intensity, a DNA internal standard such as an actin DNA fragment, in addition to a structurally unrelated internal standard, such as 5-carboxyfluorescein, is used.

The amount per area values for the PhiX digest components decrease as a function of increasing size in the range of 194 to 1353 bp. This can be corrected for by selecting a DNA internal standard of similar size to the PCR product to be quantified, but sufficiently different to be resolved on capillary electrophoresis. Alternatively, a correction factor determined

from a plot of peak areas against concentrations of PhiX fragments can
be used.

Application

The time course of BNF-induced transcription of hepatic P450 1A1
mRNA in rats was investigated, using the techniques described earlier
with GAPDH mRNA as a "housekeeping" mRNA. In addition to their
usefulness in detecting tube to tube variation, "housekeeping" mRNAs
can also be used to determine relative efficiencies of reverse transcription
reactions conducted in a number of samples. Profiles of rat liver P450 1A1
mRNA concentrations as a function of time after BNF administration are
illustrated in Fig. 3. Maximum P450 1A1 mRNA induction apparently

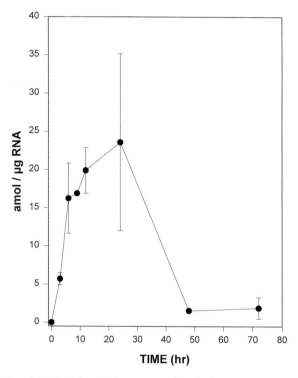

FIG. 3. Profile of P450 1A1 mRNA concentrations in liver as a function of time after
administration of a single oral dose of BNF. Conditions were as in Fig. 1, except that the
P450 1A1 assays were run in separate competitive PCR reactions. P450 1A1 values are the
mean ± SD of single assays of triplicate samples using 5 amol of the competitor/0.1-ml
PCR reaction.

occurred at 6 hr post-BNF administration, and the levels remained essentially constant for at least 18 hr before returning to essentially basal levels at 48 hr.

Acknowledgment

Supported in part by Grant ESO625601 from the National Institutes of Health, Public Health Service.

[44] Targeted Disruption of Specific Cytochromes P450 and Xenobiotic Receptor Genes

By Pedro M. Fernandez-Salguero and Frank J. Gonzalez

Introduction

Although, to date, many naturally occurring genetic diseases in humans have been identified and characterized, the use of natural animal models for the study of such diseases present several disadvantages.[1] In order to study the *in vivo* role and potential developmental relevance of a particular gene, a revolutionary new technology, called "gene knock-out" or "gene targeting," was developed several years ago.[2] Using this technique, it is possible to introduce by homologous recombination in embryonic stem (ES) cells, specific and controlled mutations in the germ line of mice that will inactivate or knock out the expression of a particular gene. Consequently, it is possible to analyze *in vivo* the effects of the lack of expression of a gene in a mouse that, with respect to any other gene, can be considered normal. Many genes have been inactivated, including housekeeping genes, oncogenes, and tumor suppressor genes as well as a large group of metabolically and physiologically relevant genes. A wide variety of phenotypes have been obtained that range from early embryonic death to apparently no obvious phenotypic change.[3-6] A large number of parameters have been found to

[1] O. Smithies, *Trends Genet.* **9,** 112 (1993).
[2] J. Travis, *Science* **256,** 1392 (1992).
[3] A. J. Copp, *Trends Genet.* **11,** 87 (1995).
[4] E. Colucci-Guyon, M.-M. Portier, I. Dunia, D. Paulin, S. Pournin, and C. Babinet, *Cell* (*Cambridge, Mass.*) **79,** 679 (1994).
[5] M. A. Rudnicki, T. Braun, S. Hinuma, and R. Jaenisch, *Cell* (*Cambridge, Mass.*) **71,** 383 (1992).
[6] Y. Saga, T. Yagi, Y. Ikawa, T. Sakakura, and S. Aizawa, *Genes Dev.* **6,** 1821 (1992).

be relevant for the success of a gene-targeting experiment that affects both the *in vitro* and the *in vivo* procedures. However, even when different modifications have been made, the basic scheme is very similar to the one developed by several pioneering teams, including O. Smithies, M. Capecchi, E. Robertson, and A. Bradley.

This chapter presents the experimental protocols established in our laboratory for the production of homozygous null mice for cytochome P450 genes (*Cypla2* and *Cyp2el*) and for xenobiotic receptor genes such as the dioxin receptor (*Ahr*) and the peroxisome proliferator-activated receptor α form (*PPARα*).[7–9] Additionally, for general methods about homologous recombination, gene targeting techniques, and manipulation of mouse embryos, the reader is referred to several of many excellent books.[10–14]

Procedures

Gene Isolation and Construction of the Targeting Vector

It has been shown that the use of isogenic DNA between the targeting vector and the genome of the ES cell line increases the frequency of homologous recombination.[15] Since most, if not all, of the ES cell lines available have been developed based on a 129/Sv mouse strain, it is desirable to use a 129/Sv mouse genomic library to isolate the genomic clone to be used in construction of the targeting vector. However, a high frequency of homologous recombination has also been reported using nonisogenic

[7] S. S. T. Lee, T. Pineau, J. Drago, E. J. Lee, J. W. Owens, D. L. Kroetz, P. Fernandez-Salguero, H. Westphal, and F. J. Gonzalez, *Mol. Cell. Biol.* **15,** 3012 (1995).

[8] T. Pineau, P. Fernandez-Salguero, S. S. T. Lee, T. McPhail, J. M. Ward, and F. J. Gonzalez, *Proc. Natl. Acad. Sci. U.S.A.* **92,** 5134 (1995).

[9] P. Fernandez-Salguero, T. Pineau, D. Hilbert, T. McPhail, S. S. T. Lee, S. Kimura, D. W. Nebert, S. Rudikoff, J. M. Ward, and F. J. Gonzalez, *Science* **268,** 722 (1995).

[10] B. Hogan, R. Beddington, F. Constantini, and E. Lact, "Manipulating the Mouse Embryo: A Laboratory Manual," 2nd ed. Cold Spring Harbor Lab., Cold Spring Harbor, NY, 1994.

[11] A. L. Joyner, "Gene Targeting: A Practical Approach." IRL Press/Oxford Univ. Press, New York, 1993.

[12] J. M. Sedivy and A. L. Joyner, "Gene Targeting." Freeman, New York, 1992.

[13] P. M. Wassarman and M. L. DePamphilis, "Guide to Techniques in Mouse Development," Chapter XI. Academic Press, San Diego, CA, 1993.

[14] F. M. Ausubel, R. Brent, R. E. Kingston, D. D. Moore, J. G. Seidman, J. A. Smith, and K. Struhl, "Current Protocols in Molecular Biology." Massachusetts General Hospital, Harvard Medical School, Wiley, New York, 1995.

[15] J. van Deursen and B. Wieringa, *Nucleic Acids Res.* **20,** 3815 (1992).

DNA.[7,16] Among the most relevant parameters to be considered is the design of the targeting vector.

1. *Coding region disruption.* The coding sequence of the gene is disrupted by inserting a positive selection marker [traditionally the phosphoribosyltransferase II ("neo" gene) that confers resistance to G418]. Whenever possible, this gene should be located in a region encoding a known functional domain of the protein (basic region involved in DNA binding for the *Ahr* or ligand-binding domain for *PPARα*). The neo gene can be simply inserted in an exon (like in the *Cyp1a2*) or a fragment of the coding sequence can be replaced with neo (*Ahr* and *PPARα*). Both approaches have proven to be quite efficient in generating inactive genes.[7–9,17,18]

2. *Length of the homologous DNA used in the targeting construct.* Several studies based on the inactivation of the *hprt* gene have shown that the frequency of homologous recombination increases as the length of the genomic sequence in the construct increases up to about 12–14 kbp.[19,20] In general, a total of 5 to 8 kbp of homologous DNA is considered enough. Additionally, since only 150–300 bp with perfect homology at the ends of the construct will be involved in the recombination event,[11,21] it is advisable to leave at least 0.5 to 1.0 kbp of genomic DNA at either end of the construct to avoid a lower frequency of recombination due to mismatches between the ES cell DNA and the targeting vector.[22]

3. *The topology of the targeting vector.* The internal positioning in the construct of both the neo gene and the negative selection marker (thymidine-kinase gene that confers sensitivity to ganciclovir), as well as the linearization site, has been shown to affect the frequency of homologous recombination.[23–25] Two promoters are commonly used to direct the expression of the neo gene: thymidine kinase (TK) or phosphoglycerate kinase (PGK). For most of the genes reported, either promoter was able to maintain a sufficient rate of transcription to keep the ES cells at a good rate of growth when G418 (300–400 μg/ml) was used in the media. Nevertheless,

[16] M. Zijlstra, E. Li, S. Subramani, and R. Jaenisch, *Nature* (*London*) **342**, 435 (1989).
[17] H. Zhang, P. Hasty, and A. Bradley, *Mol. Cell. Biol.* **14**, 2404 (1994).
[18] J. R. Dorin, P. Dickinson, E. W. Alton, S. N. Smith, D. M. Geddes, B. J. Stevenson, W. L. Kimber, S. Fleming, A. R. Clarke, M. L. Hooper, L. Anderson, R. S. Beddington, and D. J. Porteous, *Nature* (*London*) **359**, 211 (1992).
[19] C. Deng and M. Capecchi, *Mol. Cell. Biol.* **12**, 3365 (1992).
[20] M. J. Schulman, L. Nissen, and C. Collins, *Mol. Cell. Biol.* **10**, 4466 (1990).
[21] A. S. Waldman and M. R. Liskay, *Proc. Natl. Acad. Sci. U.S.A.* **84**, 5340 (1987).
[22] P. Hasty, J. Rivera-Perez, and A. Bradley, *Mol. Cell. Biol.* **11**, 5586 (1991).
[23] A. Zimmer, A. Zimmer, and K. Reynolds, *Biochem. Biophys. Res. Commun.* **210**, 943 (1994).
[24] P. Hasty, J. Rivera-Perez, and A. Bradley, *Mol. Cell. Biol.* **11**, 4509 (1991).
[25] B. H. Koller and O. Smithies, *Proc. Natl. Acad. Sci. U.S.A.* **86**, 8932 (1989).

the PGK promoter is becoming more popular since it seems to be able to transcribe the neo gene more efficiently.

4. *Positive (neo) and negative (ganciclovir or FIAU) selections.* Double selection (positive and negative) can be used versus only single, positive selection in order to reduce the background of nonhomologous recombinant clones (most of the nonhomologous recombinants will incorporate and express the thymidine kinase gene, rendering susceptibility to the toxic effects of ganciclovir). We have found that for most of our constructs, the enrichment in terms of total number of clones surviving after selection in 5 μM of ganciclovir ranged between one- and threefold. Additionally, no significant enrichment was found by the use of two thymidine kinase genes positioned at each end of the construct.[26]

5. Parameters such as the chromosomal location of the gene to be inactivated and whether the gene is actively transcribed in ES cells are also believed to influence the frequency of homologous recombination.

Preparing the DNA and Culturing Mouse Fibroblasts for Electroporation

An important factor that will dramatically influence the final outcome of generating high contribution chimeras is the ES cell line used. A number of ES cell lines have been developed in different laboratories; among the ones commonly employed are J1,[27] E14,[28] CC1,[29] R1,[30] and D3.[31] The electroporation protocol presented here is routinely used in our laboratory using J1 cells (developed in R. Jaenisch laboratory).

Preparation of the Targeting Vector DNA for Electoporation

1. Grow the bacterial clone harboring the targeting vector for 48 hr at 37° in LB media. Purify the DNA through two cesium chloride gradients. Remove the ethidium bromide by the addition of cesium chloride-saturated 2-propanol and dialyze extensively at room temperature, making the last change in pure water. Precipitate the DNA with ethanol, dry the pellet, and resuspend the DNA in 10 mM Tris–HCl, pH 7.5, 1 mM EDTA (TE) buffer.

[26] K. Horie, S. Nishiguchi, S. Maeda, and K. Shimada, *J. Biochem. (Tokyo)* **115,** 477 (1994).

[27] E. Li, E. H. Bestor, and R. Jaenisch, *Cell (Cambridge, Mass.)* **69,** 915 (1992).

[28] R. Kuhn, K. Rajewsky, and W. Muller, *Science* **254,** 707 (1991).

[29] A. Bradley, M. Evans, M. H. Kaufman, and E. Robertson, *Nature (London)* **309,** 255 (1984).

[30] A. Nagy, J. Rossant, R. Nagy, W. Abramow-Newerly, and J. C. Roder, *Proc. Natl. Acad. Sci. U.S.A.* **90,** 8424 (1993).

[31] T. Doetschman, R. G. Gregg, N. Maeda, M. L. Hooper, D. W. Melton, S. Thompson, and O. Smithies, *Nature (London)* **330,** 576 (1987).

2. Linearize with the appropriate restriction enzyme and monitor for complete digestion by agarose gel electrophoresis. Purify the DNA by phenol:chloroform and chloroform extractions, precipitate with ethanol, dry the pellet, and resuspend in DEPC-treated water at about 1 μg/μl. The DNA is now ready for electroporation.

Preparation of Feeder Cells

A monolayer of mouse fibroblasts, derived from G418-resistant homozygous mice and γ-irradiated, is used to support the growth of ES cells. These cells, together with the addition of exogenous leukemia inhibitor factor (LIF), maintain ES cell clones in an undifferentiated (pluripotential) state. This is critical since undifferentiated ES cells are required if they are to colonize the germ line in the embryo after injection. As a general step in all of the cell cultures presented here, the tissue culture flasks, petri dishes, and 24-well plates are coated with 0.2% of autoclaved gelatin at 37° for at least 1 to 2 hr. Aspirate gelatin before use.

1. *Making a primary culture of G418-resistant mouse fibroblasts.* Breed a pair of transgenic mice homozygous for G418 resistance and check for a vaginal plug the next morning (day 0.5). Between days 13.5 and 14.5, sacrifice the female and transfer the uterus to a petri dish under the cell culture hood. Isolate the embryos out of the yolk sac and place them into a new petri dish containing 1× phosphate-buffered saline (PBS) solution. Remove and discard the head and as much as possible of the internal organs (they appear as a red mass in the abdomen) which could contaminate the preparation. Wash in 1× PBS and finely mince the carcasses. Trypsinize the tissues by adding 1 ml/embryo of 2.5% trypsin (Whittaker) and incubate at 37° with shaking for 60 min. Add an equal volume of mouse embryonic fibroblasts media (MEF) which contains D-MEM media (Mediatech), 10% fetal calf serum (FCS, Hyclone), and 1% (vol/vol) antibiotic/antimycotic solution (10,000 U/ml penicillin G, 10,000 μg/ml streptomycin, and 25 μg/ml amphotericin B obtained from GIBCO-BRL). Mix and briefly spin down (for 2 to 3 min at 500 g). Carefully remove the supernatant containing the fibroblasts without disturbing the loose pellet. Plate this cell suspension using 1 × 175-cm^2 flask/embryo in 40 ml of MEF media and grow for about 3 days at 37° in a CO_2 incubator (5% CO_2) until very high confluency is reached (the culture should look very compact and dense). Aspirate the media, wash with 15 ml of 1× PBS, and trypsinize the cells by adding 3 ml of 0.25% trypsin–EDTA (GIBCO-BRL) until they detach from the bottom of the flask; dilute trypsin by adding 4 volumes of MEF media and sediment the cells for 5 min at 500 g. Aspirate the media and resuspend the cell pellet in filter-sterilized freezing media: MEF containing 10% (vol/vol) dimethyl

sulfoxide (DMSO). Divide into aliquots of 1 flask/tube and freeze (we usually put the tubes in styrofoam racks in a −80° freezer overnight to freeze slowly, then transfer to liquid nitrogen).

2. *Preparation of γ-irradiated feeder cells:* Quickly thaw one tube of frozen primary fibroblasts in a 37° water bath, add 10 ml of MEF media, and spin as indicated earlier. Plate in a single 175-cm^2 flask using 40 ml of MEF. Grow to high cell density for 3 to 4 days. Trypsinize as indicated, transfer to 8 × 175-cm^2 flasks, and grow until high density for 4 to 5 days. Trypsinize again, transfer the cells to (30–40) × 175-cm^2 flasks and grow for 7 to 8 days (the flasks should be confluent at this time). Trypsinize and collect the cells into 2 × 50-ml tubes. Inactivate cell division by γ-irradiation using a total dose of 3000 rad. We no longer use mitomycin C treatment since γ-irradiation yields good quality inactivated fibroblasts and is simpler and more convenient. Sediment and resuspend the cells in freezing media as indicated. Aliquot the "ready to use" cells at 1 flask/tube and freeze as indicated.

Electroporation and Selection of Recombinant Clones

Day 1. Thaw an aliquot of γ-irradiated, G418-resistant mouse fibroblasts and plate them into 4 × 75-cm^2 flasks using 15 ml of MEF media/flask. Put them into the cell culture incubator.

Day 2 (Step 1, Fig. 1). Aspirate the media of one of the 75-cm^2 flasks and replace it with 15 ml of ES cell media containing LIF. This media (ES + LIF) is made as follows: D-MEM containing 25 mM HEPES buffer (GIBCO-BRL), 15% (vol/vol) FCS, 1× nonessential amino acids (100× stock from Mediatech), 1% (vol/vol) antibiotics (5000 U/ml penicillin G, 5000 μg/ml streptomycin from Mediatech), 0.1 mM 2-mercaptoethanol, and 1000 U/ml LIF. Store the prepared media at 4° and use it within 1 week. Thaw one aliquot of ES cells, dilute in ES + LIF media, sediment at 500 g for 5 min, and plate into the same 75-cm^2 flask. Grow for 2 days, changing media every 24 hr. Check for the appearance of small, compact colonies that should reach around 50% density over the surface of the plate. Higher densities allow the clones to interact with each other and eventually induce ES cell differentiation. Aspirate the media, wash with 1× buffered (by the addition of 25 mM HEPES buffer) Hank's solution, and trypsinize with 2 ml of 0.25% trypsin–EDTA (buffered with 25 mM HEPES buffer) for 2 to 3 min. Add 10 ml ES + LIF media and homogenize the cell suspension by carefully pipetting up and down a few times. Sediment, resuspend, and distribute the cells equally into the remaining fibroblast-coated 75-cm^2 flasks using 12 ml of ES + LIF media/flask. Let the cells grow for 2 days until reaching 50–60% confluency.

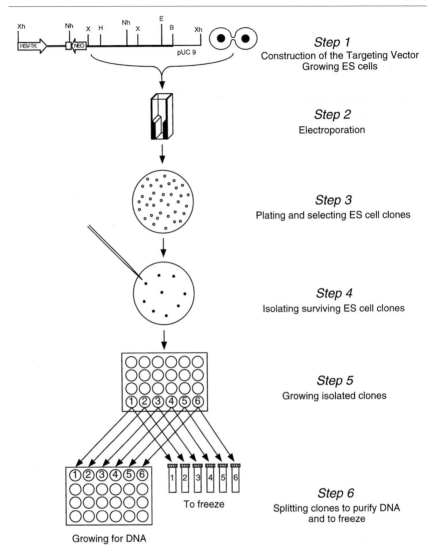

FIG. 1. Schematic representation of the overall procedure used for a gene targeting experiment. The different steps are described in the text.

Day 6. Prepare 20 × 10-cm cell culture petri dishes containing a monolayer of γ-irradiated neomycin-resistant fibroblasts in MEF media as indicated earlier.

Day 7 (Step 2). Prepare the electroporation buffer: 10 ml of HEPES-buffered Hank's solution, 100 μM 2-mercaptoethanol, and 10 μl of 1 M

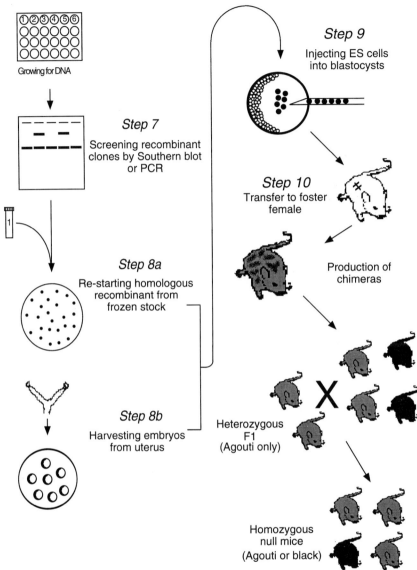

Growing for DNA

Step 7
Screening recombinant clones by Southern blot or PCR

Step 8a
Re-starting homologous recombinant from frozen stock

Step 8b
Harvesting embryos from uterus

Step 9
Injecting ES cells into blastocysts

Step 10
Transfer to foster female

Production of chimeras

Heterozygous F1 (Agouti only)

Homozygous null mice (Agouti or black)

Fig. 1. (*continued*)

NaOH and filter sterilize. The pH should be around 7.0. Trypsinize and sediment the 3×75-cm^2 flasks containing the ES cells, resuspend the pellet in 5 ml of electroporation buffer, and count the cell number. Add 1×10^7 cells to each of four electroporation cuvettes (0.5 cm gap, BioRad), add the linearized targeting vector DNA, and adjust the final volume to 1 ml with electroporation buffer. For the first electroporation, we usually test different amounts of DNA (5, 10, 20, and 40 μg) to check for the optimum transfection efficiency. Electroporate (BioRad gene pulser) at 250 V and 250 μF. Change the media of the fibroblasts-coated petri dishes to ES + LIF. Aliquot each electroporation cuvette into five petri dishes (2×10^6 cells/dish) and return them to the CO_2 incubator. Grow for 24 hr without selection.

Day 8 (Step 3). Prepare the selection media by adding 350 μg/ml G418 (positive selection) and 5 μM ganciclovir (negative selection) to the ES + LIF media. Change the media of the petri dishes by aspirating and then adding 12 ml of selection media/dish. Keep the selection for 8 days. Gradually, an increasing number of nontransfected ES cells will die and float in suspension. After 5–6 days of selection, small ES surviving clones will appear and will be fully visible by day 7–8 of selection.

Day 15 (Step 4). Surviving ES cell clones will be picked and isolated in 24-well plates. Plate γ-irradiated G418-resistant fibroblasts at a density of 7×10^4 cells/well in 24-well plates using MEF media. Prepare a total of 300–350 wells. This protocol (adapted from Dr. H. Westphal's laboratory) requires two persons in order to work efficiently and reduce stress to the ES cells.

PERSON A. This person will pick up the clones from the petri dishes. Add 10 μl/well of 0.025% trypsin–EDTA to 2×96-well plates (U bottom). Set up a dissecting microscope inside the cell culture hood. Autoclave large round tips (PGC 107-003). Take one petri dish from the CO_2 incubator, aspirate the media, and add 8 ml of HEPES-containing Hank's solution. Put the plate in the binocular and locate the ES cell clones (at this stage, when most, if not all, of the nonrecombinant cells have already died, the surviving clones should be readily seen). Select a compact, round clone with well-defined edges (avoid loose, expanding clones, if any) and lift it by suctioning with a P20 pipette connected to a large round tip. Transfer the clone to 1 well of the 96-well plate. Repeat until a row of eight clones has been completed. Double check with the binocular to make sure that all the wells contain clones. Transfer the 96-well plate to person B.

PERSON B. This person will trypsinize and transfer the clones to 24-well plates. Prepare two trays for multichannel pipettes, one containing ES + LIF media and another HEPES-buffered 0.25% trypsin–EDTA. Take yellow tips in racks suitable for multichannel pipettes and transfer every other row of 12 tips to an equivalent empty rack also in alternate rows (in this way two racks are created, each of them containing only four alternate rows of 12 tips). Additionally, complete racks of the same type of tips will also be needed. Obtain the 96-well plate with the eight isolated ES cell clones from person A and add 20 μl/well of 0.25% trypsin–EDTA using a multichannel pipette loaded with 8 tips. Wait for 2–3 min and neutralize the trypsin with 150 μl of ES + LIF media. Carefully pipette up and down a few times (using a multichannel pipette loaded with 8 tips). Now take one row of 4 tips from the racks prepared earlier (alternate rows) and aspirate four clones from the 96-well plate (either odd or even numbers). Transfer them to a row of four in one of the previously prepared 24-well plates. Do the same with the remaining four clones in the 96-well plate using a second set of 4 wells in the 24-well plate. Return the 96-well plate to person A. After a complete 24-well plate is finished, return immediately to the CO_2 incubator. Care should be taken to properly trypsinize the clones. If the clones are just broken into several large fragments, they may not attach from the feeder cells and, if they do, they may grow too fast and thus increase the chances of differentiation.

Day 16 (Step 5). Change the media of the 24-well plates to selection media daily and grow until 40–50% confluency (usually 1–2 more days depending on the size of the original clones isolated).

Day 18 (Step 6). The clones must be analyzed for homologous recombination, leaving a backup of cells to regrow them if any homologous recombinant (positive) clone is found. Using gelatin, coat as many new 24-well plates as needed and add 1 ml/well of ES-only media. Since these clones will be used just to isolate DNA, the undifferentiated state is no longer required and the ES cells can be grown without feeder cells and/or LIF. Prepare ES cell freezing media as follows: ES + LIF media containing extra 5% FCS and 20% (vol/vol) DMSO, filter sterilize, and store at 4° (we normally use it within 3–4 days). Take a 24-well plate from the incubator and, using the binocular, check the confluency in every well. Select and mark those wells having compact, isolated, and well-defined clones at around 50% confluency. Aspirate the media, wash with buffered Hank's, and add 0.2 ml of HEPES-buffered 0.25% trypsin–EDTA solution. Trypsinize for 2–3 min, dilute with 0.8 ml of ES + LIF media, and carefully pipette up and down a few times. Aliquot 0.5 ml of this cell suspension into 1 well of the 24-well plates to isolate DNA and transfer the remaining 0.5 ml to a freezing vial (make sure that the same number is written in the tube and in the 24-well plate for every clone). Repeat until a 24-well plate is completed, return it to the CO_2 incubator, and change media daily for 3–4 days until the cells grow to high density. These plates are then ready to isolate DNA. Additionally, add the same volume of ES cell freezing media (final concentration of DMSO will be 10%) to the freezing tube and slowly freeze overnight at $-80°$ (in a styrofoam rack with lid). The next day, transfer the tubes to liquid nitrogen storage.

Screening of Homologous Recombinant ES Cell Clones

Homologous recombinant clones can be identified by screening ES cell genomic DNA using either Southern blot or PCR (Step 7, Fig. 1). Although both approaches are efficient, great care should be taken when using PCR not only to detect possible false positives (which will be identified later since confirmation by Southern blot is required) but to detect false negatives due to the lack of amplification of a particular template DNA. Screening by Southern blotting is based on the use of a genomic probe located outside the fragment of genomic DNA that constitutes the targeting construct (to give specificity to the pattern obtained). This probe will hybridize with the genomic DNA isolated from the ES cell clones after digestion with the appropriate restriction enzymes (two or three different restriction enzymes, if possible). A different size pattern for the targeted versus the wild-type alleles will then be obtained.

Genomic DNA can be purified from the ES cell clones following the method previously described.[32] After the ES clones have reached a high cell density and are still in the 24-well plates, aspirate the media, wash with $1\times$ PBS, and add 0.5 ml/well of lysis solution: 100 mM Tris–HCl (pH 8.5), 5 mM EDTA, 0.2% SDS, 200 mM NaCl, and 0.15 μg/ml proteinase K (dissolved freshly from lyophilized powder). Allow cell lysis to take place for 4 to 6 hr at $37°$ (overnight lysis will not harm the DNA). To precipitate the DNA, add 0.5 ml of 2-propanol at room temperature and place the plates in an orbital shaker for 15–20 min. Prepare the same number of eppendorf tubes containing 1 ml of 80% ethanol. Transfer the precipitated DNA (it looks like a "cocoon" of fibers) from the 24-well plates to the eppendorf tubes using a pipette, do not centrifuge. Wash by inverting the tubes several times. Very briefly sediment the DNA and decant the ethanol, dry under vacuum, and resuspend in 150–250 μl of TE buffer.

Digest between 20 and 25 μl of DNA from every clone with the appropriate restriction enzyme (for enzymes with high specific activity, 5–8 hr digestion is enough) and electrophorese on agarose gels, adjusting the agarose concentration and other conditions to maximize the resolution of the bands of interest. Transfer the gels to nylon membranes (GeneScreen Plus, DuPont) as indicated elsewhere[14] and hybridize with the [^{32}P]dCTP-

---

[32] P. W. Laird, A. Zijderveld, K. Linders, M. A. Rudnicki, R. Jaenisch, and A. Berns, *Nucleic Acids Res.* **19,** 4293 (1991).

labeled genomic DNA probe. It should be emphasized that this probe must be located outside the genomic DNA included in the targeting construct and it must not be too long (anything between 0.3 and 0.8 kbp usually gives a good signal) to avoid nonspecific hybridization with repetitive sequences. It is recommended to check the probe in advance using wild-type ES cells DNA to confirm the size as well as the specificity of the band obtained. Hybridization takes place overnight at 42° in dextran–sulfate–formamide buffer.[14] Washings must be optimized for every probe, but using 2× SSC, 0.5% SDS at 65° for 20–30 min is a good start. After washing, expose the membranes to X-ray film (overnight to 2 days at −80°) or to phosphor screens for 1 to 3 hr. Identify the potential positive clones with the expected size pattern depending on the restriction map for the gene. Potential positive clones must be checked for a single, legitimate recombination event to avoid additional, nonspecific insertions within the genome of the same clone. To analyze this, the same membrane must be hybridized with neomycin gene as a probe and the pattern obtained must confirm the one corresponding to the genomic probe. Although these two size patterns are considered enough to precisely identify homologous recombinant clones, if desired, the bacterial vector and any other fragment of genomic DNA used to make the targeting construct can also be used as probes to confirm the absence of additional nonhomologous insertions resulting from these sequences.

After the homologous recombinant clones have been identified, they should be expanded and a stock of cells prepared for injection into mouse embryos (Step 8a). The day before thawing the positive ES cell clone, plate 2 wells from a 24-well plate and 2 × 75-cm² flasks with γ-irradiated G418-resistant fibroblasts using MEF media. The following day, quickly thaw and plate the ES cell positive clone into the 2 wells of the 24-well plate using ES + LIF media as indicated earlier. Grow until 50% confluency (2–3 days), changing media every 24 hr. Trypsinize and transfer each of the wells to one of the 75-cm² flasks and grow in ES + LIF the first day, adding G418 and ganciclovir during the rest of the culture until they reach about 50% confluency (3–4 days); change media every day. Trypsinize, sediment, resuspend the cells into ES cells freezing media, and freeze as indicated earlier. Normally, 5–6 tubes/flask give an average cell density.

Production of ES-Derived Chimeras by Microinjection

Two different methods are used to produce ES-derived chimeras. The most popular method consists of microinjecting single ES cells into 3.5-day mouse embryos (blastocysts). The second method is based on inducing the physical aggregation of a cluster of ES cells with an embryo at the eight-

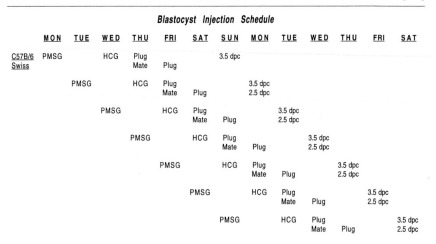

Blastocyst Injection Schedule

	MON	TUE	WED	THU	FRI	SAT	SUN	MON	TUE	WED	THU	FRI	SAT
C57B/6 Swiss	PMSG		HCG	Plug Mate	Plug		3.5 dpc						
		PMSG		HCG	Plug Mate	Plug		3.5 dpc 2.5 dpc					
			PMSG		HCG	Plug Mate	Plug		3.5 dpc 2.5 dpc				
				PMSG		HCG	Plug Mate	Plug		3.5 dpc 2.5 dpc			
					PMSG		HCG	Plug Mate	Plug		3.5 dpc 2.5 dpc		
						PMSG		HCG	Plug Mate	Plug		3.5 dpc 2.5 dpc	
							PMSG		HCG	Plug Mate	Plug		3.5 dpc 2.5 dpc

FIG. 2. Time table for preparing the mice that are used for the *in vivo* steps in the gene targeting protocol. The days of the week when the microinjection takes place are indicated by 3.5 dpc/2.5 dpc. PMSG and HCG stand for the hormones used for superovulation. C57B/6 corresponds to the mouse strain that is used to obtain the blastocysts after superovulation (C57BL/6N). Swiss stands for the strain to which the injected embryos will be transferred. It should be noted that C57BL/6N mice are mated the evening of the same day that HCG is injected. dpc, days postcoitum (as determined by the presence of a vaginal plug).

cell stage which then develops into a blastocyst *in vitro*.[33,34] This chapter focuses on microinjecting single ES cells into blastocysts.

Preparation of Mice and ES Cells for Microinjection

Two different sets of mice are used during the *in vivo* steps. Both must be synchronized in time between them and with the ES cells to be injected. The first set of mice (C57BL/6N) is used to collect the mouse embryos at the blastocyst stage for ES cell injection. The second set of mice (NIH Swiss) consists of foster mothers that will receive and develop the injected embryos to term. Because the number of fecundated embryos per mouse tends to be low by natural means,[10,13] most laboratories use a superovulation scheme to increase this number as well as the overall efficiency of the protocol; this is outlined in Fig. 2.

Day 1. At noon, using five C57BL/6N females that are 4–5 weeks of

[33] S. A. Wood, N. D. Allen, J. Rossant, A. Auerbach, and A. Nagy, *Nature* (*London*) **365**, 87 (1993).

[34] S. A. Wood, W. S. Pascoe, C. Schmidt, R. Kemler, M. J. Evans, and N. D. Allen, *Proc. Natl. Acad. Sci. U.S.A.* **90**, 4582 (1993).

age, start the superovulation by injecting 5 U/mouse of PMSG dissolved in 1× PBS ip (pregnant mare's serum gonadotropin, Sigma).

Day 3. At noon, inject the same animals with 5 U/mouse of HCG dissolved in 1× PBS (human chorionic gonadotropin, Sigma). On the evening of the same day, one female should be placed with a single C57BL/6N male for mating. The next morning check the females for the presence of a vaginal plug (appears as a yellowish mucosity blocking the vagina) and consider this time as 0.5 days postcoitum (dpc). Isolate the females until use 3 days later.

Day 4. The foster mothers are prepared by mating them with vasectomized C57BL/6N males. The use of vasectomized males maximizes the possibilities of implantation of the injected embryos since no natural fecundation of the endogenous NIH Swiss eggs will occur. This, however, does not affect induction of the apropriate hormonal levels in the recipient females to make their uteruses receptive for implantation. Additionally, because the embryos tend to be a little delayed after injection, the foster mothers should be ready 24 hr in advance to give the embryos some extra time for implantation. The presence of a vaginal plug in the foster females must be checked very carefully the next morning as a false-positive plug will prevent the mouse from carrying out the implantation of the embryos in the uterus. Also, plate γ-irradiated fibroblasts in 6 wells of a 24-well plate using MEF media.

Day 5 (Step 8a). Prepare the ES cells to be injected by thawing and plating one of the tubes from the frozen stock at different dilutions in ES + LIF media using the 24-well plate prepared the day before. Using different dilutions will ensure that one will be at the desired 40–50% confluency the day of injection. Plating 40, 20, and 10% from the frozen tube and one-half of these dilutions should give a good range of cell densities. Change ES + LIF media daily.

Day 7 (Step 8b)

1. *Harvesting the embryos:* The embryos are harvested in commercially available M2 media (Specialty Media). Sacrifice the C57BL/6N females, open the abdominal cavity, and locate both uterus horns. Dissect below the oviduct (a fine coil behind the ovary) and at the point where both horns connect. In this way, both horns can be separated from every female. Place the uteruses from the five females in a small (3 cm) petri dish containing M2 media. Transfer one horn into another 3-cm dish and put it on the dissecting microscope. Using low magnification, flush the uterus with a 25-gauge needle connected to a 1-ml syringe loaded with M2 media (we usually flush each uterus horn from both ends). Using higher magnification, search the dish for the presence of embryos and collect them using a glass pipette

(hand made by stretching and flaming regular borosilicate glass) mouth operated through connection to a rubber tubing. Put drops of the M2 media into a new 3-cm dish and cover them with light mineral oil (Fisher). Transfer the harvested embryos to one of the drops. Continue with the remaining horns. To speed the procedure up, several (usually five) horns can be flushed consecutively and all the embryos can be collected together. Although the majority of the embryos collected are usually at the blastocyst stage, a certain number of morulas (a previous stage that does not have the blastocele cavity formed yet) or even eight-cell embryos may be found due to differences in developmental timing. Morulas can be occasionally expanded to blastocysts *in vitro* by incubation in the CO_2 incubator at 37° for 1–2 hr.

2. *Preparing the ES cells:* Check the 6 wells of the 24-well plate and select one of them at 40–50% confluency. Trypsinize, sediment, and resuspend the cells in 0.5 to 1.0 ml of injection media: D-MEM containing 25 mM HEPES buffer (GIBCO-BRL) and 10% FCS. Having the cells in a 15-ml conical tube allows the fibroblasts (which might interfere during microinjection) to settle down easily due to their larger size. It is important to optimize the trypsinization process in order to obtain a single-cell suspension that will facilitate loading the injection needle.

Microinjection Procedure (Step 9)

1. The embryos are injected at 4–10° in a cooling chamber installed on the platform of the injection microscope using 200× magnification. The needle used to hold the blastocysts (holding needle) is made by stretching, cutting with a diamond pen, and flaming a borosilicate glass tube (150 mm in length with a i.d. of 0.8 mm and a o.d. of 1.0 mm). The needle used to inject the ES cells (injection needle) is made using a pipette puller (Sutter P87) to stretch the same type of glass and then cutting the conical tip obtained with a 19-gauge needle to the desired sharpness and diameter under the dissecting microscope. The optimum diameter at the tip should be wide enough to allow the cells to go in and out without major deformation and, at the same time, narrow enough not to transfer too much media when blowing the cells into the embryo. When both needles are ready, they are installed in their holders and filled with the same mineral oil used in the microinjectors (Benton Instruments).

2. When the ES cells and the embryos are ready, put a drop of injection media in the center of the cooling chamber and cover it with mineral oil. Put about 1000–2000 ES cells into the drop of injection media (usually 20–30 μl of the cell suspension). Using the dissecting microscope, check and select those embryos having both a fully expanded blastocele cavity and a complete zona pellucida (appears as a translucent membrane sur-

rounding the blastomeres that form the wall of the blastocyst) and transfer them to the drop of injection media.

3. Take one embryo with the holding needle and immobilize it by suction, being careful not to deform excessively or even break the zona pellucida. To avoid damaging the embryo, it should be held by the side containing the inner cell mass (appears as a mass of cells on the inner space of the embryo). Check that the flow in the injection needle is stable and load 40–50 single cells (avoid large cells that might break and block the needle). Place the embryo and the injection needle at the same focal distance and put the injection needle facing the space between two blastomeres in the wall of the embryo, then push forward quickly to penetrate the blastocele cavity. Slowly put 12–15 cells within the cavity (avoid releasing too much media that could increase the internal pressure and damage the embryo). Remove the injection needle from the embryo and place the injected blastocyst at a specific location in the drop of media. Repeat the same procedure with the rest of the embryos. After all of them have been injected, transfer the embryos to fresh drops of M2 media under mineral oil in a new 3-mm dish. They can be incubated at 37° for 1–2 hr before being transferred to the foster mother. Although the blastocele cavity tends to collapse after the embryo is injected, it normally reexpands during this incubation.

Transfer Injected Embryos to Foster Mothers (Step 10)

1. Prepare drops of commercially available Brinster's media (GIBCO-BRL) under mineral oil. Transfer the injected embryos from the M2 media to the Brinster's drops. Load a mouth-operated glass pipette with a few microliters of Brinster's media, a small bubble of air, 12–15 embryos, another bubble of air, and finally some more media.

2. Anesthetize the recipient mother by injecting 300–400 μl of 2.5% avertin ip (100% stock is made by dissolving 10 g tribromoethyl alcohol in 10 ml of tertiary amyl alcohol). Wash the dorsal part of the body with alcohol. Make a small longitudinal incision (about 1 cm) on the back side, just at the level of the final ribs. The fat pad attached to the ovary should be visible through the body wall. Make an incision at this level on the body wall (avoid cutting the well-defined blood vessels in that area) and grasp the fat pad connected to the ovary with a forceps. Fix a serafine clip to the fat pad to keep the exposed uterus outside the body. Using a 25-gauge needle, make a small puncture in the uterus wall 2–3 mm from the oviduct. Remove the needle and insert the mouth pipette containing the embryos 3–4 mm deep. Blow out the embryos using the first air bubble as a reference and proceed until the top air bubble is out. Remove the mouth pipette,

return the uterus and the ovary into the body, and close the incision in the skin only by applying two sterile surgical wound clips. To help in recovery, the mouse can be put on a 37° warm plate for 15–20 min. As an alternative to single horn transfer, a group of six to seven embryos can be transferred to each one of the two horns following the same procedure described earlier.

3. On some occasions, when only one to two pups are born, they may die because they cannot stimulate the mother to produce enough milk. To partially solve this problem, two or three noninjected embryos can be transferred together with the injected ones in order to increase the litter size. The foster mothers should be checked 12 to 14 days after injection for pregnancy. To minimize the possibility of the mother killing the pups, drop a few pieces of cotton-like neslets into the cage several days before birth. Once the pups are born, traces of agouti color can be seen on their coats after 7–9 days. A large variability of contribution of the ES cells in the formation of the chimeras can be obtained depending on many factors. Additionally, since the ES cell line used (diploid) was isolated from a male mouse embryo,[27] and assuming homogeneous distribution of the ES cells in the formation of all the tissues of the chimera, high contribution chimeras should be males since the ES cells will determine the sex of the animal. Occasionally, however, good contribution females chimeras can also be obtained, possibly due to karyotypic changes; these animals should not be used in the following steps.

Production of Heterozygous and Homozygous Null Mice

Assuming that the ES cells contribute approximately in the same proportion to the formation of every tissue, we consider that male chimeras having an estimated agouti coat color above 60% can be mated with females of the selected genetic background (we used C57BL/6N) to produce heterozygous animals. In the first generation, both characters, i.e., agouti coat color and the targeted mutation, segregate together, resulting in the fact that only 50% of the agouti animals are heterozygous for the targeted mutation. Agouti animals should be genotyped by Southern blot or PCR using a small piece (1 cm) of tail to isolate genomic DNA. The protocol used to isolate genomic DNA from this tissue is the same as the one described earlier for ES cells except that the incubation used to lyse the tissue takes place overnight at 50° with shaking.

Heterozygous animals are bred so as to generate a line of homozygous null mice. In this F2 generation, coat color and the targeted mutation segregate independently, resulting in any possible combination of both characters. Animals should also be genotyped using tail genomic DNA analysis as described previously.

These transgenic animals should be housed in a controlled, pathogen-free animal facility in order to keep any adverse effects on their health status to a minimum.

Phenotypic Analysis and Final Considerations

After a homozygous null mouse line has been established, a phenotypic change should be seen if the inactivated gene has a developmental role. However, in many cases, the animals are apparently normal and must be subjected to an appropriate challenge for the lack of the gene to be observed.

An important issue that has become prominent is the influence of the genetic background (mouse strain) on which the homozygous null mouse line has been established. Profound differences have been reported recently for the phenotype of mice lacking expression of the epidermal growth factor receptor, depending on the genetic background employed to cross the knockout gene. The phenotypic change ranges from lethality at an early preimplantation stage in CF-1 background,[35] to death at midgestation in the 129/Sv mouse strain,[35,36] to survival up to 8 days after birth when Swiss Webster females were used,[37] and eventually to survival up to 3 weeks after birth in either a CD-1 background[35] or a C57BL6/J × MF-1 mixed mouse strain.[36] These results clearly indicate that different factors which influence phenotype are present at variable levels in different genetic backgrounds. In some cases these factors can partially backup the lack of expression of a particular gene, allowing the developmental program to proceed to completion. These results and many others suggest that the mouse genome is highly redundant, having a compensatory system able to backup the lack of expression of certain genes previously thought to be critical for development and survival.

Additional considerations must be undertaken before attempting a gene targeting project. Many variables make the protocol unique for every gene and every laboratory. In some circumstances, alternative strategies for the production of chimeras by coculture of the ES cells with the embryos can be considered when microinjection equipment is not readily available. Additionally, the tissue-specific inactivation of genes can be achieved by the Cre-LoxP strategy in which the tissue-specific expression of the Cre

---

[35] D. W. Threadgill, A. A. Dlugosz, L. A. Hansen, T. Tennenbaum, U. Lichti, D. Yee, C. LeMantia, T. Mourton, K. Herrup, R. C. Harris, J. A. Barnard, S. H. Yuspa, R. J. Coffey, and T. Magnunson, *Science* **269,** 230 (1995).

[36] M. Sibilia and E. F. Wagner, *Science* **269,** 234 (1995).

[37] P. J. Miettinen, J. E. Berger, J. Meneses, Y. Phung, R. A. Pedersen, Z. Werb, and R. Derynck, *Nature (London)* **376,** 337 (1995).

protein leads to tissue-specific gene inactivation, leaving a normal level of expression of the gene in the nontargeted tissues of the animal.[38-40]

Finally, the ultimate goal of producing a homozygous null mouse line by gene targeting is to be able to extrapolate the information obtained to the study of human diseases.[1,2] Several human diseases, such as cystic fibrosis,[18,41,42] hemophilia A,[43] atherosclerosis,[44,45] and β-thalassemia,[46] are currently under study based on the information obtained from the corresponding null mice models.

[38] H. Gu, J. D. Marth, P. C. Orban, H. Mossmann, and K. Rajewsky, Science 265, 103 (1994).
[39] M. Barinaga, Science 265, 26 (1994).
[40] H. Gu, Y.-R. Zou, and K. Rajewsky, Cell (Cambridge, Mass.) 73, 1155 (1993).
[41] J. N. Snouwaert, K. K. Brigman, A. M. Latour, N. N. Malouf, R. C. Boucher, O. Smithies, and B. H. Koller, Science 257, 1083 (1992).
[42] L. L. Clarke, B. R. Grubb, S. E. Gabriel, O. Smithies, B. H. Koller, and R. C. Boucher, Science 257, 1125 (1992).
[43] L. Bi, A. M. Lawler, S. E. Antonarakis, K. A. High, J. D. Gearhart, and H. H. Kazazian, Jr., Nat. Genet. 10, 119 (1995).
[44] S. H. Zhang, R. L. Reddick, J. A. Piedrahita, and N. Maeda, Science 258, 468 (1992).
[45] J. A. Piedrahita, S. H. Zhang, J. R. Hanagan, P. M. Oliver, and N. Maeda, Proc. Natl. Acad. Sci. U.S.A. 89, 4471 (1992).
[46] W. R. Shehee, P. Oliver, and O. Smithies, Proc. Natl. Acad. Sci. U.S.A. 90, 3177 (1993).

Author Index

Numbers in parentheses are footnote reference numbers and indicate that an author's work is referred to although the name is not cited in the text.

A

Abdel-Razzak, Z., 381
Abe, T., 219–220, 221(22)
Abei, M., 164, 167(5, 6), 168(5, 6)
Aboula-ela, F., 277
Abrahmsen, L., 9
Abramow-Newerly, W., 415
Abuaf, N., 76
Adachi, S.-I., 362
Adachi, Y., 164, 167(5), 168(5)
Adams, N., 242
Adams, S. P., 116, 122(10)
Adedoyin, A., 171
Adele, P., 263
Adesnik, M., 25
Aggarwal, S. K., 299
Agosin, M., 287
Agui, N., 292, 303(1)
Ahlgren, R., 305
Aikens, J., 327
Aizawa, S., 412
Akiyoshi-Shibata, M., 52
Albano, E., 218–219, 221(4)
Alexander, A., 116, 118(11)
Allen, N. D., 424
Alm, C., 107, 112(6)
Alton, E. W., 414
Alvares, A. P., 170
Alvarez, J. C., 336, 338(4)
Alvinerie, P., 145–146, 146(3), 147(3, 7), 150(7), 151(7)
Amar, C., 82
Amar, M., 76
Ames, B. N., 386
Amet, Y., 116

Amichot, M., 311
Anders, E. M., 348
Anders, M. W., 334–335
Andersen, J. F., 304, 310, 311(15), 362
Andersen, K. E., 170
Anderson, D. E., 5
Anderson, J. F., 3
Anderson, J. J., 235, 236(6)
Anderson, J. W., 243–244
Anderson, L., 414
Anderson, L. W., 146
Andersson, T., 112–113, 132–133, 135(8), 136(1, 8), 137(7, 8), 138(8), 139(1, 3)
Andre, C., 76, 85
Andreassen, O. A., 199, 202(5), 204(5), 205(5)
Antonarakis, S. E., 430
Anttila, S., 220, 221(15), 232
Anzenbacher, P., 242
Aoyama, T., 199, 202(14)
Apostolakos, M. J., 402, 407(4)
Archakov, A. I., 316
Argos, P., 66
Aricò, S., 219
Arlotto, M. P., 4, 7(11), 14–15, 16(7), 21(7), 22(7), 36, 41, 48, 49(10), 50, 161, 273
Armes, L. G., 25
Arnold, G. E., 347, 354(36), 355(36), 356(36), 357(36)
Arnold, J. R., 351
Arns, P. A., 112
Ashburner, M., 298
Askergren, J., 152
Aslanian, W. S., 106–107, 109(5)
Asperger, O., 311
Asseffa, A., 44, 89
Astier, J. P., 367

432 AUTHOR INDEX

Aubert, B. S., 396
Auerbach, A., 424
Ausubel, F. M., 413, 423(14)
Axel, R., 306
Axelrod, J., 32, 160
Ayerton, A. D., 116
Ayres, M. D., 87
Azoualy, D., 179

B

Baba, T., 8–9, 35, 36(3–5), 37(3–5), 38(5), 39(3–5), 42(3–5), 43(3–5), 44, 44(3–5)
Babé, L. M., 336, 338(4)
Babinet, C., 412
Back, D. J., 144, 149(8), 150(8), 151(8)
Baertschi, S. W., 250
Bagley, J. R., 153, 157(12), 162(12)
Baim, S. B., 54
Baird, S., 78, 80(12), 116
Balbas, P., 5
Baldwin, S. J., 116
Balian, J. D., 107, 112(8), 113(8)
Balke, N. E., 236, 264
Ballet, F., 79, 81(16), 82(16)
Barbee, J. L., 87
Bardou, L. G., 123, 220, 221(20)
Bardsley, B. A., 392
Barinaga, M., 430
Barmada, S., 116, 118(13)
Barnard, J. A., 429
Barnes, H. J., 3–4, 5(12), 7(11), 8–9, 9(12), 12(12), 13, 13(12, 27), 14(12), 15, 17(2), 35–36, 36(2), 271, 273
Barrett, M., 235, 236(9)
Barrick, D., 9, 11(37)
Barry, G. F., 87
Bartoli, F., 392
Bärtsch, S., 52
Basile, A. A., 405, 407(10)
Bass, M. B., 350–351
Batard, Y., 266–267
Bates, S. E., 402
Battalle, J., 244, 245(9)
Battula, N., 171
Bauer, S., 22
Bauman, N., 340
Bautista, D., 207
Bayley, C. L., 353
Beaune, P. H., 56, 58(20), 76, 78–80, 80(12),

81(6, 16), 82, 82(16), 85, 100, 102, 102(7), 115–116, 118(3, 5), 153, 154(20), 160(20), 179, 226, 381
Bech, P., 178–179
Becker, G. E., 210
Beddington, R. S., 413–414, 424(10)
Bedoulle, H., 273
Beijnen, J. H., 149
Beke, M., 129(16), 131
Bell, D. R., 305
Bell, L. C., 44
Bell, R. P., 332
Bellamine, A., 52
Belloc, C., 76, 78, 80(12), 85, 116
Belloni, M., 99
Bellono, G., 219
Belluscio, L., 200
Beneveniste, I., 259
Benhamou, J. P., 76
Benhatter, J., 207
Bennett, P. N., 130
Benning, N. M., 367
Benveniste, I., 260, 261(2), 262–263, 263(11), 267, 271
Berberich, S. A., 7
Berendsen, H. J. C., 353
Berge, J. B., 311
Berger, J. E., 429
Bergling, H., 220, 221(13), 223(13)
Bernasconi, C. F., 333
Bernasoni, R., 99
Berns, A., 422
Bernuau, J., 76, 81(6)
Berthou, F., 115–116, 118(5, 11), 121–123, 153, 154(20), 160(20), 220, 221(20)
Bertilsson, L., 107, 111–112, 112(6, 13), 113(18), 132, 139(3), 171, 179, 199, 200(16), 202(13, 16), 206(13), 209(16)
Besler, B. H., 329
Bestervelt, L. L., 13, 25, 26(9), 27(9), 28(9), 32(9), 33(9), 365, 366(22)
Bestor, E. H., 415, 428(27)
Bhalla, K. N., 145, 146(2), 147(2), 151(2)
Bi, L., 430
Bielawska, A., 383, 384(12)
Binkley, J., 9, 11(37)
Birkett, D. J., 113, 116, 132–133, 135(8), 136(8), 137(7, 8), 138(8), 139–141, 143(1, 5), 144, 144(5, 7), 145, 145(4, 7), 149(7, 9–11), 150(7, 9–11), 151(7, 9–11), 171

Bishop, B., 367
Bjerre, M., 178
Black, S. D., 65, 350
Black, S. M., 16
Blair, I. A., 44, 171
Blaisdell, J., 105, 210, 211(5, 6), 217(5, 6)
Blake, J., 187
Blakeslee, D. M., 358
Blanchard, K., 402
Blaschke, G., 102
Blin, N., 201
Blobel, G., 9
Bloksma, N., 76
Blouin, R. A., 111, 174
Bloumer, J. L., 116
Blum, M., 199, 202(4)
Bock, K. W., 105
Böcker, R. G., 115, 118(3)
Boddupalli, S. S., 13, 315, 316(3, 5), 321(5), 322(5), 327, 350, 361, 362(11), 364(11)
Bodell, W. J., 153
Bodénez, P., 123, 220, 221(20)
Bodian, D. L., 336, 338(7)
Bodlaender, J., 7
Boeijinga, J. K., 169, 169(5), 170, 171(5)
Boeke, J., 54
Boistelle, R., 367
Boldingh, J., 252
Bolivar, F., 5
Bollenbacher, W. E., 292, 298, 303(1)
Bologa, M., 129(18), 131
Bolwell, P., 260, 261(3), 262(3)
Bonadonna, G., 149
Bonfils, C., 396
Bonierbale, E., 79, 81(16), 82(16)
Boobis, A. R., 8
Boone, P., 199, 200(7, 8), 202(7, 8), 206(8)
Bordier, C., 40
Borgna, J. L., 161
Bornheim, L. M., 3, 11(3), 13(3)
Borresen, A.-L., 199, 202(5), 204(5), 205(5)
Børresen, A.-L., 232
Botsch, S., 99–100, 102(7)
Boucher, R. C., 430
Boulikas, T., 37
Bourdi, M., 76, 81(6), 85
Bouton, M.-M., 161
Bozak, K. R., 260, 261(3), 262(3), 275, 278(2)
Braatz, J. A., 350
Bradford, J. Y., 307

Bradley, A., 414–415
Branch, R. A., 105–107, 109(5), 111–112, 121, 163, 210, 217(1)
Brash, A. R., 242, 250, 253, 256(11), 257(11), 259(1, 11)
Braun, T., 412
Breimer, D. D., 169, 169(5–8), 170–171, 171(1, 4, 5, 8), 172(24), 173, 174(24)
Brent, R., 413, 423(14)
Brian, W. R., 51, 86, 114
Bricout, N., 162
Brigman, K. K., 430
Broach, J. R., 66
Brockmeier, D., 105
Brøgger, A., 232
Broly, F., 199, 200(7, 8), 202(7, 8, 14, 15), 206(8)
Bronine, A., 51
Brooks, B. R., 353
Brøsen, K., 177, 179, 183(14), 184(15), 185, 185(8, 9, 13), 186, 186(9), 217
Brown, C. M., 12
Brownlee, R. G., 403
Bruccoleri, R. E., 353
Brugliera, F., 275, 282, 283(5), 306
Brun, A., 311
Bucheler, U. S., 7
Buck, L., 306
Budd, M., 129(19), 131
Buell, G., 9, 11(39), 12(39), 13(12)
Bühl, K., 102
Bulger, W. H., 157, 160, 161(42)
Bunn, H. F., 402
Burbott, A. J., 243
Burnett, J. P., 4
Burns, D. J., 87
Burns, J. J., 115
Burstein, S. H., 156
Busch, M. P., 403
Buswell, R. L., 336, 338(7)
Butler, M. A., 124, 128
Butlers, J. T. M., 89
Buzard, G. S., 207

C

Caldwell, J. C., 330
Caldwell, J. W., 353
Calvert, R. J., 207
Camitta, M. G., 66, 377

Campbell, M. E., 124, 127(4), 129(4, 19), 131
Camper, D. L., 347, 351, 355(35)
Capdevila, J. H., 162
Capecchi, M., 414
Capiello, E., 99
Caporaso, N. E., 128, 199, 202(9), 220, 221(21), 226
Capri, G., 149
Carlson, R. O., 387
Carlson, T. J., 328, 329(10), 330, 347, 351
Carmichael, P. L., 153
Carrière, V., 115–116, 118(5)
Carter, C. W., 370
Carter, C. W., Jr., 370
Case, D. A., 330, 353
Caslavska, J., 131
Castagnoli, N., 330
Castel, L. M., 99
Castro, N. M., 277
Catin, T., 186, 187(3), 194(3), 195(3)
Catinot, R., 79, 81(16), 82(16)
Caudle, D. L., 3, 8(9), 13(9), 14–15, 16(3), 47, 273
Cederbaum, A. I., 43
Cedermark, B., 152
Challine, D., 79, 81(16), 82(16)
Chan, G., 8, 13(27)
Chandler, M. H. H., 174
Chang, M., 112, 113(18)
Chang, S.-L., 187
Chang, Y.-T., 351, 353(34), 357(34)
Chaubert, P., 207
Chen, A. C., 302
Chen, J., 381
Chenery, R. J., 116
Cheng, H.-C., 25
Chiang, J. Y. L., 43, 74
Chiang, Y. L., 25
Chiba, K., 108, 112, 132, 217
Chmurny, G. N., 146
Chomczynski, P., 301, 384
Christiansen, J., 177–178
Christofferesen, R. E., 275, 278(2)
Chu, D. T., 348
Chui, Y. C., 153
Chun, Y.-J., 35
Chun, Y.-L., 41
Chung, B., 25
Cicalese, L., 392
Cieplak, P., 353

Cindy Lau, S.-Z., 235
Clair, P., 89
Clark, B. J., 25, 74, 86
Clarke, A. R., 414
Clarke, L. L., 430
Clarke, S. E., 116
Clot, P., 218–219, 221(4)
Cochin, J., 32
Cock, I., 35, 36(8), 37(8), 39(8), 42(8), 43(8), 44(8)
Coezy, E., 161
Coffey, R. J., 429
Cohen, F. E., 336, 338(6)
Cohen, M. B., 305
Colbert, J., 144, 149(8), 150(8), 151(8)
Coldren, J. W., 118
Cole, K. J., 153
Cole, P. E., 392, 394, 396(14, 15), 401(15)
Collins, C., 414
Collins, F. H., 310
Collins, J. M., 146, 147(9), 150(8), 151(8)
Collins, J. R., 347, 351, 353(34), 355(35), 357(34)
Colucci-Guyon, E., 412
Combest, W. L., 293
Compel, A., 162
Conn, E. E., 260
Connelly, J. A., 236
Conney, A. H., 115, 170
Connolly, M. L., 338
Constantini, F., 413, 424(10)
Cooke, N., 403
Cookman, J. R., 174
Coon, M. J., 13, 25, 26(3, 8, 9), 27(2, 3, 8, 9), 28(9), 29(2), 30, 30(8), 32, 32(3, 9), 33(3, 9), 43, 74, 153, 275, 316, 365, 366(21, 22), 372
Cooper, D. Y., 44
Copp, A. J., 412
Corbin, F. T., 235, 236(8)
Corcos, L., 381
Cornell, W. D., 353
Cornish, E. C., 275, 283(5), 306
Cosme, J., 78, 80(12)
Coulson, A. R., 77
Craigie, R., 367
Craik, C. S., 336, 338(4)
Creaser, I. I., 45
Cresteil, T., 146, 147(7), 150(7), 151(7)
Cribb, A. E., 381

Croftib-Sleigh, C., 153
Cros, S., 146, 147(6)
Croteau, R., 243–245, 245(17), 246(17, 18, 23), 247, 247(24), 248, 248(13, 17), 264
Cuany, A., 311
Cullin, C., 25, 52–53, 56(4), 58(4), 74, 77
Cupp-Vickery, J. R., 350, 363
Curtis, A. T., 302
Curtiss, J. L., 329

D

Dahl, M.-L., 112, 113(18), 199, 200(16), 202(16), 209(16), 220, 221(13), 223(13)
Dahl-Puustinen, M. L., 132, 139(3)
Dahlquist, F. W., 5
Dahlqvist, R., 171
Dale Smith, C. A., 200
Dally, S., 116, 122(9), 123(9)
Daly, A. K., 199–200, 200(11), 202(5, 11), 204(5), 205(5)
Danhof, M., 169, 169(7), 170
Danielson, P. B., 306
Dannan, G. A., 188
Dansette, P. M., 76, 79, 81(16), 82, 82(16)
Darwish, K., 361, 364(10)
Date, T., 12
Dauber-Osguthorpe, P., 353
Daujat, M., 396
Dauterman, W. C., 242, 288
Davidson, N. K., 37
Davies, A., 153
Davies, D., 367
Davies, D. S., 170
Davis, A., 153
Davis, S. N., 116, 122(8)
Dawson, S., 153
Dayer, P., 187, 193(7), 194(7), 195(7)
Dean, C., 235
Debois, J., 147(3)
DeBruin, L. S., 43
DeCamp, D. L., 336, 338(4)
de Gruyter, M., 7
Degtyarenko, K. N., 316
Dehal, S. S., 152–153
Deisenhofer, J., 13, 315, 316(3, 5), 321(5), 322(5), 350, 361–362, 362(11), 364(11)
Delaporte, E., 381
de Leede, L. G. J., 169(5), 170, 171(5)
Dellaporta, S. J., 280

de Matteis, F., 153–154, 162
de Morais, S. M. F., 210–211, 211(5, 6), 217, 217(5, 6)
Dencker, S. J., 178
Deng, C., 414
Dennis, E. S., 311
DePamphilis, M. L., 413, 424(13)
Derynck, R., 429
Desiderio, C., 131
DesJarlais, R. L., 336, 338(3, 4)
de Smith, M. H., 7
DeVito, S. C., 334
De Voss, J. J., 336, 338(2)
Dewar, M. J. S., 329
de Waziers, I., 78, 80(12), 116
Dias, C., 11
Dick, B., 101, 183
Dickinson, P., 414
Dieffenbach, C. W., 405, 407(10)
Diehl, K. E., 235, 236(9)
Dilworth, D. D., 406
DiMaio, J., 275
Dinnocenzo, J. P., 327, 328(4)
Distlerath, L. M., 188
Dixon, J. E., 28
Dixon, J. S., 336, 338(3)
Djordjevic, M. W., 152
Djordjevic, S., 66, 368, 377
Dlugosz, A. A., 429
Dobrocky, P., 130
Dobrowsky, R. T., 383, 384(12)
Doecke, C. J., 141, 144, 145(9)
Doehmer, J., 171
Doetschman, T., 415
Domergue, J., 396
Donehower, R. C., 145, 146(3), 147(3)
Dong, M.-S., 43–44
Doolittle, R. F., 33
Dorin, J. R., 414
Dorko, K., 388, 392
Doyle, M. V., 402
Drago, J., 413, 414(7)
Dragon, Y. P., 153
Dreano, P. Y., 153, 154(20), 160(20)
Dréano, Y., 122
Dressler, K. A., 384
Dreyfus, M., 5
Drobic, K., 311
Drust, F., 259
Dubbelman, A. C., 149

Dubendorff, J. W., 5
Dubois, J., 145, 146(3)
Duffaud, G., 369
Dunia, I., 412
Dunkov, B. C., 310, 311(17)
Dunn, J. J., 5
Durst, F., 57(22), 59, 235–236, 260, 261(1, 2), 262, 262(1), 263, 263(11), 265(9), 266–267, 267(7), 271, 275, 311
Dveksler, G. S., 405, 407(10)
Dvorchik, B. H., 170
Dyda, F., 367
Dykes, D. D., 222

E

Eberhart, D., 161
Eberlein, C. V., 236
Eckert, K., 11
Edwards, R. J., 8
Egeberg, K. D., 332
Egorin, M. J., 149
Eguchi, H., 220, 221(16), 224(16)
Ehrenfeucht, A., 11
Ehrlich, H. A., 228, 230(16)
Eichelbaum, M., 99–101, 101(3), 102, 102(7, 9), 105, 169(9), 170–171, 199, 200(6, 10), 202(3, 4, 6, 10), 206(6)
Ekström, G., 218
Eliasson, E., 218, 221(4)
Elswick, B., 382, 396
Engel, G., 169(9), 170
Engleman, A., 367
Epinat, J. C., 53
Erbes, D. L., 235, 236(7)
Estabrook, R. W., 3, 8(9), 13(9), 14–15, 16(3–5, 7), 21(7), 22(7), 41, 44–45, 47, 47(4), 48, 48(4), 49(10), 50, 51(3), 153, 273, 275, 316
Eugster, H. P., 52
Evans, D. H., 43
Evans, M., 415
Evans, M. J., 424
Evans, W. E., 199, 202(9)
Evert, B., 199, 200(6, 10), 202(6, 10), 206(6)

F

Fabre, G., 396
Fabre, I., 396

Fairbrother, K. S., 199
Faletto, M. B., 114, 210
Farcy, E., 275, 283(5), 306
Farr, A. L., 156
Fasco, M. J., 305, 401
Faulkner, K. M., 14–15, 16(5), 44–45, 51(3)
Fautrel, A., 381
Fecycz, I. T., 9
Fendel, K. C., 152
Feng, P., 250
Feng, R., 153
Fernandez-Salguero, P. M., 412–413, 414(7–9)
Ferrin, T. E., 339
Feyereisen, R. Y., 3, 153, 275, 304–305, 310, 311(15, 17), 316
ffrench-Constant, R., 310, 311(17)
Filipovic, D., 351–352, 353(33)
Finck, M., 76
Fingerhut, M., 129(17), 131
Fink, G. R., 54
Finn, R. F., 7
Finzel, B. C., 315, 319(1), 327, 337, 350
Fischer, C., 101, 102(9)
Fisher, B., 152
Fisher, C. W., 3, 8(9), 13(9), 14–15, 16(3–5, 7), 21(7), 22(7), 41, 44–45, 47, 47(4), 48(4), 50, 51(3), 273
Fisher, R. A., 389, 392, 392(5–7), 394, 394(5, 7), 395(6), 396(7, 14, 15), 401(7, 15)
Fitzgerald, K., 161
Flinois, J. P., 78, 80(12)
Flockhart, D. A., 107, 112(8), 113(8)
Fogleman, J. C., 306
Fojo, A. T., 402
Fonné-Pfister, R., 263
Fornander, T., 152
Förster, A., 105
Förster, E., 405
Foster, A. B., 153, 157(11), 158(11), 162(11)
Fox, T., 353
Frame, J. N., 199, 202(9)
Frampton, M. W., 402, 407(4)
Frank, R., 7
Fraser, H. S., 170
Fraser, M. J., 86
Frear, D. S., 235–236, 236(5), 237, 237(5), 242(5)
Freeswick, P. D., 392
Frei, E., 242

French, J. S., 32
French, S., 219
Fritsch, E. F., 78, 79(14), 213
Fromm, M. F., 102
Fruetel, J. A., 347, 351, 353(34), 357(34)
Frye, R., 121
Fuhr, U., 171, 348
Fujii-Kuriyama, Y., 227, 230, 275
Fujita, V. S., 30
Fukao, K., 164, 167(5, 7), 168, 168(5)
Fukushi, S., 232
Fulco, A. J., 16, 364
Funae, Y., 37, 163–164, 165(7), 167(7), 291
Funck-Brentano, C., 99
Funk, C. D., 244, 245(17), 246(17, 18), 248(17), 250, 259(1)
Furguson, D. M., 353
Furr, B. J. A., 152, 161(3)
Furth, P. S., 336, 338(4)
Furuya, H., 199, 202(12)
Furuya, K. N., 389, 390(11)

G

Gabriac, B., 260, 267, 267(7), 271
Gabriel, S. E., 430
Gaedigk, A., 199, 202(3)
Gaedigk, R., 199, 202(4)
Galteau, M. M., 199, 202(17)
Gandhi, R., 307
Garcia, P. D., 9
Gardner, H. W., 250, 253, 255(3)
Garès, M., 146, 147(6)
Garnier, J.-M., 260, 275
Garssen, G. J., 252
Gasser, R., 28
Gauffre, A., 76
Gautier, J.-C., 52, 54, 56, 56(18), 58(20), 76, 78–79, 81(16), 82(16), 85, 100, 102, 102(7), 115, 118(5), 179, 226, 381
Gearhart, J. D., 430
Geddes, D. M., 414
Gegner, J. A., 5
Gelboin, H. V., 44, 88–89, 89(14), 153, 156(14), 159(14), 160(14), 161(14), 162(14), 199, 202(14), 219
Gelfand, D. H., 10, 228, 230(16)
Genest, M., 353
Gengnagel, C., 66, 74(11)
Gerard, N., 199, 202(17)

Gerber, N. C., 351
Gershenzon, J., 264
Gertner, J. M., 3, 14(8)
Geue, R. J., 45
Geungerich, F. P., 105
Ghanayem, B. I., 114, 210
Gharbi-Benarous, J., 348
Ghrayeb, J., 9
Giani, A., 149
Giani, P., 99
Gianni, L., 149
Gibbs, A. H., 162
Gibian, M. J., 253
Gielen, J. E., 80
Gietz, D., 60
Gijzen, M., 245, 246(23), 248
Gilbert, L. I., 292–293, 294(4, 6, 7), 295(11, 12), 298, 303(1)
Gilevich, S., 45, 51(5)
Gillam, E. M. J., 35, 36(3, 6–9), 37(3, 6–9), 38(6–7, 9), 39(3, 6–9), 40(9), 42(3, 6–9), 43(3, 6–9), 44, 44(3, 6–9)
Gilland, G. L., 358
Gillette, J., 44, 89
Gillette, J. R., 332
Gilliam, E. M. J., 3, 8, 8(5), 9, 13(5), 14(5)
Gilliland, G., 402
Girard, J. E., 403
Girre, C., 116, 121, 122(9), 123, 123(9), 220, 221(20)
Glas, U., 152
Gleming, S., 414
Glenwright, P. A., 170
Globerman, H., 3, 14(8)
Glover, G. I., 7
Goasduff, T., 115, 118(5), 122
Goergoff, I., 392
Gold, L., 5, 7, 7(19), 8(19), 9, 11, 11(37)
Goldberg, M. L., 307
Golde, D. W., 381, 384(7)
Goldkorn, T., 384
Goldsmith, E. J., 361, 362(11), 364(11)
Goldstein, J. A., 105, 107, 112(8), 113(8), 114, 210–211, 211(5, 6), 217, 217(5, 6)
Gonzalez, F. J., 3, 44, 86, 88–89, 89(14), 116, 124, 146, 150(10), 151(10), 153, 199, 200(2), 202(2, 12, 14), 207, 208(23), 219, 227, 275, 316, 388, 412–413, 414(7–9)
Gordon, J., 80
Gorrod, J. W., 348

Götharson, E., 112, 113(18)
Goto, F., 107
Gotoh, O., 153, 227, 230, 316
Gotoh, S., 246
Gottesman, S., 12
Gough, A. C., 199–200, 202(3)
Gould, I. R., 353
Grace, T. D. C., 295
Graham, M. W., 278, 280(12)
Graham-Lorence, S. E., 315–316
Gram, L. F., 177–179, 185
Granger, N., 292, 303(1)
Graves, B. J., 302
Gray, R. S., 292
Grechkin, A. N., 254
Green, L., 9, 11(37)
Greenway, D., 161
Gregg, R. G., 415
Greim, H., 349
Grieneisen, M. L., 293, 294(4)
Griese, E.-U., 105, 199, 200(6, 10), 202(6, 10), 206(6)
Griffin, K. J., 3, 11(3), 13(3)
Griffith, O. M., 277
Griggs, L. J., 153, 157(11), 158(11), 162(11)
Grimberg, J., 200
Groen, K., 169, 173
Grogan, J., 146, 150(10), 151(10), 334
Gronwald, J. W., 236
Gross, B., 66, 74(11)
Gross, S. S., 16
Grubb, B. R., 430
Gu, H., 430
Gu, L., 124–125
Guénard, D., 145, 146(3), 147(3, 6)
Guengerich, F. P., 3, 8, 8(5), 9, 13(5), 14(5), 35, 36(3–9), 37, 37(3–9), 38(5, 7, 9), 39(3–9), 40(9), 41–42, 42(3–9, 13), 43, 43(3–9), 44, 44(3–9), 51, 76, 81, 81(6), 82, 85–86, 114–116, 117(17, 18), 118(3, 15), 123(17), 124, 146, 147(9), 150(8), 151(8), 153, 166, 179, 188, 191, 210, 217(1), 226, 275, 316
Guéritte-Voegelein, F., 145–146, 146(3), 147(3, 6)
Guillin, M. C., 76
Guillouzo, A., 85, 115, 118(5), 381
Gulati, R. S., 173
Gulbrandsen, A.-K., 199, 202(5), 204(5), 205(5)
Gunsalus, I. C., 153, 275, 316, 359

Guo, R., 131
Guo, Z., 3, 8, 8(5), 9, 13(5), 14(5), 35, 36(5–9), 37(5–9), 38(5–7, 9), 39(5–9), 40(9), 42(5–9), 43, 43(5–9), 44, 44(5–9), 116, 118(15)
Gurganus, T. M., 87
Guryev, O., 45, 51(5)
Gustaffsson, J.-Å., 305, 381
Gut, J., 186, 187(3), 194(3), 195(3)
Guttendorf, R. J., 174, 187
Guttman, A., 403
Guzelian, P. S., 161, 382, 389, 392(5, 6, 10), 394(5), 395(6), 396, 399(10)

H

Haack, A. E., 236, 264
Hagler, A. T., 353
Haissig, B. E., 243
Hakes, D. J., 28
Hakes, T. B., 152
Hakkola, J., 305
Halkier, B. A., 3, 9(4), 14(4), 268, 269(3), 270(2), 271(3), 273(3), 274(3)
Hall, J. L., 243
Hall, P., 171
Hall, P. D., 141
Hall, S. D., 111
Hallas, J., 179
Halperin, W., 129(17), 131
Halpert, J. R., 3, 8(6), 350
Halvorson, M., 161
Ham, J. M., 392, 394, 396(14, 15), 401(15)
Hamada, G. S., 220, 221(17)
Hamberg, M., 250
Han, X.-L., 153
Hanagan, J. R., 430
Hancock, R., 37
Hanioka, N., 199, 200(2), 202(2)
Haniskova, H., 242
Hannun, Y. A., 381, 383, 383(6), 384(6, 12)
Hansen, L. A., 429
Hansen, M. G. J., 185
Haraki, K. S., 340
Harder, P. A., 242
Hardwick, J. P., 89
Harikrishna, J. A., 16
Harris, C. C., 220, 221(21)
Harris, D., 347, 349, 351, 353(11), 355(11), 356(11), 357(11)

Harris, J. W., 107, 112(8), 113(8), 146, 147(9), 150(8, 10), 151(8, 10), 334
Harris, R. C., 429
Harris, T. M., 250
Harrowfield, J. M., 45
Hartley, R., 174
Hasemann, C. A., 13, 315, 316(3, 5), 321(5), 322(5), 327, 350, 361–362, 362(11), 364(11)
Hasenfratz, M.-P., 260, 275
Hasler, J. A., 3, 8(6), 129(16), 131
Hassett, C., 3, 11(3), 13(3), 283, 305
Hasty, P., 414
Hatanaka, T., 116
Haugen, A., 232
Haugen, D. A., 25, 32
Hay, J. V., 235, 236(7)
Hayashi, K., 66, 230
Hayashi, S., 220, 221(16), 224(16)
Hayashi, S.-I., 219, 220(11), 221(11), 226, 228, 230, 305
Hayes, R. B., 128
Hazard, E. S. III, 146
He, Y., 3, 8(6)
Healy, E. F., 329
Hefner, J., 243–244
Heidemann, H., 169(9), 170
Heim, M., 199, 202(1)
Heinkele, G., 99–100, 103
Heldrich, F. J., 146
Helvig, C., 311
Henderson, C., 102, 162
Henrich, C. P. F., 298
Henrich, V. C., 293, 298
Herlt, A. J., 45
Herrup, K., 429
Herzog, C. E., 402
Hewer, A., 153
Hewett, J., 107, 112(8), 113(8)
Hezari, M., 244–245, 247(24)
Hhispard, E., 116
Hick, J. B., 280
Hickman, A. B., 367
Hieda, K., 67
Higashi, K., 246
Higashi, T., 371
Higgins, L., 335
High, K. A., 430
Higuchi, R., 10, 200, 228, 230(16)
Higuchi, T., 243

Higushi, R., 67
Hilbert, D., 413, 414(9)
Hill, D. F., 77
Hill, W. G., 232
Hill-Perkins, M. S., 86, 87(7)
Hinuma, S., 412
Hirano, H., 246
Hirvonen, A., 232
Hirvonen, K., 220, 221(15)
Hispard, E., 122(9), 123, 123(9), 220, 221(20)
Hiwatashi, A., 361
Hjolmar, M. J., 152
Hobbs, C. A., 13
Hodgson, A. V., 305
Hodgson, E., 153, 156(21), 157(21), 158(21), 162(21), 242, 287–288
Hoffman, A. R., 160
Hoffman, K. J., 132, 139(2)
Hofman, U., 169(9), 170
Hofnung, M., 273
Hogan, B., 413, 424(10)
Högberg, J., 220, 221(13), 223(13)
Hoge, J. H., 306
Holden, H. M., 367
Holmans, P. L., 14–15, 16(4, 5, 7), 21(7), 22(7), 41, 47, 50
Holton, M. N., 153
Holton, T. A., 275, 278, 280(12), 282, 283(5), 306
Homberg, J. C., 76
Hong, G. F., 77
Hong, J., 115
Hongyo, T., 207
Honigberg, I. L., 118
Hood, B., 226
Hooper, M. L., 414–415
Hooper, W. D., 35, 36(8), 37(8), 39(8), 42(8), 43(8), 44(8)
Hope, H., 371
Horan, M. A., 173
Horecker, B. L., 368
Horie, K., 415
Horn, D. H. S., 295
Horn, G. T., 228, 230(16)
Hotta, H., 116
Hou, Z.-Y., 187
Howard, A. J., 315, 319(1), 327, 337, 350
Howard, E., 44
Howard, S. C., 87
Hsu, L.-C., 25

Hsu, M.-H., 8, 13(27)
Hu, M.-C., 25
Huang, C. C., 339
Huang, S.-L., 217
Hufschmid, E., 131
Huizing, M. T., 149
Humphrey, H. E. B., 129(19), 131
Hurwitz, C. A., 149
Husgafvel-Pursiainen, K., 220, 221(15), 232
Hutchinson, R., 362
Hutchison, J. M., 235, 236(4)
Hyland, C. D., 275, 283(5), 306

I

Iatropoulos, M. J., 152
Ibsen, I., 178
Ichikawa, Y., 361
Idle, J. R., 199–200, 200(11), 202(5, 11), 204(5), 205(5)
Iizuka, T., 230
Ikawa, S., 220, 221(22)
Ikawa, Y., 412
Ikeda, G. J., 139
Ikegami, T., 164, 167(6), 168(6)
Ikehara, M., 277
Ikeuchi, K., 246
Ile, N., 217
Ilyasov, A. V., 254
Imai, K., 219–220, 221(16), 224(16), 226, 227(6)
Imai, T., 3, 14(8)
Imaoka, S., 37, 164, 165(7), 167(7), 291
Inaba, T., 186, 186(6), 187–188, 199, 202(14)
Ingelman-Sundberg, M., 199, 200(16), 202(13, 16), 206(13), 209(16), 218–220, 220(1, 4), 221(4, 13), 223(13)
Ingram, C. D., 250
Innis, M. A., 10
Inoue, H., 280
Inouye, M., 9, 369
Inui, Y., 116, 117(18), 166, 191
Ioannides, C., 349
Iost, I., 5
Iriah, J., 124, 130(5)
Irons, G., 389, 391(3), 392(3)
Ishikawa, A., 163–164, 167(4–6), 168, 168(4–6)
Ishizaki, H., 115
Ishizaki, T., 108, 112, 132, 217

Islam, S. A., 349
Isom, H. C., 392
Israel, Y., 219
Iwahana, H., 230
Iwamoto, Y., 361
Iwasaki, K., 162
Iwasaki, M., 76, 81(6), 115, 118(3), 124
Iwasaki, Y., 164, 167(5), 168(5)
Iwase, T., 220, 221(17)
Iyanagi, T., 66
Iyer, K. R., 330

J

Jacobsen, O., 178
Jacolot, F., 153, 154(20), 160(20)
Jacqz, E., 107, 109(5), 111
Jacqz-Aigrain, E., 199, 202(17)
Jaeger, J. A., 11
Jaenisch, R., 412, 414–415, 422, 428(27)
Jaiswal, A. K., 227
Jancarik, J., 372
Jansen, E. J., 173
Janzon, L., 226
Jarman, M., 153, 157(11), 158(11), 162(11), 348
Jarvis, L. E., 339
Jayaram, M., 66
Jefcoate, C. R., 153
Jeltsch, J.-M., 260, 275
Jenkins, C. M., 3, 8(5), 13(5), 14, 14(5), 35, 36(9), 37(9), 38(9), 39(9), 40, 40(9), 42(9), 43(9, 19), 44(9)
Jenkins, T. M., 367
Jenne, J. W., 170
Jergil, B., 79
Jirtle, R. L., 389, 391(3), 392(3)
Joannet, I., 123, 220, 221(20)
Joeres, R. P., 169, 171(4)
Johansson, I., 199, 200(16), 202(13, 16), 206(13), 209(16), 218, 220, 220(4), 221(4, 13), 223(13)
John, G. H., 3, 8(6)
Johnson, E. F., 3, 8, 11(3), 13(3, 27), 35, 275
Johnson, K. S., 78
Jones, J. P., 326–328, 328(4), 329(10), 330, 332, 334–335, 347, 349, 351, 353(31, 32)
Jones, R. S., 389, 391(3), 392(3)
Jordan, V. C., 152–153, 161(3), 162(10)
Josephy, P. D., 43

Joyeux, H., 396
Joyner, A. L., 413, 414(11)
Juedes, M. J., 157
Jung, F., 3, 11(3), 13(3)
Juretzek, T., 65
Jurima, M., 186(6), 187

K

Kadar, D., 124, 126, 128(7), 129(7), 130(5)
Kadlubar, F. F., 124, 128
Kadwell, S. H., 91, 94(19)
Kagawa, N., 3, 14(8)
Kagimoto, M., 199, 202(1)
Kahn, R., 269
Kaida, S., 66
Kalb, V. F., 276
Kalhorn, T. F., 116, 122(10)
Kalow, W., 124, 126, 127(4), 128, 128(7), 129(4, 7, 9, 16, 17, 19), 130(5), 131, 186, 186(6), 187–188, 199, 202(14)
Kaltenberg, O. P., 152
Kamataki, T., 153, 219, 316
Kamin, H., 370
Kaminsky, L. S., 305, 401
Kammuller, M., 76
Kanamaru, R., 220, 221(22)
Kanazawa, H., 230
Kanazawa, I., 199, 202(12)
Kanetoshi, A., 162
Kapelner, S., 404
Kappas, A., 170
Karasaki, Y., 246
Karash, E., 122
Karbwang, J., 144, 149(8), 150(8), 151(8)
Kärgel, E., 66, 74(11)
Karjalainen, A., 220, 221(15), 232
Karki, S. B., 327, 328(4)
Karp, F., 244
Karplus, M., 353
Kasei, N., 66
Kasper, C. B., 19, 369, 371
Katahira, E. J., 292
Katki, A., 146
Kato, R., 162
Kato, S., 220, 221(21)
Kaufman, M. H., 415
Kawajiri, K., 219–220, 220(11), 221(11, 16), 224(16), 226–227, 227(6), 228, 230
Kawashima, K., 232

Kazazian, H. H., 430
Kazmaier, M., 57(22), 59, 260, 265(9), 266
Kearns, C. M., 149
Keeley, L. L., 307
Keita, Y., 171
Kellermann, G., 226
Kelly, S. L., 242, 311
Kelsell, D. P., 219
Kemler, R., 424
Kemper, B., 3, 8, 11(3), 13(3, 27)
Kempf, A. C., 13, 14(58), 43
Keohavong, P., 11
Keszenman-Pereyra, D., 67
Keung, A. C. F., 149
Kharasch, E. D., 335
Kienle, E., 116, 118(13)
Kiffel, L., 76, 179
Kikuchi, H., 219–220, 221(22)
Killackey, M. A., 152
Kim, B.-R., 8, 35, 36(3), 37(3), 39(3), 42(3), 43(3), 44(3), 146, 147(9), 150(8), 151(8)
Kim, D. H., 115
Kim, J. J., 66
Kim, J.-J. P., 368, 377
Kim, R. B., 116, 122(8)
Kim, S. G., 381
Kim, S.-H., 372
Kimber, W. L., 414
Kimua, S., 88, 89(14)
Kimura, S., 199, 200(2), 202(2, 12), 207, 208(23), 413, 414(9)
King, R. C., 299
King, X., 32
Kingma, J., 12
Kingston, R. E., 413, 423(14)
Kino, I., 220, 221(17)
Kinoshita, H., 163
Kitareewan, S., 114, 210
Kitts, P. A., 87
Kiyokawa, E., 220, 221(17)
Klein, J., 129(18), 131
Kleinman, H. K., 382, 396
Kline, M. C., 403
Klysner, R., 179
Kobayashi, K., 66, 112, 132
Kobayashi, S., 163, 167(4), 168, 168(4)
Kocarek, T. A., 389, 392(6), 395(6)
Koch, B., 3, 9(4), 14(4), 268, 269(3), 270(2), 271(3), 273(3), 274(3)
Koepp, A., 244–245, 247(24)

Koga, Y., 246
Kolesnick, R. D., 384
Koller, B. H., 414, 430
Kollman, P. A., 329–330, 353
Komel, R., 311
Kominami, S., 52
Komori, M., 219
Kondo, I., 199, 202(12)
Konno, Y., 289, 290(6)
Koop, D. R., 32, 116, 118(13), 381
Köpke, K., 66, 74(11)
Koren, G., 129(18), 131
Korsgaard, R., 226
Korzekwa, K. R., 3, 86, 89, 146, 150(10), 151(10), 326–328, 328(4), 332, 334–335, 349, 351, 353(32)
Kost, T. A., 86, 91, 94, 94(19)
Kostrubsky, V. E., 389, 392(8), 396(8)
Kotake, A. N., 129(19), 131
Kouri, R. E., 226
Kraft, R., 65, 67(3), 73(3)
Krahg-Sorensen, P., 178
Krahn, P., 217
Kramer Nielsen, K., 177, 179, 183(14), 185, 185(13)
Krautwald, O., 178
Kreibich, G., 25
Kremers, P., 80
Krishnamorthy, R., 199, 202(17)
Krishnapillai, V., 311
Kroemer, H. K., 99–101, 101(3), 102, 102(7, 9)
Kroetz, D. L., 413, 414(7)
Krom, D. P., 169
Kronbach, T., 8, 13(27), 99, 101(3), 186, 187(3), 194(3), 195(3)
Krummel, B., 10
Kudla, C., 171
Kuhn, R., 415
Kukulka, M. J., 187, 188(11), 189(11), 190(11), 191(11), 193(11), 195(11)
Kumar, G. N., 145–146, 146(2), 147, 147(2), 150(8), 151(2, 8)
Kunkel, T. A., 11
Kuntz, I. D., 336, 338, 338(3–7)
Kunze, K. L., 89
Kupfer, A., 106, 187
Kupfer, D., 152–153, 156, 156(14, 21, 22), 157, 157(21, 22), 158(21, 26), 159(14, 21), 160, 161(14, 42), 162(14, 21, 26)
Kuramshin, R. A., 254

Kuroiwa, Y., 163, 167(4), 168(4)
Kurumbail, R. G., 13, 315, 316(5), 321(5), 322(5)
Kusaka, M., 108
Kuttenn, F., 162
Kuzio, J., 87
Kvinesdal, B., 178
Kyte, J., 33

L

Lacoute, F., 54
Lact, E., 413, 424(10)
Ladam, D. J., 348
Ladner, J. E., 358
Laemmli, U. K., 37, 74, 80
LaFever, R. E., 245, 247, 247(24), 248(30)
Lafont, S., 367
Lagerström, P.-O., 132, 136(1), 139(1)
Lainé, R., 52
Laird, P. W., 422
Lake, B. G., 349
Lam, T. H., 243
Lamb, J. H., 153–154
Lambert, G. H., 129(19), 131
Landers, J. P., 403
Landolina, D., 99
Lang, N. P., 128
Langridge, R., 339
LaRosa, A. C., 334
Larrey, D., 76, 81(6), 85, 188
Larson, J. R., 13, 25, 26(3, 8), 27(2, 3, 8), 29(2), 30(8), 32(3, 9), 33(3, 9), 43, 74, 274, 365, 366(21)
Lasker, J. M., 114, 210
Latini, R., 99
Latour, A. M., 430
Latypov, S. K., 254
Lau, S. C., 242
Laughton, C. A., 348
Lawler, A. M., 430
Lawsen, M. F., 128
Leathart, J. B. S., 199, 200(11), 202(11)
Le Bot, M. H., 122
Le Breton, L., 116
Lecoeur, S., 76, 78–79, 80(12), 81(16), 82(16)
Lederer, F., 16
Lee, C. A., 86, 91, 94(19), 153, 158(26), 162, 162(26)

Lee, E. J., 413, 414(7)
Lee, G. K., 336, 338(6)
Lee, J. T., 99
Lee, K.-H., 112, 132
Lee, S. C., 87
Lee, S.-S., 293, 294(6)
Lee, S. S. T., 287, 292, 413, 414(7–9)
Lee, Y. H., 307
Leeman, T., 187, 193(7), 194(7), 195(7)
Legrand, M., 199, 200(7, 8), 202(7, 8), 206(8)
Legraverend, C., 381
LeMantia, C., 429
Lemoine, A., 179
Lennard, M. S., 349
Lenstra, R., 235, 236(2)
Lerawe-Goujon, F., 80
Leroux, J. P., 76, 179
Lesot, A., 260, 275
Lester, D. R., 275, 283(5), 306
Leto, K. J., 235, 246, 275, 278(1)
Levine, M., 66
Lewinsohn, E., 244–245, 246(18, 23), 248
Lewis, D. F. V., 348–350
Lewis, N. G., 244–245
Li, D., 382, 389, 392(5), 394(5), 396
Li, E., 414–415, 428(27)
Li, H., 351, 361, 363–364, 364(10)
Li, J. J., 160
Li, S. A., 160
Li, Y. C., 25, 43, 74
Li, Y.-Y., 66
Lichti, U., 429
Liddle, C., 381
Liehr, J. G., 153
Lietz, H., 129(19), 131
Lile, J. D., 243
Lillibridge, J., 122
Lim, C. K., 153–154, 162
Lincoln, F. H., 255
Lindberg, R. L. P., 328
Linders, K., 422
Lindros, K. O., 218, 220(1)
Lineberry, M. D., 99
Ling, L., 11
Liotta, D. C., 383–384
Lipman, D. J., 276, 280(8)
Liskay, M. R., 414
Liu, Q., 16
Lloyd, R. S., 114
Locatelli, A., 149

Lockow, V. A., 90, 91(18)
Loew, G. H., 347, 349, 351, 353(11, 34), 355(11, 35), 356(11), 357(11, 34)
Loft, S., 170
Lo-Guidice, J. M., 199, 200(7), 202(7)
Loida, P. J., 347, 351, 353(33), 354(36), 356(36), 357(36)
London, S. J., 199, 200(11), 202(11)
Lonning, P. E., 153
Looman, A. C., 7
Loomis, W. D., 243–244, 245(4, 9), 247(4)
Loper, J. C., 66, 275–276
Lopez, P. J., 5
Lopez-de-Brinas, E., 348
Lopez-Ferber, M., 87
Lopez-Garcia, M., 82
Lord, H. L., 43
Lottermoser, K., 311
Lou, Y.-C., 112, 132
Louerat, B., 51
Lowry, O. H., 156
Loyer, P., 381
Lozoya, E., 348
Lu, C.-Y., 275, 283(5), 306
Lu, J.-C., 25
Lu, J.-Y., 361, 362(11), 364(11)
Lu, Y. P., 217
Lubet, R. A., 348
Lucas, D., 115–116, 121–122, 122(9), 123, 123(9), 220, 221(20)
Luckow, V. A., 86–87
Lundborg, P., 132, 136(1), 139(1)
Lunqvist, E., 199, 200(16), 202(16), 209(16)
Luyten-Kellermann, M., 226
Lyman, S. D., 153, 162(10)

M

Machinist, J. M., 187, 188(11), 189(11), 190(11), 191(11), 193(11), 195(11)
MacIntyre, R. J., 306
Mackenzie, P. I., 141, 144, 144(7)
Madariaga, J. R., 389, 392(7), 394(7), 396(7), 401(7)
Madsen, H., 179, 185(13)
Maeda, N., 415, 430
Maeda, S., 415
Maenpaa, J., 305
Maezawa, Y., 220, 221(19)
Mahon, W. A., 186(6), 187

Mahr, G., 348
Malet, C., 162
Malouf, N. N., 430
Manaut, F., 348
Manchester, J. I., 347
Mandjes, I. M., 149
Mani, C., 153, 156, 156(14, 21, 22), 157(21, 22), 158(21, 26), 159(14, 21), 161(14), 162(14, 21, 26)
Maniatis, T., 78, 79(14), 213
Manns, M., 76
Manoharan, K., 389, 391(4)
Mansuy, D., 58, 76, 79, 81(16), 82, 82(16)
Mapoles, J., 116, 118(11)
March, P., 369
Marco, S., 367
Marez, D., 199, 200(7, 8), 202(7, 8), 206(8)
Mariel, E., 146, 147(6)
Mark, D. F., 402
Maroni, G., 293
Martasek, P., 16
Marth, J. D., 430
Martin, F. M., 277
Martin, J. C., 367
Martin, J. L., 334
Martin, M. V., 3, 8(5), 9, 13(5), 14(5), 35, 36(5, 9), 37(5, 9), 38(5, 9), 39(5, 9), 40(9), 42(5, 9), 43(5, 9), 44(5, 9)
Martin-Wixtröm, C., 3, 8(9), 13(9), 14–15, 16(3), 21(7), 22(7), 41, 47, 50, 273
Mason, H. S., 44
Massengill, J. P., 128
Masters, B. S., 66
Masters, B. S. S., 16, 37, 41(16), 368–370, 370(4), 377
Mathews, F. S., 66
Mathys, D., 186, 187(3), 194(3), 195(3)
Matsubara, K., 277
Matsukage, A., 12
Matsuki, S., 277
Matsunaga, T., 199, 202(14)
Matsuzaki, Y., 164, 167(6), 168(6)
Matteson, A., 152
Matthews, A. P., 141, 171
Matthews, B. W., 376
Mauersberger, S., 66, 74(11)
Maurakami, H., 52
Maurel, P., 161, 179, 396
Maurice, M., 396
Mauseth, J. D., 243

May, B., 396
Mayvais-Jarvis, P., 162
McAllister, B. C., 99
McAllister, C. B., 106–107, 109(5)
McBride, O. W., 219, 227
McCague, R., 153, 158(13), 348
McCarrey, J. R., 406
McCarthy, J. L., 292
McCarthy, J. E. G., 9
McCarthy, M. B., 332, 334
McFadden, J. J., 236
McFarland, J. E., 235, 236(8)
McIntosh, I. P., 5
McKee, R., 200
McKenna, R., 348
McKerrow, J. H., 336, 338(6)
McKinney, C. E., 226
McLain, J. R., 389, 391(3), 392(3)
McLemore, T. L., 226
McManus, M. E., 140–141, 143(5), 144, 144(5, 7), 171
McMillan, J. M., 147
McMillen, S. K., 161
McNamara, P. J., 111
McPhail, T., 413, 414(8, 9)
McPherson, A., 358, 370
Meese, C. O., 100–101, 102(9), 103
Meier, P. J., 101, 183
Meier, U. T., 186, 187(3)
Meijer, A. H., 306
Melton, D. W., 415
Menaldino, D., 383–384
Mendelsohn, J., 384
Meneses, J., 429
Menez, C., 122(9), 123, 123(9), 220, 221(20)
Menez, J.-F., 115–116, 118(11), 121–122, 122(9), 123, 123(9), 220, 221(20)
Meng, E. C., 338
Menting, J. G. T., 275, 283(5), 306
Merrill, A. H., 383
Merrill, A. H., Jr., 387
Merz, K. M., 329
Merz, K. M., Jr., 353
Metzger, D. A., 200
Metzger, J. D., 311
Meyer, F. P., 171
Meyer, P. S., 187
Meyer, U. A., 13, 14(58), 43, 99, 101, 101(3), 132, 137(7), 141, 171, 179, 183, 184(15), 185(8), 186, 194(3), 195(3), 199, 200(2),

202(1, 2, 4, 14, 15), 206(8), 207, 208(23),
210, 211(5, 6), 217, 217(5, 6)
Meyerhans, A., 7
Mhyre, J., 122
Michalopoulos, G., 389, 391(3), 392, 392(3),
396(14)
Michel, G., 187, 188(11), 189(11), 190(11),
191(11), 193(11), 195(11)
Michel, H., 66, 358, 365(4)
Miettinen, P. J., 429
Mignotte-Vieux, C., 266
Mihaliak, C., 244
Mihara, K., 25, 324
Miki, K., 66
Mikus, G., 99, 101(3)
Miles, J. S., 199, 202(3)
Miljkovic, B., 173
Miller, L. K., 86, 90, 91(18)
Miller, S. A., 222
Miller, W. L., 16
Mimura, M., 191
Minakuchi, T., 232
Miners, J. O., 113, 116, 132–133, 135(8),
136(8), 137(8), 138(8), 139–141, 143(1,
5), 144, 144(5, 7), 145, 145(4, 7), 149(7,
9–11), 150(7, 9–11), 151(7, 9–11), 171
Minura, M., 116, 117(18)
Mircheva, J., 85
Mirua, Y., 364
Misawa, S., 163–165, 165(7), 167(4, 7), 168(4)
Misimidrembwa, C. M., 129(16), 131
Miura, M., 166
Miura, Y., 16
Miyoshi, M., 25
Mizutani, M., 267, 275
Mode, A., 381
Moereels, H., 349–350
Moir, D. T., 199, 202(9)
Molin, J., 178
Møller, B. L., 3, 9(4), 14(4)
Moller, D. E., 208
Molowa, D. T., 389, 392(5), 394(5)
Monaharan, K., 389, 392(4)
Moncada, C., 219
Monier, S., 25
Monsarrat, B., 145–146, 146(3), 147(2, 3, 6,
7), 150(7), 151(7)
Monteith, D. L., 389, 391(4), 392(4)
Moore, A. L., 243
Moore, D. D., 413, 423(14)

Morel, F., 115, 118(5)
Moreland, D. E., 235, 236(8)
Morelle, G., 7
Morgan, E. T., 32, 381
Mörike, K., 105
Morimoto, M., 219
Morris, G. M., 350
Morris, S. C., 403
Morrow, P., 188
Moller, B. L., 268–269, 269(3), 270(2), 271(3),
273(3), 274(3)
Moss, J. E., 199, 202(4)
Mossman, H., 430
Motomiya, M., 219–220, 221(22)
Mourad, G., 396
Mourton, T., 429
Mowszowicz, I., 162
Muchmore, D. C., 5
Mucklow, J. C., 170
Muehleisen, D. P., 292
Mueller, H. K., 183
Muindi, J., 384
Müller, H.-G., 66, 74(11)
Müller, H. K., 101
Muller, W., 415
Müller-Eberhard, U., 382, 396
Mullis, K. B., 228, 230(16)
Munt, W. L., 4
Mura, C., 199, 202(17)
Murakami, H., 16, 19(13), 22(13), 57(8), 74
Murphy, L. D., 402
Murray, B. P., 8
Murray, R. I., 337
Murray, S., 170

N

Nagai, K., 65, 72
Nagata, K., 89
Nagura, K., 220, 221(17)
Nagy, A., 178, 415, 424
Nagy, R., 415
Nakachi, K., 219–220, 221(16), 224(16), 226,
227(6), 228, 230
Nakahara, K., 362
Nakamura, K., 107, 168, 210, 211(5, 6),
217(5, 6)
Nakamura, M., 164, 165(7), 167(7)
Nakanishi, K., 293, 294(6)
Nanji, M., 89

Nash, T., 165
Nataf, J., 76, 81(6)
Nawoschik, S., 200
Nebert, D. W., 153, 227, 232, 275, 413, 414(9)
Nedelcheva, V., 218
Needleman, S. B., 317
Negishi, M., 28, 328
Neidle, S., 348
Nelson, D. R., 26, 153, 275, 316, 365
Nelson, E., 139
Nelson, J. O., 348
Nelson, R., 65
Nelson, S. P., 116, 122(10)
Nguyen, D. T., 353
Nieboer, M., 12
Nielsen, H. C., 3, 9(4), 14(4)
Nielsen, H. L., 268, 269(3), 271(3), 273(3), 274(3)
Nikolova-Karakashian, M., 381
Nilakatan, R., 340
Nimkar, S., 383
Nims, R. W., 348
Nishiguchi, S., 415
Nishimura, S., 232
Nissen, L., 414
Noguchi, H., 162
Nojima, H., 280
Nolan, J. A., 403
Nomura, N., 12
Norman, A., 131
Notarianni, L. J., 130
Novak, R. F., 381
Novicki, D. L., 389, 391(3), 392(3)
Novotny, A. L., 389, 391(3, 4), 392(3, 4)
Nussler, A. K., 392
Nuwaysir, E. F., 153

O

Oatis, J. E., Jr., 146
Obara, M., 246
Obeid, L. M., 383, 384(12)
O'Connor, J. A., 295
Oda, R. P., 403
Odell, J. T., 235
Oeda, K., 25
Ohkawa, H., 16, 19(13), 22(13), 25, 52, 57(8), 74
Ohmachi, T., 219–220

Ohmori, S., 8, 35, 36(3, 7), 37(3, 7), 38(7), 39(3, 7), 42(3, 7), 43(3, 7), 44(3, 7)
Ohnachi, T., 219
Ohta, D., 267, 275
Ohtsuka, E., 277
Ohtsuka, M., 168
Okada, T., 232
Okayama, H., 280
O'Keefe, D. P., 235, 236(2), 242, 246, 275, 278(1)
Okuda, K., 153
Olafson, B. D., 353
Oldbring, J., 226
Olins, P. O., 7, 9, 87
Oliver, P., 430
Olivera, A., 387
Olken, N. M., 13, 25, 26(9), 27(9), 28(9), 32(9), 33(9), 365, 366(22)
Olson, E. R., 7
Olson, M. J., 334
Omata, Y., 89
Omer, C. A., 235, 236(2)
Omiecinski, C., 3, 11(3), 13(3), 283, 305, 382, 396
Omura, T., 25, 37, 44, 92, 324
Ono, S., 116
Orban, P. C., 430
O'Reilly, D. R., 90, 91(18)
O'Reilly, I., 139
Orita, M., 230
Orlinick, J., 162
Ornstein, R. L., 331, 347, 350–353, 353(24, 33, 38), 354(36, 38), 355(36, 38), 356(36, 38), 357(36), 364
Orron, S. V., 179
Ortiz de Montellano, P. R., 336, 338(2, 4), 347, 351, 353(34), 357(34)
Ortmann, J., 178
Osada, A., 164, 167(5, 7), 168, 168(5)
Oscarson, M., 199, 202(13), 206(13)
Osguthorpe, D. J., 353
O'Shea, D., 116, 122(8), 171
Otton, S. V., 186, 188
Overton, L. K., 94
Owens, J. W., 413, 414(7)

P

Pak, M. D., 298
Panayotatos, N., 9, 11(39), 12(39)

Panserat, S., 199, 202(17)
Pape, C. W., 153, 157(12), 162(12)
Paramsothy, V., 173
Park, B. K., 163
Park, S. S., 153, 156(14), 159(14), 160(14), 161(14), 162(14)
Parke, D. V., 349
Parkinson, A., 153, 156, 156(14), 159(14), 160(14), 161, 161(14), 162(14)
Pasanen, M., 305
Pascoe, W. S., 424
Passananti, G. T., 170
Pathak, D. N., 153
Paukstelis, J. V., 146
Paul, S. M., 160
Paulin, D., 412
Paulsen, M. D., 331, 347, 351–353, 353(24, 33, 38), 354(36, 38), 355(36, 38), 356(36, 38), 357(36), 364
Peacock, W. J., 311
Pearlman, D. A., 330
Pearlman, R. S., 334
Pearson, W. R., 276, 280(8)
Pedersen, O. L., 178
Pedersen, R. A., 429
Pelkonen, O., 169, 171(1)
Perez, M., 359
Pernecky, S. J., 13, 25, 26(8, 9), 27(8, 9), 28(9), 30(8), 32(9), 33(9), 74, 365, 366(21, 22)
Pernet, A. G., 348
Perrin, S., 402
Perry, K. M., 336, 338(5)
Persson, I., 218, 220, 221(13), 223(13)
Pessayre, D., 76, 81(6), 85
Peter, R., 115, 118(3)
Petersen, G. O., 178
Peterson, G. B., 77
Peterson, J. A., 13, 315–316, 316(3, 5), 321(5), 322(5), 350, 360–362, 362(11), 364(11)
Peterson, L., 330
Peyronneau, M. A., 58
Pfaff, G., 187
Phillips, D. H., 153
Phillips, I. R., 89
Philpot, R. M., 25, 26(8), 27(8), 28, 30(8), 74, 365, 366(21)
Phung, Y., 429
Pichard, L., 396
Pickle, L., 107, 112(8), 113(8), 187
Piedrahita, J. A., 430

Pierce, R., 153, 156, 156(14), 159(14), 160(14), 161(14), 162(14)
Pierrel, M., 266
Pike, J. E., 255
Pineau, T., 396, 413, 414(7–9)
Pinedo, H. M., 149
Pisarov, L. A., 388
Pitot, H. C., 153
Pittendrigh, B., 310, 311(17)
Pohl, L. R., 76, 334
Poitrenaud, F., 121
Polesky, H. F., 222
Polkonen, O., 305
Pompon, D., 51–54, 56, 56(4, 18), 57(22), 58, 58(4, 20), 59, 77–78, 260, 265(9)
Pond, S. M., 144, 145(9)
Pongracz, K., 153
Poon, G. K., 153
Porter, T. D., 13, 19, 25, 26(3), 27(2, 3), 29(2), 32(3), 33(3), 43, 274, 369
Portier, M.-M., 412
Posner, M. A., 392, 394, 396(14, 15), 401(15)
Possee, R. D., 86–87, 87(7)
Potts, M., 260
Poulos, T. L., 315, 319(1), 327–328, 337, 350–351, 353, 355(40), 358–359, 361, 363, 364(10)
Poulsen, H. E., 170, 186
Pournin, S., 412
Prakash, C., 171
Pratt, C. B., 149
Preisig, R., 106
Price, V. K., 152
Price Evans, D. A., 217
Priester, T. M., 235, 236(6)
Purdy, R. H., 160
Purkunen, R., 305

Q

Qian, L., 124, 130(5)
Quattrochi, L. C., 3, 8(9), 13(9), 273

R

Raag, R., 337, 353, 355(40)
Radin, N. S., 384
Raeymaekers, L., 406
Rahman, A., 146, 147(9), 150(8, 10), 151(8, 10)

Rajaonariviny, I. M., 264
Rajewsky, K., 415, 430
Ramsden, R., 283, 305
Randall, R. J., 156
Randles, D. J., 139, 145(4)
Rane, A., 305
Rangwala, S. H., 7, 9
Rankl, N. B., 87
Rannus, A., 220, 221(13), 223(13)
Rasmussen, B. B., 186
Ratanasavanh, D., 115, 118(5)
Raucy, J. L., 3, 11(3), 13(3), 114, 210
Raunio, H., 305
Ravichandran, K. G., 13, 315, 316(3), 327, 350, 361–362, 362(11), 364(11)
Ray, W. A., 107
Rayment, I., 367
Reddick, R. L., 430
Redfern, C. P. F., 293, 298
Redmond, C., 152
Reed, A. E., 329
Reeder, D. J., 403
Rees, D. L. P., 145, 149(9), 150(9), 151(9)
Regardh, C.-G., 112, 132, 136(1), 139(1, 3)
Regnier, S., 57(22), 59, 260, 265(9)
Reichhart, D., 263
Reilly, P. E. B., 35, 36(8), 37(8), 39(8), 42(8), 43(8), 44(8)
Reisby, N., 178
Reisdorf, P., 52, 56(4), 58(4), 77
Relling, M. V., 149, 199, 202(9)
Renaud, J. P., 58
Renberg, L., 132, 139(2)
Renton, K. W., 381
Rettie, A. E., 327
Reunitz, P. C., 153, 157(12), 162(12)
Reynolds, K., 414
Ribam, T. L., 305
Rice, J. W., 87
Richards, F. M., 338
Richards, W. G., 350
Richardson, T. H., 3, 8, 11(3), 13(3, 27)
Riche, C., 153, 154(20), 160(20)
Riddiford, L. M., 302
Rifkind, A. B., 153, 158(26), 162, 162(26)
Rikala, R., 244
Ring, C. S., 336, 338(6)
Ringquist, S., 9, 11(37)
Rivera-Perez, J., 414
Roberts, B., 298

Roberts, D. L., 66, 368, 377
Roberts, N. A., 173
Roberts, V. A., 353
Robertson, A., 232
Robertson, E., 415
Robson, R. A., 140–141, 143(5), 144(5), 171
Rochefort, H., 161
Roden, D. M., 99
Roder, J. C., 415
Rodopoulos, N., 131
Rodrigues, A. D., 186–187, 188(11), 189(11), 190(11), 191(11), 193(11), 194(19), 195, 195(11)
Rodriguez, J., 348
Rodriguez-Arnaiz, R., 310, 311(17)
Roman, L. J., 16
Romkes-Sparks, M., 114, 210
Ronis, M. J. J., 218, 220(1), 288
Roots, I., 127
Rosebrough, N. J., 156
Rosenberg, A., 5
Rosenberg, M. R., 389, 391(3), 392(3)
Rosenblum, W. M., 367
Rosenthal, O., 44
Rosenthal, P. J., 336, 338(6)
Rosing, H., 149
Ross, J. T., 51, 86
Rossant, J., 415, 424
Rost, K. L., 127
Rotert, G. A., 187, 188(11), 189(11), 190(11), 191(11), 193(11), 195(11)
Rougeulle, C., 53
Rountree, D. B., 293, 294(6)
Rowan, K. S., 243
Rowinsky, E. K., 145, 146(3), 147(3)
Rowlands, M. G., 153, 348
Rozman, D., 311
Rudick, J. B., 402
Rudikoff, S., 413, 414(9)
Rudnicki, M. A., 412, 422
Rudolph, M., 103
Rueff, B., 76
Rusinko, A., 340
Russell, C. B., 5
Russell, D. W., 260
Rutqvist, L. E., 152
Ryals, J., 275
Ryberg, D., 232
Ryder, D., 112
Rylander, R., 220, 221(13), 223(13)

Rypniewski, W. R., 367
Ryyppö, A., 244

S

Saarikoski, S., 305
Sabatini, D. D., 25
Sabbagh, N., 199, 200(7, 8), 202(7, 8), 206(8)
Sacchi, N., 301, 384
Saeki, Y., 14–15, 16(7), 21(7), 22(7), 41, 50
Safonova, E. Y., 254
Saga, Y., 412
Sagami, I., 219–220, 221(22)
Sagar, M., 112
Sagara, Y., 13
Sage, B. A., 295
Saiki, R. K., 10, 228, 230(16)
St. Jean, A., 60
Sakaguchi, M., 25, 324
Sakaki, T., 16, 19(13), 22(13), 25, 52, 57(8), 74
Sakakura, T., 412
Sakhova, N., 107, 112(8), 113(8)
Sakurai, S., 293, 294(6, 7)
Salaün, J.-P., 116, 262–263, 263(11), 311
Salemme, F. R., 359
Salsgiver, W. J., 7
Sambrook, J., 78, 79(14), 213
Sandhu, P., 9, 35, 36(4, 5), 37(4, 5), 38(5), 39(4, 5), 42(4, 5), 43(4, 5), 44(4, 5)
Sandström, R. P., 243
Sanger, F., 77
Sanglard, D., 66
Sansom, L. N., 144, 145(9)
Santi, D. V., 336, 338(5)
Sanyal, A. J., 392, 394, 396(14, 15), 401(15)
Sanz, E., 107, 112(6)
Sanz, F., 348
Sargeson, A. M., 45
Sari, M.-A., 51, 86
Sarker, G., 404
Sato, K., 25, 324
Sato, R., 37, 44, 92, 162, 267, 275
Satoh, K., 220, 221(22)
Satoh, T., 116
Savage, T. J., 245, 246(23)
Säwe, J., 171
Saxena, R., 199, 202(9)
Schalk, M., 259, 266
Scharf, S. J., 228, 230(16)
Schauder, B., 9

Schelin, C., 79
Schellens, J. H. M., 169(6–8), 170, 171(8)
Scheller, U., 65, 67(3), 73(3)
Scheutz, E. G., 161
Schick, M., 99
Schiestl, R. H., 60
Schipper, A. L., Jr., 243
Schirmer, R. H., 7
Schmeiser, H. H., 242
Schmid, B., 187
Schmidt, C., 424
Schmidt-Base, K., 367
Schneider, T. D., 11
Schneider, W. P., 255
Schoeller, D. A., 129(19), 131
Schonbrod, R. D., 288
Schow, G. S., 235, 236(4)
Schrenk, D., 105
Schröder, K.-L., 65, 67(3), 73(3)
Schuermann, W. H. T., 402, 407(4)
Schuetz, E. G., 382, 388–389, 390(11), 392(5–7), 394(5, 7), 395(6), 396, 396(7), 399(10), 401(7)
Schuetz, J. D., 388–389, 390(11), 392(5, 7, 9, 10), 394(5, 7), 396, 396(7), 399(10), 401(7)
Schuler, M. A., 265
Schulman, M. J., 414
Schulten, H.-R., 153, 157(11), 158(11), 162(11)
Schunck, W.-H., 65–66, 67(3), 73(3), 74, 74(11), 311
Schwartz, H. E., 403
Scott, J. A., 304, 310
Scott, J. G., 287, 289, 290(6), 292, 292(7)
Scott, S. R., 174
Seago, A., 153, 158(13)
Sedivy, J. M., 413, 414(11)
Seensalu, R., 112
Seibel, G. L., 330, 336, 338(3, 4)
Seidegård, J., 220, 221(13), 223(13)
Seidman, J. G., 413, 423(14)
Sekine, T., 219
Sekiya, T., 230
Sellers, E. M., 188
Sengstag, C., 16, 52
Serabjit-Singh, C. J., 86, 91, 94(19)
Setiabudy, R., 108
Sewer, M. B., 381
Shackleton, C. H. L., 14–15, 16(7), 21(7), 22(7), 41, 50

Shalaby, L. M., 235, 236(6)
Shaw, C. R., 226
Shaw, G. L., 199, 202(9)
Shaw, M., 243
Shea, J. P., 114
Shea, T., 66, 368, 377
Shehee, W. R., 430
Shen, A. L., 19, 369, 371
Shephard, E. A., 89
Sheridan, R. P., 336, 338(3), 340
Sherman, F., 54
Shet, M. S., 14–15, 16(3–5, 7), 21(7), 22(7), 41, 44–45, 47, 47(4), 48(4), 50, 51(3)
Sheta, E. A., 16
Shibata, M., 16, 19(13), 22(13), 52, 57(8), 74
Shields, P. G., 220, 221(21)
Shiloach, J., 22
Shimada, K., 415
Shimada, T., 42, 114, 116, 117(18), 164, 166, 167(7), 191, 226
Shin, S.-G., 112, 132
Shinedling, S., 9, 11(37)
Shinoda, N., 226, 227(6)
Shiro, Y., 362
Shoichet, B. K., 336, 338, 338(5)
Shou, M., 328, 351, 353(32)
Shoun, H., 362
Shuster, J. J., 149
Sibbesen, O., 268–269, 270(2)
Sibilia, M., 429, 430(36)
Siddoway, L. A., 99
Sigimura, H., 220, 221(17)
Silberstein, D. J., 99
Silver, P., 74
Silversward, C., 152
Simon, A., 263
Simon, F., 116, 118(11)
Simon, I., 153, 154(20), 160(20)
Simonsson, B. G., 226
Simonsson, R., 132, 139(2)
Sinclair, J. F., 389, 392(8)
Sinclair, P. R., 389, 392(8), 396(8)
Sindrup, S. H., 185
Singh, U. C., 330
Singleton, A. M., 8
Sirevag, R., 275, 278(2)
Sjöqvist, F., 171, 199, 200(16), 202(13), 206(13, 16), 209(16)
Skånberg, I., 132, 136(1), 139(1)
Skaug, V., 232

Skjelbo, E., 177, 179, 185(9), 186, 186(9)
Skoda, C., 186
Skoda, R. C., 186, 207, 208(23)
Skoog, L., 152
Slater, D. J., 353
Slattery, J. T., 116, 122(10)
Sligar, S. G., 327, 332, 337, 347, 351–352, 353(33), 354(36), 356(36), 357(36)
Sliter, T. J., 292–293
Slomainy, D. J., 226
Smith, C. E., 7
Smith, D. R., 78
Smith, G. E., 86
Smith, I. J., 174
Smith, J. A., 413, 423(14)
Smith, K. J., 140, 143(5), 144(5)
Smith, L. L., 153–154
Smith, R., 367
Smith, S. J., 44, 89
Smith, S. L., 293, 294(5), 295(5)
Smith, S. N., 414
Smith, W. A., 293, 295(12)
Smithies, O., 412, 414–415, 430
Snodgrass, D. R., 226
Snouwaert, J. N., 430
Snow, M. R., 45
Snyder, M. J., 304, 310, 311(15)
Sogawa, K., 227
Sohn, D.-R., 112, 132
Somell, A., 152
Sommer, S. S., 404
Sonderfan, A. J., 161
Song, W., 242
Song, W.-C., 250, 253, 256(11), 257(11), 259(1, 11)
Sonnichsen, D. S., 149
Soons, P. A., 169(6), 170
Sorgel, F., 348
Sondergaard, I., 178
Souer, E., 306
Sousa, R., 365
Southern, E. M., 227
Spellmeyer, D. C., 353
Spelsberg, J. A., 403
Spiegel, S., 387
Spielberg, S. P., 124, 127(4), 129(4, 19), 131
Springborg, J., 45
Spritzer, P., 162
Spurr, N. K., 199–200, 202(4), 219
Srinivasan, K., 403

Srivastava, P. K., 114
Staehelin, T., 80
Stafford, D. W., 201
Staib, A. H., 171
Stallings, W. C., 7
Stauffer, C. E., 156
Stearns, J. F., 336, 338(7)
Steele, C. L., 244, 246(18)
Steen, V. M., 199, 202(5), 204(5), 205(5)
Steinhart, H., 103
Steinmetz, J., 111
Stelle, C. L., 244
Sternberg, M. J. E., 349
Stevens, J. C., 210, 388, 389(1)
Stevens, J. L., 310, 311(15)
Stevenson, B. J., 414
Stevenson, T. W., 275, 283(5), 306
Stewart, C. B., 265
Stewart, J. T., 118, 121, 329
Stiborová, M., 242
Stiff, D., 121
Stiksa, G., 226
Stofer-Vogel, B., 244–245, 246(18), 247, 247(24), 248(30)
Stoffel, S., 228, 230(16)
Stoller, E. W., 235, 236(9)
Stoltz, C., 260, 275
Stormo, G. D., 5, 7(19), 8(19), 9, 11, 11(37)
Striberni, R., 187, 193(7), 194(7), 195(7)
Strobel, H. W., 26, 65, 305, 365
Strobl, G., 348–349
Ström, A., 381
Ström, S. C., 388–389, 391(3, 4), 392, 392(3–6, 8–9), 394, 394(5, 7), 395(6), 396(8, 14, 15), 399(10), 401(7, 15)
Stroud, R. M., 336, 338(5)
Struhl, K., 413, 423(14)
Studier, F. W., 5
Stupans, I., 171
Stura, E. A., 370
Subramani, S., 414
Suda, Y., 219
Sudhivoraseth, K., 292
Sugimura, H., 220, 221(21)
Sugiura, M., 162
Suhara, K., 25, 324, 359
Sullivan, T., 114, 210
Summers, M. D., 86
Sun, D., 389, 392(10), 399(10)
Sun, E., 336, 338(6)

Sunzeri, F. J., 403
Surber, B. W., 187, 188(11), 189(11), 190(11), 191(11), 193(11), 195(11)
Sutinen, M.-L., 244
Sutton, D. R., 161
Suzio, R. S., 116
Suzuki, I., 220, 221(17)
Swaminathan, S., 353
Swanson, H. R., 235, 236(5), 237, 237(5), 242(5)
Sweeney, M. H., 129(17), 131
Sweetman, B. J., 111
Sweetser, P. B., 235, 236(4), 236(7)
Syi, J. L., 348
Szklarz, G. D., 16, 350

T

Tabone, M., 219
Tagashira, Y., 227, 230
Takabatake, E., 163
Takada, A., 220, 221(18)
Takahashi, T., 220, 221(17)
Takahashi, Y., 277
Takase, S., 220, 221(18)
Takase, Y., 164, 167(5), 168(5)
Takeda, S., 235, 236(7)
Takemori, S., 52, 359
Tamura, S., 88, 89(14), 199, 202(12)
Tanaka, E., 163–165, 165(7), 167(4–7), 168, 168(4–6)
Tanaka, M., 220, 221(17)
Tanaka, N., 164, 167(6), 168(6)
Tanaka, Y., 275, 282, 283(5), 306
Tang, B. K., 124, 126, 128(7), 129(7, 9, 16–18), 130(5), 131
Tanihara, K., 12
Tarr, G. E., 32
Tartof, K. D., 13
Tassaneeyakul, W., 116, 132, 137(7)
Tate, W. P., 12
Tefre, T., 199, 202(5), 204(5), 205(5), 232
Teitel, C. H., 128
Temple, J. E., 157
ten Bakkel Huinink, W. W., 149
Tennenbaum, T., 429
Tepperman, J. M., 235
Terefe, H., 102
Terelius, Y., 218, 220(4), 221(4)
Teresa Landi, M., 226

Terriere, L. C., 288
Tesoro, A., 129(18), 131
Teunissen, M. W. E., 169(5), 170, 171(5)
Teutsch, H. G., 57(22), 59, 260, 265(9), 267, 267(7), 271, 275
Thalacker, F. W., 235, 236(5), 237, 237(5), 242(5)
Theoharides, A. D., 160, 161(42)
Theurillat, R., 131
Theve, T., 152
Thiel, W., 329
Thiele, D. J., 30
Thiem, S. M., 86
Thilly, W. G., 11
Thogersen, H. C., 65
Thomas, R. C., 139
Thomas, S. B., 187, 188(11), 189(11), 190(11), 191(11), 193(11), 195(11)
Thome-Kromer, B., 187, 188(11), 189(11), 190(11), 191(11), 193(11), 195(11)
Thompson, K. A., 99
Thompson, M. T., 388–389, 392, 392(5, 7), 394, 394(5, 7), 396(7, 14, 15), 401(7, 15)
Thompson, S., 415
Thorgersen, H. C., 72
Thormann, W., 131
Thornburg, K. R., 146–147
Thotassery, J. V., 389, 392(10), 399(10)
Threadgill, D. W., 429
Thummel, K. E., 116, 122, 122(10), 335
Timpkins, B., 129(17), 131
Tinel, M., 85
Tinoco, I., 277
Tjia, J. F., 144, 149(8), 150(8), 151(8)
Toda, G., 220, 221(19)
Tomasi, A., 219
Tomchick, D. R., 367
Topal, A., 242
Towbin, H., 80
Trager, W. F., 327–328, 329(10), 330, 332, 347, 351, 353(31)
Trant, J. M., 48, 49(10)
Travis, J., 412
Treanor, C., 305, 401
Trell, E., 226
Tréluyer, J. M., 146, 147(7), 150(7), 151(7)
Trivers, G. E., 220, 221(21)
Tröger, U., 171
Truan, G., 52–54, 56(4, 18), 58, 58(4), 77

Trulin, B., 381
Trump, B. F., 220, 221(21)
Tsubaki, M., 361
Tsuji, K., 164, 167(5, 7), 168, 168(5)
Tsutsui, M., 116
Tsutsumi, M., 220, 221(18)
Tucker, M. A., 220, 221(21)
Tucker, R. L., 293
Tukey, R. H., 3, 8, 8(9), 9, 13(9), 35, 36(5, 7), 37(5, 7), 38(5, 7), 39(5, 7), 42(5, 7), 43(5, 7), 44(5, 7), 273
Tunek, A., 79
Tung, M., 358
Turck, A., 200
Turner, P. H., 11
Turpen, T. H., 277
Turvey, C. G., 81, 116, 117(17), 123(17)
Tybring, G., 111–112, 112(13), 113(18)
Tyndale, R., 199, 202(14)
Tzakis, A., 392

U

Uchic, J. T., 187, 188(11), 189(11), 190(11), 191(11), 193(11), 195(11)
Uchida, E., 168
Udvardi, M. K., 311
Uematsu, F., 219–220, 221(22)
Ueng, Y.-F., 35, 36(8), 37(8), 39(8), 42, 42(8), 43(8), 44(8)
Ulfelder, K. J., 403
Umbenhauer, D. R., 114
Umeno, M., 207, 208(23), 219
Urban, P., 51–52, 54, 56, 56(4, 18), 57(22), 58, 58(4, 20), 59, 78, 260, 265(9)
Utell, M. J., 402, 407(4)
Utermohlen, J. G., 3

V

Vainio, H., 220, 221(15), 232
Valadon, P., 79, 81(16), 82, 82(16), 85
Valente, L., 145, 149(9), 150(9), 151(9)
van Bezooijen, C.F.A., 173
van Boxtel, C. J., 169
Vandenberg, P., 253
VandenBraden, M., 210
van der Graaff, M., 169, 171, 171(4), 172(24), 174(24)

van der Kuij, V., 149
van der Velde, E. A., 169(7), 170
van der Wart, J. H. F., 169(7), 170
van Deursen, J., 413
van Duin, J., 7
Van Frank, R. M., 4
van Gunsteren, W. F., 353
van Knippenberg, P. H., 7
Van Luc, P., 25
van Maanen, M. S., 153, 157(11), 158(11), 162(11)
van Zuilen, A., 170
Vapaavuori, E. M., 244
Varak, E., 307
Vaz, A. D. N., 32
Veesler, S., 367
Venitt, S., 153
Venkataraghavan, R., 336, 338(3), 340
Verbeek, R. M. A., 169
Vermeulen, N. P. E., 169, 171, 171(4), 172(24), 174(24)
Vermilion, J. L., 372
Veronese, M. E., 116, 132–133, 135(8), 136(8), 137(7, 8), 138(8), 139–141, 143(1, 5), 144, 144(5, 7), 145, 145(4, 7), 149(7, 9, 10), 150(7, 9, 10), 151(7, 9, 10), 171
Verpoorte, R., 306
Vesell, E. S., 170
Vick, B. A., 254, 256
Vidal-Puig, A., 208
Vieira, I., 146, 147(7), 150(7), 151(7)
Viganó, L., 149
Vignon, F., 161
Villen, T., 107, 112(6)
Vincent-Viry, M., 199, 202(17)
Vineis, P., 128, 226
Vitas, M., 311
Vliegenthart, J. F. G., 252
Vlietstra, T., 169, 171(4)
Vogel, F., 66, 74(11)
Vogelaar, A., 7
von Heijne, G., 9

W

Wagner, E. F., 429, 430(36)
Wagner, G. C., 359
Wakui, A., 220
Waldman, A. S., 414

Walle, T., 145–146, 146(2), 147(2), 150(8), 151(2, 8)
Walle, U. K., 145–146, 146(2), 147, 147(2), 150(8), 151(2, 8)
Wallin, H., 79
Walsh, P. S., 200
Walter, B., 153
Walter, P., 9
Wanders, B., 403
Wang, A. M., 402
Wang, E., 387
Wang, M., 66, 368, 377
Wang, P., 37
Wang, T., 99
Ward, E., 275
Ward, J. M., 413, 414(8, 9)
Warner, M., 305
Warren, J. T., 293, 294(4, 6, 7), 295(11, 12)
Wassarman, P. M., 413, 424(13)
Watanabe, J., 219–220, 220(11), 221(11, 16), 224(16), 226–227, 227(6), 228, 230
Watanabe, M., 199, 202(12), 219–220, 221(17, 22)
Waterman, M. J., 25
Waterman, M. R., 3–4, 7(11), 8, 8(9), 13, 13(9, 27), 14, 14(8), 15, 35–36, 40, 43(19), 74, 86, 273, 275
Watkins, P. B., 161
Watson, N., 7
Waxman, D. J., 153, 275, 316
Wedlund, P. J., 105–107, 109(5), 111, 174, 187, 217
Weghorst, C. M., 207
Weidolf, L., 132
Weiffenbach, B., 199, 202(9)
Weiner, P., 330
Weiner, S. J., 353
Weinhold, F., 329
Weklych, R., 260
Werb, Z., 429
Werck-Reichhart, D., 57(22), 59, 259–260, 265(9), 266–267, 267(7), 271, 275
Werner, D., 7
Wesenberg, G., 367
Wester, M., 3, 11(3), 13(3)
Weston, A., 220, 221(21)
Westphal, H., 413, 414(7)
Weyer, U., 86
Wheelock, G. D., 289, 290(6), 292(7)

White, I. N. H., 153–154, 162
White, J. M., 336, 338(7)
White, J. W., 305
White, L., 112
White, R. E., 332, 334
White, T. B., 305
Wiedmann, B., 74
Wiedmann, M., 74
Wieringa, B., 413
Wilkie, J. S., 295
Wilking, N., 152
Wilkinson, G. R., 99, 105–107, 109(5), 111–112, 116, 122(8), 171, 188, 210, 211(5, 6), 217, 217(1, 5, 6)
Williams, C. H., 370
Williams, D. C., 4
Williams, G. M., 152
Willie, K., 129(17), 131
Willye, J. C., 402, 407(4)
Wilson, I. A., 370
Wilson, T. E., 19, 369
Winkelman, D. A., 367
Winters, D. K., 43
Witholt, B., 12
Wittekindt, N. E., 16, 52
Wold, M. D., 152
Wolf, C. R., 102, 162, 199–200, 202(3), 219
Wolf, R. A., 383, 384(12)
Wolf, R. C., 349
Wölfel, C., 171
Wolff, J., 353
Wolff, T., 349
Wolfisberg, H., 131
Wood, A. J. J., 99
Wood, J., 280
Wood, S. A., 424
Wood, S. G., 389, 392(8), 396(8)
Woods, J. R. A., 60
Woosley, R. L., 99, 107, 112(8), 113(8), 187
Wray, N. P., 226
Wray, V. P., 37
Wray, W., 37
Wright, J. M., 116, 122(10)
Wright, M., 145–146, 146(3), 147(3, 6, 7), 150(7), 151(7)
Wrighton, S. A., 161, 210, 388–389, 389(1), 392(8), 396(8)
Wu, D., 188
Wu, L.-C., 66

Wunsch, C. D., 317
Wurgler, F. E., 16, 52

X

Xia, H., 217
Xiao, Z.-S., 217
Xie, H.-G., 217

Y

Yabusaki, Y., 16, 19(13), 22(13), 52, 57(8), 74
Yagi, T., 412
Yamamoto, S., 12, 164, 167(5), 168(5)
Yamamoto, T., 163
Yamamoto, Y., 168
Yamasaki, R. B., 336, 338(7)
Yamauchi, M., 220, 221(19)
Yamazaki, H., 35, 36(8), 37(8), 39(8), 42, 42(8), 43(8), 44(8), 116, 117(18), 118(15), 166, 191, 226
Yamazaki, S., 25, 324
Yaneva, H., 162
Yang, C. S., 115, 118(3)
Yang, J.-P., 220, 221(16), 224(16)
Yasuhara, H., 163, 167(4), 168, 168(4)
Yasukochi, Y., 37, 41(16), 369, 370(4)
Yasunobu, K. T., 25
Yee, D., 429
Yin, H., 218, 220(4), 221(4), 335
Yip, J., 124, 130(5)
Yokota, H., 199, 202(12)
Yoo, J. S. H., 115
Yoshida, S., 244
Yoshii, A., 226, 227(6)
Young, B., 161
Young, J. F., 128
Yu, C.-A., 359
Yu, H., 275, 278(2)
Yuan, Z.-X., 153–154
Yue, Q.-Y., 199, 202(13), 206(13)
Yun, C.-H., 226
Yuspa, S. H., 429
Yuyama, T., 235, 236(7)

Z

Zajicek, J., 245, 247(24)
Zand, R., 116, 122(10)

Zang, H., 121
Zanger, U. M., 13, 14(58), 43, 186
Zaphiropoulos, P. G., 305
Zeugin, T., 179, 184(15), 199, 202(1)
Zhang, H., 414
Zhang, S. H., 430
Zhang, Y., 111–112, 132
Zhou, H. H., 217
Zhou, Y., 126, 128(7), 129(7)
Ziegler, D., 159

Zijderveld, A., 422
Zijlstra, M., 414
Zimmer, A., 414
Zimmerlin, A., 236, 260, 261(3), 262(3)
Zimmerman, D. C., 250, 256
Zimniski, S. J., 152
Zoebisch, E. G., 329
Zou, Y.-R., 430
Zubovits, T., 128
Zuker, M., 11

Subject Index

A

N-Acetyltransferase, caffeine assay, 124, 130

Allene oxide
  degradation, 250–251
  synthesis, *see* Allene oxide synthase
Allene-oxide synthase
  assay
    high-performance liquid chromatography assay, 254–255
    spectrophotometric assay, 253–254
    substrate synthesis
      materials, 251
      purification, 253
      quantitation, 253
      soybean lipoxygenase reaction, 252
  carbon monoxide affinity, 242
  concentration in monocot microsome preparations, 241–242
  purification from flaxseed
    ammonium sulfate precipitation, 256
    anion-exchange chromatography, 257, 259
    chromatofocusing, 259
    detergent removal, 259
    hydrophobic interaction chromatography, 257
    solubilization, 256–257
    tissue extraction, 256
  substrates, 250
Antipyrine
  cytocrome P450 isoform assays
    blood sampling, 172–173
    clearance estimation, 175
    drug administration
      humans, 171–172
      rats, 172–173
    high-performance liquid chromatography

plasma, 173–174
    urine, 174–175
    urine sampling, 172–173
  metabolism, 169–170
Autoimmune drug-induced hepatitis
  clinical features, 76
  cytochrome P450 expression in heterologous systems
    autoantibody characterization, 80–82, 85
    bacteria, 78–79
    cDNA amplification and cloning, 77
    covalent binding of active metabolites, assay, 79–82
    yeast, 77
  initiation, 76

B

Baculovirus–insect cell expression system, CYP3A4 coexpression with NADPH-P450 reductase
  cell culture, 88–89, 91–92
  heme supplementation, 88–89
  host cells, 87–88, 92
  microsomal fraction isolation, 93–94
  plaque assay, 90–91
  promoters, 87
  recombinant virus formation, 87–90
  scaleup, 94
  vectors, 86–87
  virus amplification, 91

C

Caffeine
  N-acetyltransferase assay, 124, 130
  CYP1A2 assay

blood
  high-performance liquid chromatog-
    raphy, 126–127
  sampling, 126
  systemic clearance estimation, 125–128,
    131
urine
  high-performance liquid chromatog-
    raphy, 128, 130
  sampling, 128
metabolism in humans, 124
xanthine oxidase assay, 124, 130
Capillary electrophoresis, polymerase chain
  reaction products
  apparatus, 409
  electrophoresis conditions, 409
  laser-induced fluorescence detection,
    403
  peak integration, 410–411
Chlorsulfuron, detoxification in wheat,
  235–236
Chlorsulfuron-5-hydroxylase
  assay, 238–240
  induction, 237, 242
  kinetic parameters
    oat enzyme, 241
    wheat enzyme, 240–241
Chlorzoxazone, 6-hydroxylation by
  CYP2E1
  assay, in vitro
    cell culture, 118
    high-performance liquid chromatogra-
      phy, 119
    microsome incubation, 117–118
    reagents, 117–118
    thin-layer chromatography, 119–120
  assay, in vivo
    alcoholics, 122–123
    drug administration, 121
    high-performance liquid chromatogra-
      phy, 121–122
    interference, 120, 122
    reagents, 121
    sample preparation, 121
  evidence, 115–116
  phenotypes, 120, 122–123
Cinnamic acid 4-hydroxylase
  assays
    2-naphthoate hydroxylation and fluo-
      rescence detection, 266–267

radiolabeled substrate and thin-layer
    chromatography, 265–266
  distribution in plant species, 261
  immunoblotting, 267–268
  induction
    chemicals, 263
    light, 262–263
    wounding, 262
  microsome preparation
    seedlings, 261, 264–265
    tuber slices, 261–265
  Northern blot analysis, 268
  phenylpropanoid synthesis, 260–261
CPR, see NADPH-P450 reductase
Crystallization, P450
  commercial kits, 358, 363, 367
  glycerol as precipitant, 366
  light scattering, monitoring of aggrega-
    tion, 367–368
  motion in access channel, 364–365
  P450Bm-3 heme domain, 361–362, 364
  P450cam
    crystal recovery, 360
    precipitant, 360
    substrate addition and removal,
      359–361
    thiol protection, 359
  P450eryF, 362–363
  P450nor, 362
  P450scc, 361
  P450terp, 362
  protein mutagenesis and solubility,
    366–367
  solubilization from membranes, 365
CYP1A1
  alleles
    cancer association, 226, 232
    Ile-Val, 228, 230–232
    MspI, 227–228, 232
    polymerase chain reaction analysis,
      228, 230
    single-strand conformational polymor-
      phism, 230–231
    Southern blot analysis, 227–228
  cigarette smoke detoxification, 226, 232
  gene locus, 227
  induction by aromatic hydrocarbons, 396
  messenger RNA quantitation with poly-
    merase chain reaction, 411–412
  tamoxifen metabolism, 162

CYP1A2
  caffeine assay
    blood
      high-performance liquid chromatography, 126–127
      sampling, 126
      systemic clearance estimation, 125–128, 131
    urine
      high-performance liquid chromatography, 128, 130
      sampling, 128
  imipramine metabolism, 177–179
  propafenone
    dealkylation assay, 102–103
    metabolism, 100
  role in autoimmune drug-induced hepatitis, 85
CYP2B4
  assay, 31–32
  N-terminal modification
    effects
      catalytic activity, 27, 34
      subcellular localization, 26, 32, 34
    expression in *Escherichia coli*
      culture conditions, 28–29
      harvesting, 28–29
      vector construction, 27–28
    immunoblot analysis for subcellular localization, 29–30
    insertion of positive charges, 26
    purification of glutathione *S*-transferase fusion protein, 30–31
    truncation, 25–26
CYP2C8, taxol hydroxylation assay
  fecal samples, 148–149
  high-performance liquid chromatography, 147–151
  interference, 149, 151
  microsomes, 149–150
  plasma samples, 147
  quantitation, 149
CYP2C9
  role in autoimmune drug-induced hepatitis, 81–83, 85
  tolbutamide hydroxylation assay
    extraction, 141–142
    high-performance liquid chromatography, 142
    incubation conditions, 141

microsome preparation, 141
    reagents, 141
    validation, 142–145
CYP2C11, expression in cultured hepatocytes
  cell culture, 381–382
  ceramide effects
    ceramide-activated protein phosphatase, 384
    delivery of ceramide to cells, 382
    exogenous sphingomyelinase and ceramide elevation, 383
    RNA analysis of gene expression, 384–385
    Western blotting, 385
CYP2C19
  alleles
    nomenclature, 212
    polymerase chain reaction analysis
      digestion of products, 211, 214
      interpretation, 214–215
      primers, 211–212
      reaction conditions, 213–214
      sensitivity, 217
      solutions, 212–213
      specificity, 217
      troubleshooting, 215–217
    types, 210–211
  imipramine metabolism, 177–179
  mephenytoin
    assay, *in vitro*, 113–114
    assay, *in vivo*
      drug administration, 107–108
      measurement of urinary 4'-hydroxymephenytoin, 108
      measurement of urinary enantiomeric ratio, 109–111
      phenotyping, 108–109, 111–112
      principle, 107
    hydroxylation phenotypes, 105–106, 210–211
  omeprazole
    hydroxylation, 106, 112
    phenotyping, 112–113
CYP2D6
  alleles
    detection
      CYP2D6A, 203–204
      CYP2D6B, 203–204
      CYP2D6D, 204–206

CYP2D67, 206
DNA preparation, 200–201
polymerase chain reaction assay,
   201–206
single-strand conformational polymorphism analysis, 207–210
effects on activity, 199, 202
types, 200, 202
[*O-methyl*-14C]dextromethorphan assay
extraction, 190–191
inhibitor analysis, 195
microsome
   incubation, 189–190
   preparation, 189
phenotyping, 191, 193–195
principle, 187–188
reagents, 189
recombinant protein, 191
substrate synthesis, 188–189
phenotypes, 186, 200–203
propafenone
   hydroxylation assay, 101–102
   metabolism, 99–100
substrates, 186–187, 349
CYP2E1 (P4502E1)
alleles
   disease association, 220–221
   polymerase chain reaction detection
     DNA purification, 222–223
     5′-flanking region, 223–224
     intron 6, 224–225
     primers, 223–225
     reagents, 221–222
     restriction analysis, 223–225
   types, 219–221
assays, 31–32
chlorzoxazone 6-hydroxylation
assay, *in vitro*
   cell culture, 118
   high-performance liquid chromatog-
     raphy, 119
   microsome incubation, 117–118
   reagents, 117–118
   thin-layer chromatography, 119–120
assay, *in vivo*
   alcoholics, 122–123
   drug administration, 121
   high-performance liquid chromatog-
     raphy, 121–122
   interference, 120, 122

reagents, 121
sample preparation, 121
evidence, 115–116
phenotypes, 120, 122–123
induction by ethanol, 396–397
N-terminal modification
effects
   catalytic activity, 27, 34
   subcellular localization, 26, 32, 34
expression in *Escherichia coli*
   culture conditions, 28–29
   harvesting, 28–29
   vector construction, 27–28
immunoblot analysis for subcellular localization, 29–30
insertion of positive charges, 26
purification of glutathione *S*-trans-
   ferase fusion protein, 30–31
truncation, 25–26
structural alignment with P450s, 318–321
substrates, 115, 218–219
CYP3A family
induction of gene expression
   ethanol, 396–397
   reporter genes, 399–400
   species specificity of hepatocyte re-
     sponse, 389, 395
   steroids, 388–389, 394–395, 397–399
   tamoxifen metabolism, 161
CYP3A4
coexpression with NADPH-P450 reductase in baculovirus–insect cell system
   cell culture, 88–89, 91–92
   heme supplementation, 88–89
   host cells, 87–88, 92
   microsomal fraction isolation, 93–94
   plaque assay, 90–91
   promoters, 87
   recombinant virus formation, 87–90
   scaleup, 94
   vectors, 86–87
   virus amplification, 91
imipramine metabolism, 177–179
propafenone
   dealkylation assay, 102–103
   metabolism, 100
taxol hydroxylation assays
   fecal samples, 148–149
   high-performance liquid chromatogra-
     phy, 147–151

interference, 149, 151
microsomes, 149–150
plasma samples, 147
quantitation, 149
CYP4, gene isolation from cockroach using polymerase chain reaction
amplification, 309–310
cloning, 310
primer design, 308–309
sequence analysis, 310
strategy, 307–308
template isolation, 309
validation of method, 311
CYP5A1, see Thromboxane synthetase
CYP6D1, purification from house fly
anion-exchange chromatography, 291–292
buffers, 287–288
hydrophobic interaction chromatography, 290–291
microsome preparation, 287–289
yield, 292
CYP11A (P450scc)
crystallization, 361
structural alignment with P450s, 318–321
CYP17A, structural alignment with P450s, 318–321
CYP19 (P450arom), structural alignment with P450s, 318–321
CYP52A3, factor Xa recognition site in amino terminus
cytosolic domain purification
detergent removal, 73
hydrophobic interaction chromatography, 72–73
expression in Saccharomyces cerevisiae
culture conditions, 67–68, 75
factor Xa recognition site, insertion at membrane anchor region, 66–67, 74–75
strain selection, 66
vector, 66
factor Xa proteolysis
buffer, 70
comparison to cDNA truncation, 74
cytosolic domain liberation from intact microsomes, 73–74
efficiency, 70, 72
reaction conditions, 71
specificity, 72

purification
hydrophobic interaction chromatography, 69–70
hydroxyapatite chromatography, 70
microsome preparation, 68–69
solubilization, 69
CYP74, see Allene-oxide synthase
CYP79 (P450tyr)
catalytic activities, 268
expression in Escherichia coli, 273–274
isolation
recombinant protein from Escherichia coli
anion-exchange chromatography, 272
cell growth, 271
dye affinity chromatography, 272–273
extraction with glycerol, 271–272
sorghum seedling enzyme
anion-exchange chromatography, 269–270
buffers, 270
dye affinity chromatography, 269–271
solubilization, 270
CYP101 (P450cam)
crystallization
crystal recovery, 360
precipitant, 360
substrate addition and removal, 359–361
thiol protection, 359
reaction cycle, 337
structural alignment with P450s
algorithms, 317–319
core structure, 316
C-terminal alignment, 319, 321–323, 325
N-terminal alignment, 323–324
substrate specificity
active site flexibility, 350–351, 355
DOCK screening, 339, 342–347
reaction rate prediction, 328, 330
stages of substrate interaction, 337
CYP102 (P450BM-P)
structural alignment with P450s
algorithms, 317–319
core structure, 316
C-terminal alignment, 319, 321–323, 325

N-terminal alignment, 323–324
substrate specificity
  active site flexibility, 350–351, 355
  reaction rate prediction, 328, 330
CYP108 (P450terp)
  crystallization, 362
  structural alignment with P450s
    algorithms, 317–319
    core structure, 316
    C-terminal alignment, 319, 321–323, 325
    N-terminal alignment, 323–324
Cytochrome P450 (CYP), *see also specific CYPs*
  abundance of plant genes, 260, 275, 281
  assays, *see specific P450s*
  coexpression with NADPH-P450 reductase in baculovirus–insect cell system
    cell culture, 88–89, 91–92
    heme supplementation, 88–89
    host cells, 87–88, 92
    microsomal fraction isolation, 93–94
    plaque assay, 90–91
    promoters, 87
    recombinant virus formation, 87–90
    scaleup, 94
    vectors, 86–87
    virus amplification, 91
  crystallization, *see* Crystallization, P450
  electrochemical synthesis applications, *see* NADPH-P450 reductase:P450 heme domain fusion protein
  expression in *Escherichia coli*, heterologous proteins
    affinity tagging, 12–13
    amino terminus blocking, 44
    cDNA cloning, 9–10
    culture conditions, 12–13, 36–37
    fusion proteins, *see* Glutathione *S*-transferase; NADPH-P450 reductase:P450 heme domain fusion protein
    heme content, 43–44
    minimum requirements for expression, 4
    polymerase chain reaction, cDNA amplification
      3′-antisense primer, 11–12
      polymerases, 11

      5′-sense primer, 11
    promoters, 3–5, 7
    protein folding, 4
    strain selection, 12, 36
    translation, inhibitory mRNA secondary structures, 7–9, 35–36
    vectors, 5, 7, 36
  expression in *Saccharomyces cerevisiae*, heterologous proteins
    assay, 59
    cDNA amplification by polymerase chain reaction, 53–54
    cell lysis
      enzymatic procedure, 63–64
      mechanical procedure, 62–63
    culture conditions, 60–62
    humanizing of strains, 56–58
    limitations, 52
    promoter and copy number, 53
    redox environment optimization, 52
    strain selection, 58–59
    transformant selection, 60
    vectors, 54–55
  gene isolation, *see* Polymerase chain reaction
  herbicide detoxification, 235–236
  membrane association, 65
  N-terminal modification
    effects
      catalytic activity, 27, 34
      subcellular localization, 26, 32, 34
    expression in *Escherichia coli*
      culture conditions, 28–29
      harvesting, 28–29
      vector construction, 27–28
    immunoblot analysis for subcellular localization, 29–30
    insertion of positive charges, 26
    purification of glutathione *S*-transferase fusion protein, 30–31
    truncation, 25–26
  product profiling with molecular dynamics
    applications, 351–352
    force fields, 353–354
    hardware requirements, 357
    mobile binding pocket approximation, 354–356
    raw trajectory transformation to products, 356–357

solvent inclusion, 354
substrate docking, 352–353
purification, recombinant human proteins from bacteria
affinity chromatography of CYP2D6, 40
anion-exchange chromatography, 38–39
cation-exchange chromatography, 39
detergent removal, 42
functional reconstitution, 42
histidine tagging and affinity purification, 43
hydrophobic interaction chromatography, 43
membrane preparation, 37–38
P450:NADPH-P450 reductase fusion proteins, 40–41
rate predictions
AM1 Hamiltonian, 332–335
molecular dynamics approach, 326–332, 335
quantum mechanics approach, 326–327, 332–335
structural modeling with bacterial P450s, 327–332
substrate docking, 330
substrate point charges, assignment, 328–329
stereoselectivity, 326
structural alignment
algorithms, 317–319
core structure, 316
C-terminal alignment, 319, 321–323, 325
N-terminal alignment, 323–324
substrate specificity prediction, see also CYP101; DOCK
applications, 347–348
molecular orbital calculations, 349
pharmacophore models, 349
quantitative structure–activity relationships, 348–349
stages of substrate interaction, 337

inhibitor analysis, 195
microsome
incubation, 189–190
preparation, 189
phenotyping, 191, 193–195
principle, 187–188
reagents, 189
recombinant protein, 191
synthesis, 188–189
DOCK
active site modeling, 338
substrate screening
contact scoring, 341–342
CYP101, 339, 342–347
energy minimization, 340–341
limitations, 345–347
modes, 339
refinements, 347
success rate, 342–343

E

Ecdysteroid
radioimmunoassay
antiserum, 295
data analysis, 296
principle, 293, 295
protein A solution preparation, 295–296
solution preparation, 297
standards, 295–296
synthesis, see also Larval ring gland
lethal mutations relating to synthesis, 293
prothoracicotropic hormone stimulation, 292, 302–304

F

Factor Xa, see CYP52A3, factor Xa recognition site in amino terminus

D

[O-methyl-14C]Dextromethorphan
CYP2D6 assay
extraction, 190–191

G

Gas chromatography, trimethadione metabolites, 165–166
Gene knockout, see Knockout mice

Glutathione *S*-transferase, P450 fusion proteins
affinity purification, 30–31
expression in *Escherichia coli*, 28–29, 78–79

H

Hepatitis, *see* Autoimmune drug-induced hepatitis
Hepatocytes
cell culture, 381–382, 392, 394, 400–401
cryopreservation, 392, 394, 401
CYP2C11 expression in cultured hepatocytes, ceramide effects
ceramide-activated protein phosphatase, 384
delivery of ceramide to cells, 382
exogenous sphingomyelinase and ceramide elevation, 383
RNA analysis of gene expression, 384–385
Western blotting, 385
human liver donors for P450 studies, 389–391, 401
identification of P450 inducers, 394–401
isolation by perfusion, 391–392
variability of P450 expression, 389–390
Hexobarbitol
CYP isoform assays
blood sampling, 172–173
clearance estimation, 175
drug administration
humans, 171–172
rats, 172–173
high-performance liquid chromatography, 174
urine sampling, 172–173
metabolism, 171
High-performance liquid chromatography
allene-oxide synthase assay, 254–255
antipyrine metabolites, 173–175
caffeine metabolites, 126–128, 130
ceramide metabolites, 387–388
chlorzoxazone metabolites, 119, 121–122
hexobarbitol metabolites, 174
imipramine metabolites, 180–181, 183
omeprazole metabolites, 133–135
taxol metabolites, 147–151
theophylline metabolites, 173–175
tolbutamide metabolites, 142

I

Imipramine
antidepressant activity, 177
CYP isoform assays
high-performance liquid chromatography, 180–181, 183
microsomes, 183–184
plasma/urine assays
calibration, 181, 183
deconjugation, 180
extraction, 181
selectivity, 181, 183
sensitivity, 183
reagents, 179–180
thin-layer chromatography, 177–178
metabolism in humans, 177–179, 185–186

K

Knockout mice
applications to human disease, 412–413, 430
breeding, 429
embryo, transfer to foster mothers, 427–428
embryonic stem cell
cell lines, 415
feeder cell preparation, 416–417
microinjection
alternatives, 430
cell preparation, 426
embryo microinjection, 426–427
mouse preparation, 424–426
recombinant clone
screening, 422–424
selection, 417–418, 420–422
gene isolation, 413
phenotypic analysis, 429–430
targeting vector
design, 414–415
electroporation, 420
preparation for electroporation, 415–416

L

Larval ring gland
culture
Drosophila melanogaster, 298–300

RNA synthesis assay, 301–302
Sarcophaga bullata, 300
ecdysteroid synthesis, 292, 300, 302–304
Liver, *see* Hepatocytes

M

Mephenytoin
  CYP2C19 assays
    in vitro, 113–114
    in vivo
      drug administration, 107–108
      measurement of urinary 4′-hydroxy-
        mephenytoin, 108
      measurement of urinary enantio-
        meric ratio, 109–111
      phenotyping, 108–109, 111–112
      principle, 107
    hydroxylation phenotypes, 105–106
Microsome
  cytochrome P450
    assays, 117–118, 134, 141, 149–150,
      155–156, 164–165, 183–184,
      189–190
    content in insects, 290
  preparation
    house flies, 287–289
    monocots
      allene-oxide synthase content,
        241–242
      gel filtration, 238
      shoot growth, 237
      tissue homogenization, 237–238
    Saccharomyces cerevisiae, 68–69
    seedlings, 261, 264–265
    tuber slices, 261–265
    woody tissue
      buffers, 245–246
      centrifugation, 249
      difficulty of preparation, 243–244
      filtration, 249
      hammer milling and intermediate-
        scale extraction, 247–248
      phenol removal, 243–244, 247
      pulverization and small-scale extrac-
        tion, 246
      shearing and intermediate-scale ex-
        traction, 246–247
      storage, 249–250
      tissue selection, 246

Wiley mill and large-scale extraction,
  248–249

N

NADPH-P450 reductase (CPR)
  coexpression with CYP3A4 in baculovi-
    rus–insect cell system
    cell culture, 88–89, 91–92
    heme supplementation, 88–89
    host cells, 87–88, 92
    microsomal fraction isolation, 93–94
    plaque assay, 90–91
    promoters, 87
    recombinant virus formation, 87–90
    scaleup, 94
    vectors, 86–87
    virus amplification, 91
  crystallization of recombinant protein
    catalytic activity of crystals, 373, 376
    cofactor addition and removal,
      372–373
    conditions, 370–371, 373
    heavy atom derivatives, 377
    ionic strength, 373
    purity requirements, 372
  electron transfer, 368
  rat liver enzyme, *Escherichia coli* ex-
    pression
    cell growth, 369
    cloning, 369
    purification, 369–370
  X-ray analysis of crystal structure
    data collection, 371
    preliminary analysis, 376–377
NADPH-P450 reductase:P450 heme domain fusion protein
  assay in bacteria, 15
  electrochemical synthesis applications
    cobalt sepulchrate mediation, 45, 51
    examples of reactions, 46
    hydroxylation of steroids, 45, 48, 50
    NADPH as electron source, 44, 46
    reactants, 47–48
    reaction conditions, 48
    reaction monitoring, 46–47
    reaction vessel design, 45–47
    reactive oxygen species, minimization
      in vessel, 50, 51
  expression in *Escherichia coli*

amplification of cDNA by polymerase chain reaction, 17, 19–21
cell harvesting, 22
culture conditions, 21–22
examples of proteins, 17
history of expression systems, 16
plasmid construction, 17, 19–21
immobilization for bioreactor application, 16, 23
linker, effects on activity, 22–23
purification from bacteria, 15, 40–41
Nicosulfuron, detoxification in corn, 236

O

Omeprazole
assay of CYP isoforms
calibration and validation, 135
extraction, 134
high-performance liquid chromatography, 133–135
kinetic analysis, 135–136
metabolite identification, 134–135
microsome incubation, 134
CYP2C19 phenotyping, 112–113
half-life, 132
metabolism by CYP isoforms, 106, 112, 132, 137, 139

P

P450arom, see CYP19
P450Bm-3, crystallization of heme domain, 361–362, 364
P450BM-P, see CYP102
P450cam, see CYP101
P450eryF, crystallization, 362–363
P450nor, crystallization, 362
P450scc, see CYP11A
P450terp, see CYP108
P450tyr, see CYP79
Polymerase chain reaction
allele analysis
CYP1A1, 228, 230
CYP2C19, 211–217
CYP2D6, 201–206
CYP2E1, 221–225
cDNA amplification for P450 expression
Escherichia coli systems
3′-antisense primer, 11–12

NADPH-P450 reductase:P450 heme domain fusion protein, 17, 19–21
polymerases, 11
5′-sense primer, 11
Saccharomyces cerevisiae system, 53–54
insect P450 gene cloning
amplification, 309–310
applications, 305
cloning, 310
primer design, 308–309
sequence analysis, 310, 312
strategies, 305–308, 311–312
template isolation, 309
validation of method, 311
messenger RNA quantitation of P450s
accuracy, 407–409
apparatus, 404
capillary electrophoresis of products
apparatus, 409
electrophoresis conditions, 409
laser-induced fluorescence detection, 403
peak integration, 410–411
competitor DNA, 405–407
CYP1A1, 411–412
primers, 405–406
reverse transcription, 402, 404–405
RNA isolation, 403–404
standards, 402, 406–407
petunia P450 gene isolation
cloning of differentially expressed genes, 281–283
limitations, 283
single primer method
directional cDNA library construction, 277
3′-end amplification of cDNAs, 277–278
5′-end amplification of cDNAs, 278
primer design, 275–277
screening of cDNA library, 278
two primer method
amplification reaction, 280
primer design, 279
template DNA preparation, 279–280
Propafenone
CYP1A2 assay, 102–103
CYP2D6 assay, 101–102
CYP3A4 assay, 102–103

glucuronidation assay, 103–105
metabolism in man, 99–100

R

Ring gland, *see* Larval ring gland

S

Signal peptide, P450
cleavage, *see* CYP52A3, factor Xa recognition site in amino terminus
engineering in *Escherichia coli*
culture conditions, 28–29
harvesting, 28–29
vector construction, 27–28
modification effects
catalytic activity, 27, 34
subcellular localization, 26, 32, 34
Single-strand conformational polymorphism
CYP1A1 allele analysis, 230–231
CYP2D6 allele analysis, 207–210
Southern blot, CYP1A1 allele analysis, 227–228
Sphingomyelinase
assays, cell culture
high-performance liquid chromatography of ceramides, 387
lipid extraction, 385–386
microphosphate assay, 386
sphingomyelin substrate, metabolic radiolabeling, 386
effect on CYP2C11 expression in cultured hepatocytes
ceramide-activated protein phosphatase, 384
delivery of ceramide to cells, 382
exogenous sphingomyelinase and ceramide elevation, 383
RNA analysis of gene expression, 384–385
Western blotting, 385
Sphingosine, quantitation in cultured cells, 387
SSCP, *see* Single-strand conformational polymorphism
Steroid, *see also* Ecdysteroid
hydroxylation using P450 electrochemistry, 45, 48, 50

induction of P450s, 388–389, 394–395, 397–399
SYBYL, energy minimization of potential substrate structure, 340–341

T

Tamoxifen
cytochrome P450 isoform assays
extraction, 156–157
incubation conditions, 156
materials, 155, 157–160
microsome preparation, 155–156
thin-layer chromatography, 157
mechanism of action, 152, 154, 161–162
metabolism in humans, 152–154, 161–163
synthesis of metabolites
N-desmethyltamoxifen, radiolabeled, 159
3,4-dihydroxytamoxifen, 159–160
[³H]-4-hydroxytamoxifen, 158–159
tamoxifen-*N*-oxide, 157–158
[³H]-tamoxifen-*N*-oxide, 159
Taxol
CYP2C8/CYP3A4 assays
fecal samples, 148–149
high-performance liquid chromatography, 147–151
interference, 149, 151
microsomes, 149–150
plasma samples, 147
quantitation, 149
metabolism in humans, 145–147
Theophylline
CYP isoform assays
blood sampling, 172–173
clearance estimation, 175
drug administration
humans, 171–172
rats, 172–173
high-performance liquid chromatography
plasma, 173–174
urine, 175
urine sampling, 172–173
metabolism, 170–171
Thin-layer chromatography
chlorzoxazone metabolites, 119–120
cinnamic acid 4-hydroxylase assay, 265–266

imipramine metabolites, 177–178
tamoxifen metabolites, 157
Tolbutamide
  CYP2C9 assay
    extraction, 141–142
    high-performance liquid chromatography, 142
    incubation conditions, 141
    microsome preparation, 141
    reagents, 141
    validation, 142–145
  metabolism in humans, 139
Trimethadione
  CYP isoform assays
    gas chromatography, 165–166
    in vivo assays
      humans, 167–168
      pharmacokinetic parameter estimation, 168
      rat, 167
      tolerance testing for liver function, 168–169
    microsomes
      humans, 165
      rat, 164–165
      reagents, 164
      reconstituted system, 165–166
    metabolism, 163–164, 166

W

Western blot
  autoantibody characterization in autoimmune drug-induced hepatitis, 80–81
  CYP2C11 response to ceramide, 385
  subcellular localization of N-terminal-modified P450, 29–30

X

Xanthine oxidase, caffeine assay, 124, 130

ISBN 0-12-182173-0